Bioinformatics, Computational Chemistry, and AI in Drug Innovation

This book comprehensively examines the applications of bioinformatics, computational chemistry, and artificial intelligence (AI) in the field of drug discovery and innovation. It provides foundational insights into bioinformatics and its applications, from cutting-edge advancements in computational molecular phylogeny, structural bioinformatics, and metabolic computing to biological databases and data mining. In addition, this book covers critical aspects of data storage, retrieval, database structure, and annotation, providing a robust foundation for understanding the complexities in biological data management. It discusses computational protein modeling, molecular docking, pharmacophore modeling, quantitative structure–activity relationships, and innovative approaches for drug target identification. The chapters explore protein–ligand interactions, molecular dynamics simulations, density function theory, and the strategic endeavor of drug repurposing. Toward the end, this book discusses quality by design (QbD) in drug formulation, the transformative influence of AI and machine learning (ML) on drug discovery, computational tools for estimating pharmacokinetic parameters, and the conceptual underpinnings of computational genomics. This book will be useful for students, professionals, and researchers in the fields of pharmaceutical sciences, life sciences, pharmaceuticals, biotechnology, and computational biology.

Key Features:

- Discusses applications of bioinformatics, computational chemistry, and AI in drug discovery and innovation
- Covers essential aspects of data storage, retrieval, database structure, and annotation for managing complex biological data
- Reviews molecular docking, pharmacophore modeling, and quantitative structure–activity relationships in drug designing
- Explores the transformative influence of AI and ML in drug discovery
- Emphasizes QbD principles, highlighting the importance of precision and quality in drug formulation design

Bioinformatics, Computational Chemistry, and AI in Drug Innovation

Advances and Applications

Edited by
Sushil Kumar Kashaw, Anshuman Dixit,
Shivangi Agarwal, and Priyanshu Nema

CRC Press
Taylor & Francis Group
Boca Raton London New York

CRC Press is an imprint of the
Taylor & Francis Group, an **informa** business

Designed cover image: Shutterstock Image Id: 2615909837

First edition published 2026
by CRC Press
2385 NW Executive Center Drive, Suite 320, Boca Raton FL 33431

and by CRC Press
4 Park Square, Milton Park, Abingdon, Oxon, OX14 4RN

CRC Press is an imprint of Taylor & Francis Group, LLC

ISBN: 9781032733012 (hbk)
ISBN: 9781032735344 (pbk)
ISBN: 9781003464679 (ebk)

DOI: 10.1201/9781003464679

Typeset in Times
by codeMantra

Contents

Preface....vii
Editor Biographies....viii

Chapter 1 An Overview of Bioinformatics and Its Applications....1
Anushka Garhwal, Mangala Hegde, Anjana Sajeev, and Ajaikumar B. Kunnumakkara

Chapter 2 Structural Bioinformatics....13
Firdous Fatima, Sunita Dahiya, Arun K. Iyer, and Rajiv Dahiya

Chapter 3 Metabolic Computing....27
Arvindra Sahu, Devraj Rajak, and Varun Kushwah

Chapter 4 Biological Databases and Data Mining....39
Yogesh Singh, Nakkala Srinivas Mudiraj, Neha Bhatia, Satwinder Singh, and Suresh Tharej

Chapter 5 Data Storage and Retrieval, Database Structure and Annotation....56
Balaji Wamanrao Matore, Anjali Murmu, Pragya Gawande, Akshay Kumar, Souranava Jana, Jagadish Singh, and Partha Pratim Roy

Chapter 6 Computational Protein Modelling, Refinement, Validation, and Its Application....67
Sushil K. Kashaw

Chapter 7 Molecular Docking Guided Drug Design and Discovery....78
Feroz Khan, Palak Singh, and Aqib Sarfraz

Chapter 8 Pharmacophore Modeling: A Cornerstone of Modern Drug Design....89
Asif Husain, Neelima Shrivastava, and Shah Alam Khan

Chapter 9 Quantitative Structure–Activity Relationship in Drug Design....103
Satyamshyam Vishwakarma, Prashant Sahu, Pragyanshu Khare, and Mukesh Chandra Sharma

Chapter 10 Computational Approaches for Drug Target Identification....119
Shankar Gupta, Abhimanyu Shome, Moumita Saha, and Vivek Asati

Chapter 11 Protein–Ligand Interaction and Molecular Dynamics Simulation....137
Rinki Prasad Bhagat and Shovanlal Gayen

Chapter 12 Density Functional Theory and Its Applications....148
Satyam Yadav, Ayushi Gupta, Ajay Kumar, Neha Patel, and Alok Kumar Pandey

Chapter 13 Drug Repurposing ... 155

Ravichandran Veerasamy, Raghuraman Seenivasan, Prabha Thangavelu, and Harish Rajak

Chapter 14 Biodiversity Ecological and Immuno-Informatics ... 169

Shifa Khan, Akanksha gupta, Priya singh, and Varsha Kashaw

Chapter 15 Quality by Design in Drug Formulation Design ... 180

Sakshi Soni, Arpana Purohit, and Vandana Soni

Chapter 16 In Silico Designing of Vaccines: Methods, Tools, and Their Limitations ... 199

Rakibul Islam, Abhishek Samanta, Sawan Kumar, Agniv Paul, Jasmeet Kaur, Utpal Mohan, and Devendra Kumar Dhaked

Chapter 17 Artificial Intelligence and Machine Learning in Drug Discovery ... 212

Jatin Malik, Sumit Kumar, and Gopal L. Khatik

Chapter 18 Estimation of Pharmacokinetic Parameters Using Computational Tools ... 225

Samyak Bajaj, Shivam Kori, and Ratnesh Das

Chapter 19 Computational Genomics Concept and Its Utilization ... 232

Priyanshu Nema, Bineet Kumar Mohanta, Sushil K. Kashaw, and Anshuman Dixit

Index ... 247

Preface

In recent years, the convergence of bioinformatics, computational chemistry, and AI has revolutionized the landscape of drug discovery and biomedical research. These once-distinct disciplines are now merging into a dynamic and powerful triad—driven by advances in computational power, the explosion of biological data, and the rise of intelligent algorithms. As a result, this integration is not only enhancing the efficiency and precision of drug development but also unlocking entirely new possibilities in pharmaceutical innovation.

This book, *Bioinformatics, Computational Chemistry, and AI in Drug Innovation*, offers a comprehensive and forward-looking exploration of current trends, methodologies, and practical applications at this interdisciplinary frontier. Designed with clarity and purpose, the chapters aim to bridge the gap between theoretical foundations and real-world scientific challenges, empowering researchers, students, and professionals alike to navigate and leverage modern computational tools in drug discovery.

At its core, this book provides a holistic understanding of the field. It spans a broad spectrum—from foundational bioinformatics and structural biology to sophisticated computational modeling and AI-driven analytics. Readers will explore diverse topics such as molecular docking, pharmacophore modeling, metabolic simulations, ML for drug target identification, and the use of AI in vaccine design. Importantly, it also delves into sustainable healthcare approaches, ecological and immunological informatics, and innovative strategies such as drug repurposing and QbD in pharmaceutical formulation.

The intended audience for this book is equally broad. It is tailored to researchers and professionals in the pharmaceutical and biotechnology industries, graduate students in bioinformatics and cheminformatics, clinicians seeking to understand computational approaches in precision medicine, and data scientists who aspire to apply AI and ML to biomedical challenges. Whether readers are building their foundational understanding or seeking to deepen their knowledge of cutting-edge techniques, this book offers valuable insights and practical guidance.

Structurally, this book is divided into logical, interconnected sections. Beginning with fundamental principles in bioinformatics and structural biology, it progresses through advanced computational techniques, explores systems for biological data management, and examines applications in drug development. Later chapters cover topics such as ecological informatics, immunology, and the transformative role of AI in modern healthcare. Each chapter is crafted to balance theoretical clarity with applied utility—featuring real-world case studies, method evaluations, and insights into future directions.

Among the key features of this book are its interdisciplinary approach, methodological rigor, and comprehensive scope. By weaving together concepts from biology, chemistry, informatics, and AI, it provides a unified framework for understanding how data-driven science is reshaping pharmaceutical research. The inclusion of critical perspectives and emerging technologies ensures that readers not only learn current best practices but also gain a visionary outlook on where the field is headed.

We believe that we are standing at a pivotal moment in biomedical science—one that demands collaborative, innovative, and technologically empowered solutions to address global health challenges. With this book, we hope to inspire new ideas, foster meaningful collaborations, and equip the next generation of scientific leaders. May it serve as both a foundational resource and a spark for future innovation at the dynamic intersection of bioinformatics, computational chemistry, and AI.

Dr. Sushil Kumar Kashaw

Dr. Anshuman Dixit

Dr. Shivangi Agarwal

Mr. Priyanshu Nema

Editor Biographies

Prof. Sushil K. Kashaw is a Professor and Raman Postdoctoral Research Fellow in the Department of Pharmaceutical Sciences at Dr. Harisingh Gour University (A Central University) in Sagar, Madhya Pradesh (M.P), India. He has previously served as an academic and executive council member of the university. He earned his Bachelor of Pharmacy (B. Pharm, 1998), Master of Pharmacy (M. Pharm, 2000), and Ph.D. (2009) from Dr. Harisingh Gour University. His professional experience includes research positions at the Central Drug Research Institute, Lucknow, where he worked as a CSIR-SRF, and at Wayne State University, Detroit, USA, where he held the position of Raman Postdoctoral Research Fellow. He has an impressive academic record, with 162 high-impact research papers, nine books, six book chapters, and the successful completion of five sponsored research projects, along with four granted patents. He has also delivered 45 invited lectures, chaired 30 scientific sessions, presented 40 papers at conferences, and developed ten e-contents for CEC-MHRD. He has received several awards, including the Raman PDF, Outstanding Oral Presentation Award, Young Scientist Award by APP, Best Teacher Award in Pharm. Chemistry by APTI-M.P. Branch, Research Excellence Award by the Institute of Scholars, and the Best Researcher Award by Nirali Publication in association with APTI, among others. He is a life member of numerous learned societies and has organized many training programs and conferences. His area of expertise lies in molecular modeling and medicinal chemistry.

Dr. Anshuman Dixit holds a master's degree in Pharmaceutical Chemistry from Dr. Harisingh Gour University, Sagar, M.P, India, and earned his doctorate from the Central Drug Research Institute (CDRI) in 2007. Following this, he pursued postdoctoral research at the Center for Bioinformatics at the University of Kansas, Lawrence, USA, from 2007 to 2011. Currently, Dr. Dixit serves as a senior scientist at the Institute of Life Sciences in Bhubaneswar, Odisha, India. He boasts extensive research experience in drug discovery and bioinformatics and has a remarkable publication record, with more than 90 peer-reviewed publications, four book chapters, and two international patents to his name. He also contributes as an associate editor and editorial board member for various academic journals. His current research interests center around drug combinatorial strategies using AI, network pharmacology, and protein structure–function relationships.

Dr. Shivangi Agarwal currently serves as a Postdoctoral Associate at the University of Illinois at Chicago, USA, where her work primarily revolves around the analysis of next-generation sequencing data derived from various cancer types. She earned her Ph.D. in Computational Biology and Bioinformatics from the Department of Pharmaceutical Sciences at Dr. Harisingh Gour University, Sagar, M.P., India, under the guidance of Dr. Sushil K. Kashaw. She also holds a master's degree in Pharmaceutical Chemistry and was honored with a Gold Medal in 2014. Her academic achievements include 20 research publications and the publication of one book. She was granted the CSIR Foreign Travel Grant, enabling her to present her research at the FIP World Congress of Pharmacy and Pharmaceutical Sciences in the United Arab Emirates in 2019. In addition, she received the prestigious "Young Scientist Award" for an outstanding research paper during the 35th MP Young Scientist Congress held at Shri G.S. Institute of Technology and Science, Indore, in 2020. Her research interests encompass cancer genomics, computational biology, and bioinformatics, as well as drug design and discovery.

Mr. Priyanshu Nema is a UGC-Research Scholar in the Department of Pharmaceutical Sciences at Dr. Harisingh Gour University (A Central University), Sagar, Madhya Pradesh. He completed his Bachelor of Pharmacy (B. Pharm) in 2020 and Master of Pharmacy (M. Pharm) in Pharmaceutical Chemistry in 2022 from the same institution. Currently pursuing his Ph.D. since 2022, Mr. Nema has demonstrated a strong academic and research background. He qualified national-level competitive exams, securing AIR 1517 in GPAT and AIR 1004 in NIPER JEE in 2020. His research interests include molecular modeling, drug design, medicinal chemistry, and *in silico* studies focusing on diabetes-related targets. He is also proficient in protein modeling, computational biology, and bioinformatics. He has authored 16 high-impact research and review publications and contributed to seven book chapters. He has presented 20 oral and poster presentations at various national and international conferences. In recognition of his research potential, he was honored with the Young Research Scholar Award in December 2024 by the Association of Pharmacy Professionals. His work reflects a deep commitment to excellence in pharmaceutical sciences and computational drug discovery.

1 An Overview of Bioinformatics and Its Applications

Anushka Garhwal, Mangala Hegde, Anjana Sajeev, and Ajaikumar B. Kunnumakkara
Cancer Biology Laboratory, Department of Biosciences and Bioengineering, Indian Institute of Technology Guwahati (IITG), Guwahati, India

INTRODUCTION

Bioinformatics is an interdisciplinary branch of molecular biology, genetics, mathematics, and statistics with informatics [1,2]. It involves the development of methods and software tools to interpret biological data, particularly when dealing with extensive and intricate datasets [3,4]. Bioinformatics is an interdisciplinary field that integrates techniques for managing biological information, such as data storage, distribution, and analysis, supporting various scientific and research domains [5]. The main goals of using bioinformatics are data analysis and its management, disease diagnosis and treatment, personalized medicine, data visualization, etc. [3]. Various techniques, such as protein–protein and protein–ligand molecular docking, and molecular dynamics simulation, are used for several research purposes [6,7].

Moreover, databases are created to store the large volume of biological data generated and accessed worldwide by the scientific community [5,8]. Several databases were created to store vast data due to increased storage demands [9]. Moreover, databases can draw information from several sources, including raw DNA sequences, protein sequences, macromolecular structures, and genome sequences [5,10–12]. Public databases are mainly used to store large amounts of information and are generally categorized into primary and secondary databases [13]. Primary databases consist of experimental data released without prior analysis from earlier publications. Examples of primary databases include GenBank at the National Centre for Biotechnology Information (NCBI), the European Molecular Biology Laboratory (EMBL), and the DNA Data Bank of Japan (DDBJ). These databases belong to the International Nucleotide Sequence Database Collaboration (INSDC) which is responsible for data upgradation [13–15]. Subsequently, secondary databases handle compiling and interpreting data called the content curation process. Examples of secondary databases are Protein Databank (PDB), Prosite, Structural Classification of Protein 2 (SCOP), UniProtKB/Swiss-Prot, and Protein Information Resources [13,14]. In addition to these databases, there are functional databases like the Kyoto Encyclopaedia of Genes and Genomes (KEGG) and Reactome, which are used for the interpretation and analysis of metabolic maps. Currently, bioinformatics has become a crucial part of modern biological research, playing a vital role in the interpretation and analysis of proteomics, genomics, metabolomics, and transcriptomics [13,14].

HISTORICAL BACKGROUND OF BIOINFORMATICS

The history of bioinformatics (Figure 1.1) dates to the early 1960s when the first attempts were made to understand biological sequences using computational approaches [16]. In 1952, Hershey and Chase established DNA as the molecule responsible for encoding genetic information. Through their experiments, they demonstrated that DNA, not protein, was absorbed and passed on by bacterial cells infected with a bacteriophage. This crucial discovery clarified the role of DNA in heredity [16,17]. However, despite understanding its function, the precise structure of DNA remained unsolved until 1953 when Watson, Crick, and Franklin revealed its double-helix configuration. Although this discovery was recognized, it took nearly 13 more years for the genetic code to be deciphered and around two decades more for the first DNA sequencing method to be developed [18–21]. Interestingly, bioinformatics grew alongside the development of protein sequencing methods from various organisms and the availability of protein sequences following Frederick Sanger's determination of the insulin sequence in the early 1950s [20].

The term "bioinformatics" was first used by two Dutch biologists named Paulien Hogeweg and Ben Hesper for the first time in the year 1970 defining it as "the study of informatic processes in living systems" [22,23]. One of the earliest applications in bioinformatics was pattern-matching software to identify recognition sites for restriction enzymes, essential tools in DNA cloning techniques. This software uses a database of known recognition sites, such as GAATTC, to scan DNA sequences and locate these sites. This simple yet powerful tool remains one of the most widely used in bioinformatics [24]. New computational methods to analyze and compare numerous protein sequences from different organisms were necessary

DOI: 10.1201/9781003464679-1

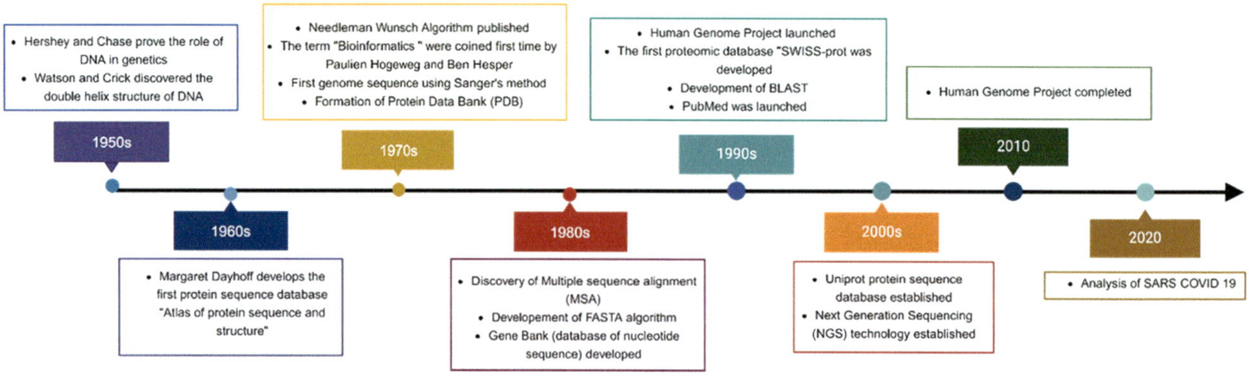

FIGURE 1.1 Illustrates the historical background and evolution of bioinformatics from 1950 to 2020s time period. The timeline highlights significant events, discoveries, and technological advancements that have shaped the field.

because manual handling of many amino acid sequences was impractical. In the late 1960s, Pehr Edman developed a fully automated protein sequencer, a machine designed for the automatic determination of amino acid sequences in proteins and peptides [25,26].

During this period, American physicist and mathematician Margaret Oakley Dayhoff developed an interest in protein and molecular evolution and began exploring mathematical approaches to analyze amino acid sequence data [16,26]. From 1958 to 1962, Dayhoff, along with Robert S. Leydley, developed the computer program COMPROTEIN. This program was designed to determine the primary structure of proteins by utilizing data from Edman peptide sequencing [27]. In the COMPROTEIN software, amino acid sequences were input and output using three-letter abbreviations, like "leu" for leucine and "met" for methionine [28]. To simplify interpretation, Dayhoff created the one-letter code for amino acids that we use today, which she introduced in her book "Atlas of Protein Sequence and Structure." This book is considered the first-ever sequence database [29]. Dayhoff is regarded as the pioneer of bioinformatics for her significant contributions to the development of computational techniques used in determining peptide sequences [30].

In 1970, Needleman and Wunsch developed a dynamic programming algorithm to perform pairwise sequence alignment of proteins [31]. Later in the 1980s, the first algorithm proposed for multiple-sequence alignment (MSA) was discovered [32]. The MSA was performed by using CLUSTAL which was introduced in 1988 [33]. Subsequently, Dayhoff developed the point mutation accepted matrix by introducing the idea of a mathematical concept for amino acid substitution [34]. Between 1970 and 1980, there was a significant shift of focus from analyzing proteins to DNA analysis. Moreover, from 1980 to 1990, more advancements in the disciplines of biology and computer science took place. Several significant events in bioinformatics, such as the establishment of the National Institute of Health (NIH) and the NCBI in 1988 and the launch of the Human Genome Project (HGP) in 1990, highlighted the importance of bioinformatics in the interpretation and management of biological data [35]. These milestones paved the way for advancements in genomics and structural bioinformatics between 1990 and 2000. Moreover, between 2000 and 2010 the high-throughput bioinformatics strategies gained the attention of scientific communities, highlighting their contribution to biology [36] Additionally, the distinctive repeat-spacer structure of clustered regularly interspaced short palindromic repeats (CRISPR) arrays was first discovered by Yoshizumi Ishino [37,38]. Later, Francisco Mojica discovered that CRISPR arrays were not only present in *Escherichia coli* but also in many archaeal and some bacterial genomes using bioinformatics [39]. Following investigations in bioinformatics, it was discovered that there were matches between bacteriophage spacers. This discovery led to the accurate conclusion that CRISPR-Cas systems operate as an acquired immune system. Through bioinformatic analysis of spacer matches, it was further revealed that the CRISPR-Cas system predominantly targets DNA instead of RNA [39]. This research was a very big success in the field of genome editing [40]. One of the notable examples from recent years is severe acute respiratory syndrome coronavirus 2 (SARS-CoV2) which was combated by using bioinformatics strategies [41]. Bioinformatics played a significant role in the discovery of therapeutics against SARS-CoV2 [42]. Recent advancements in novel algorithms, artificial intelligence, mega data technology, and deep learning have significantly improved the development of innovative bioinformatics tools for analyzing the exponentially growing SARS-CoV-2 data [42,43]. With the aid of bioinformatics tools and databases, various essential analyses were performed including genetic sequence analysis using genetic sequence databases, evolutionary and epidemiological analysis, protein structure prediction, diagnosis, and target prediction to combat the pandemic [41,44–48]. Bioinformatics has emerged as an important component of drug design, particularly in validating therapeutic targets. It can assist in the knowledge of complicated biological processes to enhance medication development [43].

Bioinformatics has a vast history background from 1950 to date; various tools and databases have been developed which are useful for understanding and managing complex biological data [49]. Handling massive datasets has become a serious difficulty nowadays, sometimes known as "Big Data." To address these issues, experienced computer scientists are required. Various tools are available for bioinformatics analysis [49,50]. The growing field of computer technology and biology led to new areas of study like synthetic biology, system biology, and whole-cell modeling [16].

GOALS OF BIOINFORMATICS

As we discussed earlier, bioinformatics is important for discovering and exploring new information in less time with the aid of computational technologies. The main aims of bioinformatics in these research projects are the development of databases, improvize available tools and software as per the requirement of research, developing platform-independent software for the analysis of biological data using computational tools and documentation of omics data, available information about animal and plant genetic source, development of graphical user interface for the integration and analysis of data [3].

APPLICATION OF BIOINFORMATICS

Bioinformatics is widely used in various fields (Figure 1.2) such as molecular biology, genomics, proteomics, metabolomics, transcriptomics, drug discovery and development, etc. In this book chapter, we will explore the role of bioinformatics in various fields mentioned below.

Molecular Biology

Molecular biology focuses on the study of biological processes at the molecular level. It mainly deals with the interaction of cells within various systems, including the correlation and regulation of DNA, RNA, and protein synthesis. Molecular biology helps us to understand the mechanisms governing replication, transcription, and translation of genetic material. The central dogma demonstrates how genetic information gets transferred from DNA to RNA and then to protein, in a simplified form. Moreover, quantitative analysis of molecular biology demands the convergence between this discipline and computer science, referred to as computational biology [51,52].

Recent technological development has led to an extraordinary revolution in science, particularly in fields like genomics, proteomics, transcriptomics, and metabolomics [53]. Further, these innovations have resulted in the development of vast amounts of "omics" data, which refers to datasets that carry detailed information about biological molecules and their interactions within organisms [53,54]. These data generated open new doors for scientists to explore more about the complexities of living systems and to tackle problems that are related to disease mechanisms, drug discovery, and personalized medicine [54].

Genomics

Genomics is the branch of genetics that applies the method of DNA sequencing and bioinformatics to sequence, assemble, and analyze the genome function and its structure [55]. Genomic analysis has shown drastic progress after the discovery of "Sanger sequencing" of DNA in the year 1977. Over the past few decades, the discovery of next-generation sequencing (NGS) has enabled faster, cost-effective, and high-throughput genome analysis. Genome analysis has a significant impact on medicine, disease diagnosis, and therapeutic response [56].

The HGP, was initiated in 1990 and completed in April 2003 [57]. It is an international collaboration of publicly supported projects that aims to sequence and map the complete human genetic makeup [58]. The US Department of Energy, NIH, and UK Welcome Trust, along with groups from Japan, France, Germany, China, and other countries, are the major collaborators of the HGP project [59]. The main goal of HGP is to accurately identify various genes and nucleotide sequences present in the human genome and to develop tools for the analysis and storage of generated data [58]. The generated sequence data from HGP are likely to be stored in large genomic repositories, the most common ones including NCBI, GenBank database, the European Bioinformatics Institute (EBI), EMBL, and DDBJ [60–62]. There are various bioinformatics tools available for sequence analysis such as basic local alignment search tool (BLAST) and FASTA to search for sequence similarity among proteins [63,64]. MOTIF, CLUSTALW, MAFFT, and TREE are used to identify sequence motifs, MSA, and phylogeny of the genome, respectively [65–67]. Moreover, the viral proteomic tree server (ViPTree), automatic genome annotation (KO assignment) (KAAS) and pathway/brite mapping), metabolic and physiological potential evaluator (MAPLE), gene network prediction by kernel-based integration of heterogenous dataset (GENIES), and drug-target interaction network interference (DINIES) tools are used for genome analysis (https://www.genome.jp/en/gn_tools.html) [68–74].

Proteomics

Proteomics is the branch of omics that explores the structure, function, and interaction of proteins, with the aid of various techniques such as mass spectrometry and protein–protein interaction networks [75]. Proteomics provides more information about the biological system by analyzing the expression of proteins compared to genomics. Bioinformatics employs a proteomic approach to handle the vast and diverse library of data involved in the process of biomarker discovery [76]. The main challenge is the management of vast data generated and its correlation with other omics technologies. Proteomics data analysis is challenging due to the complex processing parameters, quality assessment, and shortage of standardized formats. The main challenge is to interpret the complex biological data in a meaningful manner [77].

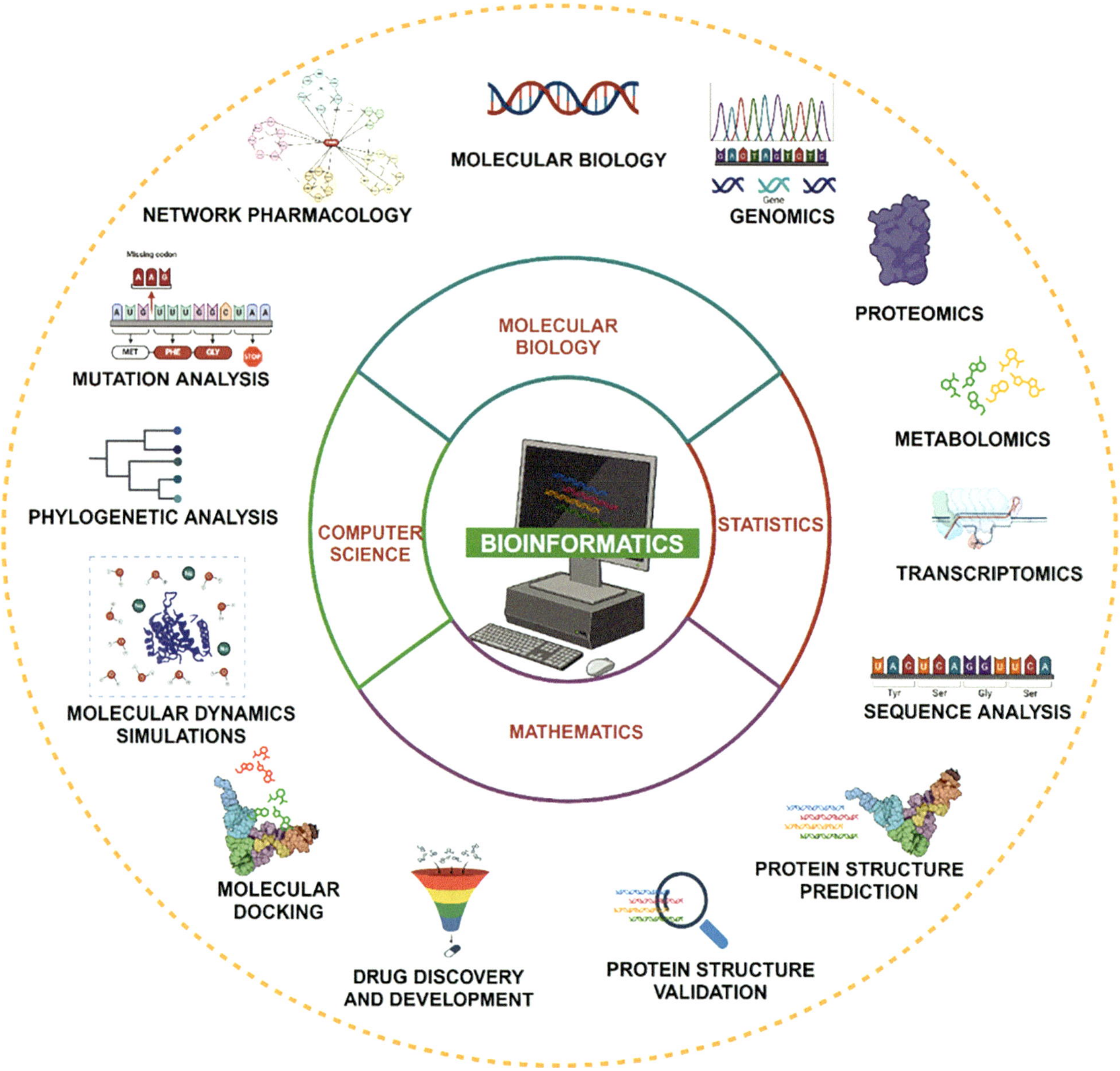

FIGURE 1.2 Depicts the role of bioinformatics in various disciplines and its application in genomics, proteomics, transcriptomics, metabolomics, molecular biology, drug discovery, protein sequence identification, validation, molecular docking, molecular dynamics simulation, etc.

Protein pathways are associated with cellular interactions that drive biological activity, and these pathways are analyzed using a proteomics approach [78,79]. Various tools and databases such as the KEGG, ingenuity pathway databases, Pathway Knowledge Base Reactome, and BioCarta provide detailed information about protein signaling, metabolism, and interactions [79–81]. Other databases such as protein analysis through evolutionary relationship and Gene MicroArray Pathway Profiler (GenMAPP) were generated to detect signal transduction pathways [82,83]. Additionally, STRING database helps to identify the interactions between genes and draw the interconnected protein networks [84,85]. InAct, MINT, BioGRID, and HRPD were used to find the details about complex protein interactions [86,87]. Netpath is another database that includes pathways related to cancer, that is used to identify proteins concerning specific cancers [79,88].

Metabolomics

Metabolomics is an evolving field that quantifies numerous small molecules known as metabolites, and the complete set of these metabolites is referred to as the metabolome [89]. This discipline aims to identify and measure metabolites present within biological systems, offering novel strategies for the discovery of active metabolites. A metabolome-based approach for identifying biological activity employs a library of metabolites created through statistical analysis of metabolomic datasets [90]. Once the peak intensity of metabolite abundance is quantified, the statistical filters can be applied based on the experiment and data. Metabolites are further selected for screening for their biological activity with a specified statistical cutoff [90]. Various alignment and peak detection tools are available, mainly MZmine2, XCMS online, Open-MS, and MS-DIAL [91–94]. Additionally, feature annotation and metabolite identification utilize various metabolite databases and spectral libraries, such as the Human Metabolome Database, METLIN, the Birmingham Metabolite Library, the Nuclear Magnetic Resonance Database, BIGG, MassBank, LipidMaps, mzCloud, the Fiehn Lab GC-MS Database, and the Golm Metabolomics Database [95–103]. The problem of data sharing and analysis was overcome by using cloud-based technologies and databases. Bioinformatics analysis of metabolic pathways and networks can reduce the complexity of data. This strategy has been merged with XCMS Online and is accessible via MetExplorer and other platforms. The primary goal of metabolomics is to assess the contributions of metabolites within various metabolic pathways, thereby helping to identify the significant roles these metabolites play in biological systems [104,105]. The selected candidates throughout the screening process are further tracked using isotope-based tracing via metabolome [106]. In conclusion, metabolomics-guided activity screening can be conducted through both in vivo and in vitro phenotypic assays, alongside chemical biology techniques and a range of omics approaches [90].

Transcriptomic Analysis

Transcriptomics quantifies the overall set of RNA transcripts present in a cell or biological system [106,107]. Transcriptomics mainly works on two approaches for the quantification of transcripts namely, hybridization-based approach, also known as microarray and RNA-sequencing (RNA-Seq)-based approaches [107]. The sequence-based strategy, which mostly involves serial gene expression analysis, is ineffective in identifying all transcript isoforms and is expensive [107]. Moreover, the microarray approach only identifies the genes and exons that are previously featured in the array and not able to identify new transcripts and genes that are low expressed. Microarray also failed to differ the genes with homology in their sequence [108,109]. Subsequently, various studies reported the high-throughput sequencing of the entire transcriptome also known as RNA-sequencing or RNA-Seq [107]. This technique is more accurate in identifying the novel genes, exons, isoforms of transcripts as well as it also detects low-expression transcripts [110]. It also helps in the identification of novel transcripts, such as long non-coding RNAs (lncRNAs), microRNAs, and allele-specific expression [111–113]. Ongoing research in transcriptomics improved the modern medicine system including diagnosis, treatment, and prognosis, by knowing the mechanism of genes in various pathways [106,114].

To generate an RNA-seq dataset, the mRNA of interest needs to be isolated, purified, fragmented into smaller pieces and then converted into complementary DNA (cDNA) using reverse transcriptase [106]. The adaptors are attached, and small fragments are selected based on size. NGS technology is utilized for cDNA sequencing [106]. Several tools present for the analysis are broadly classified into read mapping, transcript assembly, and transcript or gene quantification [111]. Tuxedo Suite Protocol (Tophat/Cufflinks) is the most widely used tool [115]. The steps involved in analysis are as follows: The read mapping of the sample genome and identification of splicing junctions are done using Tophat. Cufflinks used these alignments for transcript assembly, estimated their abundance, and identified the differentially expressed transcripts and genes which are visualized by using CummeRbund [116,117]. Further, functional enrichment analysis is performed for the genes which are selected by differential expression analysis. Furthermore, we can explore the pathways in which these genes are involved. The KEGG and Reactome databases are some of the most commonly used resources for this purpose in transcriptomics research. These databases provide comprehensive insights into the biological processes and molecular interactions associated with gene functions, facilitating a deeper understanding of cellular mechanisms [112,118,119].

Sequence Analysis

BLAST was developed by NCBI in 1990 [120]. BLAST is a tool for the alignment of sequences. It is commonly used to analyze biological sequence data, including amino acid sequences of various proteins or nucleotide sequences of DNA [120]. The BLAST algorithm is used to identify the similarity between the query sequence and the library sequence [121]. There are several types of BLAST, including Nucleotide 6-frame translation to protein (blastx), Nucleotide 6-frame translation to Nucleotide 6-frame translation (tblastx), and Protein to Nucleotide 6-frame translation (tblastn) [68,122]. The main purpose of BLAST is to identify species, locate domains in proteins, the establishment of phylogeny, and DNA mapping. Additionally, FASTA Clustal is also used as a sequence alignment tool [68,120].

Phylogenetic Analysis

Phylogenetic analysis is another key area in bioinformatics which is a scientific method used to study the evolutionary relationship between various other organisms and genes. The main objective of phylogenetic analysis is to determine the evolutionary changes through the MSA graphically represented by a Cladogram or phylogenetic tree [123]. Plethora of software and tools are available for phylogenetic analysis. The tools such as CLUSTAL_X, BLAST, Phylip, MEGA, PAUP, and PhyloDraw are used to perform phylogenetic analysis [65,120,124–126].

Protein Structure Prediction

When the experimental 3-D structure of a targeted protein is not reported in the PDB database, then we can use a computational approach to design the protein model [127]. The 3-D protein structure is determined using its amino acid sequence utilizing various bioinformatics technologies. There are three approaches for protein structure prediction namely, homology or comparative modeling, fold recognition or threading, and ab initio modeling [128]. The modeled structure is then carried out for several *in silico* analyses such as docking to target the predicted structure of a protein with ligands to identify the lead molecules in drug development [129]. SWISS-MODEL, Modeller, I-TASSER, AlphaFold, LOMETS, RaptorX, Rosetta are the most commonly used tools for estimation of protein structure [130–136].

Refinement and Validation of the Predicted Structure of a Protein

The predicted structure model of the protein is further validated and refined using various bioinformatic tools such as structural analysis and verification server (SAVES), Rampage [137], Swiss PDB viewer [138], and PROCHECK [139]. The quality of protein is estimated using these servers, and further studies are conducted using best-modeled protein structure for drug designing [140].

Analysis of Mutation

A change in one or more amino acid sequences can lead to a missense mutation, potentially resulting in disease conditions [141]. Identifying mutations in the amino acid sequence of a protein structure can aid in the prevention and treatment of various disorders. Several freely available tools and methods, such as EVmutation, PMut, SNPnexus, and DynaMut, are commonly used for mutation analysis, etc. [142–145].

Drug Discovery and Development

Computer-aided drug designing is an emerging field in the drug discovery and development process. It is a more accurate, cost-effective, and less time-consuming process as compared to the conventional drug discovery method [146]. Various techniques such as molecular docking, virtual screening, binding free energy determination, absorption, distribution, metabolism, excretion, and toxicity (ADMET) analysis, and molecular dynamics simulations (MDS) are utilized for the identification of novel lead molecules for effective treatment strategies against various diseases [146,147].

Molecular Docking

Molecular docking and virtual screening are done to identify the best ligand molecule from the library of compounds against the target protein structure based on a scoring function [148]. Moreover, MDS confirms the structural stability and conformational changes in the protein–ligand complex. These *in-silico* techniques fueled the process of investigating novel drugs or therapeutics for the treatment of several diseases in less time and experimental cost [147,148]. The tools used for molecular docking of protein–ligand are GOLD, AutoDock Vina, and SwissDock. Tools such as Clus-Pro, pepATTRACT, HDOCK, and ZDOCK are specifically used for protein–protein docking. Hybrid tools like HAD-DOCK are used to perform both protein–ligand and protein–protein docking [149–156].

Molecular Dynamics Simulations

MDS are bioinformatics techniques where Newton's classical laws of motion are applied to model the behavior of molecular systems [157]. MDS is broadly classified into quantum mechanisms/molecular mechanics (QM/MM) employed for simulating chemical reactions, conventional simulations that are widely applied for studies like protein folding, stability of complexes, and refinement of structure, and metadynamics to predict the free energy of the system [158]. The most commonly used force fields to perform MDS are Amber, CHARMM, and OPLS-AA, except GROMOS, which is particularly developed for GROMACS [158]. This software needs high-performance computational facilities for its work. Tools such as Amber, GROMACS, LAMMPS, CHARMM, NAMD, Desmond, OpenMD, ORAC, AMMP VE, ACEMD, Tinker, Abalone, and YASARA dynamics are used to perform [158–166].

Molecular Visualization

The 3-D structure elucidated for macromolecules is freely accessible in various databases like PDB which enables advanced research in the field of drug discovery, mutation, and molecular docking [159,167]. Molecular viewers help to understand the structure of various macromolecules in spatial arrangement more efficiently. 3D molecular visualization, tools like Visual Molecular Dynamics (VMD), PyMOL, Chimera, Schrödinger, MOE, and Discovery Studio (DS) Visualizer are commonly used [168–173].

Network Pharmacology

Network pharmacology (NP) is a discipline that deals with understanding the mechanism of how a drug interacts with multiple targets [174]. It uses computational tools to analyze the molecular interactions of drug molecules in living cells. NP is used to investigate novel lead therapeutics and targets, repurposing preexisting drugs for the treatment of different diseases [175,176]. The main aim of NP is to provide safer, efficacious, and more effective novel therapeutics along with drug repositioning. There are three steps involved in NP, namely, target recognition and prognostication, network building (to examine the network and confirm), and pathway enrichment analysis [177]. Various databases are employed during target selection and prediction of compounds, including the Swiss database. Further, STRING, STITCH, and KEGG databases [178–180] are used to form interactions between proteins and genes and also predict the major pathways involved [181].

ARTIFICIAL INTELLIGENCE AND MACHINE LEARNING: THE EMERGING FIELDS

Artificial intelligence (AI) and machine learning (ML) are revolutionizing numerous aspects of our lives, transforming industries and shaping the future. AI is one of the earliest and most expansive fields in computer science [182]. It encompasses all aspects of mimicking human cognitive functions for solving real-world problems and creating systems that can learn and think like humans. As a result, it is often referred to as machine intelligence [182].

AI and ML techniques are extensively utilized in the analysis of medical images, including X-rays, computed tomography (CT) scans, and magnetic resonance imaging (MRI) scans, to automate tasks such as anomaly detection, classification, and segmentation, thereby significantly assisting radiologists in diagnosing conditions such as tumors, fractures, and cardiovascular diseases [183]. These methodologies enhance the accuracy and robustness of image interpretation. In genomic research, AI and ML facilitate the analysis of complex and large-scale genomic data, driving advancements in understanding the genetic underpinnings of diseases [184]. Explainable artificial intelligence (XAI) is an emerging field aimed at increasing human trust in AI models by elucidating model construction and results [184]. The advancement in ML and AI has demonstrated promising results in the treatment of the complex, multifaceted disease, cancer [185,186]. Moreover, AI and ML facilitate personalized medicine which allows for the creation of personalized treatment plans that optimize efficacy and minimize adverse effects [187]. In oncology, for instance, AI algorithms can analyze a patient's tumor genomics to identify specific mutations and design targeted therapies [187,188]. As AI continues to evolve, it promises to improve early disease detection, streamline clinical workflows, and ultimately enhance patient outcomes, although challenges related to data privacy, regulatory approval, and model interpretability remain critical to address for widespread clinical adoption.

FUTURE PROSPECTS

Bioinformatics has gained significant attention since the HGP came into existence with a positive impact on modern research and development. In the 21st century, novel discoveries are not possible without its intervention. In today's omics era, the scientific community is increasingly turning to bioinformatics as a new frontier in science and technology. Several types of research conducted nowadays include the use of computational technology, software, and tools due to their speed, accuracy, and efficiency making bioinformatics an essential part of the lab's toolkit. In the future, a growing demand for bioinformatics will generate more biological data, which needs to be managed. The databases and other sources generated through bioinformatics directly reduce the experimental workload and cost, leading to a promising approach that can help to discover and explore more about living science. Furthermore, bioinformatics will play a vital role in the health sector and biomedical and veterinary sciences by facilitating the discovery of medicines, therapeutics, and disease management strategies.

CONCLUSION

Bioinformatics is an emerging field that combines biology, mathematics, statistics, and computer technologies to decode complex biological data. The tools of bioinformatics help to interpret, analyze, and store the data generated using various tools and databases. This chapter provides an overview of the history, introduction, and significance of bioinformatics across various disciplines. Furthermore, it places a particular emphasis on database resources and technologies, discussing their applications and accessibility. Bioinformatics will play a crucial role in managing and analyzing large data generated by omics and other breakthroughs in several fields. Overall advancement in technologies is contributing to the betterment of research and development in various fields.

REFERENCES

1. Moore, J.H., Bioinformatics. *Journal of Cellular Physiology*, 2007. 213(2): pp. 365–369.
2. Can, T., Introduction to bioinformatics. *Methods in Molecular Biology*, 2014. 1107: pp. 51–71.
3. Bayat, A., Science, medicine, and the future: Bioinformatics. *British Medical Association*, 2002. 324(7344): pp. 1018–1022.
4. Abdurakhmonov, I.Y., *Bioinformatics: Basics, Development, and Future*. 2016: InTech Rijeka, Rijeka, Croatia.
5. Luscombe, N.M., D. Greenbaum, and M. Gerstein, What is bioinformatics? A proposed definition and overview of the field. *Methods of Information in Medicine*, 2001. 40(4): pp. 346–358.
6. Pathak, R.K., D.B. Singh, and R. Singh, *Introduction to basics of bioinformatics*. In Dev, B.S. and Pathak, R.K. (Eds.), *Bioinformatics*. 2022: Elsevier, Amsterdam, Netherlands, pp. 1–15.
7. Vemula, D., et al., CADD, AI and ML in drug discovery: A comprehensive review. *European Journal of Pharmaceutical Sciences*, 2023. 181: p. 106324.
8. Pevsner, J., *Bioinformatics and Functional Genomics*. 2015: John Wiley & Sons, Hoboken, NJ.
9. Galperin, M.Y. and X.M. Fernández-Suarez, The 2012 nucleic acids research database issue and the online molecular biology database collection. *Nucleic Acids Research*, 2012. 40(D1): pp. D1–D8.

10. Sayers, E.W., et al., Database resources of the national center for biotechnology information. *Nucleic Acids Research*, 2010. 39(suppl_1): pp. D38–D51.
11. Jawdat, D. The era of bioinformatics. In *2006 2nd International Conference on Information & Communication Technologies*. 2006: IEEE, New York.
12. Diniz, W.J.S. and F. Canduri, Bioinformatics: An overview and its applications. *Genetics and Molecular Research Journal*, 2017. 16(1): p. 104238.
13. Whitfield, E.J., M. Pruess, and R. Apweiler, Bioinformatics database infrastructure for biotechnology research. *Journal of Biotechnology*, 2006. 124(4): pp. 629–639.
14. Villalba, G.C. and U. Matte, Fantastic databases and where to find them: Web applications for researchers in a rush. *Genetics and Molecular Biology*, 2021. 44: p. e20200203.
15. Brooksbank, C. and A. Cowley, *Bioinformatics for the Terrified: An Introduction to the Science of Bioinformatics*. 2020: EMBL-EBI, Cambridgeshire. https://doi. org/10.6019/TOL.BioinfTer-c_2016.
16. Gauthier, J., et al., A brief history of bioinformatics. *Briefings in Bioinformatics*, 2019. 20(6): pp. 1981–1996.
17. Hershey, A.D. and M. Chase, Independent functions of viral protein and nucleic acid in growth of bacteriophage. *Journal of General Physiology*, 1952. 36(1): pp. 39–56.
18. Maxam, A.M. and W. Gilbert, A new method for sequencing DNA. *Proceedings of the National Academy of Sciences*, 1977. 74(2): pp. 560–564.
19. Nirenberg, M. and P. Leder, RNA codewords and protein synthesis: The effect of trinucleotides upon the binding of sRNA to ribosomes. *Science*, 1964. 145(3639): pp. 1399–1407.
20. Sanger, F., S. Nicklen, and A.R. Coulson, DNA sequencing with chain-terminating inhibitors. *Proceedings of the National Academy of Sciences*, 1977. 74(12): pp. 5463–5467.
21. Watson, J.D. and F.H. Crick, Molecular structure of nucleic acids: A structure for deoxyribose nucleic acid. *Nature*, 1953. 171(4356): pp. 737–738.
22. Hogeweg, P., Simulating the growth of cellular forms. *Simulation*, 1978. 31(3): pp. 90–96.
23. Hogeweg, P. and B. Hesper, Interactive instruction on population interactions. *Computers in Biology and Medicine*, 1978. 8(4): pp. 319–327.
24. Fenstermacher, D., Introduction to bioinformatics. *Journal of the American Society for Information Science and Technology*, 2005. 56(5): pp. 440–446.
25. Edman, P. and G. Begg, A protein sequenator. *European Journal of Biochemistry*, 1967. 1(1): pp. 80–91.
26. Hagen, J.B., The origins of bioinformatics. *Nature Reviews Genetics*, 2000. 1(3): pp. 231–236.
27. Dayhoff, M.O. and R.S. Ledley. Comprotein: A computer program to aid primary protein structure determination. In *Proceedings of the December 4–6, 1962, Fall Joint Computer Conference*. 1962: National Biomedical Research Foundation, Silver Spring, MD.
28. Rules, T., IUPAC-IUB commission on biochemical nomenclature A one-letter notation for amino acid sequences 1–3. *Journal of Biological Chemistry*, 1968. 243(13): pp. 3557–3559.
29. Dayhoff, M., *Atlas of Protein Sequence and Structure*. 1978: National Biomedical Research Foundation, Silver Spring, MD, p. 345.
30. Henzy, J.E., Margaret dayhoff: Catalyst of a quiet revolution. In Whitaker, R.J. and H.A. Barton (Eds.), *Women in Microbiology*. 2018: John Wiley & Sons, Hoboken, NJ, pp. 29–36.
31. Needleman, S.B. and C.D. Wunsch, A general method applicable to the search for similarities in the amino acid sequence of two proteins. *Journal of Molecular Biology*, 1970. 48(3): pp. 443–453.
32. Feng, D.F. and R.F. Doolittle, Progressive sequence alignment as a prerequisite to correct phylogenetic trees. *Journal of Molecular Evolution*, 1987. 25(4): pp. 351–360.
33. Chenna, R., et al., Multiple sequence alignment with the Clustal series of programs. *Nucleic Acids Research*, 2003. 31(13): pp. 3497–3500.
34. Dayhoff, M., R. Schwartz, and B. Orcutt, 22 a model of evolutionary change in proteins. *Atlas of Protein Sequence and Structure*, 1978. 5: pp. 345–352.
35. Mitra, D., et al., Evolution of Bioinformatics and its impact on modern bio-science in the twenty-first century: Special attention to pharmacology, plant science and drug discovery. *Computational Toxicology*, 2022. 24: p. 100248.
36. Giani, A.M., et al., Long walk to genomics: History and current approaches to genome sequencing and assembly. *Computational and Structural Biotechnology Journal*, 2020. 18: pp. 9–19.
37. Ishino, Y., et al., Nucleotide sequence of the IAP gene, responsible for alkaline phosphatase isozyme conversion in Escherichia coli, and identification of the gene product. *Journal of Bacteriology*, 1987. 169(12): pp. 5429–5433.
38. Jiang, F. and J.A. Doudna, CRISPR-Cas9 structures and mechanisms. *Annual Review of Biophysics*, 2017. 46: pp. 505–529.
39. Mojica, F.J., et al., Biological significance of a family of regularly spaced repeats in the genomes of Archaea, Bacteria and mitochondria. *Molecular Microbiology*, 2000. 36(1): pp. 244–246.
40. Alkhnbashi, O.S., et al., CRISPR-Cas bioinformatics. *Methods*, 2020. 172: pp. 3–11.
41. Ralph, R., et al., 2019-nCoV (Wuhan virus), a novel Coronavirus: Human-to-human transmission, travel-related cases, and vaccine readiness. *The Journal of Infection in Developing Countries*, 2020. 14(01): pp. 3–17.
42. Tan, M., et al., Applying the digital data and the bioinformatics tools in SARS-CoV-2 research. *Computational and Structural Biotechnology Journal*, 2023. 21: pp. 4697–4705.
43. Ray, M., et al., Essential interpretations of bioinformatics in COVID-19 pandemic. *Meta Gene*, 2021. 27: p. 100844.
44. Sayers, E.W., et al., Database resources of the national center for biotechnology information. *Nucleic Acids Research*, 2021. 49(D1): p. D10.
45. Xiao, K., et al., Isolation and characterization of 2019-nCoV-like coronavirus from Malayan pangolins. *Nature*, 2020. 583: pp. 286–289.
46. Li, X., et al., Evolutionary history, potential intermediate animal host, and cross-species analyses of SARS-CoV-2. *Journal of Medical Virology*, 2020. 92(6): pp. 602–611.
47. Liu, P., W. Chen, and J.-P. Chen, Viral metagenomics revealed Sendai virus and coronavirus infection of Malayan pangolins (Manis javanica). *Viruses*, 2019. 11(11): p. 979.
48. Tang, X., et al., On the origin and continuing evolution of SARS-CoV-2. *National Science Review*, 2020. 7(6): pp. 1012–1023.
49. Li, Y. and L. Chen, Big biological data: Challenges and opportunities. *Genomics, Proteomics and Bioinformatics*, 2014. 12(5): pp. 187–189.
50. Greene, C.S. and O.G. Troyanskaya, PILGRM: An interactive data-driven discovery platform for expert biologists. *Nucleic Acids Research*, 2011. 39(suppl_2): pp. W368–W374.

51. Tian, J., Y. Li, and D. Zhao, Basics of molecular biology. *Molecular Imaging: Fundamentals and Applications*, 2013. 2013: pp. 541–601.
52. Nair, A.S., Computational biology & bioinformatics: A gentle overview. *Communications of the Computer Society of India*, 2007. 2: pp. 1–12.
53. Dai, X. and L. Shen, Advances and trends in omics technology development. *Frontiers in Medicine*, 2022. 9: p. 911861.
54. Babu, M. and M. Snyder, Multi-omics profiling for health. *Molecular & Cellular Proteomics*, 2023. 22(6): p. 100561.
55. Ogbe, R.J., D.O. Ochalefu, and O.B. Olaniru, Bioinformatics advances in genomics-A review. *International Journal of Current Research and Review*, 2016. 8(10): p. 5.
56. Hu, T., et al., Next-generation sequencing technologies: An overview. *Human Immunology*, 2021. 82(11): pp. 801–811.
57. Zwart, H., Human genome project: History and assessment. In Wright, J.D. (Eds.), *International Encyclopedia of Social & Behavioral Sciences*. 2nd ed. 2015: Elsevier, Oxford, pp. 311–317.
58. Collins, F.S. and L. Fink, The human genome project. *Alcohol Health and Research World*, 1995. 19(3): p. 190.
59. Wang, X., et al., The international human genome project (HGP) and China's contribution. *Protein & Cell*, 2018. 9(4): pp. 317–321.
60. Tateno, Y., et al., DNA Data Bank of Japan (DDBJ) for genome scale research in life science. *Nucleic Acids Research*, 2002. 30(1): pp. 27–30.
61. Hubbard, T., et al., Ensembl 2005. *Nucleic Acids Research*, 2005. 33(suppl_1): pp. D447–D453.
62. Wheeler, D.L., et al., Database resources of the national center for biotechnology information. *Nucleic Acids Research*, 2005. 33(suppl_1): pp. D39–D45.
63. Altschul, S.F., et al., Gapped BLAST and PSI-BLAST: A new generation of protein database search programs. *Nucleic Acids Research*, 1997. 25(17): pp. 3389–3402.
64. Pearson, W.R. and D.J. Lipman, Improved tools for biological sequence comparison. *Proceedings of the National Academy of Sciences*, 1988. 85(8): pp. 2444–2448.
65. Thompson, J.D., et al., The CLUSTAL_X windows interface: Flexible strategies for multiple sequence alignment aided by quality analysis tools. *Nucleic Acids Research*, 1997. 25(24): pp. 4876–4882.
66. Katoh, K. and D.M. Standley, MAFFT multiple sequence alignment software version 7: Improvements in performance and usability. *Molecular Biology and Evolution*, 2013. 30(4): pp. 772–780.
67. Burr, T., Phylogenetic trees in bioinformatics. *Current Bioinformatics*, 2010. 5(1): pp. 40–52.
68. Wani, M.Y., et al., Advances and applications of bioinformatics in various fields of life. *International Journal of Fauna and Biological Studies*, 2018. 5(2): pp. 03–10.
69. Teufel, A., et al., Current bioinformatics tools in genomic biomedical research. *International Journal of Molecular Medicine*, 2006. 17(6): pp. 967–973.
70. Moriya, Y., et al., KAAS: An automatic genome annotation and pathway reconstruction server. *Nucleic Acids Research*, 2007. 35(Web Server issue): pp. W182–W185.
71. Nishimura, Y., et al., ViPTree: The viral proteomic tree server. *Bioinformatics*, 2017. 33(15): pp. 2379–2380.
72. Takami, H., et al., Evaluation method for the potential functionome harbored in the genome and metagenome. *BMC Genomic s*, 2012. 13: p. 699.
73. Yamanishi, Y., J.P. Vert, and M. Kanehisa, Supervised enzyme network inference from the integration of genomic data and chemical information. *Bioinformatics*, 2005. 21(Suppl 1): pp. i468–i477.
74. Yamanishi, Y., et al., Drug-target interaction prediction from chemical, genomic and pharmacological data in an integrated framework. *Bioinformatics*, 2010. 26(12): pp. i246–i254.
75. Wilkins, M.R., et al., Progress with proteome projects: Why all proteins expressed by a genome should be identified and how to do it. *Biotechnology and Genetic Engineering Reviews*, 1996. 13(1): pp. 19–50.
76. Vihinen, M., Bioinformatics in proteomics. *Biomolecular Engineering*, 2001. 18(5): pp. 241–248.
77. Domon, B. and R. Aebersold, Challenges and opportunities in proteomics data analysis. *Molecular & Cellular Proteomics*, 2006. 5(10): pp. 1921–1926.
78. Aslam, B., et al., Proteomics: Technologies and their applications. *Journal of Chromatographic Science*, 2016. 55(2): pp. 1–15.
79. Al-Amrani, S., et al., Proteomics: Concepts and applications in human medicine. *World Journal of Biological Chemistry*, 2021. 12(5): pp. 57–69.
80. Croft, D., et al., Reactome: A database of reactions, pathways and biological processes. *Nucleic Acids Research*, 2010. 39(suppl_1): pp. D691–D697.
81. Kanehisa, M., et al., KEGG for integration and interpretation of large-scale molecular data sets. *Nucleic Acids Research*, 2012. 40(D1): pp. D109–D114.
82. Mi, H., et al., PANTHER version 6: Protein sequence and function evolution data with expanded representation of biological pathways. *Nucleic Acids Research*, 2007. 35(suppl_1): pp. D247–D252.
83. Schaefer, C.F., et al., PID: The pathway interaction database. *Nucleic Acids Research*, 2009. 37(suppl_1): pp. D674–D679.
84. Franceschini, A., et al., STRING v9. 1: Protein-protein interaction networks, with increased coverage and integration. *Nucleic Acids Research*, 2012. 41(D1): pp. D808–D815.
85. Glaab, E., et al., EnrichNet: Network-based gene set enrichment analysis. *Bioinformatics*, 2012. 28(18): pp. i451–i457.
86. Schmidt, A., I. Forne, and A. Imhof, Bioinformatic analysis of proteomics data. *BMC Systems Biology*, 2014. 8: pp. 1–7.
87. Kerrien, S., et al., The IntAct molecular interaction database in 2012. *Nucleic Acids Research*, 2012. 40(D1): pp. D841–D846.
88. Kandasamy, K., et al., NetPath: A public resource of curated signal transduction pathways. *Genome Biology*, 2010. 11: pp. 1–9.
89. Nalbantoglu, S., Metabolomics: Basic principles and strategies. *Molecular Medicine*, 2019. 10(1): 88563.
90. Rinschen, M.M., et al., Identification of bioactive metabolites using activity metabolomics. *Nature Reviews Molecular Cell Biology*, 2019. 20(6): pp. 353–367.
91. Tsugawa, H., et al., MS-DIAL: Data-independent MS/MS deconvolution for comprehensive metabolome analysis. *Nature Methods*, 2015. 12(6): pp. 523–526.
92. Pluskal, T., et al., MZmine 2: Modular framework for processing, visualizing, and analyzing mass spectrometry-based molecular profile data. *BMC Bioinformatics*, 2010. 11: pp. 1–11.
93. Smith, C.A., et al., XCMS: Processing mass spectrometry data for metabolite profiling using nonlinear peak alignment, matching, and identification. *Analytical Chemistry*, 2006. 78(3): pp. 779–787.

94. Pfeuffer, J., et al., OpenMS–a platform for reproducible analysis of mass spectrometry data. *Journal of Biotechnology*, 2017. 261: pp. 142–148.
95. Hummel, J., et al., The golm metabolome database: A database for GC-MS based metabolite profiling. In Nielsen, J. and M.C. Jewett (Eds.), *Metabolomics: A Powerful Tool in Systems Biology*. 2007: Springer, Berlin, Heidelberg, pp. 75–95.
96. Wishart, D.S., et al., HMDB 4.0: The human metabolome database for 2018. *Nucleic Acids Research*, 2018. 46(D1): pp. D608–D617.
97. Guijas, C., et al., METLIN: A technology platform for identifying knowns and unknowns. *Analytical Chemistry*, 2018. 90(5): pp. 3156–3164.
98. Ludwig, C., et al., Birmingham Metabolite Library: a publicly accessible database of 1-D 1 H and 2-D 1 HJ-resolved NMR spectra of authentic metabolite standards (BML-NMR). *Metabolomics*, 2012. 8: pp. 8–18.
99. King, Z.A., et al., BiGG models: A platform for integrating, standardizing and sharing genome-scale models. *Nucleic Acids Research*, 2016. 44(D1): pp. D515–D522.
100. Horai, H., et al., MassBank: A public repository for sharing mass spectral data for life sciences. *Journal of Mass Spectrometry*, 2010. 45(7): pp. 703–714.
101. Fahy, E., et al., Update of the LIPID MAPS comprehensive classification system for lipids1. *Journal of Lipid Research*, 2009. 50: pp. S9–S14.
102. Wang, J., et al., A platform to identify endogenous metabolites using a novel high performance Orbitrap MS and the mzCloud Library. *Blood*, 2013. 4: pp. 2–8.
103. Lai, Z., et al., Identifying metabolites by integrating metabolome databases with mass spectrometry cheminformatics. *Nature Methods*, 2018. 15(1): pp. 53–56.
104. Cottret, L., et al., MetExplore: collaborative edition and exploration of metabolic networks. *Nucleic Acids Research*, 2018. 46(W1): pp. W495–W502.
105. Huan, T., et al., Systems biology guided by XCMS Online metabolomics. *Nature Methods*, 2017. 14(5): pp. 461–462.
106. Wang, Z., M. Gerstein, and M. Snyder, RNA-Seq: A revolutionary tool for transcriptomics. *Nature Reviews Genetics*, 2009. 10(1): pp. 57–63.
107. Lowe, R., et al., Transcriptomics technologies. *PLoS Computational Biology*, 2017. 13(5): p. e1005457.
108. Fodor, S.P., et al., Light-directed, spatially addressable parallel chemical synthesis. *Science*, 1991. 251(4995): pp. 767–773.
109. Schena, M., et al., Quantitative monitoring of gene expression patterns with a complementary DNA microarray. *Science*, 1995. 270(5235): pp. 467–470.
110. Nagalakshmi, U., et al., The transcriptional landscape of the yeast genome defined by RNA sequencing. *Science*, 2008. 320(5881): pp. 1344–1349.
111. Marioni, J.C., et al., RNA-seq: An assessment of technical reproducibility and comparison with gene expression arrays. *Genome Research*, 2008. 18(9): pp. 1509–1517.
112. Hung, K.-S., et al., Functional enrichment analysis based on long noncoding RNA associations. *BMC Systems Biology*, 2018. 12: pp. 109–118.
113. Montgomery, S.B., et al., Transcriptome genetics using second generation sequencing in a Caucasian population. *Nature*, 2010. 464(7289): pp. 773–777.
114. Byron, S.A., et al., Translating RNA sequencing into clinical diagnostics: Opportunities and challenges. *Nature Reviews Genetics*, 2016. 17(5): pp. 257–271.
115. Trapnell, C., L. Pachter, and S.L. Salzberg, TopHat: Discovering splice junctions with RNA-Seq. *Bioinformatics*, 2009. 25(9): pp. 1105–1111.
116. Goff, L.A., C. Trapnell, and D. Kelley, CummeRbund: Visualization and exploration of Cufflinks high-throughput sequencing data. *R Package Version*, 2012. 2: 2.
117. Ghosh, S. and C.-K.K. Chan, Analysis of RNA-seq data using TopHat and Cufflinks. *Plant Bioinformatics: Methods and Protocols*, 2016. 1374: pp. 339–361.
118. Kanehisa, M. and S. Goto, KEGG: Kyoto encyclopedia of genes and genomes. *Nucleic Acids Research*, 2000. 28(1): pp. 27–30.
119. Haw, R., et al., Reactome pathway analysis to enrich biological discovery in proteomics data sets. *Proteomics*, 2011. 11(18): pp. 3598–3613.
120. Altschul, S.F., et al., Basic local alignment search tool. *Journal of Molecular Biology*, 1990. 215(3): pp. 403–410.
121. Pearson, W.R., BLAST and FASTA similarity searching for multiple sequence alignment. *Multiple Sequence Alignment Methods*, 2014. 1079: pp. 75–101.
122. Wheeler, D. and B. Medha, BLAST quickstart. In Bergman, N.H. (Eds.), *Comparative Genomics*. 2007: Humana Press, Totoma, NJ, p. 149.
123. Brown, D. and K. Sjölander, Functional classification using phylogenomic inference. *PLoS Computational Biology*, 2006. 2(6): p. e77.
124. Kaari, M., et al., Application of bioinformatic tools for phylogenetic analysis. *Methods in Actinobacteriology*, 2022: 2002: pp. 187–191.
125. Felsenstein, J., *PHYLIP (Phylogeny Inference Package), Version 3.5 c*. 1993: Department of Genetics, University of Washington, Seattle.
126. Tamura, K., et al., MEGA4: Molecular evolutionary genetics analysis (MEGA) software version 4.0. *Molecular Biology and Evolution*, 2007. 24(8): pp. 1596–1599.
127. Schmidt, T., A. Bergner, and T. Schwede, Modelling three-dimensional protein structures for applications in drug design. *Drug Discovery Today*, 2014. 19(7): pp. 890–897.
128. Dorn, M., et al., Three-dimensional protein structure prediction: Methods and computational strategies. *Computational Biology and Chemistry*, 2014. 53: pp. 251–276.
129. Tovchigrechko, A., C.A. Wells, and I.A. Vakser, Docking of protein models. *Protein Science*, 2002. 11(8): pp. 1888–1896.
130. Arnold, K., et al., The SWISS-MODEL workspace: A web-based environment for protein structure homology modelling. *Bioinformatics*, 2006. 22(2): pp. 195–201.
131. Webb, B. and A. Sali, Comparative protein structure modeling using MODELLER. *Current Protocols in Bioinformatics*, 2016. 54(1): pp. 5.6.1–5.6.37.
132. Zhang, Y., I-TASSER server for protein 3D structure prediction. *BMC Bioinformatics*, 2008. 9: pp. 1–8.
133. Evans, R., et al., Protein complex prediction with AlphaFold-Multimer. *biorxiv*, 2021. https://doi.org/10.1101/2021.10.04.463034.
134. Wu, S. and Y. Zhang, LOMETS: A local meta-threading-server for protein structure prediction. *Nucleic Acids Research*, 2007. 35(10): pp. 3375–3382.
135. Källberg, M., et al., Template-based protein structure modeling using the RaptorX web server. *Nature Protocols*, 2012. 7(8): pp. 1511–1522.
136. de Almeida Paiva, V., et al., Protein structural bioinformatics: An overview. *Computers in Biology and Medicine*, 2022. 147: p. 105695.

137. Carugo, O. and K. Djinović-Carugo, Half a century of Ramachandran plots. *Acta Crystallographica Section D: Biological Crystallography*, 2013. 69(8): pp. 1333–1341.
138. Kaplan, W. and T.G. Littlejohn, Swiss-PDB viewer (deep view). *Briefings in Bioinformatics*, 2001. 2(2): pp. 195–197.
139. Laskowski, R., M. MacArthur, and J. Thornton, *PROCHECK: Validation of Protein-Structure Coordinates*. 2006: John Wiley & Sons, Hoboken, NJ.
140. Xu, D., Y. Xu, and C. Uberbacher, Computational tools for protein modeling. *Current Protein and Peptide Science*, 2000. 1(1): pp. 1–21.
141. DeMillo, R.A., *Mutation Analysis as a Tool for Software Quality Assurance*. 1980: School of Information and Computer Science, Georgia Institute of Technology, Atlanta, GA.
142. Rodrigues, C.H., D.E. Pires, and D.B. Ascher, DynaMut2: Assessing changes in stability and flexibility upon single and multiple point missense mutations. *Protein Science*, 2021. 30(1): pp. 60–69.
143. Dayem Ullah, A.Z., et al., SNPnexus: Assessing the functional relevance of genetic variation to facilitate the promise of precision medicine. *Nucleic Acids Research*, 2018. 46(W1): pp. W109–W113.
144. Ferrer-Costa, C., et al., PMUT: A web-based tool for the annotation of pathological mutations on proteins. *Bioinformatics*, 2005. 21(14): pp. 3176–3178.
145. Hopf, T.A., et al., Mutation effects predicted from sequence co-variation. *Nature Biotechnology*, 2017. 35(2): pp. 128–135.
146. Marshall, G.R., Computer-aided drug design. *Annual Review of Pharmacology and Toxicology*, 1987. 27(1): pp. 193–213.
147. Sokouti, B., A review on in silico virtual screening methods in COVID-19 using anticancer drugs and other natural/chemical inhibitors. *Exploration of Targeted Anti-tumor Therapy*, 2023. 4(5): p. 994.
148. Meng, X.-Y., et al., Molecular docking: A powerful approach for structure-based drug discovery. *Current Computer-Aided Drug Design*, 2011. 7(2): pp. 146–157.
149. van Zundert, G.C.P., et al., The HADDOCK2.2 web server: User-friendly integrative modeling of biomolecular complexes. *Journal of Molecular Biology*, 2016. 428(4): pp. 720–725.
150. Pierce, B.G., et al., ZDOCK server: Interactive docking prediction of protein-protein complexes and symmetric multimers. *Bioinformatics*, 2014. 30(12): pp. 1771–1773.
151. Yan, Y., et al., The HDOCK server for integrated protein–protein docking. *Nature Protocols*, 2020. 15(5): pp. 1829–1852.
152. de Vries, S.J., et al., The pepATTRACT web server for blind, large-scale peptide–protein docking. *Nucleic Acids Research*, 2017. 45(W1): pp. W361–W364.
153. Kozakov, D., et al., The ClusPro web server for protein–protein docking. *Nature Protocols*, 2017. 12(2): pp. 255–278.
154. Jones, G., et al., Development and validation of a genetic algorithm for flexible docking. *Journal of Molecular Biology*, 1997. 267(3): pp. 727–748.
155. Tanchuk, V.Y., et al., A new, improved hybrid scoring function for molecular docking and scoring based on AutoDock and AutoDock Vina. *Chemical Biology & Drug Design*, 2016. 87(4): pp. 618–625.
156. Grosdidier, A., V. Zoete, and O. Michielin, SwissDock, a protein-small molecule docking web service based on EADock DSS. *Nucleic Acids Research*, 2011. 39(suppl_2): pp. W270–W277.
157. Hollingsworth, S.A. and R.O. Dror, Molecular dynamics simulation for all. *Neuron*, 2018. 99(6): pp. 1129–1143.
158. Guvench, O. and A.D. MacKerell, Comparison of protein force fields for molecular dynamics simulations. *Molecular Modeling of Proteins*, 2008. 443: pp. 63–88.
159. Hernández-Rodríguez, M., et al., Current tools and methods in molecular dynamics (MD) simulations for drug design. *Current Medicinal Chemistry*, 2016. 23(34): pp. 3909–3924.
160. Wang, J., et al., Development and testing of a general amber force field. *Journal of Computational Chemistry*, 2004. 25(9): pp. 1157–1174.
161. Brooks, B.R., et al., CHARMM: A program for macromolecular energy, minimization, and dynamics calculations. *Journal of Computational Chemistry*, 1983. 4(2): pp. 187–217.
162. Phillips, J.C., et al., Scalable molecular dynamics with NAMD. *Journal of Computational Chemistry*, 2005. 26(16): pp. 1781–1802.
163. Thompson, A.P., et al., LAMMPS-a flexible simulation tool for particle-based materials modeling at the atomic, meso, and continuum scales. *Computer Physics Communications*, 2022. 271: p. 108171.
164. Chow, E., et al., Desmond performance on a cluster of multicore processors. *Simulation*, 2008. 1: pp. 1–14.
165. Ma, W., *AMMP-EXTN: A User Privacy and Collaboration Control Framework for a Multi-User Collaboratory Virtual Reality System*. 2007: Georgia State University, Atlanta, GA.
166. Krieger, E. and G. Vriend, New ways to boost molecular dynamics simulations. *Journal of Computational Chemistry*, 2015. 36(13): pp. 996–1007.
167. Armstrong, D.R., et al., PDBe: Improved findability of macromolecular structure data in the PDB. *Nucleic Acids Research*, 2020. 48(D1): pp. D335–D343.
168. Dassault Systèmes, *Free Download: BIOVIA Discovery Studio Visualizer*. 2020: Dassault Systèmes, France.
169. Vilar, S., G. Cozza, and S. Moro, Medicinal chemistry and the molecular operating environment (MOE): Application of QSAR and molecular docking to drug discovery. *Current Topics in Medicinal Chemistry*, 2008. 8(18): pp. 1555–1572.
170. Hayes, J.M., An integrated visualization and basic molecular modeling laboratory for first-year undergraduate medicinal chemistry. *Journal of Chemical Education*, 2014. 91(6): pp. 919–923.
171. Pettersen, E.F., et al., UCSF Chimera—a visualization system for exploratory research and analysis. *Journal of Computational Chemistry*, 2004. 25(13): pp. 1605–1612.
172. DeLano, W.L., Pymol: An open-source molecular graphics tool. *CCP4 Newsletter on Protein Crystallography*, 2002. 40(1): pp. 82–92.
173. Humphrey, W., A. Dalke, and K. Schulten, VMD: visual molecular dynamics. *Journal of Molecular Graphics*, 1996. 14(1): pp. 33–38.
174. Hopkins, A.L., Network pharmacology. *Nature Biotechnology*, 2007. 25(10): pp. 1110–1111.
175. Zhang, B., X. Wang, and S. Li, An integrative platform of TCM network pharmacology and its application on a herbal formula, Qing-Luo-Yin. *Evidence-based Complementary and Alternative Medicine*, 2013. 2013(1): p. 456747.
176. Berger, S.I. and R. Iyengar, Network analyses in systems pharmacology. *Bioinformatics*, 2009. 25(19): pp. 2466–2472.
177. Kibble, M., et al., Network pharmacology applications to map the unexplored target space and therapeutic potential of natural products. *Natural Product Reports*, 2015. 32(8): pp. 1249–1266.
178. Kanehisa, M., et al., KEGG: New perspectives on genomes, pathways, diseases and drugs. *Nucleic Acids Research*, 2017. 45(D1): pp. D353–D361.

179. Snel, B., et al., STRING: A web-server to retrieve and display the repeatedly occurring neighbourhood of a gene. *Nucleic Acids Research*, 2000. 28(18): pp. 3442–3444.
180. Kuhn, M., et al., STITCH: Interaction networks of chemicals and proteins. *Nucleic Acids Research*, 2007. 36(suppl_1): pp. D684–D688.
181. Chandran, U., et al., Network pharmacology. *Innovative Approaches in Drug Discovery*, 2017. 14: pp. 127–164.
182. Chen, Y. and Y. Li, *Computational Intelligence Assisted Design: In Industrial Revolution 4.0*. 2018: CRC Press, Boca Raton, FL.
183. Shah, R.M. and R. Gautam, Overcoming diagnostic challenges of artificial intelligence in pathology and radiology: Innovative solutions and strategies. *Indian Journal of Medical Sciences*, 2023. 75(3): pp. 107–113.
184. Maqsood, K., H. Hagras, and N.R. Zabet, An overview of artificial intelligence in the field of genomics. *Discover Artificial Intelligence*, 2024. 4(1): p. 9.
185. Mavaddat, N., et al., Polygenic risk scores for prediction of breast cancer and breast cancer subtypes. *The American Journal of Human Genetics*, 2019. 104(1): pp. 21–34.
186. Mehrotra, R., et al., A transfer learning approach for AI-based classification of brain tumors. *Machine Learning with Applications*, 2020. 2: p. 100003.
187. Schork, N.J., Artificial intelligence and personalized medicine. *Cancer Treatment Reviews*, 2019. 178: pp. 265–283.
188. Sebastian, A.M. and D. Peter, Artificial intelligence in cancer research: Trends, challenges and future directions. *Life (Basel)*, 2022. 12(12): p. 1991.

2 Structural Bioinformatics

Firdous Fatima[1], Sunita Dahiya[2], Arun K. Iyer[3], and Rajiv Dahiya[4]

[1] Department of Pharmacy, Shri Shankaracharya Professional University, Bhilai, India

[2] Department of Pharmaceutical Sciences, School of Pharmacy, University of Puerto Rico, San Juan, Puerto Rico

[3] Department of Pharmaceutical Sciences, Use-Inspired Biomaterials and Integrated Nano Delivery (U-BiND) Systems Laboratory, School of Pharmacy, Boston, Massachusetts

[4] School of Pharmacy, Faculty of Medical Sciences, The University of the West Indies, St. Augustine, Trinidad & Tobago

INTRODUCTION

Structural bioinformatics is a subdiscipline of bioinformatics that focuses on the analysis and prediction of the three-dimensional (3D) structures of biological macromolecules—primarily proteins, nucleic acids, and complexes thereof. This sector plays a vital role in the early stages of drug development and innovation, delivering insights into molecular function, interaction specificity, and the structural basis of disease. By combining computer models, structural databases, and simulations of molecular movements, structural bioinformatics helps researchers understand the shape and behaviors of biological targets at a very detailed level. Drug innovation relies increasingly on structure-based techniques to design and optimize therapeutic medicines. Traditional drug discovery methods often involved testing large collections of chemicals on biological targets without a clear understanding of their structure. In contrast, modern methods use detailed structural information to guide structure-based drug design (SBDD), which helps scientists predict how small molecules will interact with specific protein binding sites. This technique enhances the precision of hit identification, lead optimization, and the prediction of off-target consequences. Recent improvements in techniques like X-ray crystallography, cryo-electron microscopy (cryo-EM), and nuclear magnetic resonance (NMR) spectroscopy have made it easier to obtain high-quality structural data. Simultaneously, computational methods, including homology modeling, molecular docking, and molecular dynamics simulations, have advanced to a point where they can consistently fill in the gaps when experimental structures are lacking or incomplete. Researchers now frequently use these methods to discover therapeutics against challenging targets like protein–protein interactions and inherently disordered areas. Moreover, the incorporation of artificial intelligence and machine learning into structural bioinformatics has further accelerated medication innovation. Deep-learning-based structure prediction algorithms, such as AlphaFold and RoseTTAFold, have greatly boosted the amount of reliably predicted protein structures, even for proteins previously thought "undruggable."

Over the years, DNA sequencing has gone through major improvements, showing how far science has come in understanding the blueprint of life. At the heart of this progress is next-generation sequencing (NGS), a technology that changed the game by allowing scientists to read millions of pieces of DNA all at once. This made sequencing much faster and more affordable than before. Following this, third-generation sequencing technologies, such as those developed by PacBio and Oxford Nanopore, pushed the boundaries even further. These new tools can read much longer strands of DNA in real time, giving researchers more accurate information and helping them study parts of the genome that were previously difficult to analyze. But these advances wouldn't mean much without bioinformatics—a powerful set of computer-based tools that help make sense of the massive amounts of data produced by sequencing. Bioinformatics helps scientists put the DNA pieces together, figure out what the genes do, and find important differences that might be linked to diseases [1]. Together, modern sequencing technologies and bioinformatics are giving us deeper insights into how living things work and are driving exciting new developments in medicine, genetics, and biology (Figure 2.1). This revolution in predictive structural biology is altering how pharmaceutical corporations and academic researchers approach target identification, draggability assessment, and ligand development [1,2].

DECIPHERING THE CODE OF LIFE: A FOUNDATION IN MACROMOLECULAR STRUCTURE

Proteins, the workhorses of the cellular world, represent a symphony of structure and function. These intricate biopolymers, meticulously crafted from a diverse repertoire of amino acids, engage in a mesmerizing dance with other biomolecules. Their remarkable resilience allows them to thrive in a dynamic environment, adapting to fluctuations in pH, temperature, and ionic strength. This functional versatility is a direct consequence of their exquisitely sculpted structures, a testament to the relentless optimization driven by evolution.

DOI: 10.1201/9781003464679-2

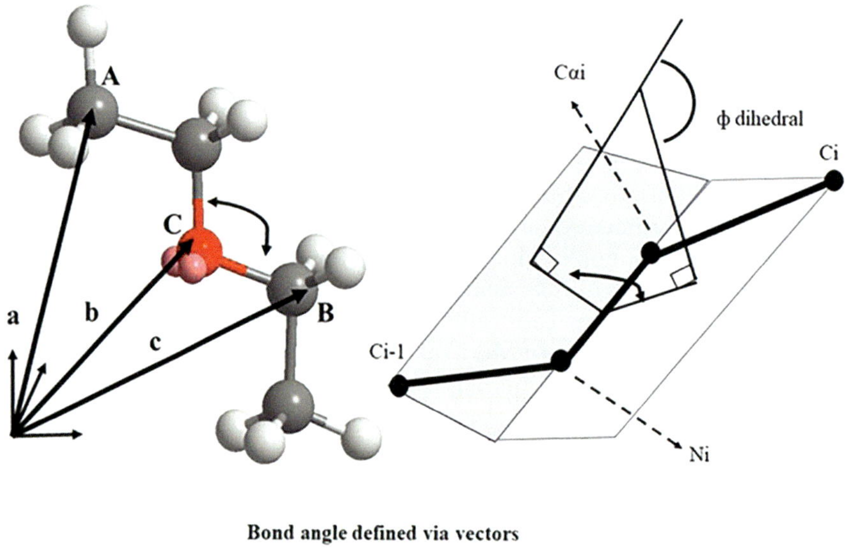

FIGURE 2.1 Diagram of bond angle and angle between these two planes is referred to as the φ (phi) angle associated with residue *i*.

Primary Structure

Proteins, the versatile biomolecules that orchestrate life processes, are meticulously constructed from a defined set of 20 amino acids. These amino acids serve as the building blocks, each featuring a central carbon atom adorned with a carboxyl group (–COOH), an amino group ($-NH_2$), a hydrogen atom, and a unique side chain (R-group). This R-group, the defining characteristic of each amino acid, can be polar, nonpolar, or charged, influencing the overall properties and function of the protein. Interestingly, with the exception of glycine, amino acids exhibit chirality, existing in mirror-image forms (D and L). However, nature has a preference, and only L-amino acids are typically incorporated into proteins by the cellular machinery.

The assembly of proteins begins with a covalent bond, the peptide bond, forged between the carboxyl group of one amino acid and the amino group of another. This elegant linkage creates a linear polymer chain, technically a polypeptide. When a polypeptide reaches a significant length and adopts a specific 3D structure. The seemingly simple peptide bond, the linchpin of protein assembly, exerts a profound influence on the 3D structure of polypeptides. Its planar and rigid nature restricts rotations within the chain, allowing flexibility only around the bonds formed by the alpha carbons (Cα). These crucial angles, designated phi (φ) and psi (ψ), dictate the possible contortions of the polypeptide backbone. However, the bulky side chains of amino acids further limit this freedom by creating steric clashes, restricting the φ and ψ angles to a narrow range of permissible values. This inherent constraint dictates the repertoire of conformations a polypeptide can adopt.

The Ramachandran plot, a graphical representation of these allowed and disallowed φ and ψ angles, serves as a map for protein folding. With a few exceptions, like the highly flexible glycine and the conformationally restricted proline, amino acid side chains dictate the accessible regions on this map. Glycine simple hydrogen side chain minimizes steric hindrance, granting it greater flexibility and expanding its allowed conformational space. Conversely, proline's unique structure, with its side chain covalently linked to the Cα carbon, significantly restricts its conformational freedom, shrinking its permissible region on the Ramachandran plot [2].

Secondary Structure

The intricate folding of a protein chain begins with its local conformation, known as the secondary structure. Pioneering work by Pauling, Corey, and Branson (1951) predicted the existence of alpha helices and beta sheets based on the inherent limitations of polypeptide chains. These secondary structures exhibit a striking degree of regularity, with a characteristic and repeating pattern of φ and ψ angles along the polypeptide backbone. The formation of hydrogen bonds between backbone atoms within these structures creates a particularly stable and energetically favorable conformation for the polypeptide chain. While α-helices and β-sheets represent the well-defined elements of protein secondary structure, irregular regions also play a crucial role [3].

The Elegant Coiling of the α-Helix

Proteins often adopt a fascinating structural motif known as the alpha helix. This regular, right-handed coil arises from the gentle curvature of the polypeptide backbone. Left-handed counterparts are disfavored due to steric clashes between bulky side chains. Among right-handed helices, the α-helix regions supreme, characterized by a pattern of 3.6 amino acids per turn. The stability of the α-helix hinges on a network of hydrogen bonds. These strategic linkages form between the carbonyl oxygen of one amino acid and the amide hydrogen four residues ahead in the sequence. This intricate arrangement creates a satisfying handshake for all backbone hydrogen bonds within the α-helix, except at the ends. Interestingly, this pattern also generates a distinct polarity within the helix. All carbonyl groups point in the same direction, while amide groups face the opposite way. This alignment of individual amino acid dipoles imbues the α-helix with a directionality, with a negative to positive charge gradient from C-terminus to N-terminus.

The propensity of amino acids to participate in α-helices varies based on their side chain complexity. Alanine, glutamate, and leucine, boasting compact side chains, are frequent residents of α-helices. In contrast, bulky side chains like those found in tryptophan are less well suited for the close quarters within the helix. Aspartate, asparagine, and serine, with their side chain hydrogen bond donors and acceptors, can form competing interactions with the backbone, disrupting the core helical structure and making them less favorable α-helix inhabitants. Glycine and proline also act as helix breakers.

3_{10}-Helix and pi (π) Helix

Beyond the ubiquitous α-helix, proteins can occasionally incorporate more exotic helical structures: the tightly coiled 3_{10}-helix and the loosely packed pi (π) helix. The 3_{10}-helix features a unique hydrogen bonding pattern, where each amino acid forms a bond with its counterpart three residues ahead, resulting in three residues per turn. In contrast, the rarer π-helix exhibits a more open conformation with hydrogen bonds established between residues five positions apart, leading to an average of 4.4 residues per turn. Interestingly, both the 3_{10}-helix and the π-helix are typically confined to the terminal regions of α-helices, suggesting a potential role in stabilizing or modulating these common protein motifs.

The Entwined Strands of β-Sheets

Beta (β) sheets represent another cornerstone of protein secondary structure, distinct from the coiled elegance of α-helices. The magic of the β-sheet lies in the alignment of its dihedral angles (φ and ψ), allowing the strands to lie nearly flat against each other (180°). This arrangement produces a pleated sheet, with the amino acid side chains projecting outward, perpendicular to the plane of the sheet. β-sheets come in two main types: parallel and antiparallel. In both, the α-carbons of adjacent strands huddle close together, with their side chains pointing in the same direction. Parallel sheets, however, are less common and require β-strands to be very distant in the overall protein sequence (like in β-α-β motifs). This parallel arrangement leads to less stable inter-strand hydrogen bonds. Here, the strands run in opposing directions, bringing the N-terminus of one strand face to face with the C-terminus of its neighbor. This perfect alignment allows for optimal formation of strong, parallel inter-strand hydrogen bonds.

Loops and Turns

Protein architecture extends beyond the well-defined α-helices and β-sheets. Disordered regions, encompassing loops (coils) and turns, play a crucial role in connecting these ordered secondary structural elements. Notably, glycine, asparagine, and proline are frequently encountered in turns, potentially due to their structural flexibility. Interestingly, loop regions in many proteins house the active site, the catalytic center responsible for enzymatic function. Hairpin loops, also known as reverse turns, constitute the most prevalent loop type, typically comprising 4–5 amino acids. These compact structures facilitate a 180° reversal of the polypeptide chain direction, often stabilized by internal hydrogen bonds and frequently incorporating proline or glycine residues.

Tertiary Structure

The biological functionality of a protein hinges on its tertiary structure, the 3D conformation adopted by the polypeptide chain. The driving force behind this folding is the hydrophobic effect, where hydrophobic side chains sequester themselves away from the surrounding hydrophilic environment by migrating toward the protein core. However, the intricate architecture of the tertiary structure is further stabilized by a multitude of intermolecular interactions, including hydrogen bonding, salt bridges, covalent disulfide bridges, and weak Van der Waals forces. This remarkable folding process facilitates the convergence of active site residues, often located at disparate regions along the polypeptide chain, enabling them to form a cohesive catalytic pocket responsible for substrate binding and subsequent catalysis. Although the overall tertiary structure might appear irregular and devoid of symmetry, a closer examination reveals a modular composition of conserved super-secondary structures known as motifs.

Motifs

Beyond the fundamental building blocks of α-helices and β-sheets, proteins assemble more complex structures called motifs. These motifs serve as the essential subunits of protein architecture, meticulously crafted by combining various secondary structures in specific arrangements. This interplay gives rise to a diverse repertoire of motifs, each with unique functional implications.

1. **Helix-Loop-Helix (HLH):** This elegant motif features two α-helices connected by a compact loop. Often found in DNA-binding proteins, HLH typically comprises a longer, basic amino-acid-rich helix responsible for DNA interaction, tethered to a shorter, secondary helix.
2. **Helix-Turn-Helix:** When two α-helices are bridged by a loop that takes a sharp turn, effectively folding back the polypeptide chain. This motif is a frequent resident in protein domains dedicated to DNA binding.

3. **Beta (β) Hairpin:** Resembling a hairpin bend, this motif consists of two β-strands linked by a loop. This loop facilitates a reversal in the direction of the polypeptide chain within the strands. β-hairpins can exist as individual motifs or contribute to the formation of continuous antiparallel β-sheets.
4. **Beta-Alpha-Beta (β-α-β) Motif:** This prevalent motif is particularly suited for proteins with parallel β-sheets. The C-terminus of the first β-strand and the N-terminus of the second are connected by a unique loop-α-helix-loop configuration. In the 3D structure, parallel β-sheets often lie flat, with the intervening α-helices positioned above this plane. Loop lengths can vary significantly within different β-α-β motifs, and some proteins even harbor catalytic sites within these loop regions.
5. **Greek Key Motif:** This visually striking motif lives up to its name, resembling an ornamental Greek key. Four adjacent antiparallel strands form the core, with three connected by two hairpin loops. The fourth strand sits close to the first, linked to the third by a longer loop.

Protein Folds and Domains

Protein folds represent intricate 3D arrangements that emerge from the assembly of simpler structural units termed motifs. Domains, defined as autonomously folding, sizable protein subunits, embody this concept. They are characterized by a conserved fold and often a dedicated function, acting as quasi-independent modules within the larger protein architecture.

The Power of Helices: Unveiling α-Domains

Alpha (α) domains represent a powerful architectural principle in proteins, built exclusively from the elegant coiling of α-helices. These domains can be further classified based on the arrangement of these helices, with two prominent examples being four-helix bundles and the globin fold.

Four-Helix Bundle: A Packing Masterclass

The four-helix bundle is a remarkably common fold found in diverse proteins. It embodies a strategy of efficiency, where four antiparallel α-helices pack tightly together. This intimate arrangement creates a hydrophobic core within the protein, effectively shielding these water-fearing residues from the surrounding aqueous environment. Meanwhile, the hydrophilic (water-loving) residues are strategically positioned on the exterior, readily interacting with the solvent. This clever segregation ensures optimal stability and function for the protein.

Globin Fold: A Haven for Heme

The globin fold is a signature structural motif found in the globin protein family, including well-known members like hemoglobin and myoglobin. This fold boasts a sophisticated arrangement of eight α-helices that come together to create a dedicated pocket for binding an essential molecule called heme. Interestingly, the C-terminal region of the globin fold features a distinctive helix-turn-helix motif, where two helices adopt an antiparallel orientation. The remaining helices in the domain huddle together, strategically positioned at an angle of roughly 50°.

The Majesty of Sheets: Unveiling β-Domains

Beta (β) domains showcase the remarkable versatility of β-sheets in protein architecture. These domains are built solely from β-sheets, each with a distinct arrangement that confers unique properties. Let's explore three prominent examples: up-and-down β-barrels, jelly roll barrels, and β-sandwiches.

Up-and-Down β-Barrel: A Transporting Powerhouse

Imagine a large antiparallel β-sheet gracefully curving around, forming a cylindrical structure like a barrel. This captivating architecture defines the up-and-down β-barrel. The edges of the sheet cleverly connect through hydrogen bonds, closing the barrel and creating a spacious void within. Interestingly, the amino acid side chains strategically alternate their orientation, pointing either above or below the sheet plane. This remarkable design is particularly well suited for membrane proteins, as the central void often serves as a dedicated channel for transporting molecules across the membrane.

Jelly Roll Barrel: Filling the Void with Loops

The jelly roll barrel shares the concept of a single β-sheet wrapping around itself, but with a twist. Here, longer loops traverse the central core, effectively filling the void seen in up-and-down β-barrels. This creates a more compact structure, with the core region typically composed of hydrophobic residues. The jelly roll barrel usually boasts eight β-strands meticulously arranged into two sets of four, forming antiparallel β-sheets.

β-Sandwich: A Stacked Sanctuary

The β-sandwich fold presents a contrasting architectural strategy. Instead of a single, curving sheet, it features two antiparallel β-sheets positioned in parallel planes, resembling a stacked sandwich (hence the name). Unlike β-barrels, this arrangement creates a well-defined hydrophobic core without internal voids.

α+β-Domains: A Blend of Helices and Strands

α+β-domains represent a prevalent class of protein structures characterized by a unique secondary structure. These domains incorporate both α-helices and β-strands arranged sequentially along the polypeptide backbone. Notably, the β-strands within these domains are predominantly antiparallel, meaning their hydrogen bond donors and acceptors point in opposite directions. Several well-characterized folds exemplify this architectural motif, including the ferredoxin fold, the DNA clamp fold, and the Src homology 2 (SH2) domain.

The Ferredoxin Fold: A Signature Motif

The ferredoxin fold stands out as a ubiquitous α+β-protein fold. It's defining characteristic lies in the distinctive βαββαβ secondary structure pattern displayed by its backbone. This fold embodies a single, elongated, and symmetrical hairpin loop that wraps around itself once. This configuration allows the terminal β-strands to establish hydrogen bonds with the central β-strands, resulting in a four-stranded, antiparallel β-sheet flanked on one side by two α-helices.

The DNA Clamp Fold: Gripping the DNA Helix

The DNA clamp fold, also referred to as the sliding clamp, embodies a distinct α+β-protein fold with a crucial function in DNA replication. This fold assembles into a multimeric structure, essentially a ring-shaped complex, that encircles the DNA double helix entirely. The toroidal (doughnut-shaped) architecture of the assembled multimer prevents dissociation from the template strand, thereby tethering the enzyme and enhancing processivity.

The SH2 Domain: Recognizing Phosphorylated Tyrosine

The SH2 domain represents a structurally conserved protein domain present within the Src oncoprotein and numerous other intracellular signaling proteins. This domain features a compact structure composed of two α-helices and seven β-strands, typically spanning around 100 amino acids in length. Notably, the SH2 domain exhibits a high affinity for phosphorylated tyrosine residues within specific peptide motifs, enabling it to recognize short amino acid sequences of three to six residues.

The Art of Duality: Unveiling α/β Domains

Alpha/beta (α/β) domains represent a captivating architectural style in proteins, characterized by a mesmerizing interplay between α-helices and β-strands along the polypeptide backbone. Let's delve into some prominent examples:

TIM Barrel: A Circular Sanctuary

The TIM barrel, famously found in triosephosphate isomerase, exemplifies the α/β barrel structure. Imagine a sheet formed by four repeating β-α-β-α units that elegantly curves around to form a circle. This ingenious design creates a central core composed of parallel β-sheets, shielded by a protective outer layer of α-helices.

Flavodoxin Fold: A Layered Masterpiece

The flavodoxin fold is a widely encountered three-layered α/β protein fold. It features a central core resembling a five-stranded parallel β-sheet, meticulously sandwiched between two distinct layers of α-helices. This intricate arrangement creates a well-defined environment for protein function.

Rossmann Fold: A Specialized Binding Site

The Rossmann fold is a renowned structural motif frequently observed in nucleotide-binding proteins. It boasts a unique architecture: an open, twisted parallel β-sheet flanked by α-helices on both sides. A specific cleft emerges between two parallel sheets connected by a helix, forming a dedicated motif for binding nucleotides.

Leucine-Rich Repeats: A Horseshoe of Strength

The leucine-rich repeat (LRR) motif exemplifies the α/β-horseshoe fold. It is a captivating structure built from repeated units of 20–30 amino acids, remarkably rich in the hydrophobic amino acid leucine. These repeating units, each adopting a β-strand-turn-α-helix fold, assemble together to form a horseshoe-shaped domain. Here, seventeen parallel β-sheets contribute to the inner concave surface, while sixteen interconnecting α-helices form the outer convex surface. This clever arrangement creates a hydrophobic core tightly packed with leucine residues, conferring stability to the entire domain. The LRR fold finds application in proteins like the placental ribonuclease inhibitor.

Quaternary Structure

Many proteins transcend the limitations of single polypeptide chains by forming intricate non-covalent associations with two or more folded subunits. Each of these individual subunits within a multimeric protein is designated as a protomer. When all protomers within a complex are identical, the protein is classified as homomeric. Conversely, if the complex harbors distinct protomers, it is categorized as heteromeric. Interestingly, the active site, the locus of enzymatic activity, can reside either at the interface between protomers or within individual protomers themselves [4].

COMPUTATIONAL METHODS FOR PROTEIN GEOMETRY ASSESSMENT

Interatomic Distance

Distance analysis allows measuring the interatomic distance between two atoms. When analyzing protein structures, determining the spatial relationships between atoms is crucial. In this system, atom A is denoted by its coordinates $(a_x, a_y, a_z)^T$, and atom B is denoted by its coordinates $(b_x, b_y, b_z)^T$. Utilizing the Pythagorean theorem, the distance between atom A and atom B can be calculated using the following formula:

$$d(A,B)=\sqrt{(a_x-b_x)^2+(a_y-b_y)^2+(a_z-b_z)^2}$$

This is the same as the normal calculation:

$$\|a-b\|=\sqrt{(a-b)^T(a-b)}$$

Bond Angle

It can be defined as the geometric angle between the two adjacent bonds of the same molecule. Since normalized vectors u and v are linearly oriented, their inner product can be understood as the cosine of the angle separating them. The following formula is used to express this:

$$\cos\theta = \langle u, v \rangle / (\| u \| \| v \|)$$

If we set $u=a-c$, $v=b-c$, the cosine of the bond angle will be given by:

$$\cos\theta = \frac{(a-c)^T (b-c)}{\| a-c \| \| b-c \|}$$

After determining this value, the necessary bond angle may be obtained with ease by taking the inverse cosine of the value on the right-hand side of the final equation (Figure 2.1).

Dihedral Angles

Unlike bond angles, which are exquisitely defined by the geometry of just three covalently bonded atoms, dihedral angles require a tetrahedral quartet. These four consecutive atoms, linked by single bonds, dictate the rotational freedom around a specific bond within the molecule (Figure 2.2). Consider atoms in the protein backbone: N1 $C\alpha 1$ C1 N2 $C\alpha 2$ C2 N3 $C\alpha 3$ C3 N4 …. As we move from residue to residue, the corresponding bond lengths do not change much.

An intriguing characteristic of protein backbones is the relatively consistent bond length between $N_i - C\alpha_i$ and $N_j - C\alpha_j$, where i and j represent any two residues in the polypeptide chain. Unlike bond lengths and bond angles, dihedral angles are defined by the rotational freedom around a single bond. This "swivel" action plays a pivotal role in determining the 3D positioning of backbone atoms. It's crucial to recognize that altering a dihedral angle has no impact on the bond angles between consecutive backbone atoms. This remarkable concept allows for the relative movement of N_i with respect to $N_{(i+1)}$ solely through a swivelling rotation around the $C_i - C\alpha_i$ bond, while maintaining consistent bond angles throughout the backbone [5].

Calculating the Phi Dihedral Angle

Labeling vectors containing these atoms coordinates with their names makes things simple. Let $u=N_i - C_{i-1}$ and $v=C\alpha_i - N_i$

Then,

$$u \times v = (u_2v_3 - u_3v_2)i + (u_3v_1 - u_1v_3)j + (u_1v_2 - u_2v_1)k$$

Dividing this by its norm will give the unit normal (call it $n(C_{i-1}, N_i, C\alpha_i)$).

Similarly, to calculate the other required normal, we let $v=C\alpha_i - N_i$ and $w=C_i - C\alpha_i$. Then, calculate

$$v \times w = \det\begin{bmatrix} i & j & k \\ v_1 & v_2 & v_3 \\ w_1 & w_2 & w_3 \end{bmatrix} = (v_2w_3 - v_3w_2)i + (v_3w_1 - v_1w_3)j + (v_1w_2 - v_2wk)$$

Finally, with both units normally computed, we can get the value of the φ dihedral for this alpha carbon using the inverse cosine function:

$$\phi = \tau(C_{i-1}, N_i, C\alpha_i, C_i) = \arccos\begin{pmatrix} n(C_{i-1}, N_i, C\alpha_i) \cdot \\ n(N_i, C\alpha_i, C_i) \end{pmatrix}$$

The concept of dihedral angles in protein structures is intricately linked to the projection of the unit normal vector. In the context of protein backbones, a positive dihedral angle translates to a projection of the unit normal vector that aligns with the direction of the final bond within the considered peptide plane. Conversely, a negative dihedral angle results in a projection of the unit normal vector that points in the opposite direction compared to the final bond.

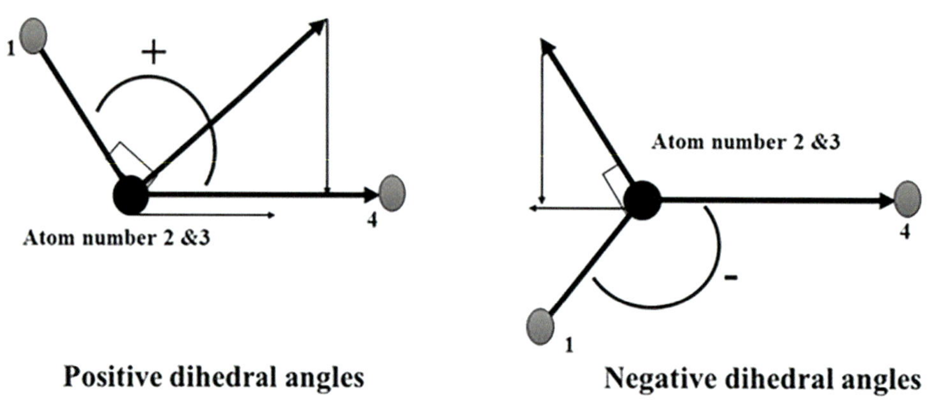

FIGURE 2.2 Diagram of dihedral angles sign.

Calculating the Psi Dihedral Angle

To calculate the ψ dihedral angle, we follow the same analysis but this time compute

$$\psi = \tau(N_i, C\alpha_i, C_i, N_{i+1}) = \arccos\begin{pmatrix} n(N_i, C\alpha_i, C_i)\cdot \\ n(C\alpha_i, C_i, N_{i+1}) \end{pmatrix}$$

INERTIAL AXES

A crucial step in analyzing helical structures involves identifying the optimal helix axis. A compelling strategy to achieve this objective involves positioning the axis such that it intersects the centroid, the geometric center of mass, calculated for all the atoms within the helix. Subsequently, the direction of this axis is meticulously chosen to minimize a specific mathematical term: the sum of the squared perpendicular distances from each atom to the axis itself. This minimization process ensures that the chosen axis represents the most accurate central core around which the helical structure coils.

More precisely, if || *di* || represents the perpendicular distance between atom **a**(*i*) and the helix axis, then we calculate the axis direction so that it minimizes the sum:

$$S = \sum_{i=1}^{N} \|di\| 2$$

After computing T (T is called the inertial tensor), we determine its eigenvalues. The eigenvector that corresponds to the least eigenvalue is the necessary w, which indicates the directions for our axis [8].

$$S = w^T T w = \lambda w^T w = \lambda$$

The inertial axis is a geometric construct that simplifies and is utilized in many research investigations to depict secondary structural elements. This is particularly crucial if one wants to compare related proteins' tertiary structures.

COMPUTATIONAL PREDICTION OF PROTEIN INTERACTIONS THROUGH EVOLUTIONARY COUPLINGS

Comparative modeling is a powerful technique for predicting the 3D structure of proteins. It hinges on the remarkable observation that protein folds and the intricate arrangements of polypeptide chains and exhibits a high degree of conservation even when the amino acid sequences themselves diverge over evolutionary time. This underlying principle allows researchers to transfer atomic coordinates from a protein with a known structure (the template) to a target protein with a similar sequence [6]. The evolutionary pressure to maintain crucial interactions between amino acids within a protein structure plays a vital role. This phenomenon, known as coevolution (CE), dictates that residues in physical contact tend to coevolve, meaning their sequences change in a coordinated fashion throughout evolution. Numerous studies have demonstrated, often through anecdotal evidence, that evolutionary coupling models can effectively identify functionally constrained residues—those essential for protein function—beyond what can be achieved by simply analyzing single-residue conservation patterns. Recently, these models have been further refined to make quantitative predictions of the impact of mutational changes in proteins [7,8].

DETERMINATION METHOD

This study introduces a novel computational method that leverages the concept of evolutionary couplings to predict the impact of disease variants (DVs) at sites with low sequence conservation. The proposed method hinges on a composite score termed the CE score. This score is calculated by multiplying two distinct coevolutionary matrices: the coupling number (CN) and the cost of coupling (CC). A high CN value indicates a functionally important residue, suggesting that a mutation at this position might be detrimental.

In contrast, the CC score assesses the impact of the specific amino acid change introduced by the variant on these evolutionary couplings. It essentially evaluates the evolutionary tolerance for altered amino acid pairs within coupled residues. A high CC score signifies that the variant introduces an uncommon pairing compared to wild-type sequences, suggesting potential functional disruption. Conversely, the CC quantifies the impact of the variants' amino acid alteration on evolutionary couplings and signifies the modified amino acid pairs' evolutionary resilience in the related residues [9,10].

CALCULATING COST OF COUPLING

To understand how altering amino acid types within evolutionarily coupled sites might affect the protein function, we devised a metric called the cost of coupling (CC). This score reflects the evolutionary tolerance for changes in amino acid pairs at coupled residue positions. To calculate the CC score, researchers leverage the concept of entropy, a measure of uncertainty applied to the distribution of amino acid pairs within two aligned columns representing evolutionarily coupled residues (*i* and *j*). The mathematical formula for this entropy is as follows:

$$Si, j = \ln \frac{N!}{\prod \alpha, \beta \; ni, \; j\,(\alpha, \beta)}$$

- N is the total number of amino acid pairs in the aligned columns of the residues *i* and *j*.
- α and β are wild-type amino acids on the residues *i* and *j*.
- $n_{i,j}$ (α, β) the number of α-β amino acid pairs in aligned columns *i* and *j*.

The entropy changes resulting from a variant in residue I paired with its associated residues were averaged to get the coefficient of correlation. If a variation from α to γ arises in residue i, its complementary chain looks like this:

$$\Delta Si, j\left(\alpha \rightarrow \gamma\right) = -\ln \frac{ni,\ j\left(\gamma,\ \beta\right)+1}{ni,\ j\left(\alpha,\ \beta\right)}$$

$$CCi\left(\alpha \rightarrow \gamma\right) = \frac{\Sigma j \int c\ \Delta Si, j\left(\alpha \rightarrow \gamma\right)}{|C|}$$

- γ is the altered amino acid of residue i, and β is the wild-type amino acid for the residue j.
- C is a group of evolutionarily coupled partners of the residue i.

Thus, evolutionary coupling suggests that the two residues are linked to carry out important structural or functional roles.

MOTIF

Motifs: These are relatively short, recurring segments of a protein 3D structure. Notably, motifs are conserved across various, unrelated proteins, implying a shared function. **Domains:** In contrast to motifs, domains represent larger and more intricate regions within a protein. Each domain typically folds into a unique structure and accomplishes a specific function. Remarkably, domains often possess the ability to fold and function independently of the rest of the protein.

Methods Using 3D Motif Finding

The practice of identifying patterns within sets of sequences is known as motif discovery. This task is a "needle in a haystack" problem as, in many circumstances, we do not know what the pattern looks like. It entails sorting through a large number of *K*-mers that aren't in the pattern but happen frequently.

Sequence Complexity

Motif discovery can be hindered by prevalent, uninformative sequences in biological data. The Wootton–Federhen (WF) complexity score helps address this issue by quantifying the complexity of a sequence [11]. The WF complexity measures how many different sequences may be created with the same amount of as n_A, number of Cs n_C, number of Gs n_G, and number of Ts n_T. This can be calculated using a multinomial coefficient that appears in a log in the formula:

$$CW\ F = \frac{1}{N}\log D\left(\frac{N!}{nA!nC!nGnT!}\right)$$

The factor of $1/N$, where N is the number of characters in the sequence, is to ensure the value is between 0 and 1.

Weight Matrices

Weight matrices represent the most versatile approach for capturing motifs in sequences. They function as a probabilistic model, allowing calculation of the likelihood that a given sequence segment truly represents the motif compared to a random "background" sequence.

Probabilistic Models of Motifs

The concept of the position-specific scoring matrix (PSSM), also known as a weight matrix, was developed by Stormo et al. [12]. Under this model, the overall probability of a specific sequence is calculated by multiplying the probabilities of each individual nucleotide within that sequence. This can be mathematically expressed as the product shown in the formula:

$$P(x|R) = \prod^{|x|}_{i=1} p_x[I]$$

Entropy and Information Content

Shannon entropy is a valuable concept in motif finding. It essentially quantifies the uncertainty associated with a particular model, reflecting how unpredictable the sequences generated by that model would be [13]. In the context of the simple background model (using individual nucleotide probabilities), the Shannon entropy is given by the formula:

$$H = -\sum_{b=A}^{T} pb\ \log 2\ pb$$

Relative Entropy

The idea of "relative entropy," sometimes referred to as "information gain" or the Kullback–Leibler divergence, is another helpful tool for calculating the amount of information in a motif [14]. The expression for relative entropy for biological motifs at a specific position i is given by the equation:

$$D(fi \parallel p) = \sum_{b=A}^{T} fib\ \log\left(\frac{fib}{pb}\right)$$

This expression quantifies how different two discrete probability distributions are, in this case, the background frequencies *pb* and one position of our motif probability matrix *fib*.

De novo Motif Finding

Gibbs Sampling

Generally, statistical physics has various uses for Gibbs sampling and is occasionally referred to as a Markov Chain Monte Carlo (MCMC) method. It may come as a surprise to learn how effective this strategy is at identifying motifs [15].

$$P(u) \propto \prod_{C}\prod_{j=1}^{W}\left(\frac{p_{c,j}}{p_{c,0}}\right)^{n_{c,j}(u)}$$

SEQUENCE ALIGNMENT

Cracking the Code: Predicting Gene Function through Sequence Analysis

Understanding a gene function is a crucial step in gene annotation. The first approach involves sequencing the gene and then using sequence alignment to compare it with genes of known function. Sequence alignment arranges these gene sequences side by side, highlighting similarities and differences. Each position in the alignment can be a match (identical or similar residues), a mismatch (substitution), or a gap (insertion/deletion in one sequence). The presence of conserved residues (exact matches or similar substitutions) across multiple sequences indicates a functionally or structurally important region within the gene (Figure 2.3).

Sequence Identity, Sequence Similarity, and Sequence Homology

Percent identity or sequence identity refers to the proportion of identical bases (amino acids) or bases (DNA) in an alignment at the same places in relation to the length of the sequence:

$$\text{Sequence identity} = \frac{\text{Number of identical bases or residues} \times 100}{\min(\text{length } i,\ \text{length } j)}$$

Identity computations do not account for gaps, making identity a less accurate and sensitive metric. Another argument is that if $A = B$ and $B = C$, then A does not necessarily equal C because the shorter of the two sequences is used to measure identity, making it non-transitive [16].

Demystifying Sequence Similarities: Not All Matches Are Created Equal

When analyzing sequences, it is important to distinguish between sequence similarity and sequence homology.

- **Sequence Similarity:** This refers to the presence of similar residues (amino acids in proteins, nucleotides in DNA/RNA) at corresponding positions in an alignment. In proteins, similar substitutions (e.g., chemically similar amino acids) may occur without affecting the protein function or structure. These substitutions are often tolerated by the protein and receive less penalty during alignment scoring. Sequence similarity can be used to estimate evolutionary distance, but it does not necessarily imply a shared ancestor.
- **Sequence Homology:** This is a stricter concept that indicates a common evolutionary origin for the compared sequences. Homology is an absolute term—sequences are either homologous (share a common ancestor) or not.

Key Components of Sequence Alignment

Query Sequence: This is the sequence of interest that serves as the starting point for finding and comparing similar sequences. It can be the entire sequence or a shorter segment.

Homologous Sequences: These are sequences identified as having a common ancestor with the query sequence. Finding homologs is often the first step in sequence analysis.

Substitution Scores: These scores represent the degree of similarity between aligned pairs of residues (amino acids in proteins, nucleotides in DNA/RNA). Substitution matrices provide a reference for assigning these scores.

Gaps: These are insertions or deletions of characters within sequences that are introduced during the alignment process. Gaps are represented by dashes (-) and can be placed in any sequence (query or homolog) to optimize the alignment.

Sequence i	G	A	A	T					
Sequence j	G	A	A	T	A	T	T	G	C
Sequence k	A	C	G	T	G	A	T	A	G

(a)

Sequence 1	G	A	A	T	G	T	C	A	C
Sequence 2	-	-	-	T	G	T	C	A	-

(b)

FIGURE 2.3 (a) Sequence identity considers the smaller, potentially untrustworthy of the two sequences compared. Three fictitious nucleotide sequences, *i*, *j*, and *k*, are shown on the left. (b) Introducing gaps.

Gap Penalty: This score penalizes the introduction of gaps in the alignment. It typically consists of two components: a gap opening penalty (for introducing a new gap) and a gap extension penalty (for extending an existing gap).

Achieving an Optimal Alignment

The choice between global and local sequence alignment depends on the nature of the sequences being compared:

Global Alignment: This method aligns the entire length of both sequences. Global alignments are useful for identifying homologous sequences (sharing a common ancestor) and potentially similar functions (DNA) or structures (proteins).

Local Alignment: This method focuses on aligning specific, similar segments within sequences. It is well suited for sequences with different lengths or that share only certain regions of similarity. Local alignments help identify conserved motifs or domains within sequences that may appear globally dissimilar.

Role of Substitution Matrices

Substitution matrices are like cheat sheets for aligning biological sequences. They provide scores for pairing different residues (amino acids in proteins, nucleotides in DNA/RNA). These scores reflect the likelihood of one residue mutating into another based on evolutionary data.

For Nucleotides: Nucleotide substitution matrices consider the varying probabilities of different substitutions. Transitions (between purines or pyrimidines) are generally more common than transversions (between purines and pyrimidines). These matrices typically have two parameters: mismatch penalty and gap penalty.

For Amino Acids: Amino acid substitution matrices go a step further. They express scores as log odds ratios, which represent the observed likelihood of one amino acid substituting for another, compared to what would be expected by pure chance.

$$\text{Log odds ratio} = \log 2\left(\frac{\text{observed}}{\text{expected}}\right)$$

Gap Penalties

Sequence alignments often involve insertions (indels) in one sequence compared to another. Since indels are thought to be less frequent than substitutions, a penalty is applied whenever a gap is introduced in an alignment. This penalty discourages unnecessary gaps and promotes more realistic alignments [17].

$$G_{\text{total}} = G_o + G_e \bullet L$$

G_{total} = Total gap penalty, G_o = Gap opening penalty, G_e = Gap extension penalty, L = total number of gaps – 1.

There are two components to gap penalties:

Gap Opening Penalty: This penalty applies when a new gap is first introduced in a sequence.

Gap Extension Penalty: This penalty applies when an existing gap is extended by one position.

Fast and Efficient: Word-Based Methods for Large Databases

Word-based methods, also known as k-tuple methods, prioritize speed over finding the absolute best alignment. These methods work by breaking the query sequence into short, fixed-length subsequences called "words" or "k-tuples." They then search the database for sequences containing identical matches to these words. FASTA and BLAST are two prominent examples of word-based methods widely used for database searches [18]. FASTA, developed by Lipman and Pearson, is a landmark sequence alignment software that introduced the FASTA file format (denoted by ">" at the start). FASTA searches sequence databases for matches to a query DNA or protein sequence using local alignment.

Balancing Speed and Sensitivity: The key parameter in FASTA is the k-tuple size, which defines the length of word segments used for initial search. Larger k-tuple values lead to fewer, more specific matches (higher specificity), while smaller k-tuples provide a more sensitive search but with potentially more irrelevant hits.

Beyond the Initial Search: FASTA refines the initial search results using a more detailed local alignment algorithm based on the Smith–Waterman approach. This step provides a more accurate assessment of the similarity between the query sequence and each database sequence.

FASTA, originally designed for protein sequence similarity searches, has evolved into a suite of programs. These programs can now identify regions of local or global similarity between protein or DNA sequences, while also evaluating the statistical significance of the matches.

Basic local alignment search tool (BLAST) is another prominent algorithm developed for sequence comparison. Like FASTA, BLAST can compare a query sequence (DNA or protein) to a vast database of sequences. Its goal is to identify homologs (sequences sharing a common ancestor) or sequences with a certain level of similarity to the query [19].

Multiple Sequence Alignment

Multiple sequence alignment (MSA) is a powerful tool used to compare more than two biological sequences, often focusing on homologs (related genes). By meticulously aligning these sequences, scientists can identify conserved regions—stretches of genetic code that remain remarkably similar across different species. MSAs also serve as a springboard for constructing phylogenetic trees, which map out the evolutionary relationships between different organisms [20,21]. There are two main strategies for generating MSAs:

Progressive MSA: This method starts with a single reference sequence and gradually adds others based on pairwise similarities.

Iterative MSA: This method takes a more holistic approach, realigning all sequences multiple times to refine the overall alignment.

Applications of MSA

- **Evolutionary Detective Work:** MSAs reveal patterns that shed light on the evolutionary history of related sequences.
- **Protein Family Secrets:** By identifying conserved and variable regions within a protein family, MSAs help us understand functional differences and critical residues.
- **Structural and Functional Inferences:** MSAs provide a foundation for making informed predictions about protein structure and function.
- **Phylogenetic Launchpad:** MSAs often serve as the first step in reconstructing phylogenetic trees, which illustrate evolutionary relationships between species.

Clustal Omega is a novel, high-quality aligner that creates alignments for numerous sequences using hidden Markov model profile–profile approaches and seeded guide trees. GCG, FASTA, EMBL, and GenBank are examples of input, and output is FASTA, MSF, etc. [22].

BRIDGING THE GAP: FROM SEQUENCE TO STRUCTURE WITH PROTEIN PREDICTION

Unlocking the secrets of protein structure is a multifaceted endeavor. Scientists employ a diverse arsenal of techniques, each with its own strengths:

Homology Modeling (Template-Based Methods): This approach leverages known protein structures as templates to build models for similar proteins. Imagine using a familiar blueprint to design a new house.

Fold Recognition (Threading): When a close relative isn't available for homology modeling, fold recognition steps in. Here, the protein sequence is threaded through different known folds, identifying the most compatible one.

De Novo (Ab Initio) Methods: These methods build protein structures entirely from scratch, using only the amino acid sequence as a starting point. Imagine constructing a house from its building blocks alone, without any existing blueprints.

Protein-Folding Forces

The information is encoded within the protein amino acid sequence itself, like a secret instruction manual. Scientists believe the most stable folded state (native state) has the lowest overall energy. While some interactions guide overall folding, others refine the precise structure. Imagine building a house: Some forces ensure the basic structure is stable, while others fine-tune details like windows and doors. Predicting protein structures requires a multistep approach. Initially, scientists use simplified models to capture the broad strokes of folding. Then, they refine the model with more complex calculations to achieve atomic-level accuracy. This allows them to see the precise interactions of all the protein building blocks.

Protein Structure Prediction

Protein structure prediction can be tackled through two main approaches:

Template-Based Modeling: This method leverages the known structure of a related protein (template) as a guide to model the unknown structure of the target protein. It focuses on finding and aligning the target sequence to the template structure.

Template-Free Modeling: This approach does not rely on preexisting structural similarities. It can handle proteins with novel folds not present in the Protein Data Bank (PDB). Template-free methods employ large-scale sampling of possible protein conformations and utilize physics-based energy functions to evaluate their stability.

While traditionally distinct, these methods are increasingly incorporating elements from each other. Template-based methods have begun to include energy-based refinement steps to improve model accuracy. Conversely, template-free methods are increasingly using machine learning and fragment-based sampling techniques that exploit information from the structural database. This cross-pollination of ideas is leading to more powerful and versatile protein structure prediction methods (Figure 2.4).

Building on Blueprints: Template-Based Protein Modeling

Template-based modeling is a powerful technique for predicting protein structures. It works in three key steps:

1. **Finding the Right Blueprint:** Scientists search for existing protein structures (templates) that closely resemble the target protein.
2. **Aligning the Pieces:** The target protein amino acid sequence is meticulously aligned to the template structure, ensuring a good match.
3. **Accounting for Differences:** Any variations between the target and template protein, such as insertions or deletions of amino acids, are factored into the final model.

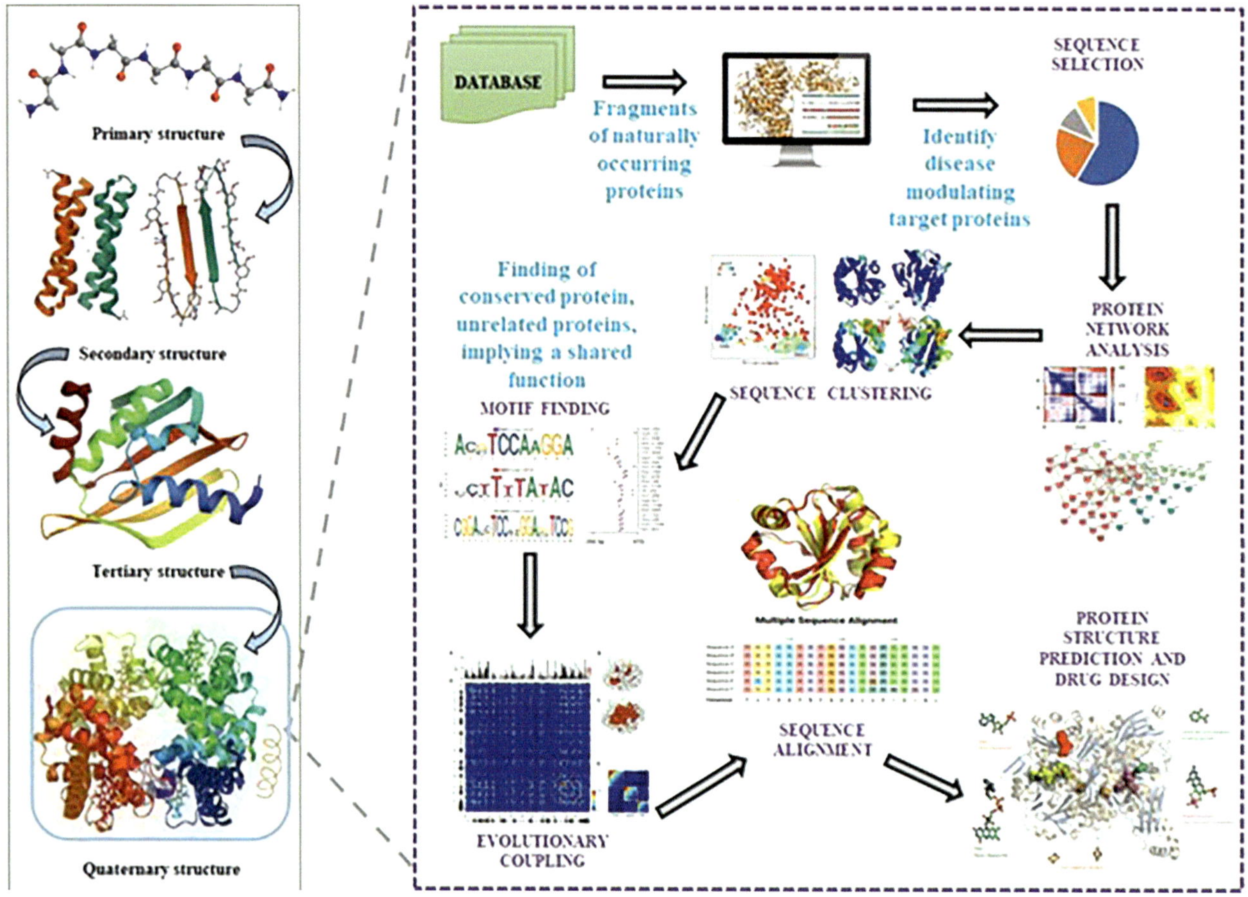

FIGURE 2.4 Schematic representation of protein structure and protein structure prediction.

Template-Free Protein Structure Prediction: Building from Scratch

Template-free modeling tackles protein structures that lack close relatives in the PDB. Here is how it works:

1. **Generating Candidate Models:** Since there is no template to copy, the method needs a strategy to explore a vast number of possible protein shapes (conformational sampling).
2. **Identifying the Best Fit:** A scoring system is employed to rank these candidate models and identify ones that most resemble a real protein structure (native-like conformations).
3. **Leveraging Sequence Information:** The process starts with aligning the target protein sequence with similar ones. This combined information helps predict local features like secondary structure (shape) and backbone angles, as well as nonlocal features like interactions between distant parts of the protein chain.
4. **Building and Refining Models:** Based on these predictions, the method constructs 3D models of the target protein. These models are then further refined, compared against each other, and finally, the most promising candidates are selected as the predicted structures.

In essence, template-free modeling builds protein structures from the ground up, relying on sequence information and computational techniques to explore a vast landscape of possibilities and identify the most likely native conformation.

Refining the Rough Draft: Model Refinement in Protein Structure Prediction

While initial protein structure models can be generated, their accuracy often needs improvement for practical applications. Here is why refinement is crucial:

Limited Detail: The initial models may lack the fine-grained details that differentiate the true structure (e.g., precise side-chain arrangements and hydrogen bonding).

Real-World Applications: These models often lack the atomic-level accuracy needed for tasks like SBDD.

A Booming Database Fuels Progress in Protein Structure Prediction

A breakthrough in template-free protein structure prediction is the ability to predict contacts between amino acids based on how their mutations are correlated across vast protein sequence databases. This progress is driven by two key factors:

1. **The Data Deluge:** The explosion of protein sequence data from next-generation sequencing and metagenomics provides a richer pool of information for analysis.
2. **Statistical Savvy:** Sophisticated statistical methods have been developed to distinguish between direct contacts (due to physical proximity in the folded protein) and indirect correlations arising from other factors.

Designing Proteins from Scratch: The Inverse Folding Challenge

Protein design flips the script on protein folding. This field leverages various protein structure prediction techniques like template-based modeling (homology modeling) and fold recognition (threading). The key difference is that homology modeling relies solely on sequence similarity, while threading considers both sequence and structure of the template protein.

Template-Based Design

Protein design often starts with using existing proteins, like naturally occurring ones, as a foundation. This approach is useful for creating proteins that bind to other proteins, molecules (ligands), or act as enzymes. Each starting protein (template) has unique folds and surfaces, making some a better fit for the desired outcome than others. Scientists can initially pick templates with pockets roughly the size of the target molecule they want to bind. However, to find the best fit, they typically run simulations with various templates to see which ones can form the strongest bond with the target molecule. In the first step of protein design, scientists often choose existing proteins (templates) with desirable qualities. These templates might be stable, easy to produce, or have a specific function, like an enzyme or antibody. The goal is then to modify this starting point to achieve the desired outcome, such as improved stability, function, or activity [23].

Fold Recognition Method

Fold recognition methods help predict the 3D structure of a new protein, even when it has no close relatives with known structures. These methods come in handy when the similarity between the new protein and existing ones falls within a fuzzy area. Two common methods are:

3D Profile Matching: This approach analyzes the chemical properties of the protein building blocks (amino acids) to see if they could fit within the environment of a known protein structure. Imagine checking if the puzzle pieces (amino acids) of the new protein would fit the overall shape (environment) of an existing protein.

Threading: This method "threads" the amino acid sequence of the new protein onto the backbone of known protein structures, one by one. It then checks how well the new protein sequence fits the existing structure, like trying different combinations on a necklace to see which one looks best [24,25].

Ab-Initio or De Novo Design

De novo protein design takes protein creation to a whole new level. Instead of relying on existing structures, it allows researchers to design proteins entirely from scratch. Here is a breakdown of the process:

1. **Choosing the Blueprint (Fold):** The first step involves selecting a desired protein fold or shape. This choice can be driven by curiosity about a specific fold or the intended function of the new protein. For example, beta-barrel folds are known to create pockets for binding small molecules.
2. **Building the Backbone:** Once the fold is chosen, a model of the protein backbone is constructed. However, not all possible backbone conformations are stable or functional. De novo design focuses on methods to create realistic backbones that:
   - Allow for tight packing of amino acid side chains.
   - Facilitate hydrogen bonding through secondary structures (like alpha helices or beta sheets).
   - Avoid strain in the backbone's torsion angles.
3. **Fragment Assembly:** A common technique for building stable backbones involves stitching together small fragments of naturally occurring proteins. These fragments not only provide well-formed secondary structures but can also encode favorable motifs for connecting these structures.
4. **Shaping the Protein:** To achieve the desired 3D structure (tertiary fold), fragment-based folding is often combined with user-defined constraints. These constraints specify the spatial arrangement of the secondary structural elements. This approach has successfully generated various protein folds, including all-helical, repeat proteins, and mixed alpha-beta proteins.
5. **Unveiling Design Principles:** Building idealized folds from fragments provides valuable insights into the physical and structural constraints that govern a protein shape. For instance, designing an all-beta-sheet protein revealed key connections between loop geometry and length in beta-sheet connections. These principles can be applied to design connections in other protein fold types as well.
6. **Exploring Possibilities:** De novo design allows researchers to rapidly generate a multitude of protein backbones, systematically varying specific parameters. This exploration helps identify regions in the design space that promote strong interactions between structural elements, leading to exceptionally stable proteins. Recent successes include designing transmembrane proteins and alpha-helical barrels using this approach.

Finding the Perfect Fit: Sequence Optimization in Protein Design

After generating a protein backbone model, the next step is to find the optimal amino acid sequence that will bring it to life. Sequence optimization software acts like a skilled chef, using two key tools:

Energy Scorecard: An energy function evaluates the favorability of different sequences, like judging the taste of a dish.

Search Strategy: A search protocol explores the sequence landscape, seeking even better combinations of amino acids.

This strategy produced better results in conventional benchmarks for protein design, including structure prediction, free-energy change prediction by mutation, and native-sequence recovery, which attempts to recreate native-like sequences when redesigning naturally occurring proteins [26].

CONCLUSION

Protein structure prediction tools have gotten much better in recent years, but there are still hurdles to overcome. The calculations used to predict and design protein structures are simplified to save time, but this makes it difficult to accurately account for how different parts of the protein interact with each other and with water. This is especially true for protein interfaces, which are important for functions like docking and enzyme activity. One promising solution involves including a small number of specific water molecules in the calculations, but this is complex and expensive to do. As we collect more data on protein structures, we will have more building blocks to work with. This opens up the possibility of designing new protein functions and binding sites. Additionally, machine learning techniques that have been successful in protein structure prediction might also be useful for protein design.

REFERENCES

1. Heather JM, Chain B. The sequence of sequencers: the history of sequencing DNA. *Genomics*. 2016;107(1):1–8.
2. Ramachandran GN, Ramakrishnan C, Sasisekharan V. Stereochemistry of polypeptide chain configurations. *Journal of Molecular Biology*. 1963;7:95–99.
3. Pauling L, Corey RB, Branson HR. The structure of proteins; two hydrogen-bonded helical configurations of the polypeptide chain. *Proceedings of the National Academy of Sciences of the United States of America*. 1951;37(4):205–211.
4. Stollar EJ, Smith DP. Uncovering protein structure. *Essays in Biochemistry*. 2021;65(2):407.
5. Döker R, Maurer T, Kremer W, Neidig K, Kalbitzer HR. Determination of mean and standard deviation of dihedral angles. *Biochemical and Biophysical Research Communications*. 1999;257(2):348–350.
6. Webb B, Sali A. Comparative protein structure modeling using modeller. *Current Protocols in Bioinformatics*. 2016;54:5.6.1–5.6.37.
7. Qian B, Raman S, Das R, et al. High-resolution structure prediction and the crystallographic phase problem. *Nature*. 2007;450(7167):259–264.
8. Lindorff-Larsen K, Piana S, Dror RO, Shaw DE. How fast-folding proteins fold. *Science*. 2011;334(6055):517–520.
9. Kim D, Han SK, Lee K, Kim I, Kong J, Kim S. Evolutionary coupling analysis identifies the impact of disease-associated variants at less-conserved sites. *Nucleic Acids Research*. 2019;47(16):e14.
10. Green AG, Elhabashy H, Brock KP, Maddamsetti R, Kohlbacher O, Marks DS. Large-scale discovery of protein interactions at residue resolution using co-evolution calculated from genomic sequences. *Nature Communications*. 2021;12(1):1396.
11. Wootton JC, Federhen S. Analysis of compositionally biased regions in sequence databases. *Methods in Enzymology*. 1996;266:554–571.
12. Stormo GD, Schneider TD, Gold L, Ehrenfeucht A. Use of the 'Perceptron' algorithm to distinguish translational initiation sites in E. coli. *Nucleic Acids Research*. 1982;10(9):2997–3011.
13. Shannon CE. A mathematical theory of communication. *ACM Sigmobile Mobile Computing and Communications Review*. 2001;5(1):3–55.
14. Kullback S, Leibler RA. On information and sufficiency. *The Annals of Mathematical Statistics*. 1951;22(1):79–86.
15. Geman S, Geman D. Stochastic relaxation, Gibbs distributions, and the bayesian restoration of images. *IEEE Transactions on Pattern Analysis and Machine Intelligence*. 1984;6:721–741.
16. Pearson WR. An introduction to sequence similarity ("homology") searching. *Current Protocols in Bioinformatics*. 2013;3:3.1.1–3.1.8.
17. Smith T, Waterman M. Identification of common molecular subsequences. *Journal of Molecular Biology*. 1981;147:195–197.
18. Lipman DJ, Pearson WR. Rapid and sensitive protein similarity searches. *Science*. 1985;227:1435–1441.
19. Altschul SF, Gish W, Miller W, et al. Basic local alignment search tool. *Journal of Molecular Biology*. 1990;215:403–410.
20. Feng DF, Doolittle RF. Progressive sequence alignment as a prerequisite to correct phylogenetic trees. *Journal of Molecular Evolution*. 1987;25:351–360.
21. Sievers F, Wilm A, Dineen D et al. Fast, scalable generation of high-quality protein multiple sequence alignments using clustal omega. *Molecular Systems Biology*. 2011;7:539.
22. Higgins DG, Sharp PM. Clustal: a package for performing multiple sequence alignment on a microcomputer. *Gene*. 1988;73(1):237–244.
23. Bagchi A. Molecular modeling techniques and in-silico drug discovery. *Methods in Molecular Biology*. 2024;2719:1–11.
24. Bowie JU, Lüthy R, Eisenberg D. A method to identify protein sequences that fold into a known three-dimensional structure. *Science*. 1991;253(5016):164–170.
25. Jones DT, Taylor WR, Thornton JM. A new approach to protein fold recognition. *Nature*. 1992;358(6381):86–89.
26. Kuhlman B, Bradley P. Advances in protein structure prediction and design. *Nature Reviews Molecular Cell Biology*. 2019;20(11):681–697.

3 Metabolic Computing

Arvindra Sahu[1], *Devraj Rajak*[1], *and Varun Kushwah*[2]
[1]Integrated Drug Discovery Research Laboratory, Department of Pharmaceutical Sciences, Dr. Harisingh Gour University (A Central University), Sagar, India
[2]CSIR-Central Drug Research Institute, Lucknow, India

INTRODUCTION

Metabolomics is really the art and science of cell exploration by monitoring the entire range or at least a very large fraction—of the small molecules produced by a cell, called metabolites. They indicate the dynamic chemical processes within a cell and provide researchers with a snapshot of the physiological condition of a cell. Metabolomics has become an important protagonist in the biological sciences over the last decade, with the assistance of high-throughput technologies and sophisticated computational capabilities. What makes metabolomics so useful is the fact that it can study body fluids such as blood, urine, and even feces to see the small molecule byproducts of different metabolic pathways. This gives important information on how our bodies react to disease, treatment, and changes in the environment. In combination with bioinformatics and artificial intelligence (AI), metabolomics becomes even more potent—AI can identify subtle patterns in metabolite data that are perhaps too intricate or masked to conventional analysis, allowing researchers to expose possible drug targets or biomarkers with greater accuracy and efficiency. Being a relatively young discipline after genomics, proteomics, and transcriptomics, metabolomics is regarded as a pillar of systems biology and phenomics. It not only considers individual metabolites but investigates how they interact with each other and with genes or the environment. These interactions create complex metabolic networks that can be modeled and simulated by AI and accelerate drug discovery by predicting the impact of candidate compounds before they reach the laboratory [1–5].

Eventually, the metabolome—the total of metabolites within a biological organism—constitutes an honest-to-goodness real-time measurement of health, disease, and response to therapy. Due to its dynamic and changing character relative to the state of cellular development or disease, it presents an invaluable wealth from which new medicines can be identified. Here again, bioinformatics is crucially involved, helping to process and interpret the plethora of information on metabolite abundances arising through metabolomics experimentation. With the use of AI, scientists are now able to process this data quicker, better, and at a level that was unthinkable before, opening doors to more intelligent, more precise drug development [6,7].

HISTORICAL BACKGROUND

The origins of metabolomics date back to 2000–1500 B.C. This is the earliest application of body fluids in establishing a biological status and thus can be considered one of the first applications of metabolomics. Further early attempts at metabolomics were made in 300 B.C. by Galen. When ancient Greeks first identified that it was necessary to study body fluids (then referred to as humor). From here, the next step in the journey of metabolomics was in 131 A.D. When Galen developed a pathology system that blended the humoral theories of Hippocrates with the Pythagorean theory, the first Metabolomics paper, although not named as such then, was by Robinson and Pauling in 1971. It was entitled "Quantitative Analysis of Urine Vapor and Breath by Gas Liquid Partition Chromatography" and appeared in Proceedings of the National Academy of Sciences [8]. Metabolomics is an important one among the omics technologies [9]. The combination of omics information with computational tools employed in systems biology characterizes systems metabolic engineering. This systems-level metabolic engineering integrational lows the development of novel metabolic pathways and products and rewiring of regulatory circuits which, in turn, facilitates enhanced strain development [10]. Three main approaches are employed in metabolomics research: targeted analysis, metabolite profiling, and metabolic fingerprinting [11,12].

Targeted Analysis

Targeted analysis is the most advanced analytical method in metabolomics. Targeted analysis is applied to quantify a small number of known metabolites accurately. Targeted analysis is a real quantitative method and gives extremely low limits of detection for known metabolites. Targeted analysis can also be performed in a high-throughput manner.

Metabolite Profiling

Metabolite profiling is the quantification of the concentration of a set of metabolites within a sample. Metabolite profiling is a standard protocol in biomedical sciences that is routinely applied to bodily fluids as a diagnosis that helps to describe the patient's state of health [13]. This use of

DOI: 10.1201/9781003464679-3

metabolite profiling for functional genomics is analogous to transcript and protein profiling, and as with those technologies, it will be valuable for determining the full complement of the cell in terms of metabolites—the metabolome [14].

Metabolic Fingerprinting

Metabolic fingerprinting finds its greatest utility in biomarker discovery and diagnostics [15,16]. Fingerprinting is typically conducted with spectroscopic methods like nuclear magnetic resonance (NMR) [17–19], Fourier transform infrared spectroscopy (FT-IR), Fourier transform ion cyclotron resonance mass spectroscopy [20], or mass spectrometry (MS) [21] by simply recording physical spectra without pre-separation procedures such as chromatography or electrophoresis. The drawback of metabolic fingerprinting is that direct comparison of the metabolic pathways is not possible, as the compounds in the samples are not easily recognizable. Metabolic fingerprinting has been extensively employed to identify distinctive metabolic patterns of diseases [22,23].

METABOLOMICS PROCESS

The basis of metabolomic strategies in biomarker discovery (Table 3.1) is rooted in the hypothesis that a given biological stimulus—e.g., the onset of illness—tunes the body's normal biochemical processes. These changes result in measurable differences in the metabolome, or metabolic phenotype, distinguishable from that of healthy non-diseased subjects. In essence, the presence of disease upsets the normal metabolic balance, producing a characteristic chemical fingerprint that metabolomics seeks to identify and define. To accurately identify such disease-associated metabolic changes, a metabolomics experiment needs to be carefully designed and conducted. Reproducibility is paramount, particularly when dealing with intricate biological samples and high-dimensional data. This necessitates not only the choice of reliable and sensitive analytical methods with a broad range of metabolites that can be covered, but also the incorporation of sophisticated computational tools. The application of bioinformatics and chemometrics is necessary at this phase. These technologies enable the extraction of informative patterns from raw data, comparison of differences between patient groups, and determination of biological significance. With this integrated strategy—elaborate data capture followed by clever data analysis and interpretation—scientists have an improved understanding of the extent to which metabolic profiles differ between healthy and diseased subjects and, ultimately, the pursuit of robust biomarkers (Figure 3.1) [24–27].

Sample Preparation

The metabolomics pipeline starts with one of its most essential steps: sample preparation. This is the basis for guaranteeing the reliability and quality of downstream analysis. A standard pipeline encompasses a number of key steps, such as the instant quenching of metabolic functions to lock the original composition of metabolites, gentle extraction of these metabolites, and usually chemical derivatization to make them easier to detect during analysis. This is then followed by instrumental analysis, computational data processing with or without recourse to existing databases, and ultimately statistical evaluation and biological interpretation [28]. At the center of this operation is the biological sample itself—blood plasma, serum, urine, saliva, tissue biopsies, or cultured cells. The protocol employed for sample collection, handling, and extraction of metabolites from them is a determinant of the outcome of a metabolomics study [29]. Though some standardized methods have been devised for various sample types, there remains a need for more harmonization in the collection and processing methods to reduce variability and increase reproducibility between studies. For sample preparation prior to analysis, physical procedures like homogenization or pulverization are utilized. These tend to reduce the material to particle form, provide greater surface area, and facilitate interaction with the selected extraction buffer. The choice of this buffer is optimized based on the chemical type of target metabolites. Combined with AI-facilitated metabolic computing platforms, this procedure becomes all the more influential. Sophisticated algorithms are now able to evaluate the quality of sample preparation in real time, forecast possible sources of variability, and even propose optimized extraction strategies from historical datasets. These smart systems

TABLE 3.1
Major Approach in Metabolomics

Approach	Advantage	Disadvantage
Target analysis	Quantitative	Limited number of compounds can be targeted
	Low limit of detection	Does not detect compounds that were not targeted
	High throughput	Targeted compounds must be available purified for calibration
Metabolite profiling	Global (not targeted)	Semi-quantitative
Metabolic fingerprinting	Global (not targeted)	Majority of peaks are not identifiable
	Directly applicable to pattern recognition	Difficult informatics
	Highest throughput	Medium throughput

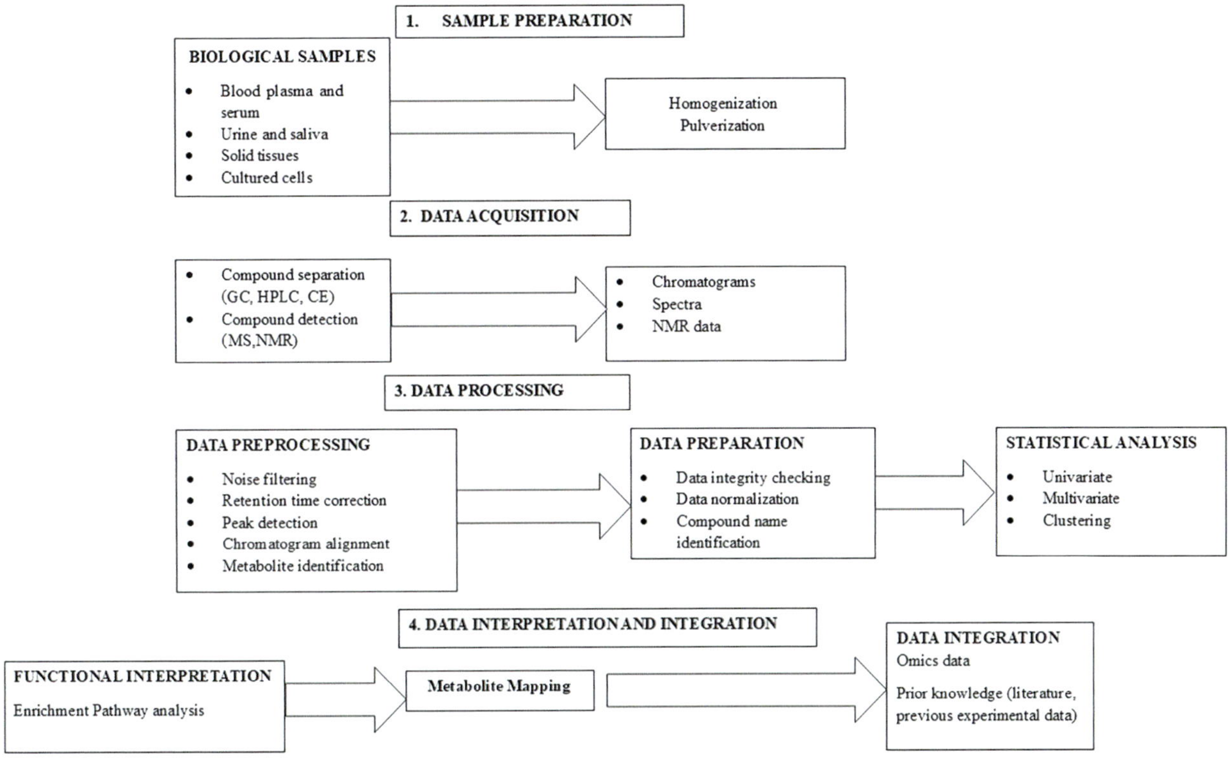

FIGURE 3.1 Flowchart of a typical metabolomic study. After sample preparation, specific metabolic signals are acquired using heterogeneous analytical platforms (data acquisition). Raw signals are then preprocessed to produce data in a suitable format for univariate and multivariate statistical analyses. For untargeted studies, metabolites have first to be identified from spectral information (data processing). Significantly expressed metabolites are then linked to the biological context, through enrichment and pathway analysis, and mapped into networks. Finally, metabolomic data are integrated with other "omics" data and with prior knowledge to gain a comprehensive view of the molecular processes involved (data interpretation and integration).

improve the reproducibility and stability of metabolomic studies, which is particularly critical in drug development, where even minor inconsistencies can cause significant downstream effects. By combining standardized sample handling with AI-driven computational analysis, metabolic computing provides a strong foundation for precision biomarker discovery and therapeutic innovation.

Data Acquisition in AI-Driven Metabolomics

After the biological sample has been adequately prepared, the subsequent key step in the metabolomics pipeline is data acquisition. During this step, sophisticated analytical methodologies are employed to isolate and detect the chemically heterogeneous collection of metabolites contained within the sample. Separation methods such as gas chromatography, high-performance liquid chromatography, ultra-high-performance liquid chromatography, and capillary electrophoresis are used extensively to separate various classes of metabolites. These are subsequently paired with very sensitive detection systems like MS or NMR spectroscopy for the identification and quantification of the compounds of interest [30–32]. MS provides excellent sensitivity and a wide dynamic range, while NMR gives unparalleled structural information and excellent reproducibility, although it is usually less sensitive. In drug innovation, AI and bioinformatics technologies are transforming this phase by maximizing both data acquisition and quality control. Machine learning methods can now support real-time decision-making during acquisition—e.g., selecting the best analytical approach for a particular sample type or dynamically adjusting instrument settings according to data quality indicators. In addition, AI can assist in spectral deconvolution and pattern recognition to facilitate the faster and more accurate identification of new drug candidates or disease-specific biomarkers. By combining these robust analytical tools with intelligent computational algorithms, scientists can optimize metabolome coverage and technical variability. This combination drastically improves the validity of metabolomics data in drug discovery and enables more accurate, data-driven therapeutic approaches.

Data Processing

Data processing in metabolomics encompasses a number of important steps to ready raw analytical outputs—like chromatograms, spectra, or NMR data—for significant interpretation. These involve noise reduction, retention time correction, peak detection, and chromatogram alignment [33]. These processes are carried out using commercial software (e.g., SIEVETM [34]) as well as open-source software such as XCMS [35], MAVEN, and MZmine 2 [36,37].

Following preprocessing, metabolite identification is performed with databases like HMDB and METLIN [38,39]. Statistical analysis is usually initiated with univariate and multivariate approaches to emphasize important metabolic changes, followed by data mining approaches to identify biologically meaningful metabolite patterns [40]. AI and bioinformatics augment this process by automating pattern detection, enhancing precision, and speeding up the identification of drug-relevant biomarkers.

Data Interpretation and Integration in Metabolomics

The last step of the metabolomics pipeline is essential for converting analytical data into actionable biological knowledge. After statistically significant metabolites are detected, scientists need to put them into context within relevant biochemical pathways and biological functions. This process includes three core elements: functional interpretation, metabolite mapping and visualization, and data integration. Collectively, they are a bridge connecting raw data and actionable knowledge, especially useful in drug target discovery and disease biomarker identification.

Functional interpretation incorporates the application of enrichment analysis and pathway analysis, frequently in tandem as pathway-enrichment analysis. These procedures contrast the ranked list of statistically significant metabolites with predetermined biological pathways or functional sets to ascertain their possible roles in cellular processes [41]. Packages like MetaCore™ are frequently applied in such a scenario, providing powerful functionalities for enrichment and pathway analysis, which are drug discovery and disease research oriented. MetaCore™ enables researchers to utilize its complete database or apply filters to narrow down to certain compound subsets. Statistical significance is calculated with the hypergeometric mean, and results are ranked and presented in histogram form, showing the most impacted pathways [42,43]. MetaboAnalyst is another widely used platform that consists of more than one tool for pathway analysis. It has metabolite set enrichment analysis (MSEA) and metabolomics pathway analysis (MetPA) [44]. MSEA offers three modes of enrichment: over-representation analysis (ORA), single sample profiling (SSP), and quantitative enrichment analysis (QEA) [34]. All of them generate interactive outputs, i.e., graphs and tables, that are linked directly to extended pathway diagrams and disease information. MetPA, in contrast, provides statistical utilities like Fisher's exact test, hypergeometric test, global test, and GlobalAncova for more detailed pathway analysis [45]. These tools, upon being augmented by AI, can compare datasets at high speed to large pathway databases and allow for hypothesis-driven searching in drug discovery.

Metabolite mapping involves the assignment of identified metabolites to known pathways of biology. This is critical in organizing unmapped metabolite data into readable visual forms which are easier to interpret. The metabolites are represented as nodes in biological networks, where the interactions and relationships between them can be seen [46]. InCroMAP facilitates such mapping by providing two forms of visual output: integrated visualization, which superimposes omics data onto chosen pathways, and a global metabolic overview, which combines multi-omics data and emphasizes pathway significance through color-coded metrics [47]. These visualizations reveal concealed patterns, relationships, and potential drug targets that may be overlooked in tabular data.

The final, and possibly most challenging, stage of the interpretation process is **data integration.** In contemporary bioinformatics pipelines, this is normally done by integrating metabolomics data with other omics levels—e.g., genomics, transcriptomics, or proteomics—to give a better overall view of the biological system. Integration is done mainly through network-based visualization by high-throughput platforms, aligning different data types in a common biological context. Such integration is possible with tools such as MetaCore™, InCroMAP, and 3Omics, which present gene expression, protein abundance, and metabolite levels within the same pathway or interaction map [48]. Although this integrated view maximizes the chances of discovering key regulatory pathways and biomarkers, caution is required by users since visualizations have the potential to oversimplify intricate interactions or introduce misinterpretations. Data integration and interpretation are the pinnacle of the metabolomics pipeline, converting numerical information into biological stories. With the power of AI and bioinformatics, this becomes not only more effective but also more informative, allowing researchers to detect disease mechanisms and therapeutic targets with unprecedented accuracy.

POTENTIAL BIOMARKERS UNVEILED BY METABOLOMICS

Metabolomics has been a revolutionary technology in identifying and validating new biomarkers, providing in-depth insights into disease mechanisms and therapeutic value. By using untargeted metabolomics, scientists have identified vast numbers of potential metabolite biomarkers in a variety of diseases. These biomarkers, individually or in combination, can provide high diagnostic utility by indicating disease-specific metabolic derangements. For instance, initial metabolomic analyses have been able to identify unique metabolic signatures of diseases such as myocardial ischemia, and this implies the potential of using certain metabolites for early diagnosis and disease state differentiation [49]. Yet the discovery of a promising biomarker is merely the first step toward a rigorous validation process. For a biomarker to be useful in the clinic, it must prove to offer added value over existing diagnostic norms. This entails providing new information regarding the pathophysiology of a disease and enhancing clinical decision-making through greater sensitivity, specificity, or prognostic significance [50,51]. Such an example is the research into sarcosine as a biomarker for prostate cancer diagnosis. Early results indicated its potential in not only detecting prostate cancer but also predicting its progression and perhaps directing treatment options. Still, future validation trials in larger groups of patients are necessary to demonstrate these advantages and determine their clinical usefulness [52]. Here, bioinformatics and AI tools are of critical importance to speed up biomarker discovery and validation. Machine learning algorithms have the ability to dig out discriminative patterns in complicated metabolomic data, whereas network-based methods assist in integrating these biomarkers into larger regulatory and metabolic networks. As the field of metabolomics evolves further, it holds great potential for providing precision biomarkers that will revolutionize diagnostics and inform new drug development.

ANALYTICAL METHODS IN METABOLOMICS: IMPLICATIONS FOR AI AND DRUG DISCOVERY

Metabolomics has emerged as a key part of systems biology, providing information on cellular metabolism with important implications for drug discovery and personalized medicine. Analytical methods (Table 3.2) like NMR, MS, FTIR, gas chromatography–mass spectrometry (GC-MS), liquid chromatography–mass spectrometry (LC-MS), and capillary electrophoresis–mass spectrometry (CE-MS) play a central role in detecting and quantifying metabolites. When combined with bioinformatics and AI, such technologies facilitate large-scale, high-throughput analysis of data to hasten biomarker discovery, therapeutic response prediction, and identification of drug targets [53,54].

Nuclear Magnetic Resonance

NMR spectroscopy is a building block of metabolomics owing to its nondestructive and reproducible nature as well as its ability to determine simultaneously a wide range of organic molecules present in biological samples [54]. It provides a global overview of metabolic alterations through the ability to quantify metabolites present in concentrations of micromoles, making it extremely useful in the clinical arena where sample availability is not always optimal [55,56]. The most frequently used formats are ^{1}H-NMR, ^{31}P-NMR, and ^{13}C-NMR, with distinct advantages each depending on the issue of interest to biological research. High-resolution magic angle spinning (HR-MAS) advances the application of the technique to solid and semisolid materials. The incorporation of AI within NMR data analysis facilitates automated deconvolution of spectra, detection of anomalies, and pattern identification—thus facilitating

TABLE 3.2
Comparative Overview of Analytical Techniques in Metabolomics

Analytical method	Advantages	Disadvantages	Applications in AI & drug discovery
NMR (nuclear magnetic resonance)	• Rapid and nondestructive analysis • High spectral resolution • No need for derivatization • Reusable samples	• Low sensitivity (micromolar range) • Complex and overlapping spectra • Multiple peaks per metabolite • Limited library utility in complex matrices	• AI-enabled spectral deconvolution • Biomarker discovery in cancer and neurology • Monitoring metabolic flux and drug metabolism
GC-MS (gas chromatography-mass spectrometry)	• High sensitivity and robustness • Wide linear dynamic range • Extensive public/commercial libraries	• Time consuming • Requires derivatization • Not suitable for thermally unstable or high-MW compounds	• Disease stratification models • Metabolic profiling of volatile organics • Drug toxicity screening with machine learning
LC-MS (liquid chromatography-mass spectrometry)	• No derivatization typically required • Multiple separation modes (HILIC, RP) • High sensitivity for a wide polarity range	• Slower throughput • Limited standardized libraries • Complex data interpretation	• Deep phenotyping using AI • Lead compound identification • Pharmacometabolomics and personalized therapy
CE-MS (capillary electrophoresis-mass spectrometry)	• High resolution and separation efficiency • Requires small sample volumes • Can profile cations, anions, and neutrals in one run • Usually no derivatization	• Limited commercial libraries • Low reproducibility of retention times • Susceptible to matrix effects	• AI-assisted profiling of charged metabolites • Ideal for limited clinical samples • Identifying ionic biomarker panels for disease
FTIR (Fourier transform infrared spectroscopy)	• Very rapid analysis • Whole-sample metabolic fingerprint • Minimal preparation and no derivatization	• Highly convoluted spectra • Requires dry samples • Lower specificity compared to MS-based methods	• Pattern recognition with AI/ML • Useful in preliminary screening and sample triage • Environmental exposure assessment in population health

biomarker discovery pipelines and enabling drug development approaches [57–59]. NMR has been found especially useful in the metabolomic analysis of cancer and other diseases of complexity, which facilitates early diagnosis and personalized treatment plans [60,61].

Mass Spectrometry

MS is among the most sensitive and flexible metabolomic analysis tools, with the ability to detect a broad variety of metabolites over a wide range of polarity and molecular weight spectra. MS-based profiling can be achieved using direct infusion techniques or in conjunction with separation methods like gas chromatography, liquid chromatography, ion chromatography, or capillary electrophoresis [62,63]. MS identification relies on the characteristic fragmentation patterns of metabolites, which are separated according to their mass-to-charge (m/z) ratio [64]. High-performance high-resolution MS instruments such as Orbitrap®, time-of-flight (mass spectrometry (TOF-MS), and FT-ICR MS offer improved mass accuracy and are perfectly suited for characterizing complex biological mixtures [65,66]. When combined with machine learning algorithms, MS information can be used to design predictive models for disease diagnosis and drug efficacy. These methodologies are indispensable in contemporary drug discovery, especially in lead compound identification and pharmacodynamic marker validation [12].

Fourier-Transform Infrared Spectroscopy

FTIR spectroscopy offers fast, reproducible metabolic fingerprinting and is therefore amenable to high-throughput experiments in clinical metabolomics. The method requires little or no sample preparation and can detect molecular vibrations specific to certain metabolites. Although spectral interpretation is challenging, chemometric algorithms—those augmented by AI included—can identify minute differences between biological conditions, allowing for effective classification and biomarker identification [8]. FTIR simplicity and amenability to automation make it an attractive tool for point-of-care diagnostic use and drug screening in the early stages.

Gas Chromatography-Mass Spectrometry

GC-MS has been a workhorse in metabolomics for the profiling of volatile and derivatized low-molecular-weight compounds. It is especially useful for detecting organic acids, amino acids, sugars, sugar alcohols, and fatty acids [67,68]. Although derivatization increases separation, it can introduce artifacts that influence data interpretation. GC-MS does not offer spatial information regarding metabolite distribution, thus limiting its use in tissue-level diagnostics. However, with the integration into AI tools, GC-MS information can serve to build accurate disease classifiers and pharmacometabolomic models with increased therapeutic monitoring accuracy and predicted drug response [69].

Liquid Chromatography-Mass Spectrometry

LC-MS is found in common applications across metabolomics owing to high sensitivity and generality over a wide range of metabolite polarity and molecular masses [70–72]. In contrast to GC-MS, LC-MS does not typically involve derivatization, retaining the original chemical structure of metabolites. It can be tailored for targeted or untargeted analysis, based on the research goal. Soft ionization techniques such as electrospray ionization (ESI) have enhanced LC-MS use in biomedical research. When combined with machine learning methods, LC-MS data help in pharmacokinetic modeling, toxicity profiling, and the generation of predictive diagnostics, which are all key elements of precision medicine and drug discovery [73].

Capillary Electrophoresis-Mass Spectrometry (CE-MS)

CE-MS is becoming an increasingly promising metabolomics technique because it has outstanding resolving power, minimal sample consumption, and fast analysis time. It is especially applicable to the profiling of ionic and polar metabolites, allowing for the concurrent detection of cations, anions, and neutral molecules within a single run [74–77]. This broad profiling is critical for delineating disease-related metabolic changes and drug-induced modifications. AI-fortified CE-MS platforms support intricate pattern recognition and classification operations, enabling the discovery of therapeutic biomarkers and the individualized optimization of treatment protocols [78,79].

All these analytical methods, though potent by themselves, are most effective when combined with AI and bioinformatics. Together, they enable researchers to traverse the vast complexity of metabolomics information, converting raw signals into usable knowledge. For drug discovery, this translates to quicker identification of drug targets, better biomarker validation, and ultimately, more personalized and targeted therapies.

APPLICATION OF METABOLOMICS

Metabolomics methods have been used to study many diseases, including breast cancer, kidney cancer, cardiovascular disease, bladder cancer, esophageal cancer and gastric cancer, kidney disease, natural metabolic errors, toxicological effects, and nutrition [80,81]. Metabolomics is used in several areas, including plant biotechnology, toxicology, pharmacology, and medicine, where it is employed to study numerous human diseases, to improve diagnosis, identify biomarkers, clarify pathological mechanisms, develop new drugs, and evaluate its action mechanism and efficacy [82]. In medicine, metabolomics is helpful to understand, diagnose, and treat different disorders, such as endocrine problems or cancer. Oncology is one of the scientific fields that most uses metabolomics. Carcinogenic cells present a great proliferative rate and a high demand for energy and usually lose some regulatory functions. Thus, carcinogenic cells have different metabolic requirements than those of normal cells [83]. Metabolomics is useful to quickly identify and measure hundreds of metabolites, allowing to recognize of several diseases and evaluate their state much earlier. Finally, metabolomics has also been applied combined with genome-wide application studies (mGWAS) for their interest in epidemiological and precision medicine scopes [84]. Key applications are discussed as follows:

Toxicity Assessment/Toxicology: Metabolic profiling (especially of urine or blood plasma samples) can be used to detect the physiological changes caused by toxic insult of a chemical (or mixture of chemicals). In many cases, the observed changes can be related to specific syndromes, e.g., a specific lesion in liver or kidney. This is of particular relevance to pharmaceutical companies wanting to test the toxicity of potential drug candidates: If a compound can be eliminated before it reaches clinical trials on the grounds of adverse toxicity, it saves the enormous expense of the trials.

Functional Genomics: Metabolomics can be an excellent tool for determining the phenotype caused by a genetic manipulation, such as gene deletion or insertion. Sometimes this can be a sufficient goal in itself—for instance, to detect any phenotypic changes in a genetically modified plant intended for human or animal consumption. More exciting is the prospect of predicting the function of unknown genes by comparison with the metabolic perturbations caused by the deletion/insertion of known genes. Such advances are most likely to come from model organisms such as *Saccharomyces cerevisiae* and *Arabidopsis thaliana*.

Nutrigenomics: It is a generalized term which links genomics, transcriptomics, proteomics, and metabolomics to human nutrition. In general, a metabolome in a given body fluid is influenced by endogenous factors such as age, sex, body composition, and genetics as well as underlying pathologies. The large bowel microflora are also a very significant potential confounder of metabolic profiles and could be classified as either an endogenous or exogenous factor. The main exogenous factors are diet and drugs. Diet can then be broken down to nutrients and non-nutrients. Metabolomics is one means to determine a biological endpoint, or metabolic fingerprint, which reflects the balance of all these forces on an individual's metabolism [8].

Cancer Research: Cancer is still a major problem, with tens of millions of people dying of cancer every year; therefore, how to treat and conquer cancer is a contemporary issue of the times. Understanding the causes of cancer and its influencing factors is important for understanding cancer. Because cancer is affected by many factors, such as environment, irrational diet, and genetic factors, researchers use metabolomics to describe the effects of various factors on cancer [85]. MS used to explore the metabolic changes of ovarian cancer and identified 18 metabolites closely related to ovarian cancer. LC/MS was used to study the metabolic changes of pancreatic cancer tissues and found that seven metabolites may be potential biomarkers. A combination of LC-MS and GC-MS was used to study the effect of macranthoidin B on the total metabolites of colorectal cancer cells. It was found to inhibit colorectalcancercellproliferationbyinducingreactive-oxygen-species-mediated apoptosis. Metabolomics technology is used to discover the mechanism of action in cancer; cancer is a special disease that is difficult to cure, and clinical samples of cancer patients are difficult to obtain. Therefore, further development has been hindered [86].

Nutrition: The application of metabolomics in the field of nutrition is called nutritional metabolomics and refers to the systematic study of the interaction between diet and organism metabolism using metabolomics in different health and disease states of organisms. Vegetables and fruits are full of trace elements required by the human body, which can help supplement the trace elements required by the human body. There are many factors affecting the content of trace elements in fruit. These include geographical factors and biological stress. Characteristics of metabolomics in nutrition can be described by a number of examples. ^{1}H NMR and GC-MS techniques are used to examine the changes of cucumber fruit metabolites under nano-Cu stress, which found that nano-copper exposure had an effect on the metabolic profile of fruit metabolites [87]. A UHPLC-MS/MS-based metabolomics strategy was used to study the nutrient profiles of Chinese kale. It was found that the contents of components in different cabbage varieties were different [88]. Quadrupole time-of-flight mass spectrometry (Q/TOF-MS) technology was used to explore the difference in secondary metabolites and heavy metal content in dried ginger from different habitats. It is speculated that abiotic stress due to metals may lead to differences in metabolite abundance [89].

Agronomy and Plant Biology: Metabolomics based on MS is increasingly used in different scientific fields, especially agronomy and plant biology, to understand the behavior of plants under different stress conditions [90]. In the process of plant growth, a large number of elements are often required, and it is essential to explore the influence of elements on the growth laws of plants. Metabolomics based on MS used to investigate the effects of sulfur deficiency on metabolites, which laid the foundation for agricultural research. This is helpful to the management of fertilization in agriculture [91]. Metabonomic analysis was carried out on two wheat varieties with different induced nitrogen levels and found that avoids and their related derivatives were considered as potential biomarkers under low nitrogen stress [92].

In addition to their significant application in agriculture, the plant and biological aspects still have a very wide range of applications. Gas chromatography-mass metabolomics methods were used to characterize the effects of drought stress on metabolites in different growth stages of eggplants. Drought stress was found to have a significant effect on metabolites such as amino acids, sugars, and organic acids [93]. UPLC/MS was used to investigate the effect of in vitro sucrose on the quality components of tea plants and found that sucrose regulates polyphenol biosynthesis in tea plants by changing the expression of transcription factor genes and pathway genes. Metabolomics technology is widely used in crop information measurement, disease diagnosis, etc., and contributes to increasing agricultural income. The mutual integration of the two has opened up a new direction for the development of the discipline of metabolomics [94].

Environmental Metabolomics: It is the application of metabolomics to characterize the interactions of organisms with their environment. This approach has many advantages for studying organism—environment interactions and for assessing organism function and health at the molecular level. As such, metabolomics is finding an increasing number of applications in the environmental sciences, ranging from understanding organism responses to abiotic pressures, to investigating the responses of organisms to other biota. These interactions can be studied from individuals to populations, which can be related to the traditional fields of ecophysiology and ecology [95].

Organ Transplantation: Metabolite measurements have been part of organ transplant monitoring for more than 40 years. While most metabolite measurements have been restricted to just a few well-known compounds (creatinine, glucose), there is a surprisingly large body of literature describing injury-dependent changes for a large number of lesser-known metabolites. Until recently, most of these measurements have been done using "classical" clinical chemistry methods such as GC-MS, but since 1999, a growing number of reports have described the use of true metabolomics methods (NMR, LC–MS, spectral pattern analysis, etc.) to monitor organ function. Regardless of the technique or speed by which these compounds were measured, it is still quite instructive to explore what these metabolite measurements have revealed and what compounds are proving to be particularly good diagnostic or prognostic biomarkers. In general, metabolite measurements have been performed to monitor two key aspects of organ physiology: (i) Organ reperfusion injury and (ii) Organ function (or dysfunction). Both types of measurements have been performed and reported on almost all solid organs that can be transplanted including kidneys, liver, lung, and heart. The frequency of these reports for specific organs closely reflects the reported frequency of the corresponding organ transplant, with the kidney leading the way, followed by liver (21%), heart (10%), pancreas (5%), and lung (4%). Most metabolite measurements associated with organ transplant analysis have been performed ex vivo, using biofluids such as urine, serum, or bile. More recently, a few measurements have been performed in vivo using NMR chemical shift imaging techniques. These primarily measure inorganic phosphate or phosphorylated metabolites phosphocreatine) [96].

Metabolomics in Alcohol Research and Drug Development Metabolomics: A characterizing systems metabolites biology approach to produce in biochemical pathways—is contributing to many studies of disease progression and treatment, although it has not yet been extensively applied in research on metabolic perturbations associated with alcohol abuse. However, numerous metabolomics approaches may contribute to alcohol-related research, as illustrated by studies on alcohol-related metabolic dysfunctions such as alterations in fat metabolism and thiamine deficiency. By further increasing the number and types of metabolites that can be measured in a given biological sample, metabolomic approaches may be able to help define the role of the many different metabolic pathways affected by alcohol abuse and support the discovery and development of novel medications for the treatment of alcoholism and related conditions [97].

Life Science and Medical Science: Metabolome is an important component of phenome. Metabolomics conducts qualitative and quantitative analysis of all small molecule metabolites in organisms and searches for the relative relationship between metabolites and physiological and pathological changes. The subjects are mostly small molecules with molecular weights of less than 1,000. With the development of high-throughput technology, the study of living organisms has developed from single small molecules to multi-omics such as genomics, transcriptomics, proteomics, and metabolomics. Multi-omics reflects molecular changes in a disease or biological process, and molecules that can be identified can be used as valuable biomarkers. Metabolites are substances produced or consumed through the metabolic process. Metabolites are the final expression products subject to genetic control and environmental influence. Imprints with genomic, transcriptomic, epigenetic, and environmental effects are called "associations between genotypes and phenotype" [11].

In Clinical Pharmacology: Pharmacogenomics focuses on the identification of genome variants that influence drug response, while metabolomics investigates the biochemical variations associated with varied drug responses and disease heterogeneity. As a pathophysiological state often involves metabolic perturbation, metabolomics is increasingly recognized as a powerful tool for its ability to profile and delineate metabolic perturbations and, thus, link metabolic phenotypes to genotypic variations. Clinical applications of metabolomics involve the identification of diagnostic biomarkers, elucidation of disease mechanisms, discovery of novel drug targets, and prediction of drug responses. Patients often show great heterogeneity at the molecular level due to their diverse genotypes, gut microbiota, nutrition status, and lifestyles. All these factors contribute to interindividual variability in drug response and treatment outcome. Therefore, it is vital to investigate these differences from a systematic perspective using multiplex analytical and computational/data mining tools (e.g., genomics and metabolomics) [98–103].

CONCLUSION

Metabolomics has become an essential platform in drug discovery, providing in-depth information on metabolic pathways, disease mechanisms, and biomarker identification. This chapter emphasized some of the important analytical methods—NMR, MS (GC-MS, LC-MS, CE-MS), and FTIR—and their applications in high-resolution metabolite profiling. Each method offers distinct advantages, such as NMR's nondestructive character and LC-MS's extensive analyte coverage, although constraints such as sensitivity and sample complexity remain. Combining these approaches with AI and machine learning facilitates effective data interpretation, pathway mapping, and predictive modeling. The chapter also highlighted the need for multi-omics data integration and metabolic computing to drive precision medicine and strain development. High-throughput platforms and microfluidics, while still in their development stages, hold potential for rapid and automated metabolite analysis. In general, metabolomics coupled with computational capabilities offers a broad platform for streamlining drug discovery, enhancing diagnostics, and designing biological systems for therapeutic development. Future progress is dependent on continued advances in analytics and bioinformatics.

REFERENCES

1. Oliver SG, Winson MK, Kell DB, et al. Systematic functional analysis of the yeast genome. *Trends Biotechnology* 1998;16:373–8.
2. Mehrotra B, Mendes P. Bioinformatics approaches to integrate metabolomics and other system biology data. *Biotechnology in Agriculture and Forestry. Plant Metabolomics* 2006;57:105–15.
3. Ju R, Liu X, Zheng F, et al. Removal of false positive features to generate authentic peak table for high resolution mass spectrometry-based metabolomics study. *Analytica Chimica Acta* 2019;1067:79–87.
4. Zhang A, Sun H, Wang X. Urinary metabolic profiling of rat models revealed protective function of scopa one against alcohol induced hepatotoxicity. *Scientific Reports* 2014;4:6768.
5. Zhan X, Long Y, Lu M. Exploration of variations in proteome and metabolome for predictive diagnostics and personalized treatment algorithms: innovative approach and examples for potential clinical application. *Journal of Proteomics* 2018;188:30–40.
6. Gomase VS, Changbhale SS, Patil SA, Kale KV. Metabolomics. *Current Drug Metabolism* 2008;9(1):89–98.
7. Zhan X, Yang J, Zheng S. Introductory chapter: metabolomics. In Zhan X (eds.), *Metabolomic: Methodology and Applications in Medical Sciences and Life Sciences.* InTech, London.
8. Kasture VS, Musmade DS, Vakte MB, et al. Metabolomics: current technologies and future trends. *International Journal of Research and Development in Pharmacy & Life Sciences* 2012;2(1):206–17.
9. Griffin JL. The Cinderella story of metabolic profiling: does metabolomics get to go to the functional genomics ball? *Philosophical Transactions of the Royal Society B: Biological Sciences* 2004;359(1446):857–71. https://doi.org/10.1098/rstb.2005.1734
10. Lee JW, Na D, Park JM, Lee J, Choi S, Lee SY. 2012. Systems metabolic engineering of microorganisms for natural and non-natural chemicals. *Nature Chemical Biology* 8(6):536–46.
11. Fiehn O. Metabolomics–the link between genotypes and phenotypes. *Plant Molecular Biology* 2002;48:155–71.
12. Halket JM, Waterman D, Przyborowska AM, et al. Chemical derivatization and mass spectral libraries in metabolic profiling by GC/MS and LC/MS/MS. *Journal of Experimental Botany* 2005;56:219–43.
13. Weckwerth W, Loureiro ME, Wenzel K, et al. Differential metabolic networks unravel the effects of silent plant phenotypes. *Proceedings of the National Academy of Sciences of the United States of America* 2004;101:7809–14.
14. Villas-Boas SG, Moxley JF, Akesson M, et al. High-throughput metabolic state analysis: the missing link in integrated functional genomics of yeasts. *Beachem* 2005;388: 669–77.

15. Florian CL, Preece NE, Bhakoo KK, et al. Characteristic metabolic profiles revealed by 1H NMR spectroscopy for three types of human brain and nervous system tumours. *NMR in Biomedicine* 1995;8:253–64.
16. Florian CL, Preece NE, Bhakoo KK, et al. Cell type-specific fingerprinting of meningioma and meningeal cells by proton nuclear magnetic resonance spectroscopy. *Cancer Research* 1995;55:420–7.
17. Kim HK, Choi YH, Erkelens C, et al. Metabolic fingerprinting of Ephedra species using 1H-NMR spectroscopy and principal component analysis. *Chemical and Pharmaceutical Bulletin* 2005;53:105–9.
18. Choi HK, Choi YH, Verberne M, et al. Metabolic fingerprinting of wild type and transgenic tobacco plants by 1H NMR and multivariate analysis technique. *Phytochemistry* 2004;65:857–64.
19. Lamers RJ, Wessels EC, van de Sandt JJ, et al. A pilot study to investigate effects of inulin on Caco-2 cells through in vitro metabolic fingerprinting. *Journal of Nutrition* 2003;133:3080–4.
20. Johnson HE, Broadhurst D, Kell DB, et al. High-throughput metabolic fingerprinting of legume silage fermentations via Fourier transform infrared spectroscopy and chemometrics. *Applied and Environmental Microbiology* 2004;70:1583–92.
21. Allen J, Davey HM, Broadhurst D, et al. High-throughput classification of yeast mutants for functional genomics using metabolic foot printing. *Nature Biotechnology* 2003;21:692–6.
22. Chung Y-L, Stubbs M, Griffiths JR. Metabolic profiling in tumors by in vivo and in vitro NMR spectroscopy. In Harrigan GG, Goodacre R (eds.), *Metabolic Profiling: Its Role in Biomarker Discovery and Gene Function Analysis.* Boston, Dordrecht and London: Kluwer Academic Publishers, 2003: pp. 83–94.
23. Fiehn O, Spranger J. Use of metabolomics to discover metabolic patterns associated with human disease. In: Harrigan GG, Goodacre R (eds.), *Metabolic Profiling: Its Role in Biomarker Discovery and Gene Function Analysis.* Boston, Dordrecht and London: Kluwer Academic Publishers, 2003: pp. 199–215.
24. Roessner U, Bowne J. What is metabolomics all about? *Biotechniques* 2009;46(5):363–5.
25. Lindon JC, Nicholson JK. Analytical technologies for metabonomics and metabolomics, and multi-omic information recovery. *Trends in Analytical Chemistry* 2008;27(3):194–204.
26. Patti GJ, Yanes O, Siuzdak G. Innovation: metabolomics: the apogee of the omics trilogy. *Nature Reviews Molecular Cell Biology* 2012;13(4):263–9.
27. Griffin JL. Metabolic profiles to define the genome: can we hear the phenotypes? *Philosophical Transactions of the Royal Society B: Biological Sciences* 2004;359(1446):857–71.
28. da Luz JA, Hans E, Zeng AP. Automated fast filtration and on-filter quenching improve the intracellular metabolite analysis of microorganisms. *Engineering in Life Sciences* 2014;14(2):135–42.
29. Vuckovic D. Current trends and challenges in sample preparation for global metabolomics using liquid chromatography mass spectrometry. *Analytical and Bioanalytical Chemistry* 2012;6:1523–4.
30. Zhang A, Sun H, Wang P, et al. Modern analytical techniques in metabolomics analysis. *Analyst* 2012;137:293.
31. Krastanov A. Metabolomics - The state of art. *Biotechnology & Biotechnological Equipment* 2010;24:1537–43.
32. Castle AL, Fiehn O, Kaddurah-Daouk R, et al. Metabolomics standards workshop and the development of international standards for reporting metabolomics experimental results. *Briefings in Bioinformatics* 2006;7:159–65.
33. Sugimoto M, Kawakami M, Robert M, et al. Bioinformatics tools for mass spectroscopy-based metabolomic data processing and analysis. *Current Bioinformatics* 2012;7(1):96–100.
34. Thermo Scientific. SIEVETM Software for Differential Analysis. https://www.thermoscientific.com/en/product/sieve-soft ware-differential-analysis.html (25 February 2016, last accessed).
35. Smith CA, Want EJ, O'Maille G, et al. XCMS: processing mass spectrometry data for metabolite profiling using nonlinear peak alignment, matching, and identification. *Analytical Chemistry* 2006;78:779–87.
36. Clasquin MF, Melamud E, Rabinowitz JD. LC-MS data processing with MAVEN: a metabolomic analysis and visualization engine. *Current Protocols in Bioinformatics* 2012;14:11.
37. Pluskal T, Castillo S, Villar-Briones A, et al. MZmine 2: modular framework for processing, visualizing, and analyzing mass spectrometry-based molecular profile data. *BMC Bioinformatics* 2010;11:395.
38. Wishart DS, Jewison T, Guo AC, et al. HMDB 3.0- The human metabolome database in 2013. *Nucleic Acids Research* 2013;41:801–7.
39. Smith CA, O'Maille G, Want EJ, et al. METLIN: a metabolite mass spectral database. *Therapeutic Drug Monitoring* 2005;27:747–51
40. Sugimoto M, Kawakami M, Robert M, et al. Bioinformatics tools for mass spectroscopy-based metabolomic data processing and analysis. *Current Bioinformatics* 2012;7:96–108.
41. Cambiaghi A, Ferrario M, Masseroli M. Analysis of metabolomic data: tools, current strategies and future challenges for omics data integration. *Briefings in Bioinformatics* 2017;18:498–510. https://doi.org/10.1093/bib/bbw031
42. GeneGo. *MetaCore Training Manual - Version 5.0.* St. Joseph MI (USA), 2008.
43. GeneGo. MetaCore: Integrated pathway analysis for all omics data, 2014.
44. Reshetova P, Smilde AK, Van Kampen AHC, et al. Use of prior knowledge for the analysis of high-throughput transcriptomics and metabolomics data. *BMC Systems Biology* 2014;8(Suppl. 2):S2.
45. Xia J, Wishart DS. MetPA: a web-based metabolomics tool for pathway analysis and visualization. *Bioinformatics* 2010;26:2342–4.
46. To¨pfer N, Kleessen S, Nikoloski Z. Integration of metabolomics data into metabolic networks. *Frontiers in Plant Science* 2015;6:49.
47. Eichner J, Rosenbaum L, Wrzodek C, et al. Integrated enrichment analysis and pathway-centered visualization of metabolomics, proteomics, transcriptomics, and genomics data by using the InCroMAP software. *Journal of Chromatography B: Analytical Technologies in the Biomedical and Life Sciences* 2014;966:77–82.
48. Xia J, Sinelnikov IV, Han B, et al. MetaboAnalyst 3.0-making metabolomics more meaningful. *Nucleic Acids Research* 2015;43:W251–7.
49. Sabatine MS, Liu E, Morrow DA, et al. Metabolomic identification of novel biomarkers of myocardial ischemia. *Circulation* 2005;112(25):3868–75.
50. Mamas M, Dunn WB, Neyses L, Goodacre R. The role of metabolites and metabolomics in clinically applicable biomarkers of disease. *Archives of Toxicology* 2011;85(1):5–17.

51. Bennett MR, Devarajan P. Characteristics of an ideal biomarker of kidney diseases. In Charles LE (eds.), *Biomarkers in Kidney Disease*. San Diego, CA: Academic Press, 2011, pp 1–24.
52. Sreekumar A, Poisson LM, Rajendiran TM, et al. Metabolomic profiles delineate potential role for sarcosine in prostate cancer progression. *Nature* 2009;457(7231):910–4.
53. Shulaev V. Metabolomics technology and bioinformatics. *Briefings in Bioinformatics* 2006;7(2):128–39.
54. Spratlin JL, Serkova NJ, Eckhardt SG. Clinical applications of metabolomics in oncology: a review. *Clinical Cancer Research* 2009;15(2):431–40.
55. Le GG, Colquhoun IJ, Davis AL, et al. Metabolite profiling of tomato (Lycopersicon esculentum) using 1H NMR spectroscopy as a tool to detect potential unintended effects following a genetic modification. *Journal of Agricultural and Food Chemistry* 2003;51:2447–56.
56. Krishnan P, Kruger NJ, Ratcliffe RG. Metabolite fingerprinting and profiling in plants using NMR. *Journal of Experimental Botany* 2005;56:255–65.
57. Park VY, Yoon D, Koo JS, et al. Intratumoral agreement of high-resolution magic angle spinning magnetic resonance spectroscopic profiles in the metabolic characterization of breast cancer. *Medicine* 2016;95(15):e3398.
58. Precht C, Diserens G, Oevermann A, et al. Visibility of lipid resonances in HR-MAS spectra of brain biopsies subject to spinning rate variation. *Biochimica et Biophysica Acta* 2015;1851(12):1539–44.
59. Martinez-Bisbal MC, Martinez-Granados B, Rovira V, et al. Magnetic resonance spectroscopy and imaging on fresh human brain tumor biopsies at microscopic resolution. *Analytical and Bioanalytical Chemistry* 2015;407:6771–80.
60. Monleo′n D, Morales JM, Gonzalez-Darder J, et al. Benign and atypical meningioma metabolic signatures by high-resolution magic-angle spinning molecular profiling. *Journal of Proteome Research* 2008;7(7):2882–8.
61. Palmnas MS, Vogel HJ. The future of NMR metabolomics in cancer therapy: towards personalizing treatment and developing targeted drugs? *Metabolites* 2013;3(2):373–96.
62. Junot C, Madalinski G, Tabet JC, et al. Fourier transform mass spectrometry for metabolome analysis. *Analyst* 2010;135(9): 2203–19.
63. Junot C, Fenaille F, Colsch B, et al. Hig resolution mass spectrometry based techniques at the crossroads of metabolic pathways. *Mass Spectrometry Reviews* 2014;33(6):471–500.
64. Davis VW, Bathe OF, Schiller DE, Slupsky CM, Sawyer MB. Metabolomics and surgical oncology: potential role for small molecule biomarkers. *Journal of Surgical Oncology* 2011;103(5):451–9.
65. Nordstrom A, Lewensohn R. Metabolomics: moving to the clinic. *Journal of Neuroimmune Pharmacology* 2010;5(1):4–17.
66. Dettmer K, Aronov PA, Hammock BD. Mass spectrometry-based metabolomics. *Mass Spectrometry Reviews* 2007;26(1):51–78.
67. Griffiths WJ, Koal T, Wang Y, Kohl M, Enot DP, Deigner HP. Targeted metabolomics for biomarker discovery. *Angewandte Chemie International Edition* 2010;49(32):5426–45.
68. Roessner U, Wagner C, Kopka J, et al. Technical advance: simultaneous analysis of metabolites in potato tuber by gas chromatography–mass spectrometry. *Plant Journal* 2000;23:131–42.
69. Scalbert A, Brennan L, Fiehn O, et al. Mass spectrometry-based metabolomics: limitations and recommendations for future progress with particular focus on nutrition research. *Metabolomics* 2009;5(4):435–58.
70. Maurer HH. Advances in analytical toxicology: the current role of liquid chromatography–mass spectrometry in drug quantification in blood and oral fluid. *Analytical and Bioanalytical Chemistry* 2005;381:110–8.
71. Maurer HH. Liquid chromatography–mass spectrometry in forensic and clinical toxicology. *Journal of Chromatography B: Biomedical Sciences and Applications* 1998;713:3–25.
72. Maurer HH, Peters FT. Toward high-throughput drug screening using mass spectrometry. *Therapeutic Drug Monitoring* 2005;27:686–8.
73. Tolstikov VV, Lommen A, Nakanishi K, et al. Monolithic silica-based capillary reversed-phase liquid chromatography/ electrospray mass spectrometry for plant metabolomics. *Analytical Chemistry* 2003;75:6737–40.
74. Terabe S, Markuszewski MJ, Inoue N, et al. Capillary electrophoretic techniques toward the metabolome analysis. *Pure and Applied Chemistry* 2001;73:1563–72.
75. Perrett D, Ross G. Capillary electrophoresis—a powerful tool for biomedical analysis and research. *Trends in Analytical Chemistry* 1992;11:156–63.
76. Perrett D, Alfrzema L, Hows M, et al. Capillary electrophoresis for small molecules and metabolites. *Biochemical Society Transactions* 1997;25:273–8.
77. Perrett D. Capillary electrophoresis in clinical chemistry. *Annals of Clinical Biochemistry* 1999;36:133–50.
78. Soga T, Imaizumi M. Capillary electrophoresis method for the analysis of inorganic anions, organic acids, amino acids, nucleotides, carbohydrates and other anionic compounds. *Electrophoresis* 2001;22:3418–25.
79. Soga T, Ueno Y, Naraoka H, et al. Simultaneous determination of anionic intermediates for Bacillus subtilis metabolic pathways by capillary electrophoresis electrospray ionization mass spectrometry. *Analytical Chemistry* 2002;74:2233–9.
80. Zhang A, Sun H, Qiu S, et al. NMR-based metabolomics coupled with pattern recognition methods in biomarker discovery and disease diagnosis. *Magnetic Resonance in Chemistry* 2013;51(9):549–56.
81. Feng CC, Zhang AH, Miao JH, et al. Recent advances in understanding cross-talk between bile acids and gut microbiota. *Open Journal of Proteomics and Genomics* 2018;3(1):024–034.
82. Gomez-Casati DF, Zanor, MI, Busi MV. Metabolomics in plants and humans: applications in the prevention and diagnosis of diseases. *BioMed Research International* 2013;2013:792511–27.
83. Hanahan D, Weinberg RA. The hallmarks of cancer. *Cell* 2000;100:57–70.
84. Liu X, Zhou L, Shi X, Xu G. New advances in analytical methods mass-spectrometry-based large-scale metabolomics study. *Trends in Analytical Chemistry* 2019; 121(115665):1–11.
85. Yang W, Mu T, Jiang J, et al. Identication of potential biomarkers and metabolic proling of serum in ovarian cancer patients using UPLC/Q-TOF MS. *Cellular Physiology and Biochemistry* 2018;51(3):1134–48.
86. Fan X, Rao J, Zhang Z, et al. Macranthoidin B modulates key metabolic pathways to enhance ROS generation and induce cytotoxicity and apoptosis in colorectal cancer. *Cellular Physiology and Biochemistry* 2018;46(4):1317–30.

87. Zhao L, Hu J, Huang Y, et al. 1H NMR and GC-MS based metabolomics reveal nano-Cu altered cucumber (Cucumis sativus) fruit nutritional supply. *Plant Physiology and Biochemistry* 2017;110:138–46.
88. Wang YQ, Hu LP, Liu GM, et al. Evaluation of the nutritional quality of chinese kale (brassica alboglabra bailey) using UHPLC-quadrupole-orbitrap MS/MS-based metabolomics. *Molecules* 2017;22(8):E1262.
89. Rahman MM, AbdEl-Aty AM, Kabir MH, et al. Aquick andeffective methodologyforanalyzing dinotefuran andits highly polar metabolites in plum using liquid chromatography-tandem mass spectrometry. *Food Chemistry* 2018;239:1235–43.
90. do Prado RM, Porto C, Nunes E, et al. Metabolomics and agriculture: what can be Done?, *Systems* 2018;3(2):e00156–17.
91. Ghosson H, Schwarzenberg A, Jamois F, et al. Simultaneous untargeted and targeted metabolomics profiling of underivatized primary metabolites in sulfur deficient barley by ultra-high performance liquid chromatography-quadrupole/time-of-fight spectrometry. *Plant Methods* 2018;14:62.
92. Du P, Wu X, Xu J, et al. Determination and dissipation of mesotrione and its metabolites in rice using UPLC and triple-quadrupole tandem mass spectrometry. *Food Chemistry* 2017;229:260–7.
93. Mibei EK, Owino WO, Ambuko J, et al. Metabolomic analyses to evaluate the effect of drought stress on selected African eggplant accessions. *Journal of the Science of Food and Agriculture* 2018;98(1):205–16.
94. Qian Y, Zhang S, Yao S, et al. Effects of vitro sucrose on quality components of tea plants (Camellia sinensis) based on transcriptomic and metabolic analysis. *BMC Plant Biology* 2018;18(1):121.
95. Ravenzwaay BV, Cunha GCP, Leibold E, et al. The use of metabolomics for the discovery of new biomarkers of effect. *Toxicology Letters* 2007;172:21–8.
96. Lewis GD, Asnani A, Gerszten RE. Application of metabolomicsto cardiovascular biomarker and pathway discovery, metabolomics in biomarker discovery. *JACC* 2008;52(2):117–23.
97. Sumner LW, Mendes P, Richard A. Dixon plant metabolomics: large-scale phytochemistry in the functional genomics era. *Phytochemistry* 2003;62:817–36.
98. Corona G, Rizzolio F, Giordano A, Toffoli G. Pharmaco-metabolomics: an emerging "omics" tool for the personalization of anticancer treatments and identification of new valuable therapeutic targets. *Journal of Cellular Physiology* 2012;227:2827–31.
99. Kowalczyk T, Ciborowski M, Kisluk J, Kretowski A, Barbas C. Mass spectrometry based proteomics and metabolomics in personalized oncology. *Biochimica Et Biophysica Acta-Molecular Basis of Disease* 2020;1866:165690. https://doi.org/10.1016/j.bbadis.2020.165690.
100. Di Ottavio F, Gauglitz JM, Ernst M, et al. A UHPLC-HRMS based metabolomics and chemoinformatics approach to chemically distinguish 'super foods' from a variety of plant-based foods. *Food Chemistry* 2020;313:126071. https://doi.org/10.1016/j.foodchem.2019.126071.
101. Castro-Puyana M, Perez-Mıguez R, Montero L, Herrero M. Reprint of: application of mass spectrometry-based metabolomics approaches for food safety, quality and traceability. *TrAC - Trends in Analytical Chemistry* 2017;96:62–78. https://doi.org/10.1016/j.trac.2017.08.007.
102. Camacho DM, Collins KM, Powers RK, Costello JC, Collins JJ. Next- generation machine learning for biological networks. *Cell* 2018;173(7):1581–92.
103. Wang J, Chen G. Microchip capillary electrophoresis with electrochemical detector for fast measurements of aromatic amino acids. *Talanta* 2003;60:1239–44.

4 Biological Databases and Data Mining

Yogesh Singh[1], *Nakkala Srinivas Mudiraj*[2], *Neha Bhatia*[1], *Satwinder Singh*[2], *and Suresh Tharej*[1]

[1]Department of Pharmaceutical Sciences and Natural Products, School of Pharmaceutical Sciences, Central University of Punjab, Bathinda, India

[2]Department of Computer Science and Technology, Central University of Punjab, Bathinda, India

INTRODUCTION

The advancement in the field of technology significantly impacts the lives of individuals, primarily through business data processing and scientific computing. It opens the path for exponential growth in collecting data. The concept of data mining arises since traditional data mining systems fail to process large amounts of data. Data mining is the process called knowledge discovery from databases (KDD). Data mining has roots in databases, machine learning (ML), deep learning, and statistics and has contributions toward many other areas [1]. In this chapter, we explore the biological database, which is a repository used to store biological information.

Biology and biological data are computed in the field of bioinformatics, an interdisciplinary science of analyzing biological data using information technology and computer science. The collection of data from the biological world is known as biological data, and the system that is used to store, search, and retrieve biological data is known as a biological database. Biological database design, development, and long-term management are core areas of the discipline of bioinformatics (http://en.wikipedia.org/wiki/Biological_database). It is the repository of biological information [2].

A database is a platform where all the data is stored systematically and can be accessed and updated occasionally. It stores a massive amount of data generated in experiments of genome, transcriptome, and exome sequences of various organisms in current times; the biological data is stored exponentially [3]. The availability of enormous amounts of biological data (sequences and structural data) has generated a need for managing, storing, and retrieving these vast data. These data have played a fundamental role in bioscience, particularly in bioinformatics, where we can access sequence and structure data for tens of thousands of protcins/amino acids from a broad range of organisms and use these data for drug innovation [2]. It is an invaluable resource that also helps to support biological research with the help of a huge database pattern and their relationship to finding a meaningful correlation, pattern, and trends for discoveries. The process by which a correlation between the huge data is established is known as data mining [4].

DATA MINING

The practice of utilizing statistical analysis and ML to find anomalies, correlations, and hidden patterns in big datasets is known as data mining [5]. In 1989, Gregory Piatetsky–Shapiro first used "Knowledge Discovery in Databases." Nonetheless, the press and business groups started to use the term "data mining" more frequently. Data mining and knowledge discovery are currently used synonymously. Various key steps in data mining are [6]:

a. Define problem	f. Select model
b. Collect data	g. Train model
c. Prep Data	h. Evaluate model
d. Explore data	i. Deploy model
e. Select predictors	j. Monitor and maintain

Techniques of Data Mining

i. **Classification:** The process of classifying data into preset classes or categories according to the characteristics of the data instances is known as classification. In order to predict the class labels of new, unknown data instances, a model must be trained on labeled data.
ii. **Regression:** The data mining technique known as regression analysis is used to find and examine the relationship between variables when another component is present. It's employed to specify the likelihood of a certain variable. Regression is mostly used in modeling and planning. We may use it, for instance, to forecast costs based on variables like availability, customer demand, and competition.
iii. **Clustering:** Information is divided into groups of related objects through the process of clustering. Describing the data by a small number of clusters primarily results in the loss of certain important information. Data is modeled with the help of clustering. Clustering is viewed historically through the lens of data modeling, which is grounded in mathematics, statistics, and numerical analysis.
iv. **Association Rule:** This kind of data mining aids in identifying a connection between two or more objects. In the dataset, a hidden pattern is uncovered. If-then statements, known as association

DOI: 10.1201/9781003464679-4

rules, are used to assist the display of the likelihood of interactions between data items in huge datasets across various database formats [7].

Biological Databases

These life-science information libraries can be gathered from several sources, including published literature, high-throughput experiment technologies, computational analysis, and experimental studies. Biological data are typically collected from diverse sources spanning several structural and functional boundaries [8]. The biological data might take several forms, such as inquiries about ecology, medicinal discoveries, protein activities, genetic sequences, and more. Therefore, the availability of biological databases facilitates the handling of gathered biological data by many stakeholders in a way that is well organized, accessible, centrally maintained, and flexible in terms of updating [9] (Figure 4.1).

Classification of Biological Databases

A. **Data Source:** A data source is defined as a location from where the data is being originated or first digitalized from the physical information. Databases are the most common data sources stored in relational database management systems (RDBMS) [10]. It is further classified into three categories:
   i. **Primary Databases:** These databases are derived experimentally and contain information related to the raw sequenced data with sequence and structural information. These are well-organized portals with large amounts of biological data that are convenient for research across the globe. Primary databases were first utilized for storing DNA and protein sequences in the 1980s and 1990s. Various primary databases include Swiss-Prot, UniProtKB, and Protein Information Resource (PIR) for protein sequence, as well as Gene bank [11]. Further primary databases are classified into two categories:
      a. **Sequence Database:** These databases contain DNA/nucleotide and protein information sequences [12].
         – **DNA/Nucleotide Database** stores DNA and RNA sequence information. Although each database keeps its own collection of submission and retrieval tools, they interchange data daily to ensure that every database has the same set of sequences. Examples of DNA/nucleotide databases are:
            i. **GenBank:** It is an open-access database with a huge collection of nucleotide sequences, including mRNA. It is in the USA and can be accessed through NCBI portal.
            ii. **DNA Data Bank of Japan (DDBJ):** It is a nucleotide sequence database that collects data from researchers from Japan. It is in NIG in Mishmia, Japan.
            iii. **European Molecular Biology Laboratory-European Bioinformatics Institute (EMBL-EBI):** It is the first nucleotide sequence database and provides

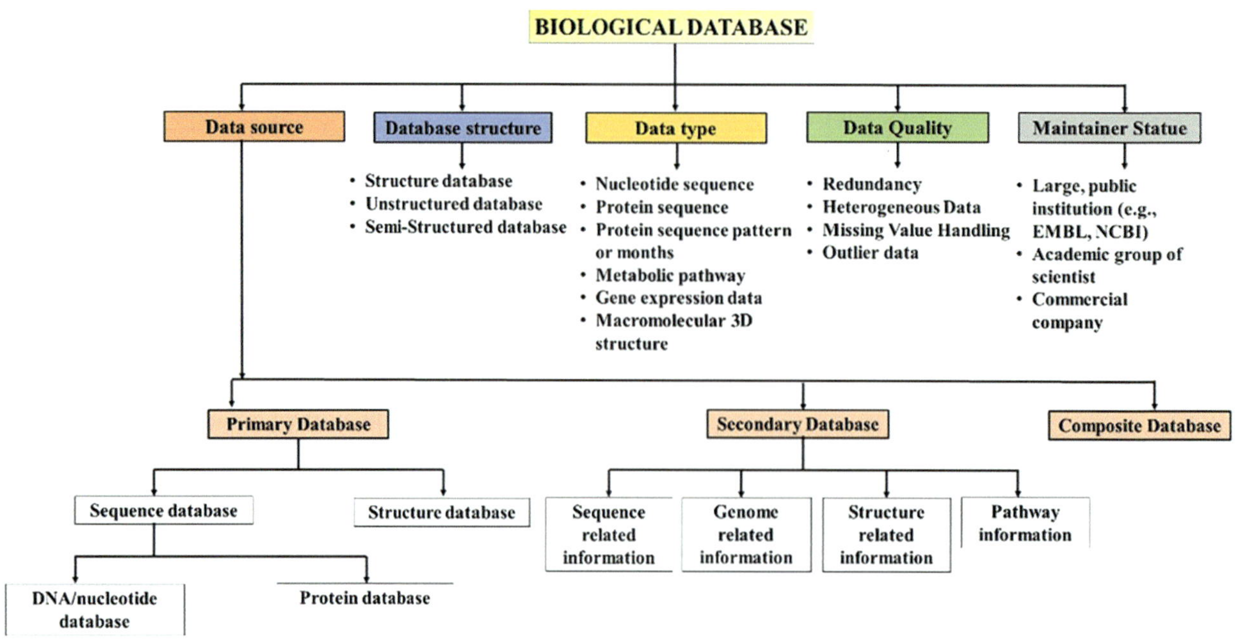

FIGURE 4.1 Classification of biological database.

data belonging to life science, computational biology, and training for researchers. It is in the UK and was established in 1980 [13].

iv. **EnsEMBL:** It annotates the genome automatically, combines it with other biological data that is already available, and makes the information available to researchers online. It is a joint venture of EMBI, EMBL, and Welcome Trust Sanger Institute. EnsEMBL creates genomic databases for eukaryotic organisms, including vertebrates, and makes them publicly accessible on the internet [14].

– **Protein Database:** It is a database that stores information related to protein sequences.

i. **Protein Information Resource:** It was developed by the National Research Foundation (NBRF) for the identification of protein sequence information. It consists of three sections: PIR1, PIR2, and PIR3. PIR1 consists of classified and annotated entries, PIR2 consists of redundancy entries, and PIR3 consists of unverified entries.

ii. **Swiss Institute of Bioinformatics (SWISS-PROT):** It is a protein sequence database started in 1986. It is a part of the UniProt consortium and provides information about the functions of a protein, domain structure post-translational modifications, etc.

iii. **Translation of EMBL Nucleotide Sequence Database (TrEMBL):** It is composed of computer-annotated entries derived from the translation of all coding sequences in the nucleotide databases. The data are unreviewed and automatically annotated benefits from SWISS-PROT and comprises translations of all coding sequences (CDS) in EMBL [15].

b. **Structure Databases:** These contain structure information of biological macromolecules, e.g., nucleic acids, proteins, and nucleic acids.

– **Protein Data Bank:** It contains 3D structures of biological molecules maintained by X-ray crystallography and NMR. It was established in 1971 at Brookhaven National Laboratories (BNL). It has been maintained by the Research Collaboratory for Structural Bioinformatics (RCSB) at Rutgers University since 1998. Today, 162,041 structures in the PDB have a structure factor file. 11,242 structures have an NMR restraint file. 5,774 structures in the PDB have a chemical shift file.

– **Macromolecular Structure Database:** It is the gathering, administration, and dissemination of macromolecular structures by Europe, fusing modern IT and database technology with a strong foundation of structural biology knowledge. It collaborates closely with PDBj in Japan and RCSB in the USA.

– **Electron Microscopy Database:** It is overseen by MSD and arranges biological macromolecule structures determined by three-dimensional electron microscopy.

– **Molecular Modeling Database:** NCBI's structural database containing experimentally determined 3D biomolecular structures. The database contains information on the biological function, the mechanisms associated with the function, and the evolutionary history of macromolecules and their relationships.

– **Nucleic Acid Database:** It distributes and holds nucleic acid structures. It was established in 1992 by David Beveridge of Wesleyan University, Helen M. Berman of Rutgers University, and Wiliam K. Olson of Olson University.

– **Cambridge Crystallographic Data Centre/Cambridge Structural Database (CCDB/CSD):** A nonprofit organization compiles and distributes organic and metal-organic structure data. It is situated at Cambridge University, England [16].

ii. **Secondary Databases:** These databases contain data derived from the primary databases, organized in the form of structural classification of protein class, folds, superfamily, etc. The data generally includes secondary structures, plots, conserved sequences, signature sequences, and active sites of residues of protein families that arise due to multiple sequence alignments of a set of related proteins and various domains. SCOP, CATH, PROSITE, and eMOTIF are individual databases created by universities such as Cambridge, University College of London, Swiss Institute of Bioinformatics, and Stanford respectively.

a. **Sequence-Related Database:** These databases include the protein sequence patterns, which are preserved amino acid residues rather than the complete sequence. These amino acid patterns were identified using particular functional and structural features in the protein. These patterns are aligned using multiple alignment techniques.
   - **ProSite:** It is a database that consists of data related to families, domains, functional sites, amino acid patterns, and protein profiles. These data were arranged by the Swiss Institute of Bioinformatics. It was developed in 1988 by Amos Bairoch. It is of two types: the pattern and the relative descriptive texts. Identification of new sequences, protein family identification, biological sites as well as models can be made.
   - **Pfam:** It is a library of protein families that offers hidden Markov models (HMM) and multiple sequence alignment for protein domains found in all kinds of living things. HMM is used to generate information about multiple sequence alignment. There are two parts to the Pfam. Pfam-A stores carefully selected premium entries by hand. An alignment of the protein sequence and an HMM are saved for every entry. Pfam-B contains automatically created lower-quality entries. Since not all known proteins are covered by Pfam-A entries, Pfam-B families of lesser quality can be utilized instead.
   - **ExploreEnx:** It is an open-access database that is peer reviewed and curated manually. In order to access the data from the International Union of Biochemistry and Molecular Biology (IUBMB) enzyme nomenclature list, Andrew McDonald of the School of Biochemistry and Immunology at Trinity College, Dublin, Ireland, devised, developed, and maintains ExplorEnz.
   - **Restriction Enzyme Database REBase:** Richard J. Roberts has been in charge of its upkeep since before 1980. A variety of data are stored in it, including recognition and cleavage sites, isoschizomers, commercial availability, methylation sensitivity, crystal, genome, sequence data, DNA methyltransferases, homing endonucleases, nicking enzymes, specificity subunits, and control proteins. It also contains information about restriction enzymes and related proteins.
   - **PRIDE:** A database of protein and peptide identifications that has been reported in scientific literature, along with post-translational modifications and spectral evidence in support of those identifications, is called proteomics identifications.
   - **InterPro:** By grouping protein sequences into families and forecasting the existence of domains, Biological Databases - I 9 and significant locations, this database offers functional analysis of protein sequences. InterPro uses prediction models, or signatures, from many member databases to classify proteins in this manner.
   - **PRINTS:** It is a library of conserved motifs known as protein fingerprints, which are used to describe a family of proteins. Compared to single motifs, they are more flexible and potent in encoding protein folds and capabilities.
   - **ProDom:** The TrEMBL and Swiss-Port sequence databases are automatically used to create this protein domain database. It has automatic homologous domain clustering, which is a logical method of arranging protein sequence data.
   - **BLOCKS:** It's a sequence analysis server at the Fred Hutchinson Cancer Research Center in Seattle, Washington, in the United States. A protein or DNA sequence can be compared to known protein families using the Blocks Database, which is a database of blocks with protein families.
   - **Clusters of Orthologous Groups of Proteins (COGs):** It is a database that

helps in protein functions and evolution with genome-scale analysis [17].

b. **Genome-Related Database**
   - **Genome:** The complete genomes of approximately 1000 species, including bacteria, archaea, and eukaryotes, as well as several viruses, phages, viroids, plasmids, and organelles, are organized in this database based on sequence, map, chromosomes, assemblies, and annotations.
   - **The Saccharomyces Genome Database (SGD):** This database contains information on the genetics and molecular biology of the yeast *Saccharomyces cerevisiae.*
   - **EBI Genomes:** It offers access to finished genomic statistics as well as details on current research.
   - **JGI Genome Portal:** Databases of numerous eukaryotic and microbial genomes are available. Additionally, it offers a number of specialized analytical tools for organizing and deciphering intricate genetic datasets.
   - **Vertebrate Genome Annotation Database (VEGA):** It serves as a repository of high-quality gene models generated through manual annotation of vertebrate genome.
   - **UCSC Genome Browser:** For a sizable number of genomes, it includes working draft assemblies and the reference sequence. It also offers links to the Neanderthal project and data from the Encyclopedia of DNA Elements (ENCODE).
   - **PlantGDB:** It is a database that includes molecular sequence information for every species of plant that has undergone extensive sequencing. EST sequences are arranged in the database into contigs, which stand for putatively unique genes.
   - **Gene Expression Omnibus:** It is a freely accessible, functional genomic data repository that accepts information based on sequences and arrays and offers resources to assist users in finding and downloading curated gene expression profiles and experiments.
   - **Gene Expression Database:** Gene expression data for laboratory mice can be found in this community resource. Information is given regarding which genes create particular transcripts and proteins, where and how much their products are expressed, and how their expression changes in various mouse strains and mutants. The Mouse Genome Database (MGD) and Gene Expression Database (GXD) are connected.
   - **Microarray Gene Expression Data (MGED):** It constitutes the functional genomics and proteomics data in the form of microarray [18].

c. **Structure-Related Database:** It consists of the PDB information in a detailed and organized way, which is protein class classification on structural bases.
   - **Database of Secondary Structure Assignments (DSSP):** It is a database for the storage of Protein Data Bank protein entries. It was designed by Wolfgang Kabsch and Chris Sander to assign it to a secondary structure.
   - **Homology-Derived Secondary Structure of Proteins (HSSP):** It consists of aligned sequence, secondary structure, sequence variability, and sequence profile for every protein structure, which is known as the tertiary structure of the protein.
   - **Distance Matrix Alignment (DALI):** It comprises the 3D protein structures from the Protein Data Bank.
   - **Families of Structurally Similar Proteins (FSSP):** It consists of similar protein structures from DALI and helps in protein structure comparison.
   - **Structural Classification of Protein Database (SCOP):** It is a database which, on the basis of hierarchal order, classifies protein structures. This takes data from PDB and gives detailed protein information on their structural and evolutionary relationships.
   - **DBAli Database:** This database is similar to SCOP; it arranges the structural alignments in a pairwise manner.
   - **Class, Architecture, Topology, Homologous (CATH):** This database presents data in a hierarchal [19].

d. **Pathway Information Database:** It is a database that is specially designed for information related to various biological processes, genes-and-proteins-related information, chemical compounds and their structures.
   - **KEGG PATHWAY Database (Kyoto Encyclopedia of Genes and Genomes):** It contains graphical pathway maps for all known metabolic

pathways from various organisms. It is a collection of databases dealing with genomes, biological pathways, diseases, drugs, and chemical substances.
- **EcoCyc:** It is a bioinformatics database – *E. coli* database, stores information regarding the genome and biochemical machinery of *E. coli* K-12 MG1655.
- **LIGAND:** It is a chemical database for enzyme reactions at the Institute for Chemical Research, Kyoto. It is a composite database currently consisting of the COMPOUND, DRUG, GLYCAN, REACTION, RPAIR, and ENZYME databases.
- **PathDB:** It is a public server. It encodes the contents of over 10,000 original publications on the topics of enzymology and metabolism. This information is transformed into a queryable database. It plays an important role in the interpretation of genetic sequence data.
- **Database of Enzymes and Metabolic Pathways (EMP):** It is a public server with more than 10,000 original publications related to enzymology and metabolism. It has an essential role in genetic sequence data interpretation, and data is converted into a queryable database [20].

iii. **Composite Database:** These databases consist of various database sources, which help compile data in a single platform to avoid searching multiple resources. These databases combine different primary databases using several algorithms. Various examples of composite databases include:
   a. **Nonredundant Database**: It uses sequences from GenBank, PDB, Swiss-Prot, PIR, and Protein Research Foundation (PRF).
   b. **International Nucleotide Sequence Databases:** These databases combine nucleotide sequences from EMBL, GenBank, and DDBJ.
   c. **Universal Protein Sequence databases:** These databases combine PIR-PSD, Swiss-Prot, and TrEMBL protein sequences [21].

B. **Database Structure**
   a. **Structured Database:**
      i. Relational Databases
      ii. Columnar databases
   b. **Unstructured Database:**
      i. Document store
      ii. Key-value store
   c. **Semi-Structured**
      i. Database: Graph Databases
      ii. Document stores [22]

C. **Data Type**
   i. **Nucleotide Sequence:** A series of bases within the nucleotides that create alleles within a DNA (using GACT) or RNA (GACU) molecule is known as a nucleic acid sequence. The most basic level of understanding of a gene or genome is its nucleotide sequence. Without a comprehensive grasp of genetic function and evolution, the blueprint containing the instructions for developing a creature would be incomplete.
   ii. **Protein Sequence:** The practical technique of ascertaining the amino acid sequence of a protein or peptide, in whole or in part, is known as protein sequencing. This could be used to describe or identify the post-translational modifications of the protein.
   iii. **Protein Sequence Pattern:** Conserved sequence patterns or motifs are frequently used to identify protein families. It is common for a researcher to want to assess the importance of a certain pattern found in a protein or to use information about recognized motifs to help identify highly divergent but homologous family members.
   iv. **Metabolic Pathway:** The series of processes that are catalyzed by enzymes and result in the transformation of a material into a finished product is known as a metabolic pathway. A set of reactions known as metabolic cycles involves the constant regeneration of intermediate metabolites and the ongoing reformation of the substrate.
   v. **Macromolecular 3D Structure:** These databases contain data regarding the spatial organization of atoms within proteins, DNA, and RNA molecules, which is essential for comprehending the function of biomolecules, their interactions with other molecules, and their potential involvement in diseases [23].

D. **Data Quality**
   i. **Redundancy:** Biological databases may contain replicated records of the same gene, protein, or organism. This duplication can clutter the database and hinder the search for accurate information. Minimizing duplication ensures data integrity and simplifies searches.
   ii. Heterogeneous Data
   iii. Missing Value Handling
   iv. Outlier data [24]

E. **Maintainer Statue**
   i. **A large, public institution** typically employs a dedicated team to ensure data quality and

accessibility, often funded by government grants or public donations for open access.

ii. **An academic group of scientists**, maintained by a research consortium, fosters collaboration within its community, often funded by research grants, with potentially limited resources, and may restrict data access to collaborators or require a user account.

iii. **Commercial Company:** Developed and maintained by for-profit companies, with data as a potential revenue stream, requiring a subscription or license fee for data access, and offering additional features or services for a cost, such as advanced data analysis tools [25].

COMPUTATIONAL CHEMISTRY IN DRUG DISCOVERY

Computational chemistry has significance in drug design as it provides data on molecular interactions, predicts molecular characteristics, and optimizes potential drugs before they are synthesized and evaluated in the lab. Below is the manner in which computational chemistry helps the drug design technique:

Virtual Screening: Computational approaches are used to evaluate massive chemical databases for possible therapeutic candidates. By modeling the relationships among compounds and protein targets, researchers may identify molecules that are predicted to bind efficiently and display the required biological function [26].

Structure-Based Drug Design (SBDD): Molecular docking and molecular dynamics (MD) simulations are two computational methods that aid in comprehending the atomic-level interactions between possible drug compounds and their intended target proteins. The logical design of novel compounds with enhanced binding capacity and selectivity is made easier by this information [27].

Ligand-Based Drug Design (LBDD): To find common structural traits or pharmacophores, computational techniques may also be used to examine the characteristics and structure of previously identified ligands or compounds that interact with target proteins [28].

Quantitative Structure-Activity Relationship (QSAR): These QSAR models link a compound's structure to its biological function or other characteristics. Using computational tools, researchers may create QSAR models from experimental data. These models may be utilized later to forecast the activity of novel compounds and rank them in order of significance for synthesis and evaluation [29].

Estimation of ADME/T Properties: The pharmacokinetics and toxicology data of potential drugs are significantly influenced by the features of ADME/T, which stands for absorption, distribution, metabolism, excretion, and toxicity. Earlier in the drug designing, computational approaches may anticipate these features, which can assist discover compounds with favorable ADME/T profiles and lower the risk of loss in later phases of development [30].

Fragment-Based Drug Design: The discovery and optimization of small molecule fragments that interact with certain areas of target proteins is made easier by computational approaches. Afterward, the individual fragments can be further developed into bigger molecules with improved efficacy and binding affinity [31].

Artificial Intelligence and Machine Learning: In the field of computational chemistry, sophisticated ML algorithms are being employed more and more for tasks like virtual screening, de novo drug creation, and attribute prediction. To speed up the drug development process, these techniques can uncover intricate patterns from massive databases and analyze them [32].

All things considered, computational chemistry is essential to rational drug development because it combines data-driven methods, theoretical models, and calculations to find and optimize drug candidates with better safety and effectiveness profiles more quickly.

Principles and Applications in Drug Design and Optimization

To study molecular interactions to the comprehensive stage, drug design, and optimization uses a variety of computational approaches such as MD simulations, quantum mechanics (QM), and additional computational methods. The following are the concepts and implications of these approaches.

A. **MD Simulation:** The principle underlying MD simulations uses basic mechanics to explain the periodic development of a molecular entity. At the core of their work, MD simulations seek to computationally solve Newton's laws of motion for every atom in a system's structure, hence revealing details about its dynamic behavior over time, which helps them to understand a variety of atomic-level characteristics and processes. An outline of MD simulations is provided here:

Atomistic Description: MD simulations examine the precise positions and velocities of every single atom in an entire system as they change over time. MD enables computing the solutions of Newton's laws of motion, allowing for great spatial and temporal precision in molecular motion modeling.

Force Fields: Force fields represent atom interactions by approximating systems potential energy as an indicator of atomic positions. It includes phrases for both bonded and non-bonded events (e.g., bonds, angles, dihedrals).

Simulation Setup: Before running an MD simulation, experts must provide the system of concern, including its molecular structure, starting atomic positions, and other applicable external parameters (such as pressure and temperature). To simulate the biological surroundings, the simulation box could additionally contain ions or solvent as well.

Integration Algorithm: To move the system ahead in time, MD simulations incorporate equations of motion. Atomic locations and velocities are updated at specified time intervals (time steps) using various numerical integration techniques, including the Varlet and leapfrog procedures.

Applications: MD simulations are used in many different domains, including drug development, materials biology, biochemistry, and biophysics. Protein folding, protein–ligand interaction, membrane dynamics, chemical responses, and several other processes in biology and chemistry. In general, MD simulations are an effective means of investigating MD and comprehending the fundamental workings of intricate chemical and biological systems. They support experimental methods and offer atomic-level discoveries that are frequently hard or impossible to get through experiments [33].

B. **Quantum Mechanics (QM):** The area of physics known as QM explains how matter and energy behave on their smallest sizes, which are usually in the subatomic and atomic range. QM controls how units like electrons, protons, and photons behave; traditional mechanics deals with larger things. An overview of QM, its principles, and its uses is given below:

Principle: The wave-particle duality, which postulates that particles may behave both like waves and like fragments, is one of the essential concepts of quantum physics. Wave functions, which express the probability amplitude of obtaining a particle in a specific state, perfectly capture this duality.

Techniques in QM:

Schrödinger Equation: This basic formula of QM describes the evolution of a quantum system's wave function as time passes. A system's energy and wave function are related by the use of differential calculations.

Quantum Operators: In QM, mathematical equations acting in the wave function are used to express physical observables like position, momentum, and energy.

Density Functional Theory: Density functional theory (DFT) is a computer technique that helps many-body systems, like molecules and atoms, resolve the Schrödinger equation. In big networks, it is computationally efficient since it corresponds to the electron density instead of the wave function.

Applications of QM: The mathematical basis for comprehending the composition, motion, and interconnections of atoms and molecules is provided by QM. It provides an explanation for things like spectroscopy, bonding of chemicals, and electronic structure. The concepts of overlap and superposition are used in quantum computing, which was made possible by QM, and enables the execution of calculations that would be impossible with traditional computers. Quantum computing holds the potential to transform domains including material science, encryption, and optimization. In quantum chemistry, chemical processes are modeled, molecular characteristics are predicted, and novel materials and medications are designed using the principles of QM [34].

C. **Additional Computational Methods:**

Docking Studies: Molecular docking facilitates structure-based drug discovery by predicting the binding strengths and mechanisms of ligands to target [35].

Free Energy Calculations: These calculations help optimize drug candidates by estimating the binding free energy of ligand for specific proteins [36].

Machine Learning and Data Mining: To find patterns and forecast the characteristics of novel compounds, ML algorithms examine sizable datasets containing biological activity and chemical structure information [37].

Discussion on the Integration of Chemical/Biological Databases and Molecular Modeling Tools in Drug Discovery

The drug discovery approach has been transformed by integrating molecular modeling techniques with biochemical and biological databases. This has sped up the generation of new drugs and made it easier to identify novel drug candidates. The following is an explanation of the way drug discovery has changed as a result of the melding of various resources:

Access to Extensive Data: Pharmaceutical and biological databases have a wealth of knowledge on targets in biology, pharmacological actions, clinical data, and molecular frameworks. These databases may be integrated with molecular modeling tools to provide researchers with a plethora of details that can help guide the course of drug discovery.

Machine Learning and Data Mining: Database-driven drug discovery methodologies are made possible by the combination of ML tools with information pertaining to biology and chemistry. ML algorithms can prioritize potential drug ideas for experimental validation, find trends, and anticipate biological activities by analyzing massive databases containing chemical and biological information.

Collaborative Research and Knowledge Sharing: Scholars from universities, pharmaceutical corporations, and other organizations may work together more easily when they use combined systems that merge biomedical and chemical data with molecular modeling tools. This encourages the sharing of knowledge, information, and ideas which results in pipelines for drug development that are more effective [38].

Applications of Data Mining in Drug Discovery and Development

The use of data mining techniques can reveal underlying trends in the chemical and pharmacological properties space that are essential for the development and discovery of new drugs. ML techniques and visualization are two of the most popular strategies. Dimensionality reduction techniques are used by visualization approaches to minimize information loss while reducing multidimensional data into 2D or 3D depictions. Using recurrent mathematical models, ML looks for relationships between particular actions or categories for a group of chemicals and their characteristics. Both models make use of the many and complex interactions that may exist between compound properties, and they are useful in classifying compounds according to these features or, in the case of visualization techniques, in revealing underlying trends of the feature space.

a. **Target Identification and Validation** Data mining tools are essential for determining targets and validation, two crucial phases in the drug development process. Following is how determining targets and validation using data mining is done:

   i. **Target Identification**
      - **Proteomic and Genomic Data Analysis**: Large-scale genomic and proteomic datasets are analyzed using methods such as data mining to find possible therapeutic targets linked to certain ailments. This entails using methods like analysis of pathway enrichment and variable gene expression analysis to discover genes, proteins, or networks that are dysregulated or linked to disease states.
      - **Network Analysis**: To find important nodes or proteins which could be suitable targets for drugs, data mining techniques are used to build and examine biological networks, such as networks involving protein–protein interactions and signaling cascades. Network analysis reveals how biological processes are interrelated and helps pinpoint major nodes that could be essential for the development of a disease.

   ii. **Validation of the Target:**
      - **Integrating Multi-Omics Data**: To verify the applicability of putative therapeutic targets, data mining incorporates multi-omics data, such as transcriptomics, proteomics, metabolomics, and genomics. Data mining enables to recognize and prioritize targets that are frequently dysregulated in disease settings by establishing a correlation between molecular alterations across many omics' levels and disease phenotypes.
      - **Clinical and Practical Data Interpretation:** Potential drug targets' clinical significance is confirmed by data mining from clinical and practical data sources, including patient databases with electronic health records (EHRs).

b. **Compound Screening and Design**

   i. **High-Throughput Virtual Screening:**
      a. **Database Screening**: High-throughput virtual screening (HTVS) uses computational techniques to screen enormous chemical databases with millions of molecules for a biological target of concern. Compounds that are marketed commercially, compound libraries, or digital collections of compounds created via molecular docking or synthetic chemistry may be found in these databases.
      b. **Fast and Effective Screening**: HTVS quickly screens hundreds to millions of chemicals against the target protein by utilizing quick computational methods

and procedures. HTVS frequently uses ligand-based simulations, pharmacophore-based assessment, and molecular docking to forecast a compound's binding efficiency and specificity.

c. **Prioritization and Scoring**: Estimated binding affinities, interaction energies, and match to the targeted binding site are used to provide scores to compounds. The screened compounds are ranked using prioritization algorithms to determine which top results have the best chance of interacting with the target and displaying the required pharmacological activity.

d. **Hit Identification**: Strong affinity and specificity of molecules who bind to the protein of interest can be identified using HTVS. These hit molecules are used as building blocks for additional optimization via structure–activity relationship (SAR) and medicinal chemistry research.

c. **Pharmacogenomics and Personalized Medicine**

By personalizing medical procedures to each patient's unique genetic and genomic description, pharmacogenomics and personalized medicine provide the promise of more secure and more efficient therapeutic interventions. An outline is provided below:

Utilizing *Genomic* and *Genetic Information* to Provide Personalized Care:

Genetic Variability: The field of pharmacogenomics is concerned with comprehending how medication response and metabolism are impacted by genetic variations, including single-nucleotide polymorphisms (SNPs) and copy number variations (CNVs). Through the analysis of genetic data, scientists can find genetic variations connected to the effectiveness, toxicity, and side effects of drugs.

Pharmacogenetic Biomarkers: Individual reactions to certain medications or pharmacological classes are predicted using genetic biomarkers. By using genomic data mining, pharmacogenetic biomarkers that are based on a person's genetic profile may be found to inform therapy choices, dose modifications, and medication selection.

Gene Expression Mapping: Next-generation sequencing (NGS) and microarray analysis (MA) are two genomic technologies that make it possible to evaluate hundreds of genes' activities at once through gene expression profiling. Personalized treatment techniques can be aided by the revelation of molecular signatures linked to medication response or illness progression through the use of gene expression data.

Personalized medicine has both opportunities and challenges such as data integration and interpretation, clinical implementation, ethical and legal considerations, cost and accessibility, therapeutic innovation [39].

INTEGRATION OF ARTIFICIAL INTELLIGENCE IN BIOLOGICAL DATABASES FOR DRUG INNOVATION

Artificial intelligence (AI) in medicine refers to the application of AI techniques, algorithms, and technologies in healthcare. Medical image analysis, medication discovery, customized treatment planning, illness diagnosis and prediction, virtual health assistants, EHR management, and patient monitoring are just a few of the many fields in which AI is being applied. One of the major areas of AI in medicine that is developing quickly and having an impact is drug discovery, and the process of development involves more efficiency.

ML techniques have been used in the fields of drug discovery and the development of novel drug candidates. ML and deep learning algorithms are now commonly used in drug target design and novel drug discovery techniques to improve productivity, effectiveness, and output quality. There are two main types of ML algorithms: supervised and unsupervised learning. In supervised learning, new samples' labels are determined by using training samples with predetermined labels. Unsupervised learning, which typically occurs without sample labels, finds patterns in a collection of samples [40].

Supervised ML Algorithms

A. **Support Vector Machine:**

The support vector machine (SVM) algorithm is a supervised ML algorithm used for both regression and classification tasks. SVM can be applied to real-world tasks, such as text classification, image classification, spam detection, handwriting identification, gene expression analysis, face detection, and anomaly detection. The utility of these algorithms in drug discovery is in assisting in virtual screening, drug–target interaction prediction, and the identification of new drug targets. It also predicts drug similarity by quantitative structure–activity relationship (QSAR) analysis and detecting activity cliffs, pairs of structurally similar compounds with significant activity variations. SVMs use ML techniques to analyze and interpret large-scale chemical and biological datasets, which speeds up the drug discovery process. SVMs are useful tools for computational drug discovery and medicinal chemistry because of their capacity to simulate intricate relationships between chemical structures and biological activities.

B. **Naïve Bayes – Simplicity and Versatility in Probabilistic Modeling:**

Naive Bayes is a collection classification algorithm based on the Bayes theorem. The Naïve Bayes classifier, one of the most straightforward and efficient classification algorithms, facilitates the quick creation of ML models with quick prediction capabilities. This algorithm is used for many of the classification problems, mostly in text classification. In text classification tasks, the data contains high dimensions. It is used in spam filtering, sentiment detection, rating classification, etc. Naive Bayes algorithms may not be as complex or versatile as SVMs in handling high-dimensional and nonlinear data, but they can still offer value in certain aspects of drug discovery, particularly in probabilistic modeling, text analysis, and classification tasks involving structured and categorical data.

C. **Random Forest – The Power of Ensemble Learning:**

Random forest (RF) is a powerful ML technique that belongs to supervised learning. It can be applied to ML problems involving both classification and regression. Its foundation is the idea of ensemble learning, which is the process of merging several classifiers to solve a challenging issue and enhance the model's functionality. Random forest works in two phases: The first is to create the RF by combining N decision trees, and the second is to make predictions for each tree created in the first phase. Random forest is used in various aspects of drug discovery, ranging from compound screening and prediction to optimization of compound properties and understanding SARs. Thus, in the evolving landscape of drug discovery, supervised learning techniques, including random forest, are indispensable assets, shaping a future of enhanced efficiency and precision in drug development [41].

Unsupervised Learning

A. **K-means Clustering: Unraveling Molecular Mysteries:**

K-means is an unsurprised ML algorithm and is a process of training the machines to use unlabeled, unclassified data and enabling the algorithm to operate on that data without supervision. It is used to solve clustering problems in ML or data science. The unlabeled dataset is fed into the algorithm, which then divides it into k clusters and keeps going until it finds no optimal clusters. In this algorithm, the value of k ought to be predetermined. K-means clusters play an important role in drug discovery in defining molecular descriptors, numeric representations of a compound's physicochemical properties, and analyzing large datasets of chemical compounds, biological data, and clinical information. K-means clustering is used in compound screening to prioritize lead compounds for additional research and identify chemical classes by grouping compounds based on bioactivity profiles or structural similarity.

B. **T-Distributed Stochastic Neighbor Embedding: Navigating the High-Dimensional Wilderness:**

T-distributed stochastic neighbor embedding (t-SNE) is a dimensionality reduction algorithm. This algorithm employs a randomized strategy to nonlinearly reduce the dimensionality of the current dataset. This is primarily concerned with keeping the dataset's local structure in the lower dimension. Principal Component Analysis (PCA) and t-SNE belong to the unsupervised algorithms, where PCA is a deterministic algorithm to reduce the dimensionality of the algorithm and t-SNE is a randomized nonlinear method to map the high-dimensional data to the lower-dimensional data. t-SNE takes on a crucial role in drug discovery. It helps with molecular representation, drug target exploration, compound clustering, and drug design itself. Through the process of reducing high-dimensional data to a lower-dimensional canvas, t-SNE facilitates a thorough comprehension of intricate biological data and compound similarities. t-SNE is utilized to analyze and interpret complex biological and chemical datasets, aiding in data exploration, feature visualization, pattern recognition, and knowledge extraction in the context of drug discovery.

C. **Hierarchical Clustering:**

Hierarchical clustering is an unsupervised machine. Learning algorithms are used to group the unlabeled datasets into a cluster, also known as hierarchical cluster analysis. The dendrogram is the structure that results from developing the hierarchy of clusters in this algorithm in the shape of a tree. In the hierarchal clustering, unlike the K-means algorithm, there is no need to predetermine the number of clusters. It uses two approaches: agglomerative and divisive. This algorithm facilitates the investigation, arrangement, and comprehension of intricate datasets spanning various phases of the pharmaceutical development process. Compound analysis, target identification, patient stratification, adverse event profiling, and knowledge extraction from various data sources are just a few of its many uses [42].

Category	Algorithm	Key Applications
Supervised machine learning	SVM	• Virtual screening • Activity prediction • Toxicity prediction • Compound optimization • ADME (absorption, distribution, metabolism, excretion) prediction • Drug–target interaction prediction • Structure–Activity relationship (SAR) analysis • Multi-target drug design • de novo drug design
	Naïve Bayes	• *Bioactivity prediction* • Toxicity prediction • Compound similarity and clustering • Text mining and literature analysis • Multi-label classification • Integration of omics data • Data preprocessing and feature selection
	Random forest	• Compound bioactivity prediction • Toxicity prediction • Compound property prediction • Compound similarity and clustering • Multi-target activity prediction • ADME (absorption, distribution, metabolism, excretion) prediction • Virtual screening and compound prioritization • Structure–activity relationship (SAR) analysis
Unsupervised machine learning	K-means clustering	• Bioactivity profiling and compound prioritization • Target profiling and similarity analysis • Disease subtyping and biomarker discovery • Virtual screening and compound selection • Analysis of high-dimensional omics data • Identification of drug combinations and synergistic effects • Adverse event profiling and pharmacovigilance • Text mining and literature analysis
	T-distributed stochastic neighbor embedding	• Structure–activity relationship (SAR) analysis • Molecular similarity and clustering • Integration of multi-omics data • Cell population analysis and single-cell data visualization • Patient stratification and disease subtyping • Drug–target interaction analysis • Text mining and literature analysis
	Hierarchical clustering	• Activity landscape analysis and SAR exploration • Gene expression clustering and omics data analysis • Target profiling and protein family analysis • Patient stratification and disease subtyping • Biomarker discovery and molecular signature identification • Integration of multimodal data sources • Adverse event profiling and safety assessment • Text mining and literature analysis

D. **Performance Metrics for Supervised Algorithms [43]:**

- **Precision:** To evaluate the performance of the system, precision is the average ratio of positive observation successfully predicted to the total positive observation predicted.

$$\text{Precision} = \frac{\text{True Positive}}{(\text{True Positive} + \text{False Positive})}$$

- **Recall:** Recall is the average ratio of positive (or correct) results to the total number of the actual relevant results.

$$\text{Recall} = \frac{\text{True Positive}}{(\text{True Positive} + \text{False Negative})}$$

- **F1-score:** F1-score combines precision and recall.

$$F1-\text{Score} = \frac{2 \times \text{Precision} \times \text{Recall}}{(\text{Precision} + \text{Recall})}$$

E. **Performance metrics for unsupervised Algorithms [44]:**

- **Silhouette Score**: It is a method of interpretation and validation of consistency within clusters of data.

$$S_i = \frac{b_i - a_i}{\max(b_i, a_i)}$$

where; a=average intra-cluster distance, *i.e.*, the average distance between each point within a cluster; b=average inter-cluster distance, *i.e.*, the average distance between all clusters.

- **Davies–Bouldin Index:** To calculate the average similarity measure of each cluster with the cluster most similar to it.

$$DB = \frac{1}{k}\sum_{i=1}^{k} \max\left(\frac{\Delta(X_i) + \Delta(X_j)}{\delta(X_i, X_j)}\right)$$

where; ΔX_k is the intracluster distance within the cluster X_k; $\delta(X_i, X_j)$is the intercluster distance between the clusters X_i and X_j.

- **Dunn Index:** To identify sets of clusters that are compact, with a small variance between members of the cluster, and well separated, where the means of different clusters are sufficiently far apart, as compared to the within cluster variance.

$$\text{Dunn index } (U) = \min_{1 \le i \le c}\left\{\min_{1 \le j \le c,\, j \ne i}\left\{\frac{\delta(X_i, X_j)}{\max\{\Delta(X_k)\}}\right\}\right\}$$

Where; $\delta(X_i, X_j)$ is the intercluster distance, i.e., the distance between clusters X_i and X_j; ΔX_k is the intracluster distance of cluster X_k, i.e. distance within the clusters X_k.

CHALLENGES AND FUTURE PERSPECTIVE/DIRECTIONS

Biological databases and data mining play a pivotal role in advancing our understanding of biological research, fostering groundbreaking discoveries and innovations across diverse fields such as biology, enabling breakthroughs in medicine, agriculture, biotechnology, and beyond. Addressing the challenges such as data integration, quality control, scalability, interpretability, and ethical issues; scientists can use the abundance of biological data to gain a deeper understanding of the workings of complex biological systems.

Challenges

Data Integration: Biological data are often vast and heterogeneous in nature of biological information. Data are kept in different databases, each with its own formats, standards, and identifiers, across a variety of domains, including proteomics, metabolomics, and genomics. Integrating this kind of diverse data requires sophisticated computational methods. Inconsistencies in data formats, missing annotations, and discrepancies in nomenclature are a challenge in integration and analysis. Overcoming this hurdle is essential for gaining a holistic understanding of biological systems.

1. **Data Quality**: Data quality is one of the fundamental challenges, and ensuring the quality and reliability of data in biological databases is crucial. Biological datasets encompass a wide range of information that is susceptible to issues such as incomplete annotations, errors, and biases that can affect the results of data mining analyses.
2. **Scalability**: As the volume of biological data continues to grow exponentially, scalable data mining algorithms and computational resources are essential to handle large-scale datasets efficiently. The exponential growth in data production has outpaced the capacity of traditional computational resources, necessitating scalable solutions to process, store, and analyze large-scale datasets efficiently.
3. **Interpretability**: While ML techniques can uncover complex patterns in biological data, interpreting these results in biologically meaningful ways can be challenging, especially for nonexperts as the complexity and multidimensionality of biological systems processing requires expert domain knowledge and contextual understanding (Figure 4.2).
4. **Ethical and Privacy Concerns**: With the increasing availability of personal genomic data, ethical considerations regarding data privacy, consent, and potential misuse are paramount. Because biological data are sensitive and frequently contain genetic information about specific individuals, there are serious risks to informed consent, privacy, and confidentiality. Furthermore, privacy risks are further increased by the possibility of reidentification and the connection of genomic data with other personal information [45].

Future Perspectives/Directions

1. **Multi-Omics Integration**: Through the integration of data from various omics layers, such as transcriptomics, proteomics, metabolomics, and

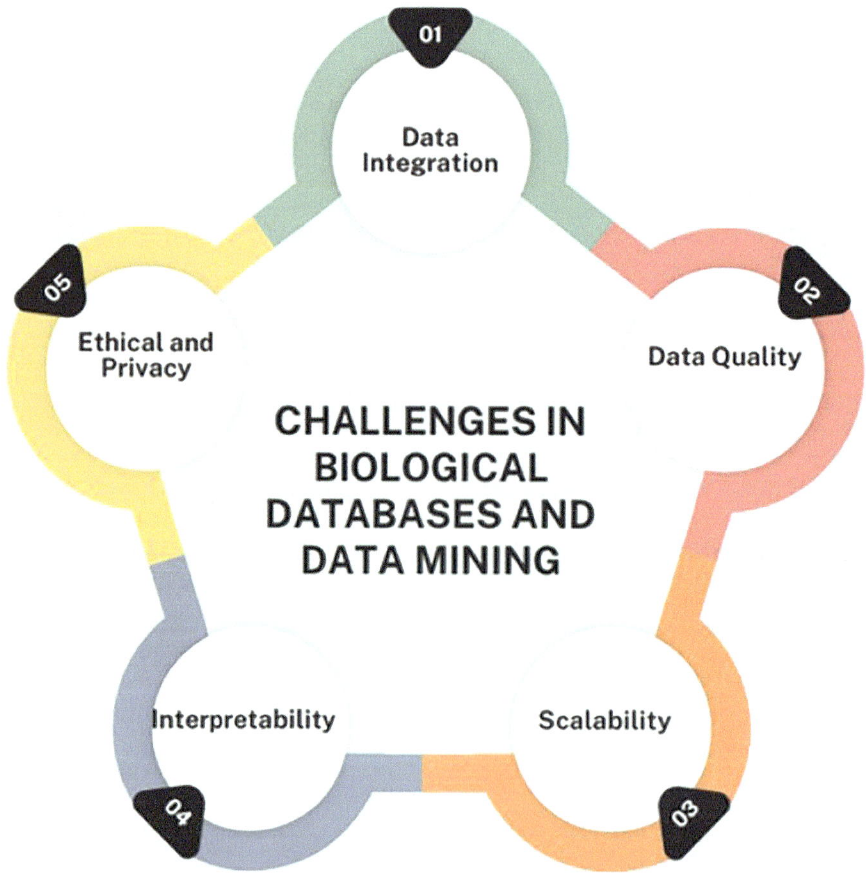

FIGURE 4.2 Challenges in biological databases and mining.

epigenomics, scientists can acquire a more profound understanding of the molecular processes that underlie both health and illness. Further, precision medicine approaches catered to individual patients could be made possible by the potential for multi-omics integration to uncover novel biomarkers, therapeutic targets, and personalized treatment strategies for a variety of diseases [46].

2. **Single-Cell Analysis**: Advancements in single-cell technologies enable the study of individual cells, uncovering cellular heterogeneity and dynamics. Furthermore, privacy risks are further increased by the possibility of reidentification and the connection of genomic data with other personal information. Researchers can clarify intricate cellular behaviors – like differentiation, proliferation, and stimulus response – that are frequently obscured in bulk cell analyses by examining individual cells [47].
3. **Artificial Intelligence and Machine Learning**: Continued advancements in AI and ML will drive the building of more sophisticated algorithms for analyzing complex biological data. As a promising future direction in biological data analysis, the combination of AI and ML will transform our capacity to glean meaningful insights from intricate biological datasets [48].
4. **Precision Medicine**: Leveraging biological data to personalized medical treatments based on individual genetic profiles and biomarkers holds immense promise for improving healthcare outcomes. Precision medicine seeks to provide more individualized and efficient healthcare strategies that maximize patient outcomes while minimizing side effects by integrating diverse biological data, including genetic, clinical, environmental, and lifestyle information [49].
5. **Network Biology**: Understanding biological systems as interconnected networks of genes, proteins, and other molecules will provide insights into emergent properties and system-level behaviors. Network biology will be shaped by the creation of novel computational approaches, experimental strategies, and interdisciplinary teams that will use network analysis to address important biological and medical issues, eventually leading to breakthroughs in biotechnology, healthcare, and other fields [50].

6. **Data Sharing and Collaboration**: Encouraging open access to biological data and fostering collaboration between researchers and institutions will accelerate discoveries and innovation in the field. In order to maximize the value of these datasets and speed up research, collaborative efforts are crucial as the volume and complexity of biological data continue to grow exponentially.
7. **Ethical and Responsible Data Use**: Establishing guidelines and frameworks for ethical and responsible use of biological data, including privacy protection and informed consent, is crucial for maintaining public trust and advancing research ethically. The creation of strong ethical standards, legal frameworks, and best practices to protect people's rights, privacy, and autonomy while advancing accountability, transparency, and fairness in data use will define the future of ethical and responsible data use [51] (Figure 4.3).

ETHICAL AND LEGAL CONSIDERATIONS IN DATA-DRIVEN DRUG DISCOVERY

With the use of large datasets and sophisticated computational techniques, data-driven drug discovery has the potential to significantly speed up the creation of new medications. To ensure responsible and moral behavior, it also raises a number of ethical and legal issues that need to be properly negotiated. Here are some crucial things to remember:

Data Privacy and Security: It is imperative for pharmaceutical companies and research institutions to guarantee the confidentiality and integrity of the data they gather and examine. This entails safeguarding patient data from unwanted access, making sure rules like the US's Health Insurance Portability and Accountability Act (HIPAA) are followed, and putting strong cyber security measures in place to stop data breaches.

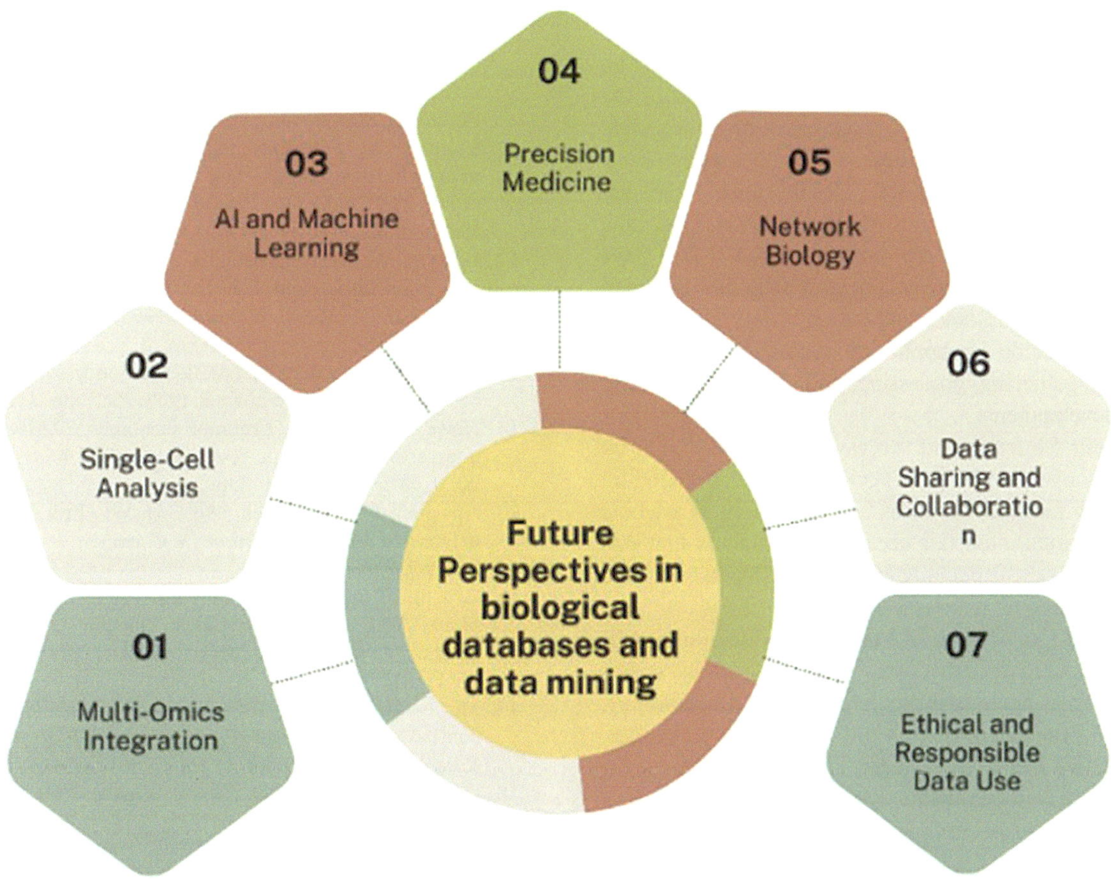

FIGURE 4.3 Future perspectives of biological databases and mining.

Informed Consent: Getting informed consent is crucial when using patient data for medication discovery. Patients ought to be fully informed about the alternatives available to them, the possible risks and benefits, and how their data will be used. In order to preserve patient privacy, researchers must also make sure that data is de-identified or anonymized.

Bias and Fairness: Dataset biases, such as the underrepresentation of particular demographics, can produce discriminatory results and biased algorithms. To ensure fairness and equity in healthcare outcomes, researchers must be diligent in identifying and mitigating bias throughout the drug discovery process.

Transparency and Reproducibility: In data-driven drug discovery, transparency is essential to guarantee reproducibility of outcomes and allow the scientific community to review. Scholars ought to furnish comprehensive records of data sources, methodologies, and analytical techniques in order to enable autonomous validation and replication of discoveries.

Intellectual Property Rights: Data-driven drug discovery may give rise to intellectual property rights issues, such as those pertaining to patents and licensing contracts. In order to guarantee that intellectual property is properly protected and to foster innovation and collaboration within the scientific community, researchers must negotiate these legal complexities.

Regulatory Compliance: Organizations like the European Medicines Agency (EMA) and the U.S. Food and Drug Administration (FDA) have strict regulatory oversight over drug discovery. To ensure compliance and patient safety, researchers must abide by applicable regulations governing the gathering, processing, and use of data in drug development.

Benefit Sharing and Access: As data-driven drug discovery develops, it's critical to take into account how medical advances benefit society as a whole and guarantee fair access to treatments that can save lives, especially in underprivileged areas and low-income nations.

Ethical Use of AI and ML: The application of ML and A) in drug discovery raises new ethical issues such as accountability, algorithmic transparency, and the possibility of unforeseen consequences. When developing, implementing, and overseeing AI-driven healthcare systems, researchers have a duty to respect ethical norms and values [51].

REFERENCES

1. Fayyad, U. and P. Stolorz, Data mining and KDD: Promise and challenges. *Future Generation Computer Systems*, 1997. 13(2–3): pp. 99–115.
2. Lacroix, Z. and T. Critchlow, *Bioinformatics: Managing Scientific Data*. Vol. 6. 2003: Academic Press, New York.
3. Chen, J. and G. Coppola, Bioinformatics and genomic databases. *Handbook of Clinical Neurology*, 2018. 147: pp. 75–92.
4. Gorunescu, F., *Data Mining: Concepts, Models and Techniques*. Vol. 12. 2011: Springer Science & Business Media, Berlin.
5. Dogan, A. and D. Birant, Machine learning and data mining in manufacturing. *Expert Systems with Applications*, 2021. 166: p. 114060.
6. Pohle, C., *Integrating and updating domain knowledge with knowledge discovery*. Knowledge Creation Diffusion Utilization, https://www.hhl.de/fileadmin/LS/micro/Download/Pohle_2003_IntegratingAndUpdatingA.pdf, 2002.
7. Olson, D.L. and D. Delen, *Advanced Data Mining Techniques*. 2008: Springer Science & Business Media, Berlin.
8. Cannataro, M., P.H. Guzzi, and A. Sarica, Data mining and life sciences applications on the grid. *Wiley Interdisciplinary Reviews: Data Mining and Knowledge Discovery*, 2013. 3(3): pp. 216–238.
9. Dubitzky, W., M. Granzow, and D.P. Berrar, *Fundamentals of Data Mining in Genomics and Proteomics*. 2007: Springer Science & Business Media, Berlin.
10. Mandapuram, M. and M.F. Hosen, The object-oriented database management system versus the relational database management system: A comparison. *Global Disclosure of Economics and Business*, 2018. 7(2): pp. 89–96.
11. Lazarova, M., S. Markov, and A. Vassilev. On some classes of growth functions and their links to reaction network theory. In *AIP Conference Proceedings*. 2020: AIP Publishing, Melville, NY.
12. Ludwig, W., Sequence databases. In Akkermans, A.D.L., J.D. Van Elsas, and F.J. De Bruijn (Eds.), *Molecular Microbial Ecology Manual*. 1995: Springer, Berlin, pp. 309–330.
13. Stoesser, G., et al., The EMBL nucleotide sequence database. *Nucleic Acids Research*, 1997. 25(1): pp. 7–13.
14. Howe, K.L., et al., Ensembl Genomes 2020—enabling non-vertebrate genomic research. *Nucleic Acids Research*, 2020. 48(D1): pp. D689–D695.
15. Apweiler, R., A. Bairoch, and C.H. Wu, Protein sequence databases. *Current Opinion in Chemical Biology*, 2004. 8(1): pp. 76–80.
16. Groom, C.R., et al., The Cambridge structural database. Acta Crystallographica Section B: Structural Science. *Crystal Engineering and Materials*, 2016. 72(2): pp. 171–179.
17. Mewes, H.W., et al., MIPS: Curated databases and comprehensive secondary data resources in 2010. *Nucleic Acids Research*, 2011. 39(suppl_1): pp. D220–D224.
18. Courteau, J., Genome databases. *Science*, 1991. 254(5029): pp. 201–207.

19. Lo Conte, L., et al., SCOP: A structural classification of proteins database. *Nucleic Acids Research*, 2000. 28(1): pp. 257–259.
20. Schaefer, C.F., Pathway databases. *Annals of the New York Academy of Sciences*, 2004. 1020(1): pp. 77–91.
21. Kim, W., et al., Composite object support in an object-oriented database system. In *Conference Proceedings on Object-Oriented Programming Systems, Languages and Applications*. 1987: Austin, Texas.
22. Chang, K.C.-C., et al., Structured databases on the web: Observations and implications. *ACM Sigmod Record*, 2004. 33(3): pp. 61–70.
23. Codd, E.F., Data models in database management. In *Proceedings of the 1980 Workshop on Data Abstraction, Databases and Conceptual Modeling*. 1980: Austin, Texas.
24. Wang, Y.-Y.R., et al., *Data Quality*. Vol. 23. 2001: Springer Science & Business Media, Berlin.
25. El-Attar, N.E., Perspectives review and challenges in biological databases integration. *Alfarama Journal of Basic & Applied Sciences*, 2022. 3(2): pp. 346–362.
26. Cosconati, S., et al., Virtual screening with AutoDock: Theory and practice. *Expert Opinion on Drug Discovery*, 2010. 5(6): pp. 597–607.
27. Wang, X., et al., Structure-based drug design strategies and challenges. *Current Topics in Medicinal Chemistry*, 2018. 18(12): pp. 998–1006.
28. Yadav, V. and R.K. Tonk, Ligand-based drug design (LBDD). In *Computer Aided Drug Design (CADD): From Ligand-Based Methods to Structure-Based Approaches*. 2022: Elsevier, Amsterdam, Netherlands, pp. 57–99.
29. Isarankura-Na-Ayudhya, C., et al., A practical overview of quantitative structure-activity relationship. *EXCLI Journal*, 2009. 8: pp. 74–88.
30. Shen, J., et al., Estimation of ADME properties with substructure pattern recognition. *Journal of Chemical Information and Modeling*, 2010. 50(6): pp. 1034–1041.
31. Kumar, A., A. Voet, and K.Y. Zhang, Fragment based drug design: From experimental to computational approaches. *Current Medicinal Chemistry*, 2012. 19(30): pp. 5128–5147.
32. Das, S., et al., Applications of artificial intelligence in machine learning: Review and prospect. *International Journal of Computer Applications*, 2015. 115(9): pp. 31–41.
33. Hansson, T., C. Oostenbrink, and W. van Gunsteren, Molecular dynamics simulations. *Current Opinion in Structural Biology*, 2002. 12(2): pp. 190–196.
34. Ballentine, L.E., *Quantum Mechanics: A Modern Development*. 2014: World Scientific Publishing Company, London.
35. Scotti, L., et al., Docking studies for multi-target drugs. *Current Drug Targets*, 2017. 18(5): pp. 592–604.
36. Chipot, C. and A. Pohorille, *Free Energy Calculations*. Vol. 86. 2007: Springer, Berlin.
37. Mitchell, T.M., Machine learning and data mining. *Communications of the ACM*, 1999. 42(11): pp. 30–36.
38. Oprea, T.I., et al., Computational systems chemical biology. *Chemoinformatics and Computational Chemical Biology*, 2011. 672: pp. 459–488.
39. Agatonovic-Kustrin, S. and D. Morton, Data mining in drug discovery and design. In *Artificial Neural Network for Drug Design, Delivery and Disposition*. 2016: Elsevier, Amsterdam, Netherlands, pp. 181–193.
40. Parvatikar, P.P., et al., Artificial intelligence: Machine learning approach for screening large database and drug discovery. *Antiviral Research*, 2023. 220: p. 105740.
41. Singh, A., N. Thakur, and A. Sharma, A review of supervised machine learning algorithms. In *2016 3rd International Conference on Computing for Sustainable Global Development (INDIACom)*. 2016: IEEE, New York.
42. Barlow, H.B., Unsupervised learning. *Neural Computation*, 1989. 1(3): pp. 295–311.
43. Suthaharan, S. and S. Suthaharan, Supervised learning algorithms. *Machine Learning Models and Algorithms for Big Data Classification: Thinking with Examples for Effective Learning*, 2016. 36: pp. 183–206.
44. Nguyen, T.T. and U.Q. Nguyen, An evaluation method for unsupervised anomaly detection algorithms. *Journal of Computer Science and Cybernetics*, 2016. 32(3): pp. 259–272.
45. Doan, A., A. Halevy, and Z. Ives, *Principles of Data Integration*. 2012: Elsevier, Amsterdam, Netherlands.
46. Subramanian, I., et al., Multi-omics data integration, interpretation, and its application. *Bioinformatics and Biology Insights*, 2020. 14: p. 1177932219899051.
47. Stuart, T. and R. Satija, Integrative single-cell analysis. *Nature Reviews Genetics*, 2019. 20(5): pp. 257–272.
48. Moroney, L., *AI and Machine Learning for Coders*. 2020: O'Reilly Media, Sebastopol, CA.
49. Kosorok, M.R. and E.B. Laber, Precision medicine. *Annual Review of Statistics and Its Application*, 2019. 6: pp. 263–286.
50. Barabasi, A.-L. and Z.N. Oltvai, Network biology: Understanding the cell's functional organization. *Nature Reviews Genetics*, 2004. 5(2): pp. 101–113.
51. Papale, D., et al., Database maintenance, data sharing policy, collaboration. In Aubinet, M., Vesala, T., and Papale, D. (Eds.), *Eddy covariance: A practical guide to measurement and data analysis*, 2012: Springer, Dordrecht, pp. 399–424.

5 Data Storage and Retrieval, Database Structure and Annotation

Balaji Wamanrao Matore[1], Anjali Murmu[1], Pragya Gawande[1], Akshay Kumar[1], Souranava Jana[1,2], Jagadish Singh[1], and Partha Pratim Roy[1]
[1]Department of Pharmacy, Guru Ghasidas Vishwavidyalaya (A Central University), Bilaspur, India
[2]Department of Pharmaceutical Chemistry, Haldia Institute of Pharmacy, Haldia, India

INTRODUCTION

In the modern era of drug discovery, bioinformatics and cheminformatics played a crucial role inspeeding up drug discovery in an economical way [1,2]. In this landscape of information technology, where data serves as the cornerstone of modern operations, the importance of effective data storage and retrieval, the significance of database structure, and the pivotal role of annotation in data management cannot be overstated [3,4]. These elements collectively form the foundation of efficient data management systems, facilitating the organization, storage, retrieval, and manipulation of vast amounts of data [5,6]. The fundamental concepts, methodologies, challenges, and advancements in data storage and retrieval, database structure, and annotation, exploring their interconnected roles in drug discovery, need to be detailed [7,8]. Nowadays, these practices are the daily routine for bioinformaticians or drug designers to enhance the productivity of work.

With the expansion of computers, artificial intelligence (AI), and machine learning (ML) technologies, the volume and diversity of data generated globally continue to escalate exponentially. This phenomenon, often referred to as the "data deluge," encompasses diverse data types such as structured, semi-structured, and unstructured data, posing challenges for traditional methods of data management [2,9]. In the field of drug design, development, and discovery, vast data are gathered and processed to optimize the results. The most common data information includes physicochemical properties, ADMET parameters, different biological responses, preclinical and clinical trial results, etc [10]. The traditional model of storing data in structured databases often falls short when dealing with this type of complex data, necessitating innovative approaches to data storage, retrieval, and management.

At the core of effective data management lie the fundamental processes of data storage and retrieval, which form the backbone of data associated drug design and discovery [11]. A well-designed database structure serves as the roadmap for organizing data in a systematic and coherent manner, facilitating efficient data retrieval, processing, and analysis [12]. By optimizing performance, ensuring data integrity, and supporting scalability and adaptability, a robust database structure enables organizations to harness the full potential of their data assets and drive innovation and growth [13,14].

In parallel, annotation emerges as a crucial mechanism for enhancing data usability, accessibility, and understanding. By adding metadata, tags, or labels to data, annotation enriches its context and semantic meaning, making it more interpretable and discoverable [15,16]. Annotation serves as a bridge between raw data and human understanding, providing additional context and insights that empower users to navigate, search, and analyze data more effectively [17]. Moreover, annotation facilitates data integration, interoperability, and knowledge discovery, enabling organizations to unlock hidden patterns, correlations, and insights that drive strategic decision-making and innovation [18].

In overview, the interconnected roles of data storage and retrieval, database structure, and annotation are indispensable in today's data-driven world. In the context of the above facts, this chapter explores three interconnected pillars of data management: data storage and retrieval, database structure, and annotation. We also highlighted different types of data and databases useful in rational drug design and, further, different AI tools and software used for the data storage, retrieval, database structure, and annotation.

DATA STORAGE AND RETRIEVAL

In the realm of science, data have emerged as the Fourth Paradigm alongside experimental, theoretical, and computational sciences, with data storage playing a pivotal role in fields like drug discovery and bioinformatics [19]. Efficient storage systems are essential for managing and analyzing the vast datasets generated. Historically, access to chemistry-related data was monopolized, but recent trends toward openness have democratized method development, fostering the creation of robust computational algorithms. Large-scale public repositories now offer extensive data sources for scientific exploration. Technological advancements have led to a surge in data volume and complexity, encompassing molecular structures, genomic sequences, experimental findings, and clinical trial results [5]. Reliable

DOI: 10.1201/9781003464679-5

storage solutions are crucial in drug discovery, serving as the foundation for analyses, interpretations, and therapeutic breakthroughs. Effective storage mechanisms not only facilitate seamless data management but also drive collaboration, innovation, and the acceleration of drug discovery processes [20]. Similarly, data retrieval plays a crucial role in accessing and analyzing diverse datasets related to compounds, targets, assays, and clinical trials. Several tools, servers, and software are used for data retrieval in drug discovery, catering to the specific needs of researchers and organizations in the pharmaceutical and biotechnology industries.

NEED FOR DATA STORAGE AND RETRIEVAL

Data storage and retrieval plays a crucial role in drug discovery by facilitating secure retention and efficient retrieval of experimental data across various studies, including high-throughput screening and molecular modelling. It enables the integration and thorough analysis of diverse data types such as chemical structures, biological assays, and clinical trial results, fostering insights that might be overlooked individually. Centralized storage promotes collaboration among multidisciplinary teams, while computational techniques extract patterns and identify potential drug candidates, expediting the discovery process. Compliance with regulatory standards ensures data integrity, while knowledge management informs decision-making and preserves valuable intellectual property for future innovation [3,21]. Similarly, data retrieval is pivotal for managing, accessing, analyzing, and processing extensive data in drug discovery, driving scientific advancement and therapy development.

DIFFERENT TYPES OF DATA INFORMATION AND DATABASES

Drug discovery involves the storage and retrieval of diverse data types to facilitate research and development endeavors. It's essential to carefully scrutinize different data types before assessing the suitable data sources and methodologies to generate fresh insights as shown in Figure 5.1 [3,6]. The common types of data used in drug discovery are discussed below.

CHEMICAL STRUCTURE

The core backbone of chemistry data analysis lies in understanding chemical structures, which dictate all associated properties and data. These chemical frameworks primarily structure chemistry databases, typically store structures in topological form, and advanced ones like the Cambridge Structural Database (CSD) accommodate experimentally derived 3D structures. Similarly, databases such as the Protein Data Bank (PDB) facilitate the linkage between the 3D structures of compounds and biological targets through co-crystal structures, providing insights into potential bioactive conformations. Researchers have focused their efforts on creating chemical structure representations for computational analysis, highlighting SMILES and InChI as essential tools for computational input.

CHEMICAL PROPERTIES

Chemical properties are decisive in drug discovery, as they determine the drug interactions within biological systems and their impact on the intended target. These properties include molecular structure, lipophilicity, solubility, ionization behavior, reactivity, metabolism, toxicity, stability, binding affinity, hydrogen bonding, conformational flexibility, etc. By understanding and optimizing these properties, researchers can enhance a drug candidate's efficacy, safety, and pharmacokinetic profile, ultimately advancing its potential for clinical application.

BIOLOGICAL ASSAY DATA

The collection of information from experiments and assays aimed at comprehending the interactions of potential drug compounds with biological systems constitutes biological data in drug discovery. This data includes target identification and validation, screening assays, binding affinity, specificity, pharmacokinetics, pharmacodynamics, toxicity and safety assessment, and genomic and proteomic data. It plays a crucial role in different phases of drug development, aiding in the selection of promising drug candidates and advancing them through preclinical and clinical evaluation.

ADMET PROPERTIES

ADMET, which stands for Absorption, Distribution, Metabolism, Excretion, and Toxicity, is a key area of focus in drug discovery. It encompasses important biological data that plays a crucial role in the development of new drugs. The properties discussed have implications for biological data, making them relevant to drug discovery programmes on a global scale. The drug discovery process requires the conduct of assays to evaluate the ADMET properties of each compound generated. As a result, researchers widely study these assays and have a significant amount of data available in databases.

Until the early 2000s, chemical and biological data were primarily accessible through extensive curated datasets sold by publishing companies, vast unstructured data from journal papers, or within extensive pharmaceutical databases. However, since then, the availability of high-quality public data has increased significantly, allowing for more frequent exploration. This includes large-scale extraction from literature and patents, as well as smaller but meticulously curated datasets tailored to specific inquiries. We highlight several notable community-driven databases that facilitate data submission and retrieval for drug design and discovery.

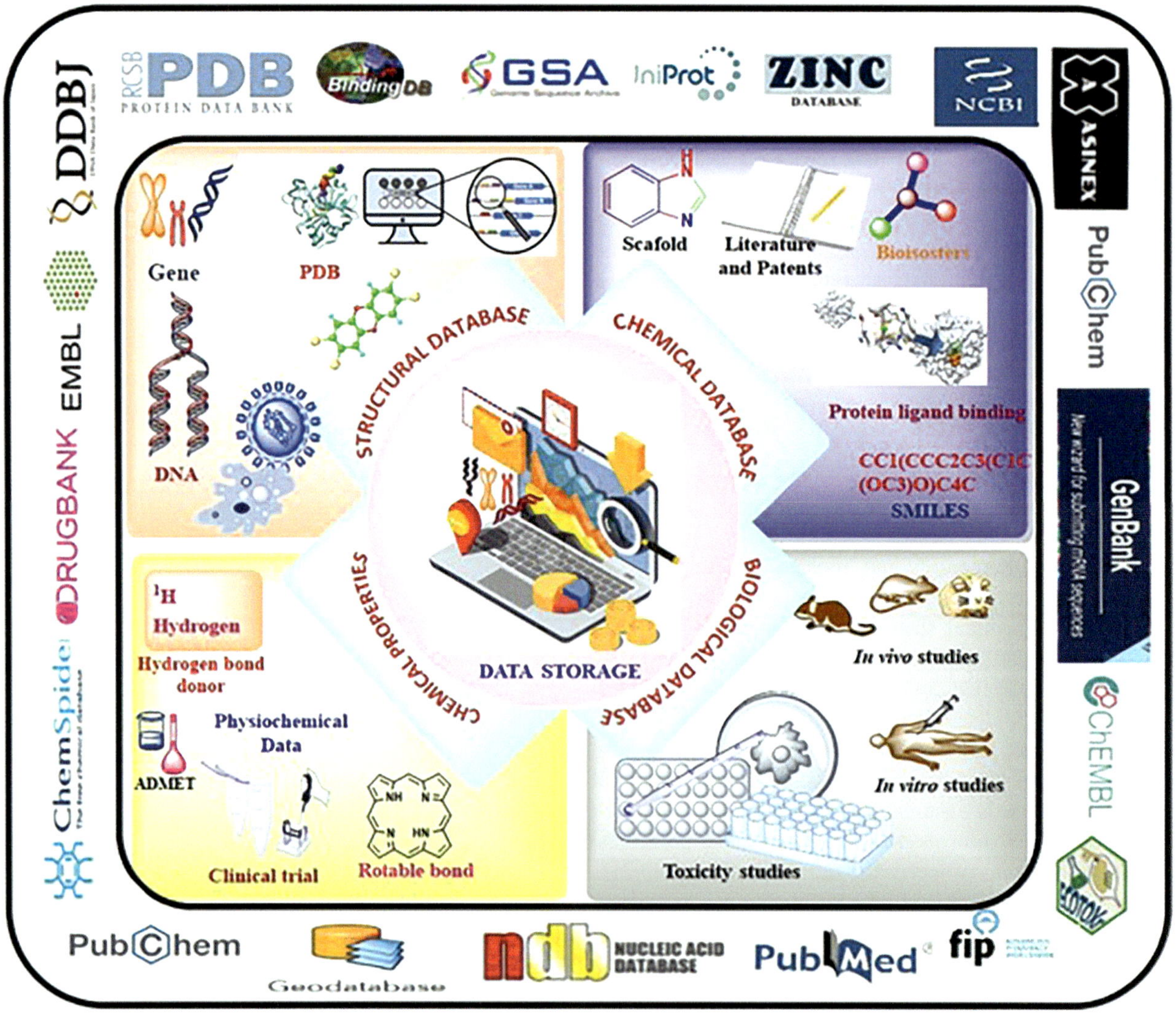

FIGURE 5.1 Overview of the common data types and databases used in drug discovery.

PubChem: The NCBI under the NIH operates the PubChem Database, a widely used resource for researchers in chemistry, bioinformatics, and drug discovery (https://pubchem.ncbi.nlm.nih.gov/). It hosts a vast collection of data on chemical substances, including structures, properties, biological activities, and experimental findings. With millions of compounds sourced from various repositories like literature, patents, and vendors, PubChem offers researchers a comprehensive platform. Users can utilize its search and analysis tools to explore compounds, identify potential drugs, and study structure–activity relationships. Additionally, PubChem facilitates data sharing and integration, fostering collaboration across fields such as pharmacology, toxicology, and chemical biology [22].

ChEMBL: Similar to PubChem, it is a leading database in drug discovery and chemical biology, houses a wide array of bioactive molecules with detailed information on their activities, properties, and structures. It focuses on small molecules and includes targets such as proteins, enzymes, and disease receptors. Researchers utilize ChEMBL to probe compound–target interactions, prioritize drug candidates, and comprehend structure–activity relationships. With supporting analysis tools, ChEMBL empowers users to glean valuable insights, aiding in the discovery and optimization of potential therapeutics [23].

BindingDB: It stores experimentally measured *in vitro* and *in vivo* results for different biological targets. It has IC50, EC25, EC50, LD50, Ki, etc data for millions of compounds, which is widely used in drug discovery and bioinformatics. It is managed by UC San Diego researchers, and it compiles data from diverse sources, offering a reliable dataset. Encompassing various protein targets and small-molecule ligands, BindingDB aids in exploring critical molecular interactions for drug development [24].

ChemDiv: It is a vital tool in drug discovery because it provides access to a wide range of chemical compounds from the ChemDiv library. It serves as a comprehensive repository for commercially available small molecules with diverse chemical structures. Researchers utilize the database for lead identification, virtual screening, and accelerating drug discovery. With its various search options and filters, the ChemDiv Database streamlines compound selection, aiding in easy retrieval.

ZINC: A vast array of chemical structures, each representing commercially available compounds, is accessible for purchase. The ZINC database stands out as one of the largest freely available repositories for such compounds. Its latest version, ZINC 15, boasts over 120 million purchasable compounds with drug-like properties. Approximately a quarter of these compounds are ready for immediate purchase, while the rest may require lead time for synthesis. The ZINC compound collection enables extensive virtual screening of pertinent chemistry [24].

Protein Data Bank (PDB): It offers free access to a huge repository of over 20 billion macromolecular structures, primarily X-ray crystal structures, but also structures from methods like solution NMR and electron microscopy. This resource is invaluable in drug discovery, particularly for structure-based drug design (SBDD), where structures serve to dock virtual chemical structures, aiding in identifying potential ligands. These structural repositories, beyond SBDD, facilitate understanding protein structural relationships and uncovering new binding opportunities through large-scale binding site analysis [25].

DrugBank: The comprehensive data on US FDA-approved drugs is now compiled in an open-access repository known as DrugBank. This repository contains extensive biochemical and pharmacological information, including details on targets and mechanisms, drug interactions, and structures. Hand-curated, DrugBank encompasses a wealth of data on drugs, including synonyms, identifiers, clinical indications, marketed formulations, pharmacodynamics, and ADMET data [26].

ChemSpider: The Royal Society of Chemistry manages ChemSpider, an accessible chemical database that serves as an extensive storehouse of chemical information. With a collection of over 100 million chemical compounds, it provides detailed access to their properties, structures, spectra, and related data. By collating information from diverse sources such as public databases, literature, and user inputs, ChemSpider emerges as a valuable resource for those engaged in chemistry research, education, and learning [27].

UniProt: It is a vital resource in drug discovery, offering extensive information about proteins, including their sequences, functions, interactions, and structures. Drawing from diverse sources such as scientific literature, experimental results, and computational predictions, UniProt provides a centralized platform for researchers and scientists. Its role encompasses target identification, validation, and characterization, aiding in the design and optimization of drug candidates. Furthermore, UniProt contributes to pharmacogenomics by incorporating data on genetic variations that influence drug responses [28].

AlphaFold: It is a creation of DeepMind that harnesses advanced AI to predict protein structures with exceptional accuracy. Although it is not a database per se, its predictions help to create databases housing protein structures. In drug discovery, these accurate predictions are critical. They aid in understanding protein functions, interactions, and dynamics, which are essential for identifying potential drug targets and designing effective therapeutics. Researchers can utilize this database to explore structures relevant to specific diseases or biological processes, facilitating the discovery and development of new drugs [29].

GenBank: GenBank, administered by the National Center for Biotechnology Information (NCBI), is a repository of nucleotide sequences and their corresponding translated proteins. As of April 2024, it hosts around 27.9 trillion bases across over 4.20 billion sequences. Each submission to GenBank receives a unique identifier or accession number [30].

Nucleic Acid Database (NDB): It was established in 1991 with the purpose of collecting and disseminating structural data pertaining to nucleic acids. The Nucleic Acid Database (NDB) is a web portal accessible at http://ndbserver.rutgers.edu, offering comprehensive information on 3D nucleic acid structures and their complexes. Alongside primary data, the NDB features derived geometric data, classifications of structures and motifs, and standards for describing nucleic acid features. Additionally, it provides various tools for nucleic acid analysis. The website offers diverse search capabilities and generates different types of reports [31].

In drug discovery, data retrieval plays a crucial role in accessing and analyzing diverse datasets related to compounds, targets, assays, and clinical trials. The process of data retrieval at different stages of drug discovery is shown in Figure 5.2. Here are some commonly used data retrieval systems:

Sequence Retrieval System (SRS) typically refers to a computer program or a set of algorithms designed to search for and retrieve sequences of data from a

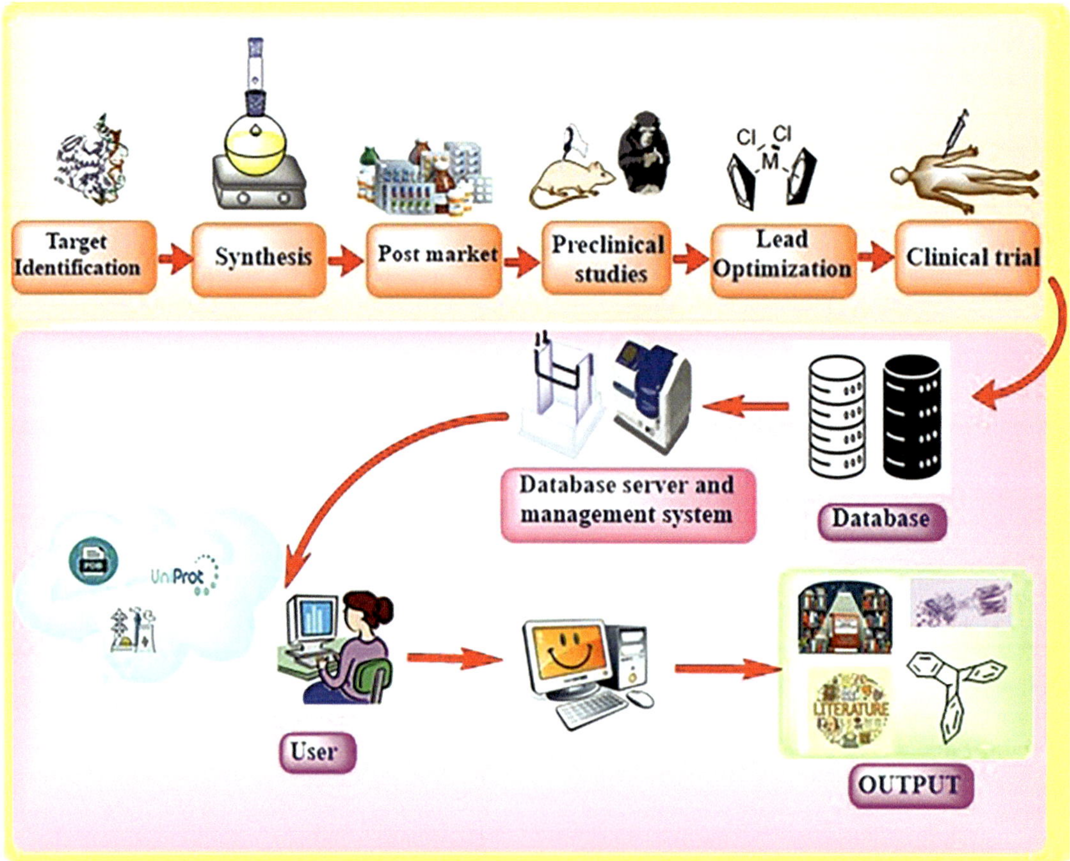

FIGURE 5.2 Process of data retrieval at different stages of drug discovery.

database or collection [32]. These sequences could be in the form of DNA, RNA, protein, or any other biological or non-biological data represented as a sequence of symbols.

Entrez Molecular Sequence Database System: It is an integrated platform offering access to diverse biomedical data, with a particular focus on nucleotide and protein sequences (https://www.ncbi.nlm.nih.gov/search/). It encompasses databases like GenBank, DNA sequences worldwide, Sequence for curated gene and protein sequences, 3D protein structures, the Sequence Read Archive (SRA), and Genome for annotated genomes. Through Entrez, users can explore, retrieve, and analyze sequence data using a variety of search tools and algorithms, facilitating research in genomics, bioinformatics, molecular biology, and related fields [33].

The different databases like PubChem, ChEMBL, DrugBank, and others mentioned in the database section can be directly accessed for data retrieval. Along with these, various cheminformatics and bioinformatics software tools are available for retrieving and analyzing chemical data in drug discovery. The most common tools include RDKit, Open Babel, BLAST, KNIME, TIBCO, BIOVIA Pipeline Pilot, and ChemAxon's Marvin suite. These tools provide functionalities for working with chemical structures, performing similarity and sequence searches, and analyzing chemical and biological properties [22-32,34]. These tools, servers, and software provide researchers and organizations in drug discovery with the capabilities to retrieve, analyze, and visualize diverse datasets, accelerating the discovery and development of new therapeutic compounds and treatments. The choice of tools depends on factors such as the nature of the data, the specific research objectives, and the preferences of the users.

DATA RETRIEVAL TECHNIQUES IN DRUG DISCOVERY

1. **Structured Query Language (SQL):** SQL queries are commonly used to retrieve data from relational databases in drug discovery. Researchers can use SQL to retrieve information such as compound structures, assay results, target interactions, and clinical trial data from databases containing drug-related information [35].
2. **Full-Text Search:** Full-text search engines allow researchers to search for specific keywords or phrases within textual documents such

as research articles, patents, and clinical trial reports. This technique enables efficient retrieval of relevant literature and information related to drug targets, compounds, and therapeutic indications [36].

3. **Data Warehousing:** Data warehousing techniques involve consolidating and integrating data from various sources into a centralized repository. In drug discovery, data warehouses can store diverse datasets, including chemical structures, biological assays, genomics data, and clinical trial results, facilitating comprehensive data retrieval and analysis [37,38].
4. **Cheminformatics and Bioinformatics Tools:** Specialized software tools and libraries in cheminformatics and bioinformatics enable researchers to retrieve and analyze molecular data, such as chemical structures, physicochemical properties, biological sequences, protein–ligand interactions, and metabolic pathways. These tools provide functionalities for searching, filtering, and analyzing molecular data to identify potential drug candidates and targets [39].
5. **Data Mining and Machine Learning:** Data mining and machine learning techniques are increasingly used in drug discovery to retrieve insights from large-scale datasets. These techniques can identify patterns, correlations, and predictive models from diverse data sources, including compound libraries, high-throughput screening data, and omics data, to aid in target identification, lead optimization, and drug repurposing [40,41].

CHALLENGES IN DRUG DISCOVERY DATA STORAGE AND RETRIEVAL

1. **Data Heterogeneity:** Drug discovery data come from diverse sources, formats, and types, making integration, storage, and retrieval difficult due to issues like harmonization, normalization, and interoperability.
2. **Data Volume and Complexity:** The sheer amount and complexity of drug discovery data, including compound libraries and omics data, require efficient storage and retrieval techniques.
3. **Data Quality and Consistency:** Ensuring reliable and reproducible research findings necessitates addressing challenges such as data curation, annotation, validation, and handling inconsistencies, errors, and missing values.
4. **Privacy and Security:** Protecting sensitive information like patient data and proprietary research findings is crucial, requiring robust data access controls, encryption, and compliance with privacy regulations.
5. **Computational Resources:** Drug discovery often demands significant computational resources for storage, retrieval, and analysis, posing challenges for smaller research groups or organizations with limited infrastructure and funding.

DATABASE STRUCTURE AND ANNOTATION

Database structure and annotation play critical roles in drug discovery, enabling researchers to organize, analyze, and interpret vast amounts of chemical and biological data efficiently. In the realm of drug discovery, databases serve as invaluable repositories of chemical and biological information [2]. Contemporary efforts in drug discovery followed by development rely heavily on accumulated knowledge amassed over years, particularly from studies on potential therapeutics. Nowadays, this wealth of information is stored in databases, serving as structured collections for text, numerical data, visuals, and structural details. These databases facilitate efficient management of up-to-date information and ensure swift access to relevant data. Annotating data and structure databases are key components of efficiently arranging, storing, and interpreting information. The importance of database structure and annotation in drug discovery lies in their function as essential elements of intricate biomedical data management [42,43].

THE IMPORTANCE OF DATABASE STRUCTURE

Effective data management and organization are essential to the success of drug discovery projects. In addition to making data retrieval and analysis easier, a well-designed database structure speeds up decision-making, promotes smooth researcher collaboration, and increases the reproducibility of research findings. Furthermore, strong database architectures set the stage for the implementation of machine learning algorithms and predictive modeling strategies to advance drug discovery in an era defined by big data and advanced analytics [44]. It encompasses the schema, tables, relationships, and constraints that govern how data are stored and accessed. The key components of a database structure are reported in Table 5.1.

FUNDAMENTALS OF DATABASE STRUCTURE

1. **Data Models:** Relational or NoSQL data models are commonly used in drug discovery database structures. NoSQL databases, like MongoDB or Cassandra, provide greater flexibility in managing unstructured or semi-structured data, while relational databases, like MySQL or PostgreSQL, arrange data into tables with predefined relationships [14,45-47].
2. **Entity-Relationship Diagrams (ERDs):** A comprehension of the data schema is made possible by ERDs, which graphically depict the relationships between various entities within the database.

TABLE 5.1
The Key Components of a Database Structure

Sl. No.	Components	Application/Description
1	Schema	Schema serves as an architectural blueprint to outline the underlying structure, data categories, and constraints of the database system.
2	Table	Tables are essential for arranging and storing structured data, as they demonstrate. It describes how rows and columns in tables stand for specific data instances and attributes, respectively.
3	Relationships	This efficiently manages and organizes data for one-to-one, one-to-many, and many-to-many relationships. It also helps to understand how relationships create connections between tables, data integrity, and enable complex queries across multiple entities.
4	Indexes	They function as literary guides, enabling users to efficiently navigate vast amounts of stored data and act as organized signposts. Indexes enhance query performance and streamline data retrieval, making them indispensable tools in database management, ultimately saving time and enhancing efficiency.
5	Constraints	Database management constraints serve as vital guardians, guaranteeing the consistency and integrity of data. Primary key constraints allow for efficient data retrieval by preventing duplication. Primary key, foreign key, and check constraints provide robustness and dependability to databases, protecting the consistency and integrity of the data they hold.

Entities in drug discovery databases can include, among other things, compounds, targets, assays, and experimental outcomes. Maintaining data consistency and integrity requires clearly defining the relationships between these entities.

3. **Data Integration Tools:** A key component of drug discovery databases is the integration of various data sources, such as chemical, biological, and pharmacological data. Data integration tools, namely APIs (Application Programming Interfaces) and ETL (Extract, Transform, Load) processes, enable the smooth transfer of data from multiple sources into the database, guaranteeing data accessibility and homogeneity [9].
4. **Metadata Management**: Data about data, or metadata, offers descriptions and crucial context to help with data analysis and interpretation. In order to improve data discoverability and usability, effective metadata management entails defining standardized metadata attributes, such as data provenance, experimental conditions, and quality metrics [47].
5. **Query and Analysis Tools**: Database structures in drug discovery often incorporate query and analysis tools to facilitate ad-hoc querying, data visualization, and statistical analysis. These tools empower data-driven decision-making in drug discovery research by enabling researchers to create intricate queries, produce insightful conclusions, and visualize data trends [48].

ESSENTIAL COMPONENTS OF DESIGNING AND MANAGING RELATIONAL DATABASES

1. **Entity-Relationship Model (ERM):** A conceptual framework for illustrating connections between different system components, aiding in understanding interactions and facilitating data manipulation and analysis.
2. **Data Normalization:** Minimizing redundancy, upholding data integrity, and streamlining storage and retrieval processes by organizing data into logical tables and eliminating duplicates.
3. **Flexibility and Scalability:** Adaptability of database structures to changes without extensive alterations, achieved through flexible schema designs, and the ability of the architecture to manage rising data volumes and user requirements without compromising performance, often utilizing distributed systems or cloud infrastructure [48].
4. **Indexing and Query Optimization:** Improving system performance by creating specialized data structures for swift data access (indexing) and refining query execution plans to minimize resource usage and maximize efficiency. Effective implementation requires selective indexing, regular maintenance, and continuous monitoring to sustain optimal performance [49,50,51].

Database annotations are all about labeling or tagging information/metadata in a dataset to let the machine understand what they are. The dataset is composed of any set of forms, i.e., an image, an audio file, video footage, or even text. In the realm of databases, annotations play a significant role in providing additional information about the data stored within. They serve as metadata, offering insights into the structure, relationship, and constraints of the database objects [52].

Annotations are descriptive labels or notes attached to database elements such as tables, columns, constraints, and relationships. They provide contextual information that aids in understanding the database's schema and its

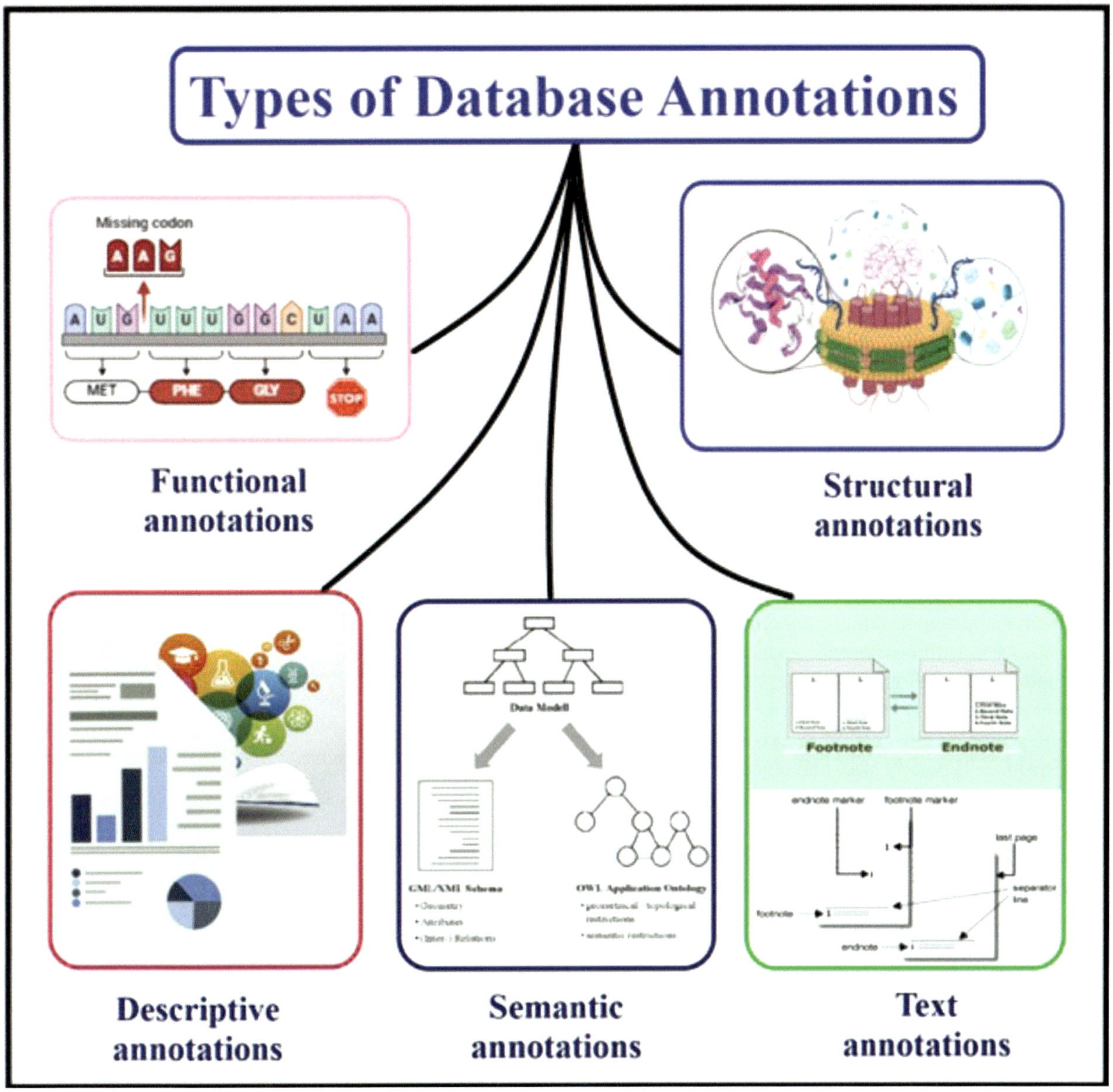

FIGURE 5.3 Different types of database annotations used.

components. Annotations are commonly used in both relational and non-relational databases to document various aspects of the data model [53].

Table and Column Comments: Annotations can be added to the tables. It is easy for developers to understand the databases and interact easily, as shown in Figure 5.3.

Index Comment: Comments can be added to indexes to explain why they were created, their intended use, or any other relevant details. This helps in understanding the purpose and impact of indexes on database performance.

Query Comments: In query parts, comments can be added to SQL to explain the purpose of a specific part which helps to understand the query.

Stored Procedure and Function Comments: Annotations can be added to the stored procedure and functional comments to describe functionality, parameters, and return value.

Database Objects Comments: Comments can also be added to objects such as views, triggers, or sequences to provide additional information.

TYPES OF DATABASE ANNOTATIONS

1. **Structural Annotations**: Structural annotations are vital for understanding gene and protein functions, covering experimental and computational techniques, ontology, databases, and multi-omic data integration. They assist in decoding pathogenesis, pathways, and biological processes, aiding

in gene prioritization, drug target identification, disease mechanism understanding, and regulatory network unveiling. Continuous refinement and integration of diverse data types are crucial for keeping pace with expanding biological information, advancing biomedical research, and drug discovery [4].

2. **Functional Annotations**: Functional annotations are essential for deciphering genome complexities, shedding light on gene functions, regulatory mechanisms, and disease pathways. They drive genome annotation, gene expression analysis, and comparative genomics, transforming raw genomic data into actionable insights. Despite challenges in accuracy and completeness, interdisciplinary collaborations and methodological improvements are advancing their utility. Vital for medical, agricultural, and biotechnological innovation, functional annotations pave the way for transformative discoveries encoded within the genome [54].
3. **Descriptive Annotations**: Descriptive annotations are succinct summaries and evaluations of resources, helping readers gauge relevance and quality. They inform, evaluate, and aid in selection by outlining main topics, arguments, and key points, while assessing credibility and significance. Typically brief and objective, they highlight key features like methodology and audience. Found in library catalogs, databases, and academic publications, they streamline information retrieval and scholarly communication, enhancing resource selection and engagement [55,56].
4. **Semantic Annotations**: Semantic annotations enhance data with explicit semantic metadata, improving search, interoperability, reasoning, personalization, and knowledge representation. They include entities, attributes, relations, and ontology, utilizing standards like RDF, Schema.org, and OWL. Implementation methods, whether manual or automatic, face challenges of ambiguity, scalability, interoperability, and quality control. Despite these hurdles, semantic annotations are crucial for maximizing data utilization, enabling precise search and interpretation across applications and domains. They unlock the full potential of data assets in the Semantic Web era and beyond [57].
5. **Text Annotations:** Text annotations provide commentary and context to enhance understanding for readers. They come in various forms such as footnotes, endnotes, highlighting, links, and markup languages, serving to offer additional information, citations, or explanations. Found across literary works, academic papers, digital content, and programming code, they enrich the reading experience and aid comprehension, playing a crucial role in facilitating deeper engagement with text [58].

CONCLUSION

In the contemporary landscape of drug discovery and bioinformatics, the seamless integration of data management principles is paramount. The symbiotic relationship between efficient data storage and retrieval, database structure, and annotation elucidates the intricate tapestry of modern information management. As the realm of data expands exponentially, traditional methodologies falter in addressing the complexities inherent in diverse data types. However, innovative strategies, propelled by advancements in computer, AI, and ML technologies, offer a promising avenue for overcoming these challenges. It is important to note that while these tools are powerful, they are just one part of the larger drug discovery process, which also involves experimental validation and clinical trials. Through meticulous database structuring and comprehensive annotation, organizations can enhance data usability, accessibility, and understanding, thus fostering informed decision-making and strategic innovation. This chapter underscores the indispensable role of integrated data management approaches in optimizing performance, ensuring data integrity, and unlocking the latent potential within vast data repositories. By embracing these principles, organizations can navigate the evolving landscape of drug discovery and bioinformatics with agility and efficacy, ultimately driving transformative advancements in the field.

REFERENCES

1. Velmurugan, D.; Pachaiappan, R.; Ramakrishnan, C. Recent Trends in Drug Design and Discovery. *Curr. Top. Med. Chem.*, 2020, 20 (19), 1761–1770.
2. Paul, D.; Sanap, G.; Shenoy, S.; Kalyane, D.; Kalia, K.; Tekade, R. K. Artificial Intelligence in Drug Discovery and Development. *Drug Discov. Today*, 2021, 26 (1), 80–93.
3. Duke, R.; Bhat, V.; Risko, C. Data Storage Architectures to Accelerate Chemical Discovery: Data Accessibility for Individual Laboratories and the Community. *Chem. Sci.*, 2022, 13 (46), 13646–13656.
4. Borujeni, P. M.; Salavati, R. Functional Domain Annotation by Structural Similarity. *NAR Genomics Bioinforma.*, 2024, 6 (1), lqae005.
5. Brown, N.; Cambruzzi, J.; Cox, P. J.; Davies, M.; Dunbar, J.; Plumbley, D.; Sellwood, M. A.; Sim, A.; Williams-Jones, B. I.; Zwierzyna, M.; et al. Big Data in Drug Discovery. *Prog Med Chem.*, 2018, 57, 277–356.
6. Liu, X.; Xu, Y.; Li, S.; Wang, Y.; Peng, J.; Luo, C.; Luo, X.; Zheng, M.; Chen, K.; Jiang, H. In Silico Target Fishing: Addressing a "Big Data" Problem by Ligand-Based Similarity Rankings with Data Fusion. *J. Cheminform.*, 2014, 6 (1), 1–14.
7. Vamathevan, J.; Clark, D.; Czodrowski, P.; Dunham, I.; Ferran, E.; Lee, G.; Li, B.; Madabhushi, A.; Shah, P.; Spitzer, M.; et al. Applications of Machine Learning in Drug Discovery and Development. *Nat. Rev. Drug Discov.*, 2019, 18 (6), 463–477.
8. Reisberg, B.; Franssen, E. H.; Bobinski, M.; Auer, S.; Monteiro, I.; Boksay, I.; Wegiel, J.; Shulman, E.; Steinberg, G.; Souren, L. E. M.; et al. Overview of Methodologic

Issues for Pharmacologic Trials in Mild, Moderate, and Severe Alzheimer's Disease. *Int. Psychogeriatrics*, 1996, 8 (2), 159–193.
9. Dara, S.; Dhamercherla, S.; Jadav, S. S.; Babu, C. M.; Ahsan, M. J. Machine Learning in Drug Discovery: A Review. *Artif. Intell. Rev.*, 2022, 55 (3), 1947–1999.
10. Kim, S.; Thiessen, P. A.; Bolton, E. E.; Chen, J.; Fu, G.; Gindulyte, A.; Han, L.; He, J.; He, S.; Shoemaker, B. A.; et al. PubChem Substance and Compound Databases. *Nucleic Acids Res.*, 2016, 44 (D1), D1202–D1213.
11. Wruck, W.; Peuker, M.; Regenbrecht, C. R. A. Data Management Strategies for Multinational Large-Scale Systems Biology Projects. *Brief. Bioinform.*, 2014, 15 (1), 65–78.
12. Maurer, M.; Molidor, R.; Sturn, A.; Hartler, J.; Hackl, H.; Stocker, G.; Prokesch, A.; Scheideler, M.; Trajanoski, Z. MARS: Microarray Analysis, Retrieval, and Storage System. *BMC Bioinform.*, 2005, 6 (1), 101.
13. Zarour, M.; Alenezi, M.; Ansari, M. T. J.; Pandey, A. K.; Ahmad, M.; Agrawal, A.; Kumar, R.; Khan, R. A. Ensuring Data Integrity of Healthcare Information in the Era of Digital Health. *Healthc. Technol. Lett.*, 2021, 8 (3), 66–77.
14. Aniceto, R.; Xavier, R.; Guimarães, V.; Hondo, F.; Holanda, M.; Walter, M. E.; Lifschitz, S. Evaluating the Cassandra NoSQL Database Approach for Genomic Data Persistency. *Int. J. Genomics*, 2015, 2015, 1–7.
15. Gromiha, M. M.; Ou, Y.-Y. Bioinformatics Approaches for Functional Annotation of Membrane Proteins. *Brief. Bioinform.*, 2014, 15 (2), 155–168.
16. Cao, H.; Meyer-Lindenberg, A.; Schwarz, E. Comparative Evaluation of Machine Learning Strategies for Analyzing Big Data in Psychiatry. *Int. J. Mol. Sci.*, 2018, 19 (11), 3387.
17. Mitchell, B. R.; Cohen, M. C.; Cohen, S. Dealing with Multi-Dimensional Data and the Burden of Annotation: Easing the Burden of Annotation. *Am. J. Pathol.*, 2021, 191 (10), 1709–1716.
18. Humann, J. L.; Lee, T.; Ficklin, S.; Main, D. Structural and Functional Annotation of Eukaryotic Genomes with GenSAS. *Methods Mol. Biol. (Clifton, N.J.)*, 2019, 1962, 29–51.
19. Tolle, K. M.; Tansley, D. S. W.; Hey, A. J. G. The Fourth Paradigm: Data-Intensive Scientific Discovery [Point of View]. *Proc. IEEE*, 2011, 99 (8), 1334–1337.
20. Matore, B. W.; Banjare, P.; Singh, J.; Roy, P. P. In Silico Selectivity Modeling of Pyridine and Pyrimidine Based CYP11B1 and CYP11B2 Inhibitors: A Case Study. *J. Mol. Graph. Model.*, 2022, 116, 108238.
21. Merelli, I.; Pérez-Sánchez, H.; Gesing, S.; D'Agostino, D. Managing, Analysing, and Integrating Big Data in Medical Bioinformatics: Open Problems and Future Perspectives. *Biomed Res. Int.*, 2014, 2014, 1–13.
22. Wang, Y.; Xiao, J.; Suzek, T. O.; Zhang, J.; Wang, J.; Bryant, S. H. PubChem: A Public Information System for Analyzing Bioactivities of Small Molecules. *Nucleic Acids Res.*, 2009, 37, W623–W633.
23. Gaulton, A.; Bellis, L. J.; Bento, A. P.; Chambers, J.; Davies, M.; Hersey, A.; Light, Y.; McGlinchey, S.; Michalovich, D.; Al-Lazikani, B.; et al. ChEMBL: A Large-Scale Bioactivity Database for Drug Discovery. *Nucleic Acids Res.*, 2012, 40 (D1), D1100–D1107.
24. Gilson, M. K.; Liu, T.; Baitaluk, M.; Nicola, G.; Hwang, L.; Chong, J. BindingDB in 2015: A Public Database for Medicinal Chemistry, Computational Chemistry and Systems Pharmacology. *Nucleic Acids Res.*, 2016, 44 (D1), D1045–D1053.
25. Burley, S. K.; Berman, H. M.; Kleywegt, G. J.; Markley, J. L.; Nakamura, H.; Velankar, S. Protein Data Bank (PDB): The Single Global Macromolecular Structure Archive. *Methods Mol. Biol.*, 2017, 1607, 627–641.
26. Wishart, D. S.; Knox, C.; Guo, A. C.; Cheng, D.; Shrivastava, S.; Tzur, D.; Gautam, B.; Hassanali, M. DrugBank: A Knowledgebase for Drugs, Drug Actions and Drug Targets. *Nucleic Acids Res.*, 2008, 36 (suppl_1), D901–D906.
27. Pence, H. E.; Williams, A. ChemSpider: An Online Chemical Information Resource. *J. Chem. Educ.*, 2010, 87 (11), 1123–1124.
28. Bateman, A. UniProt: A Worldwide Hub of Protein Knowledge. *Nucleic Acids Res.*, 2019, 47 (D1), D506–D515.
29. Varadi, M.; Anyango, S.; Deshpande, M.; Nair, S.; Natassia, C.; Yordanova, G.; Yuan, D.; Stroe, O.; Wood, G.; Laydon, A.; et al. AlphaFold Protein Structure Database: Massively Expanding the Structural Coverage of Protein-Sequence Space with High-Accuracy Models. *Nucleic Acids Res.*, 2022, 50 (D1), D439–D444.
30. Benson, D. A.; Cavanaugh, M.; Clark, K.; Karsch-Mizrachi, I.; Lipman, D. J.; Ostell, J.; Sayers, E. W. GenBank. *Nucleic Acids Res.*, 2012, 41(D1), D36–D42.
31. Coimbatore Narayanan, B.; Westbrook, J.; Ghosh, S.; Petrov, A. I.; Sweeney, B.; Zirbel, C. L.; Leontis, N. B.; Berman, H. M. The Nucleic Acid Database: New Features and Capabilities. *Nucleic Acids Res.*, 2014, 42 (D1), D114–D122.
32. Etzold, T.; Argos, P. SRS--An Indexing and Retrieval Tool for Flat File Data Libraries. *Comput. Appl. Biosci.*, 1993, 9 (1), 49–57.
33. Baxevanis, A. D. Searching the NCBI Databases Using Entrez. *Curr. Protoc. Hum. Genet.*, 2006, 51 (1), Unit 6.10.
34. Petitjean, M.; Camproux, A. In Silico Medicinal Chemistry: Computational Methods to Support Drug Design. *ChemMedChem.*, 2016, 11 (13), 1480–1481.
35. Jamison, D. C. Structured Query Language (SQL) Fundamentals. *Curr. Protoc. Bioinforma.*, 2003, 9 (1), Unit 9.2.
36. Almeida, T.; Antunes, R. F.; Silva, J.; Almeida, J. R.; Matos, S. Chemical Identification and Indexing in PubMed Full-Text Articles Using Deep Learning and Heuristics. *Database*, 2022, 2022, baac047.
37. Gedela, S. Integration, Warehousing, and Analysis Strategies of Omics Data. *Methods Mol. Biol. (Clifton, N.J.)*, 2011, 719, 399–414.
38. Nusrat, S.; Harbig, T.; Gehlenborg, N. Tasks, Techniques, and Tools for Genomic Data Visualization. *Comput. Graph. Forum*, 2019, 38 (3), 781–805.
39. Sukumar, N.; Krein, M.; Breneman, C. M. Bioinformatics and Cheminformatics: Where Do the Twain Meet? *Curr. Opin. Drug Discov. Devel.*, 2008, 11 (3), 311–319.
40. Lan, K.; Wang, D.-T.; Fong, S.; Liu, L.-S.; Wong, K. K. L.; Dey, N. A Survey of Data Mining and Deep Learning in Bioinformatics. *J. Med. Syst.*, 2018, 42 (8), 139.
41. Masoudi-Sobhanzadeh, Y.; Omidi, Y.; Amanlou, M.; Masoudi-Nejad, A. Drug Databases and Their Contributions to Drug Repurposing. *Genomics*, 2020, 112 (2), 1087–1095.
42. Ishaq, M., Abid, A.; Farooq, M. S.; Manzoor, M. F.; Farooq, U.; Abid, K.; Helou, M. A. Advances in Database Systems Education: Methods, Tools, Curricula, and Way Forward. *Educ. Inf. Technol.*, 2023, 28 (3), 2681–2725.
43. Ferrero, E.; Brachat, S.; Jenkins, J. L.; Marc, P.; Skewes-Cox, P.; Altshuler, R. C.; Gubser Keller, C.; Kauffmann, A.; Sassaman, E. K.; Laramie, J. M.; et al. Ten Simple Rules to Power Drug Discovery with Data Science. *PLoS Comput. Biol.*, 2020, 16 (8), e1008126.

44. Zhu, H. Big Data and Artificial Intelligence Modeling for Drug Discovery. *Annu. Rev. Pharmacol. Toxicol.*, 2020, 60 (1), 573–589.
45. da Silva, L. F.; Lima, J. V. F. An Evaluation of Relational and NoSQL Distributed Databases on a Low-Power Cluster. *J. Supercomput.*, 2023, 79 (12), 13402–13420.
46. Deari, R.; Zenuni, X.; Ajdari, J.; Ismaili, F.; Raufi, B. Analysis And Comparision of Document-Based Databases with Relational Databases: MongoDB vs MySQL. In *2018 International Conference on Information Technologies (InfoTech)*; IEEE, New York, 2018; pp. 1–4.
47. Oussous, A.; Benjelloun, F.-Z.; Ait Lahcen, A.; Belfkih, S. Big Data Technologies: A Survey. *J. King Saud Univ. - Comput. Inf. Sci.*, 2018, 30 (4), 431–448.
48. Falatah, M. M.; Batarfi, O. A. Cloud Scalability Considerations. *Int. J. Comput. Sci. Eng. Surv.*, 2014, 5 (4), 37–47.
49. Corwin, J.; Silberschatz, A.; Miller, P. L.; Marenco, L. Dynamic Tables: An Architecture for Managing Evolving, Heterogeneous Biomedical Data in Relational Database Management Systems. *J. Am. Med. Informatics Assoc.*, 2007, 14 (1), 86–93.
50. Ramu, V. B. Optimizing Database Performance: Strategies for Efficient Query Execution and Resource Utilization. *Int. J. Comput. Trends Technol.*, 2023, 71 (7), 15–21.
51. Sliwoski, G.; Kothiwale, S.; Meiler, J.; Lowe, E. W. Computational Methods in Drug Discovery. *Pharmacol. Rev.*, 2014, 66 (1), 334–395.
52. Zhao, Y.; Kuang, Z.; Wang, Y.; Li, L.; Yang, X. MicroRNA Annotation in Plants: Current Status and Challenges. *Brief. Bioinform.*, 2021, 22 (5), bbab075.
53. Rebolledo, C.; Silva, J. P.; Saavedra, N.; Maracaja-Coutinho, V. Computational Approaches for CircRNAs Prediction and in Silico Characterization. *Brief. Bioinform.*, 2023, 24 (3), bbad154.
54. Singh, N. K. MiRNAs Target Databases: Developmental Methods and Target Identification Techniques with Functional Annotations. *Cell. Mol. Life Sci.*, 2017, 74 (12), 2239–2261.
55. Scala, G.; Serra, A.; Marwah, V. S.; Saarimäki, L. A.; Greco, D. FunMappOne: A Tool to Hierarchically Organize and Visually Navigate Functional Gene Annotations in Multiple Experiments. *BMC Bioinform.*, 2019, 20 (1), 79.
56. Cooper, D. A.; Cortés-López, M.; Miura, P. Genome-Wide CircRNA Profiling from RNA-Seq Data. *Methods Mol. Biol.*, 2018, 1724, 27–41.
57. Yu, T.; Xu, N.; Haque, N.; Gao, C.; Huang, W.; Huang, Z. Popular Computational Tools Used for MiRNA Prediction and Their Future Development Prospects. *Interdiscip. Sci. Comput. Life Sci.*, 2020, 12 (4), 395–413.
58. Wang, Y.; Zhu, S.; Liu, C.; Deng, H.; Zhang, Z. Text Mining and Hub Gene Network Analysis of Endometriosis. *Biomed Res. Int.*, 2021, 2021, 1–10.

6 Computational Protein Modelling, Refinement, Validation, and Its Application

Sushil K. Kashaw
Integrated Drug Discovery Research Laboratory, Department of Pharmaceutical Sciences, Dr. Harisingh Gour University (A Central University), Sagar, India

INTRODUCTION

Current drug discovery has moved from empirical trial-and-error techniques to more methodical, data-driven methods that use bioinformatics, artificial intelligence (AI), and computational chemistry. Drugs in the past were found either by chance or through high-throughput screening (HTS) of libraries of synthetic and natural compound collections. Such "random" approaches, which were typically steered by combinatorial chemistry, delivered chemical diversity breadth but not mechanistic understanding [1,2]. Conversely, the "rational" strategy—based on structure-based drug design (SBDD)—is dependent on the determination of the three-dimensional (3D) structure of biological targets to identify and optimize therapeutic candidates through molecular modelling methods [3]. Figure 6.1 shows a historical timeline of how machine learning entered and changed Protein modelling. Although structural databases like the Protein Data Bank (PDB) are growing rapidly, a minority (~1%–2%) of all proteins whose structures have been elucidated by experiment because it is too expensive and difficult to do X-ray crystallography, NMR spectra, and cryo-electron microscopy [4]. To fill this gap, computational methods like homology modelling have become crucial tools for protein tertiary structure prediction. Such approaches deduce the 3D structure of a target protein based on experimentally known template, yielding useful information about active sites, ligand-binding interactions, and draggability [5]. Computational modelling approaches to proteins can be broadly divided into heuristic (e.g., ab initio, Monte Carlo, molecular dynamics) and deterministic methods (e.g., homology modelling), depending on whether pre-existing structural knowledge is utilized [6]. Breakthroughs in AI, specifically machine learning (ML), have improved the reliability and accuracy of homology modelling by a considerable extent, allowing prediction systems to be trained to identify structural patterns and energetics from vast databases without specific programming [7–9]. These AI-based platforms have demonstrated exceptional improvement in predictive accuracy and form the core of contemporary in silico drug design workflows. With the combination of AI, structural bioinformatics, and computational chemistry, homology modelling is an effective tool in target identification, ligand docking, and functional annotation in drug discovery. The chapter presents an overview of comparative protein modelling and its contribution to speeding up drug innovation across research areas such as molecular biology, physics, and systems pharmacology [10–29].

HOMOLOGY MODELLING OR COMPARATIVE MODELLING

Homology modelling, or comparative modelling, is a key computational tool used to predict the 3D protein structure, which may be a beginning point in rational drug design. As the experimental methods such as X-ray crystallography, NMR, and cryo-electron microscopy are limited by their cost and the enormous amount of time they consume, homology modelling provides a time-efficient and economical solution, particularly for those proteins that are hard to crystallize or too big for structure determination [5,29–35]. This approach relies on the principle that proteins with common evolutionary heritage tend to have similar structural conformations. By superposing a target sequence (query) onto a structurally known homolog (template), a 3D model can be inferred if the sequence identity is greater than a certain level—usually more reliable above 35% identity, with models above 50% being most appropriate for drug development purposes [36,37]. The modelling process consists of four main steps: template selection, sequence alignment, construction of the first model, and improvement and validation of the final structure [37]. Though useful, homology modelling is hampered by limitations like poor template selection with low quality, misalignment, and errors in modelling flexible parts such as loops. Furthermore, although sequence identity is related to model accuracy, local structural variations can still be observed even for highly homologous sequences, possibly restricting the usefulness of the model for detailed drug–target interaction research [38–40]. Increased integration of machine learning and AI software into homology modelling pipelines in recent times has enhanced the accuracy of alignment and model prediction, further entrenching its

DOI: 10.1201/9781003464679-6

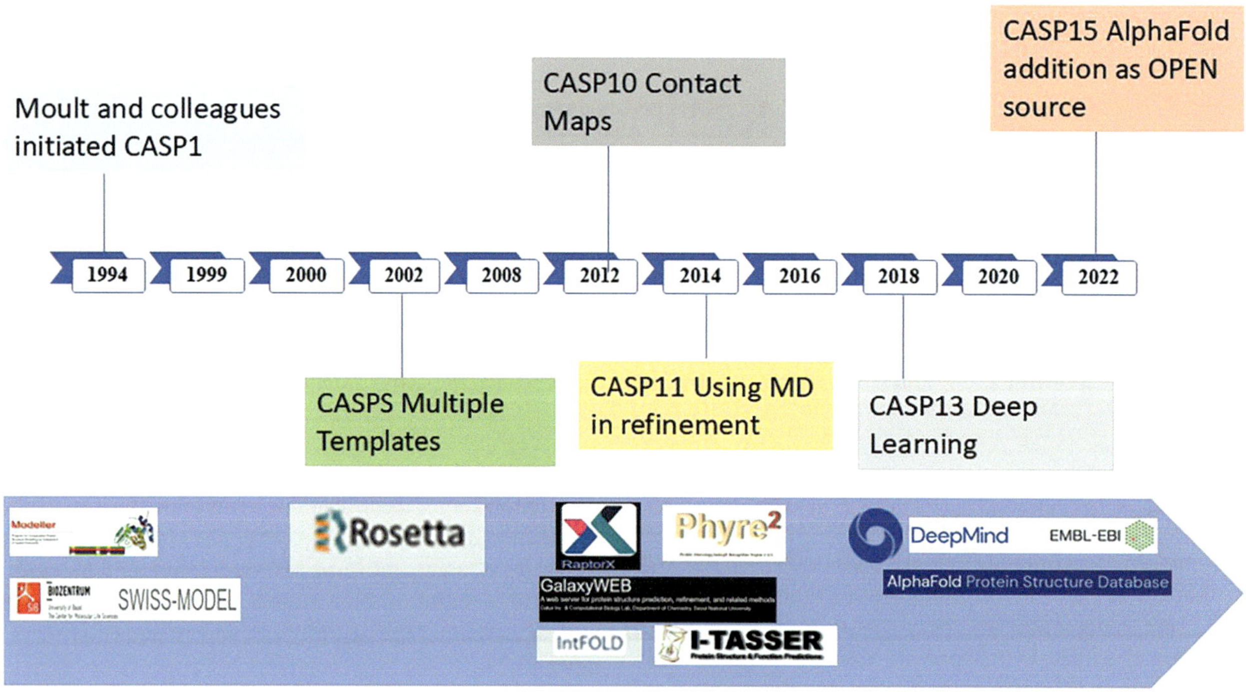

FIGURE 6.1 Historical timeline of major developments in homology modelling. CASP, Critical Assessment of Protein Structure Prediction.

position in structure-based drug discovery and target validation pipelines [41,42].

BASIC STEPS IN HOMOLOGY MODELLING

Homology modelling is a multistep computational procedure for predicting the 3D structure of a protein from its primary amino acid sequence. This approach is widely used in rational drug design, particularly when experimental structures are unavailable. Integration of AI, machine learning algorithms, and bioinformatics tools has significantly enhanced the speed and accuracy of each stage in the modelling pipeline [43,44].

A. **Searching for a Query Sequence**: The process begins by obtaining the target protein sequence in FASTA format from publicly available databases such as UniProt, NCBI, or Swiss-Prot. Using similarity search tools (e.g., BLAST), the sequence is scanned against known structures to identify suitable homologs [45].
B. **Template Selection:** The query sequence is compared to structurally characterized proteins in databases like PDB. High sequence identity (>30%) ensures better model accuracy. Tools such as BLAST, CARD, and 3D-PSSM assist in identifying the best-fit templates. AI-enhanced pattern recognition can improve low-homology template detection [46–54].
C. **Sequence Alignment:** Once a template is selected, precise alignment with the query sequence is critical. Errors in this step can significantly affect model quality. Modern AI-driven algorithms (e.g., HMMs, PSI-BLAST) and advanced alignment tools such as Clustal, MAFFT, and Modeller facilitate accurate pairwise and multiple sequence alignments, even under low identity conditions [30,55–61].
D. **3D Model Building:** Model construction relies on structural data from the selected template. Methods include rigid-body assembly, segment matching, spatial restraints, and artificial evolution. Software like SWISS-MODEL, MODELLER, and JACKAL automate this step. AI and deep learning approaches are increasingly used to predict protein backbone and side-chain conformations from sequence data [29,45,62–68].
E. **Model Refinement and Loop Modelling:** Refinement ensures structural accuracy by correcting misalignments and adjusting loops and side chains. Loops, which often mediate ligand binding, are modelled using either knowledge-based or energy-based methods. AI tools, including neural networks and Monte Carlo simulations, optimize loop conformations and reduce energetically unfavourable structures [69–86].
F. **Side-Chain Prediction:** Accurate side-chain placement is essential for evaluating binding

site geometry and protein–ligand interactions. Rotamer libraries and energy scoring functions are employed by tools like SCWRL and OPUS-Rota2. AI-driven rotamer prediction has improved side-chain accuracy and model realism [87–92].

G. **Model Optimization:** The final model is refined using molecular mechanics force fields and energy minimization techniques. Molecular dynamics (MD) and Monte Carlo simulations help reduce steric clashes and improve thermodynamic stability. Recent integration of machine learning methods accelerates this process and improves model robustness [93].

H. **Model Validation:** Validation of the model (Table 6.1) is the last and most important step in the homology modelling process and guarantees that the modelled protein structure is not only biologically significant but also computationally trustworthy. Because each modelling step—particularly sequence alignment and template selection—is prone to cumulative errors, rigorous assessment is needed to validate accuracy and direct structural refinement [37,94,95]. Several computational methods evaluate the stereochemical quality and atomic-level accuracy of the model by comparing predicted residue environments with those seen in high-resolution experimental structures. Methods originally developed for the validation of X-ray crystallography data, including Ramachandran plot analysis, Verify3D, and ProSA, are frequently used on homology models to identify misfolded areas or improper alignments [96–99]. Sophisticated AI and statistical tools are increasingly being integrated into such validation platforms for improving model assessment. These are neural network-based local error predictors, energy scoring functions such as DOPE and QMEAN, and consensus-based evaluators such as MolProbity. Together, these programs

TABLE 6.1
Model Validation Parameters in Homology Modelling

No.	Validation Parameter	Description	References
1	Ramachandran Plot	Evaluates backbone dihedral angles to identify residues in allowed/disallowed regions.	[140–143]
2	ERRAT Plot	Assesses non-bonded atom interactions; identifies structural error zones.	[140,144]
3	ProSA Folding Energy	Evaluates overall folding energy; negative scores imply stable conformations.	[145,146]
4	G-Factor	Measures stereochemical quality; low values (< -0.5) suggest unusual torsion angles.	[147]
5	Verify 3D	Assesses compatibility between 3D structure and amino acid sequence.	[148–150]
6	Z-Score	Measures how native-like a structure is compared to misfolds; used in ProSA.	[150]
7	RMSD	Quantifies deviation between template and model backbone atoms.	[29,151]
8	WHAT_CHECK	Identifies crystallographic anomalies and assesses conformational energy.	[152,153]
9	PROCHECK	Evaluates stereochemistry via bond lengths, angles, and torsions.	[98,153]
10	Local–Global Alignment (LGA)	Measures local–global similarity between model and PDB structures.	[151]
11	GDT_TS	Global Distance Test Total Score measures model accuracy across different distance cutoffs.	[154,155]
12	GDT_HA	High-accuracy version of GDT; more sensitive to finer structural deviations.	[155,156]
13	TM-Score & TM-align	Measures overall structural similarity; independent of protein size.	[157]
14	MaxSub	Identifies the largest subset of well-predicted Cα atoms; returns normalized score.	[158]
15	SphereGrinder	Visualizes accuracy using colour-coded 2D3D landscapes; CASP-approved.	[159]
16	CAD-Score	Measures contact area differences; reflects both backbone and side-chain packing.	[160,161]
17	MolProbity	Evaluates hydrogen bonding, sterics, and atomic clashes; offers detailed structural analysis.	[162]
18	WHAT IF	Performs comprehensive structure analysis, docking, and mutation prediction.	[163]
19	QMEAN	Composite energy function for selecting the best-quality model from decoys.	[164]
20	DOPE Score	Statistical potential to assess and compare model quality; lower values are better.	[165]
21	MetaMQAP	Meta-predictor combining various MQAPs to predict local–global model accuracy.	[166]
22	Machine Learning Methods	AIML models (e.g., SVMs) used for RMSD prediction and error detection in structure modelling.	[167]
23	Experimental Validation (MD Sim)	Molecular dynamics simulations validate the flexibility, motion, and stability of the model.	[168–171]
24	GRASP2	Assesses geometric feasibility and assists with manual inspection of structural models.	[172,173]

assess structural parameters including backbone torsion angles, side-chain conformations, residue environment compatibility, and quality of packing [100–103]. Templates with sequence identities less than 35% give rise to models that are especially prone to errors and should be carefully examined. It is crucial that various validation approaches are merged to guarantee confidence in homology-derived structures employed in drug design, target binding prediction, and ligand improvement.

SOFTWARE AND SERVERS HOMOLOGY MODELLING

Homology modelling is facilitated by a wide variety of computational algorithms and web-servers that design good-quality 3D protein structures from a query sequence. They utilize various strategies for modelling such as rigid-body assembly, segment matching, spatial restraints, and artificial evolution to develop precise tertiary structure using homologous templates. MODELLER, Swiss-Model, RAMP, PrISM, COMPOSER, CONGEN+2, and DISGEOConsensus are the essential tools with their specific abilities towards prediction, refinement, and structure analysis. Molecular graphics programs are generally employed in combination with these modelling packages to view, manipulate, and check protein structures, augmenting the analysis of functional and structural characteristics. Commercial and freely available platforms are available, and the choice frequently hinges on the particular objectives of the research, including structural accuracy, efficiency, or compatibility with docking pipelines. Comparative analyses have established that the majority of homology modelling software act similarly when there is high query-template sequence identity. At low identity levels, however, stark differences are established, with superior performance by more sophisticated software incorporating powerful alignment programs and refinement schemes. For example, MODELLER has universally been noted to be reliable and versatile in template-based modelling. With AI and machine learning increasingly incorporated into modelling protocols, these tools are now well-equipped to make high-accuracy structural predictions even under constrained experimental conditions, thus becoming an essential part of structure-based drug discovery pipelines [104–125].

MODELLING CHALLENGES

A. **Intrinsically Disordered Proteins (IDPs) in Drug Innovation:** Intrinsically disordered proteins (IDPs) and intrinsically disordered protein regions (IDPRs) represent a unique class of biomolecules that lack a stable 3D conformation under physiological conditions, particularly in the absence of binding partners such as proteins or nucleic acids. Despite their structural uncertainty, IDPs are involved in essential cellular regulation, signalling, and disease functions, particularly in advanced organisms where their abundance rises with biological complexity. They contain enrichment with disorder-favouring amino acids such as Arg, Pro, Gln, Gly, Glu, Ser, Ala, and Lys, making them highly structurally flexible and low in sequence complexity. IDP structure and function understanding and prediction need integrative methods combining experimental and computational techniques. Sophisticated methods such as multidimensional nuclear magnetic resonance (NMR) and small-angle X-ray scattering (SAXS) are commonly used to investigate their conformational landscapes. Experimental information can be supplemented by molecular dynamics simulations and predictive computational methods. Computational methods of IDP characterization are broadly classified into three categories. The first is based on intrinsic physicochemical properties and disorder propensity scales (e.g., IsUnstruct). The second is based on ML methods, as represented by tools such as SPINE-D. The third type consists of meta-predictor approaches, which combine several algorithms to enhance prediction accuracy. Template-based meta-predictors, like GSmetaDisorder3D, augment these abilities by superposing disordered segments onto homologous structures if present, enhancing functional interpretation and assisting in the discovery of drug targets and inhibitor design for disordered binding sites [126–129].

B. **Modelling Multi-Domain Proteins**: A large majority of proteins, roughly estimated at more than 70%, consist of several domains, and each domain acts as an independent structural or evolutionary unit. Recombination of such domains frequently occurs via genomic rearrangement to generate intricate quaternary structures that modulate various biological functions. Structural analysis of multi-domain proteins is crucial for deciphering domain–domain interactions, allosteric control, and the functional effects of mutations, especially for disease-causing variants. Modelling of multi-domain proteins is challenging because of domain flexibility and dynamic inter-domain orientations. To overcome this, structure prediction approaches increasingly utilize cryo-electron microscopy (cryo-EM), SAXS, and high-resolution homology templates data. AI-based homology modelling software and domain assembly algorithms have appeared to handle these complexities. These include AIDA (Ab Initio Domain Assembly), which builds individual domain models into entire structures without the need for global templates, and the Multidomain Assembler (MDA) module in UCSF CHIMERA, which provides hybrid modelling from experimental constraints and template information [126–129]. These computational developments, coupled with AI-aided insights, allow for a more comprehensive structural elucidation of multi-domain proteins and IDPs—crucial for the identification of new druggable sites, interaction network mapping, and structure-based drug design in the post-genomic era.

APPLICATION OF HOMOLOGY MODELLING

Homology modelling has become a vital tool in drug discovery, particularly when the 3D structure of target proteins is not available or cannot be easily determined. The technique applies the amino acid sequence of a protein to calculate its 3D structure based on comparisons with homologous proteins with known structures. This strategy is becoming ever more vital for targets that do not have experimental structural information but are very important for drug development. Its uses in drug discovery are many, and progress in AI and ML has further increased the accuracy and speed of homology modelling in the past few years. One important usage of homology modelling is in drug design and discovery. Computer algorithms driven by AI can accurately predict the 3D structure of proteins, which can help identify potential drug targets. Modelling the protein structure allows researchers to simulate how small molecules bind to the target, thereby helping discover new therapeutics [5]. Machine learning methods, including deep learning, are being used more and more to improve homology modelling by learning from big sets of protein structures to make predictions about the most likely conformations of new targets [130]. During the construction of 3D protein models, homology modelling is very important in supplying detailed structural information for proteins where the structure has not been experimentally determined. This is especially so at the beginning stages of drug discovery, when possessing a credible 3D model is pivotal to in silico screening of possible drug leads. Advances in AI have more recently enabled more efficient and more precise prediction of protein structures even when there are no close homologs available, greatly enhancing drug development workflows significantly [131].

A further critical use of homology modelling is in determining the function of proteins without a known function. When a protein's function is not known, homology modelling can be used to predict its role within cellular functions by comparing its sequence to that of already characterized proteins. AI-based tools can complement this by determining functional motifs and areas of interest within the protein sequence, helping to elucidate its biological function. In addition, homology modelling is utilized to calculate mutations in protein structures and forecast how they affect biological activity, which is helpful in the comprehension of genetically caused diseases and the design of therapeutic drugs [132].

The discovery of active sites and recognition of binding modes is one more vital use of homology modelling in drug discovery. By precisely simulating the 3D protein structure, scientists can determine its active sites, where binding interactions with small molecules or other proteins take place. New AI methods, including deep learning models on large protein–ligand interaction datasets, have enhanced the accuracy of these predictions, resulting in more effective drug screening and design [133].

Homology modelling is also extensively employed in structure-based pharmacophore modelling and virtual screening, two key methods of lead compound identification in drug discovery. In such processes, AI algorithms can quickly scan large databases of compounds and predict which are most likely to bind to a target protein. This speeds up the drug discovery process by focusing on potential candidates for further experimental verification [134]. Likewise, docking-based virtual screening, in which small molecules are docked into the 3D protein model structure to assess binding affinity, is significantly aided by AI algorithms that optimize docking protocols and accurately predict binding modes [135].

For database mining, homology modelling enables the generation of thorough structural databases for use in high-throughput screening of compounds. This application is particularly crucial while searching the enormous chemical space of drug candidates because AI-driven search algorithms are capable of detecting new compounds with therapeutic utility based on predicted interactions with target proteins [136]. Furthermore, protein–protein docking simulations, employed for the study of protein–protein interactions, are also enhanced through the use of AI-driven homology modelling methods. These simulations assist in targeting potential drug intervention areas, especially in diseases caused by aberrant protein–protein interactions [137].

Homology modelling has also been used in X-ray crystal structure refinement, in which case it assists in the improvement of low-resolution crystal structures determined by X-ray crystallography. AI-supported approaches can enhance the quality of the improved models by combining experimental observations with predicted structural data, a necessity in drug development [138].

Homology modelling, within the context of cancer drug discovery, is a breakthrough tool as it offers structural information regarding oncogenic proteins and their interaction with possible inhibitors. AI and ML codes have been used to envisage the potency of small molecules as cancer-related protein inhibitors and thus design more potent cancer therapeutics [139]. In addition, structure-based protein engineering, in which homology modelling is employed to design proteins with pre-defined characteristics, has come to the forefront of developing new biotherapeutics and enzyme inhibitors.

Homology modelling has a very important role to play in vaccine design, especially for the development of vaccines against viral proteins. Through the development of models of viral proteins, researchers can pinpoint potential epitopes for use in vaccine development. AI-based approaches are increasingly being applied to predict the consequences of mutations in viral proteins on immune responses, enhancing the design of vaccines that can offer broad protection against quickly changing pathogens.

CONCLUSION

Homology modelling has emerged as a necessity in contemporary drug discovery, especially when experimentally known structures are not available for most pharmacological targets. It makes it possible to predict the 3D structures of proteins from amino acid sequences, allowing fundamental applications like virtual screening, ligand docking, enzyme–substrate interaction examination, and mutagenesis studies. The model building process is particularly important for intricate multi-domain or multi-protein targets that are hard to solve experimentally. Technological advances in bioinformatics andAI have dramatically enhanced the accuracy of homology modelling. Machine learning techniques now enable improved template recognition, sequence alignment, loop modelling, and structural refinement. Computational intelligence (CI) has also improved model evaluation and integration processes, making overall in silico predictions more reliable. Although these advances are notable, difficulties persist—most notably in modelling low-homology targets and flexible regions of proteins. Effective alignment and model valthe idation at sites of binding are key to success in drug discovery. Integration of AI, large-scale structural databases, and high-throughput screening tools will continue to drive utility and predictivity of models. Homology modelling—enhanced through AI and bioinformatics—provides a strong, affordable platform for structure-based drug design, speeding up discovery and enhancing accuracy in the creation of novel therapeutics.

REFERENCES

1. Giersiefen H, Hilgenfeld R, Hillisch A. Modern methods of drug discovery: an introduction. *EXS*. 2003;(93):1–18.
2. Congreve M, Murray CW, Blundell TL. Structural biology and drug discovery. *Drug Discov Today*. 2005;10(13):895–907.
3. Hillisch A, Pineda LF, Hilgenfeld R. Utility of homology models in the drug discovery process. *Drug Discov Today*. 2004;9(15):659–69.
4. Muhammed MT, Aki-Yalcin E. Homology modeling in drug discovery: overview, current applications, and future perspectives. *Chem Biol Drug Des*. 2019;93(1):12–20.
5. Cavasotto CN, Phatak SS. Homology modeling in drug discovery: current trends and applications. *Drug Discov Today*. 2009;14(13–14):676–83.
6. Hatfield M, Lovas S. Conformational sampling techniques. *Curr Pharm Des*. 2014;20(20):3303–13.
7. Nichols JA, Herbert Chan HW, Baker MAB. Machine learning: applications of artificial intelligence to imaging and diagnosis. *Biophys Rev*. 2019;11(1):111–8.
8. Kourou K, Exarchos TP, Exarchos KP, Karamouzis MV, Fotiadis DI. Machine learning applications in cancer prognosis and prediction. *Comput Struct Biotechnol J*. 2015;13:8–17.
9. AlQuraishi M. ProteinNet: a standardized data set for machine learning of protein structure. *BMC Bioinform*. 2019;20(1).
10. Blundell TL, Sibanda BL, Montalvão RW, Brewerton S, Chelliah V, Worth CL, Harmer NJ, Davies O, Burke D. Structural biology and bioinformatics in drug design: opportunities and challenges for target identification and lead discovery. *Philos Trans R Soc Lond B Biol Sci*. 2006;361(1467):413–23.
11. Halgren TA. Identifying and characterizing binding sites and assessing druggability. *J Chem Inf Model*. 2009;49(2):377–89.
12. Orry AJW, Abagyan RA, Cavasotto CN. Structure-based development of target-specific compound libraries. *Drug Discov Today*. 2006;11(5–6):261–6. https://doi.org/10.1016S1359-6446(05)03717-7
13. Cavasotto CN, Orry AJW. Ligand docking and structure-based virtual screening in drug discovery. *Curr Top Med Chem*. 2007;7(10):1006–14.
14. Cavasotto C, Singh N. Docking and high throughput docking: successes and the challenge of protein flexibility. *Curr Comput Aided Drug Des*. 2008;4(3):221–34.
15. Lundstrom K. Structural genomics and drug discovery. *J Cell Mol Med*. 2007;11(2):224–38.
16. Anderson AC. The process of structure-based drug design. Chem Biol. 2003;10(9):787–97.
17. Weigelt J, McBroom-Cerajewski LDB, Schapira M, Zhao Y, Arrowmsmith CH. Structural genomics and drug discovery: all in the family. *Curr Opin Chem Biol*. 2008;12(1):32–9.
18. Manjasetty BA, Turnbull AP, Panjikar S, Büssow K, Chance MR. Automated technologies and novel techniques to accelerate protein crystallography for structural genomics. *Proteomics*. 2008;8(4):612–25.
19. Mooij WTM, Hartshorn MJ, Tickle IJ, Sharff AJ, Verdonk ML, Jhoti H. Automated protein–ligand crystallography for structure-based drug design. *ChemMedChem*. 2006;1(8):827–38.
20. Villar HO, Yan J, Hansen MR. Using NMR for ligand discovery and optimization. *Curr Opin Chem Biol*. 2004;8(4):387–91.
21. Jung JW, An JH, Na KB, Kim YS, Lee W. The active site and substrates binding mode of malonyl-CoA synthetase determined by transferred nuclear Overhauser effect spectroscopy, site-directed mutagenesis, and comparative modeling studies. *Protein Sci*. 2000;9(7):1294–303.
22. De Rienzo F, Fanelli F, Menziani MC, De Benedetti PG. Theoretical investigation of substrate specificity for cytochromes P450 IA2, P450 IID6 and P450 IIIA4. *J Comput Aided Mol Des*. 2000;14(1):93–116.
23. Ginalski K, Rychlewski L, Baker D, Grishin NV. Protein structure prediction for the male-specific region of the human Y chromosome. *Proc Natl Acad Sci USA*. 2004;101(8):2305–10.
24. Talukdar AS, Wilson DL. Modeling and optimization of rotational C-arm stereoscopic X-ray angiography. *IEEE Trans Med Imaging*. 1999;18(7):604–16.
25. Ceulemans H, Russell RB. Fast fitting of atomic structures to low-resolution electron density maps by surface overlap maximization. *J Mol Biol*. 2004;338(4):783–93.

26. Teichmann SA, Chothia C, Gerstein M. Advances in structural genomics. *Curr Opin Struct Biol.* 1999;9(3):390–9.
27. Capener CE, Shrivastava IH, Ranatunga KM, Forrest LR, Smith GR, Sansom MS. Homology modeling and molecular dynamics simulation studies of an inward rectifier potassium channel. *Biophys J.* 2000;78(6):2929–42.
28. Kairys V, Gilson MK, Fernandes MX. Using protein homology models for structure-based studies: approaches to model refinement. *Sci World J.* 2006;6:1542–54.
29. Agarwal S, Verma E, Kashaw SK. *Homology Modelling: A Computational Tool in Drug Design and Discovery.* Sharjah: Bentham Science Publishers; 2022, pp. 1–32.
30. Bourne PE, Weissig H. *Structural Bioinformatics: Bourne Structural Bioinformatics.* Nashville, TN: John Wiley & Sons; 2003.
31. Eswar N, Webb B, Marti-Renom MA, Madhusudhan MS, Eramian D, Shen MY, Pieper U, Sali A. Comparative protein structure modeling using MODELLER. *Curr Protoc Protein Sci.* 2007;Chapter 2:Unit 2.9. doi: 10.10020471140864.ps0209s50.
32. Bordoli L, Schwede T. Automated protein structure modeling with SWISS-MODEL workspace and the protein model portal. *Methods Mole Biol.* 2011;857:107–36.
33. Marti-Renom MA, Madhusudhan MS, Eswar N, Pieper U, Shen M-Y, Sali A, Fiser A, Mirkovic N, John B, Stuart A. Modeling protein structure from its sequence. *Curr Protoc Bioinform.* 2003;3(1):5.1.1–5.1.33.
34. Chandonia J-M, Brenner SE. Implications of structural genomics target selection strategies: Pfam5000, whole genome, and random approaches. *Proteins.* 2004;58(1):166–79.
35. Vitkup D, Melamud E, Moult J, Sander C. Completeness in structural genomics. *Nat Struct Biol.* 2001;8(6):559–66.
36. Tramontano A, Morea V. Assessment of homology-based predictions in CASP5. *Proteins.* 2003;53 Suppl 6(S6):352–68.
37. Martí-Renom MA, Stuart AC, Fiser A, Sánchez R, Melo F, Sali A. Comparative protein structure modeling of genes and genomes. *Annu Rev Biophys Biomol Struct.* 2000;29(1):291–325.
38. Burley SK, Joachimiak A, Montelione GT, Wilson IA. Contributions to the NIH-NIGMS protein structure initiative from the PSI production centers. *Structure.* 2008;16(1):5–11.
39. Ginalski K. Comparative modeling for protein structure prediction. *Curr Opin Struct Biol.* 2006;16(2):172–7.
40. Larsson P, Wallner B, Lindahl E, Elofsson A. Using multiple templates to improve quality of homology models in automated homology modeling. *Protein Sci.* 2008;17(6):990–1002.
41. Jalily Hasani H, Barakat K. Homology modeling: an overview of fundamentals and tools. *Int Rev Model Simul (IREMOS).* 2017;10(2):129.
42. Haddad YA. The syntax of the addressee in imperatives: what levantine Arabic attitude datives bring to the table. *Proc Linguist Soc Am.* 2020;5(1):1.
43. Kopp J, Schwede T. Automated protein structure homology modeling: a progress report. *Pharmacogenomics.* 2004;5(4):405–16.
44. Joo K, Lee J, Lee J. Methods for accurate homology modeling by global optimization. *Methods Mol Biol.* 2012;857:175–88.
45. Mishra S, Saxena A, Sangwan RS. Fundamentals of homology modeling steps and comparison among important bioinformatics tools: an overview. *Sci Int.* 2013;1(7):237–52.
46. Fiser A, Sali A. Modeller: generation and refinement of homology-based protein structure models. *Methods Enzymol.* 2003;374:461–91.
47. Wong W-C, Maurer-Stroh S, Eisenhaber F. Not all transmembrane helices are born equal: towards the extension of the sequence homology concept to membrane proteins. *Biol Direct.* 2011;6(1):57.
48. Lindahl E, Elofsson A. Identification of related proteins on family, superfamily and fold level. *J Mol Biol.* 2000;295(3):613–25.
49. Park J, Karplus K, Barrett C, Hughey R, Haussler D, Hubbard T, Chothia C. Sequence comparisons using multiple sequences detect three times as many remote homologues as pairwise methods. *J Mol Biol.* 1998;284(4):1201–10.
50. Sauder JM, Arthur JW, Dunbrack RL Jr. Large-scale comparison of protein sequence alignment algorithms with structure alignments. *Proteins.* 2000;40(1):6–22.
51. Feng DF, Doolittle RF. Progressive sequence alignment as a prerequisite to correct phylogenetic trees. *J Mol Evol.* 1987;25(4):351–60.
52. Thompson JD, Higgins DG, Gibson TJ. CLUSTAL W: improving the sensitivity of progressive multiple sequence alignment through sequence weighting, position-specific gap penalties and weight matrix choice. *Nucleic Acids Res.* 1994;22(22):4673–80.
53. Thompson JD, Gibson TJ, Plewniak F, Jeanmougin F, Higgins DG. The CLUSTAL_X windows interface: flexible strategies for multiple sequence alignment aided by quality analysis tools. *Nucleic Acids Res.* 1997;25(24):4876–82.
54. Ashkenazy H, Unger R, Kliger Y. Hidden conformations in protein structures. *Bioinformatics.* 2011;27(14):1941–7.
55. Remmert M, Biegert A, Hauser A, Söding J. HHblits: lightning-fast iterative protein sequence searching by HMM-HMM alignment. *Nat Methods.* 2011;9(2):173–5.
56. Johnson LS, Eddy SR, Portugaly E. Hidden Markov model speed heuristic and iterative HMM search procedure. *BMC Bioinform.* 2010;11(1):431.
57. Altschul SF, Madden TL, Schäffer AA, Zhang J, Zhang Z, Miller W, Lipman DJ. Gapped BLAST and PSI-BLAST: a new generation of protein database search programs. *Nucleic Acids Res.* 1997;25(17):3389–402.
58. Lam SD, Das S, Sillitoe I, Orengo C. An overview of comparative modelling and resources dedicated to large-scale modelling of genome sequences. *Acta Crystallogr D Struct Biol.* 2017;73(8):628–40.
59. Larkin MA, Blackshields G, Brown NP, Chenna R, McGettigan PA, McWilliam H, Valentin F, Wallace IM, Wilm A, Lopez R, Thompson JD, Gibson TJ, Higgins DG. Clustal W and Clustal X version 2.0. *Bioinformatics.* 2007;23(21):2947–8.
60. Li W, Cowley A, Uludag M, Gur T, McWilliam H, Squizzato S, Park YM, Buso N, Lopez R. The EMBL-EBI bioinformatics web and programmatic tools framework. *Nucleic Acids Res.* 2015;43(W1):W580 4.
61. Manuel M. A new semi-subterranean diving beetle of the Hydroporus normandi-complex from south-eastern France, with notes on other taxa of the complex (Coleoptera: Dytiscidae). *Zootaxa.* 2013;3652(4):453.

62. Collura V, Higo J, Garnier J. Modeling of protein loops by simulated annealing. *Protein Sci.* 1993;2(9):1502–10.
63. Browne WJ, North ACT, Phillips DC. A possible three-dimensional structure of Bovine a-Lactalbumin based on that of Hen's EGG-white lysozyme. *J Hoi Bid.* 1969;42:65–86.
64. Greer J. Comparative modeling methods: application to the family of the mammalian serine proteases. *Proteins.* 1990;7(4):317–34.
65. Levitt M. Accurate modeling of protein conformation by automatic segment matching. *J Mol Biol.* 1992;226(2):507–33.
66. Rost B. Twilight zone of protein sequence alignments. *Protein Eng.* 1999;12(2):85–94.
67. Xiang Z. Advances in homology protein structure modeling. *Curr Protein Pept Sci.* 2006;7(3):217–27.
68. Fiser A. Template-based protein structure modeling. *Methods Mol Biol.* 2010;673:73–94.
69. van Gelder CW, Leusen FJ, Leunissen JA, Noordik JH. A molecular dynamics approach for the generation of complete protein structures from limited coordinate data. *Proteins.* 1994;18(2):174–85.
70. Zhu J, Fan H, Periole X, Honig B, Mark AE. Refining homology models by combining replica-exchange molecular dynamics and statistical potentials. *Proteins.* 2008;72(4):1171–88.
71. Cornell WD, Cieplak P, Bayly CI, Gould IR, Merz KM, Ferguson DM, Spellmeyer DC, Fox T, Caldwell JW, Kollman PA. A second generation force field for the simulation of proteins, nucleic acids, and organic molecules. *J Am Chem Soc.* 1995;117(19):5179–97.
72. Das R, Qian B, Raman S, Vernon R, Thompson J, Bradley P, Khare S, Tyka MD, Bhat D, Chivian D, Kim DE, Sheffler WH, Malmström L, Wollacott AM, Wang C, Andre I, Baker D. Structure prediction for CASP7 targets using extensive all-atom refinement with Rosetta@home. *Proteins.* 2007;69 Suppl 8(S8):118–28.
73. Han R, Leo-Macias A, Zerbino D, Bastolla U, Contreras-Moreira B, Ortiz AR. An efficient conformational sampling method for homology modeling. *Proteins.* 2008;71(1):175–88.
74. Misura KMS, Baker D. Progress and challenges in high-resolution refinement of protein structure models. *Proteins.* 2005;59(1):15–29.
75. Lee J, Lee D, Park H, Coutsias EA, Seok C. Protein loop modeling by using fragment assembly and analytical loop closure. *Proteins.* 2010;78(16):3428–36.
76. Michalsky E, Goede A, Preissner R. Loops in proteins (LIP)--a comprehensive loop database for homology modelling. *Protein Eng Des Sel.* 2003;16(12):979–85.
77. Hameduh T, Haddad Y, Adam V, Heger Z. Homology modeling in the time of collective and artificial intelligence. *Comput Struct Biotechnol J.* 2020;18:3494–506.
78. Jones TA, Thirup S. Using known substructures in protein model building and crystallography. *EMBO J.* 1986;5(4):819–22.
79. Blundell TL, Sibanda BL, Sternberg M, Thornton JM. Knowledgebased prediction of protein structures and the design of novel molecules. *Nature.* 1987;326:347–52.
80. Sutcliffe MJ, Haneef I, Carney D, Blundell TL. Knowledge based modeling of omologous proteins, part I: three-dimensional frameworks derived from the simultaneous superposition of multiple structures. *Prot Eng.* 1987;5:377–84.
81. Moult J, James MN. An algorithm for determining the conformation of polypeptide segments in proteins by systematic search. *Proteins.* 1986;1(2):146–63.
82. Pellequer JL, Chen SW. Does conformational free energy distinguish loop conformations in proteins? *Biophys J.* 1997;73(5):2359–75.
83. Fine RM, Wang H, Shenkin PS, Yarmush DL, Levinthal C. Predicting antibody hypervariable loop conformations. IL Minimization and molecular dynamics studies of MCPC603 from many randomly generated loop conformations. *Proteins.* 1986;1:342–62.
84. Sowdhamini R, Rufino SD, Blundell TL. A database of globular protein structural domains: clustering of representative family members into similar folds. *Fold Des.* 1996;1(3):209–20.
85. Shenkin PS, Yarmush DL, Fine RM, Wang HJ, Levinthal C. Predicting antibody hypervariable loop conformation. I. Ensembles of random conformations for ringlike structures. *Biopolymers.* 1987;26(12):2053–85.
86. Shirai H, Nakajima N, Higo J, Kidera A, Nakamura H. Conformational sampling of CDR-H3 in antibodies by multicanonical molecular dynamics simulation. *J Mol Biol.* 1998;278(2):481–96.
87. Mendes J, Baptista AM, Carrondo MA, Soares CM. Improved modeling of side-chains in proteins with rotamer-based methods: a flexible rotamer model. *Proteins.* 1999;37(4):530–43.
88. Liang S, Grishin NV. Side-chain modeling with an optimized scoring function. *Protein Sci.* 2002;11(2):322–31.
89. Canutescu AA, Shelenkov AA, Dunbrack RL Jr. A graph-theory algorithm for rapid protein side-chain prediction. *Protein Sci.* 2003;12(9):2001–14. https://doi.org/10.1110ps.03154503
90. Xu G, Ma T, Du J, Wang Q, Ma J. OPUS-Rota2: an improved fast and accurate side-chain modeling method. *J Chem Theory Comput.* 2019;15(9):5154–60.
91. Krivov GG, Shapovalov MV, Dunbrack RL Jr. Improved prediction of protein side-chain conformations with SCWRL4: side-chain prediction with SCWRL4. *Proteins.* 2009;77(4):778–95.
92. Huang X, Pearce R, Zhang Y. FASPR: an open-source tool for fast and accurate protein side-chain packing. *Bioinformatics.* 2020;36(12):3758–65.
93. Hong SH, Joung I, Flores-Canales JC, Manavalan B, Cheng Q, Heo S, Kim JY, Lee SY, Nam M, Joo K, Lee IH, Lee SJ, Lee J. Protein structure modeling and refinement by global optimization in CASP12. *Proteins.* 2018;86:122–35.
94. Petrey D, Honig B. Protein structure prediction: inroads to biology. *Mol Cell.* 2005;20(6):811–9.
95. Gerstein M, Levitt M. Comprehensive assessment of automatic structural alignment against a manual standard, the SCOP classification of proteins: structural alignment. *Protein Sci.* 1998;7(2):445–56.
96. Chothia C, Lesk AM. The relation between the divergence of sequence and structure in proteins. *EMBO J.* 1986;5(4):823–6.

97. Sippl MJ. Knowledge-based potentials for proteins. *Curr Opin Struct Biol.* 1995;5(2):229–35.
98. Baumann G, Froömmel C, Sander C. Polarity as a criterion in protein design. *Protein Eng Des Sel.* 1989;2(5):329–34.
99. Holm L, Sander C. Evaluation of protein models by atomic solvation preference. *J Mol Biol.* 1992;225(1):93–105.
100. Gregoret LM, Cohen FE. Novel method for the rapid evaluation of packing in protein structures. *J Mol Biol.* 1990;211(4):959–74.
101. Haddad Y, Adam V, Heger Z. Ten quick tips for homology modeling of high resolution protein 3D structures. *PLoS Comput Biol.* 2020;16(4):1–19.
102. Munsamy G, Soliman MES. Homology modeling in drug discovery-an update on the last decade. *Lett Drug Des Discov.* 2017;14(9):27.
103. Kryshtafovych A, Monastyrskyy B, Fidelis K. CASP prediction center infrastructure and evaluation measures in CASP10 and CASP ROLL: CASP prediction center. *Proteins.* 2014;82(Suppl 2):7–13.
104. Sali A, Blundell TL. Comparative protein modelling by satisfaction of spatial restraints. *J Mol Biol.* 1993;234(3):779–815.
105. Sánchez R, Sali A. Evaluation of comparative protein structure modeling by MODELLER-3. *Proteins.* 1997;Suppl 1:50–8.
106. Webb B, Sali A. Comparative protein structure modeling using MODELLER. *Curr Protoc Bioinform.* 2014;47(1).
107. Guex N, Peitsch MC. SWISS-MODEL and the Swiss-PdbViewer: an environment for comparative protein modeling. *Electrophoresis.* 1997;18(15):2714–23.
108. Arnold K, Bordoli L, Kopp J, Schwede T. The SWISS-MODEL workspace: a web-based environment for protein structure homology modelling. *Bioinformatics.* 2006;22(2):195–201.
109. Biasini M, Bienert S, Waterhouse A, Arnold K, Studer G, Schmidt T, Kiefer F, Gallo Cassarino T, Bertoni M, Bordoli L, Schwede T. SWISS-MODEL: modelling protein tertiary and quaternary structure using evolutionary information. *Nucleic Acids Res.* 2014;42(Web Server issue):W252–8.
110. Schwede T, Kopp J, Guex N, Peitsch MC. SWISS-MODEL: an automated protein homology-modeling server. *Nucleic Acids Res.* 2003;31(13):3381–5.
111. Peitsch MC. ProMod and Swiss-Model: internet-based tools for automated comparative protein modelling. *Biochem Soc Trans.* 1996;24(1):274–9.
112. Roy A, Kucukural A, Zhang Y. I-TASSER: a unified platform for automated protein structure and function prediction. *Nat Protoc.* 2010;5(4):725–38.
113. Yang J, Zhang Y. Protein structure and function prediction using I-TASSER. *Curr Protoc Bioinform.* 2015;52(1):5.8.1–5.8.15.
114. Yang J, Zhang Y. I-TASSER server: new development for protein structure and function predictions. *Nucleic Acids Res.* 2015;43(W1):W174–81.
115. Kim DE, Chivian D, Baker D. Protein structure prediction and analysis using the Robetta server. *Nucleic Acids Res.* 2004,32(W1).W526–31.
116. Du Z, Su H, Wang W, Ye L, Wei H, Peng Z, Anishchenko I, Baker D, Yang J. The trRosetta server for fast and accurate protein structure prediction. *Nat Protoc.* 2021;16(12):5634–51.
117. Wang S, Li W, Liu S, Xu J. RaptorX-Property: a web server for protein structure property prediction. *Nucleic Acids Res.* 2016;44(W1):W430–5.
118. Källberg M, Wang H, Wang S, Peng J, Wang Z, Lu H, Xu J. Template-based protein structure modeling using the RaptorX web server. *Nat Protoc.* 2012;7(8):1511–22.
119. Jumper J, Evans R, Pritzel A, Green T, Figurnov M, Ronneberger O, Tunyasuvunakool K, Bates R, Žídek A, Potapenko A, Bridgland A. Highly accurate protein structure prediction with AlphaFold. *Nature.* 2021;596(7873):583–9.
120. Senior AW, Evans R, Jumper J, Kirkpatrick J, Sifre L, Green T, Qin C, Žídek A, Nelson AW, Bridgland A, Penedones H. Improved protein structure prediction using potentials from deep learning. *Nature.* 2020;577(7792):706–10.
121. Qian B, Raman S, Das R, Bradley P, McCoy AJ, Read RJ, Baker D. High-resolution structure prediction and the crystallographic phase problem. *Nature.* 2007;450(7167):259–64.
122. Keedy DA, Williams CJ, Headd JJ, Arendall III WB, Chen VB, Kapral GJ, Gillespie RA, Block JN, Zemla A, Richardson DC, Richardson JS. The other 90% of the protein: assessment beyond the Cαs for CASP8 template-based and high-accuracy models. *Proteins: Struct Func Bioinform.* 2009;77(S9):29–49.
123. Callaway E. 'It will change everything': deepmind's AI makes gigantic leap in solving protein structures. *Nature.* 2020;588(7837):203–4.
124. Callaway E. After AlphaFold: protein-folding contest seeks next big breakthrough. *Nature.* 2023;613(7942):13–4.
125. AlQuraishi M. AlphaFold at CASP13. *Bioinformatics.* 2019;35(22):4862–5.
126. Xu D, Jaroszewski L, Li Z, Godzik A. AIDA: Ab initio domain assembly for automated multi-domain protein structure prediction and domain–domain interaction prediction. *Bioinformatics.* 2015;31(13):2098–105.
127. Hertig S, Goddard TD, Johnson GT, Ferrin TE. Multidomain assembler (MDA) generates models of large multidomain proteins. *Biophys J.* 2015;108(9):2097–102.
128. Hou J, Adhikari B, Tanner JJ, Cheng J. SAXSDom: modeling multidomain protein structures using small-angle X-ray scattering data. *Proteins.* 2020;88(6):775–87.
129. Zhou X, Hu J, Zhang C, Zhang G, Zhang Y. Assembling multidomain protein structures through analogous global structural alignments. *Proc Natl Acad Sci USA.* 2019;116(32):15930–8.
130. França TCC. Homology modeling: an important tool for the drug discovery. *J Biomol Struct Dyn.* 2015;33(8):1780–93.
131. Rezácová P, Borek D, Moy SF, Joachimiak A, Otwinowski Z. Crystal structure and putative function of small Toprim domain-containing protein from Bacillus stearothermophilus. *Proteins.* 2008;70(2):311–9.
132. Cohen NC. Structure-based drug design and the discovery of aliskiren (Tekturna): perseverance and creativity to overcome a R&D pipeline challenge: discovery of aliskiren. *Chem Biol Drug Des.* 2007;70(6):557–65.
133. Park H, Bahn YJ, Jung SK, Jeong DG, Lee SH, Seo I, Yoon TS, Kim SJ, Ryu SE. Discovery of novel Cdc25 phosphatase inhibitors with micromolar activity based on the structure-based virtual screening. *J Med Chem.* 2008;51(18):5533–41.

134. Nguyen TL, Gussio R, Smith JA, Lannigan DA, Hecht SM, Scudiero DA, Shoemaker RH, Zaharevitz DW. Homology model of RSK2 N-terminal kinase domain, structure-based identification of novel RSK2 inhibitors, and preliminary common pharmacophore. *Bioorg Med Chem.* 2006;14(17):6097–105.
135. Hellmuth K, Grosskopf S, Lum CT, Würtele M, Röder N, von Kries JP, Rosario M, Rademann J, Birchmeier W. Specific inhibitors of the protein tyrosine phosphatase Shp2 identified by high-throughput docking. *Proc Natl Acad Sci USA.* 2008;105(20):7275–80.
136. Xu Q, Canutescu A, Obradovic Z, Dunbrack RL Jr. ProtBuD: a database of biological unit structures of protein families and superfamilies. *Bioinformatics.* 2007;23(15):2028–2028.
137. Ring CS, Sun E, McKerrow JH, Lee GK, Rosenthal PJ, Kuntz ID, Cohen FE. Structure-based inhibitor design by using protein models for the development of antiparasitic agents. *Proc Natl Acad Sci USA.* 1993;90(8):3583–7.
138. Ruback E, Lobo LA, França TCC, Pascutti PG. Structural analysis of Pla protein from the biological warfare agentYersinia pestis: docking and molecular dynamics of interactions with the mammalian plasminogen system. *J Biomol Struct Dyn.* 2013;31(5):477–84.
139. Howell PL, Almo SC, Parsons MR, Hajdu J, Petsko GA. Structure determination of turkey EGG-white lysozyme using Laue diffraction data. *Acta Crystallogr B.* 1992;48(2):200–7.
140. MacArthur MW, Laskowski RA, Thornton JM. Knowledge-based validation of protein structure coordinates derived by X-ray crystallography and NMR spectroscopy. *Curr Opin Struct Biol.* 1994;4(5):731–7.
141. Morris AL, MacArthur MW, Hutchinson EG, Thornton JM. Stereochemical quality of protein structure coordinates. *Proteins.* 1992;12(4):345–64.
142. Laskowski RA, MacArthur MW, Moss DS, Thornton JM. PROCHECK: a program to check the stereochemical quality of protein structures. *J Appl Crystallogr.* 1993;26(2):283–91.
143. Hooft RW, Vriend G, Sander C, Abola EE. Errors in protein structures. *Nature.* 1996;381(6580):272.
144. Khor BY, Tye GJ, Lim TS, Noordin R, Choong YS. The structure and dynamics of BmR1 protein from Brugia malayi: in silico approaches. *Int J Mol Sci.* 2014;15(6):11082–99.
145. Wiederstein M, Sippl MJ. ProSA-web: interactive web service for the recognition of errors in three-dimensional structures of proteins. *Nucleic Acids Res.* 2007;35(Web Server):W407–10.
146. Sippl MJ. Recognition of errors in three-dimensional structures of proteins. *Proteins.* 1993;17(4):355–62.
147. Kumar V, Saravanan P, Arvind A, Mohan CG. Identification of hotspot regions of MurB oxidoreductase enzyme using homology modeling, molecular dynamics and molecular docking techniques. *J Mol Model.* 2011;17(5):939–53.
148. Lüthy R, Bowie JU, Eisenberg D. Assessment of protein models with three-dimensional profiles. *Nature.* 1992;356(6364):83–5.
149. Eisenberg D, Lüthy R, Bowie JU. VERIFY3D: assessment of protein models with three-dimensional profiles. *Methods Enzymol.* 1997;277:396–404.
150. Zhang L, Skolnick J. What should the Z-score of native protein structures be? *Protein Sci.* 1998;7(5):1201–7.
151. Mande SC, Kumar A, Ghosh P. Analysis of dihedral angle variability in related protein structures. In: Bansal M, Srinivasan N (Eds.), *Biomolecular Forms and Functions.* India: World Scientific Indian Inst of Science; 2013, pp. 107–15.
152. Petrey D, Honig B. Free energy determinants of tertiary structure and the evaluation of protein models. *Protein Sci.* 2000;9(11):2181–91.
153. Huang ES, Subbiah S, Levitt M. Recognizing native folds by the arrangement of hydrophobic and polar residues. *J Mol Biol.* 1995;252(5):709–20.
154. Zemla A. LGA: a method for finding 3D similarities in protein structures. *Nucleic Acids Res.* 2003;31(13):3370–4.
155. Kryshtafovych A, Venclovas Č, Fidelis K, Moult J. Progress over the first decade of CASP experiments. *Proteins.* 2005;61(S7):225–36.
156. Read RJ, Chavali G. Assessment of CASP7 predictions in the high accuracy template-based modeling category. *Proteins.* 2007;69(Suppl 8):27–37.
157. Zhang Y. TM-align: a protein structure alignment algorithm based on the TM-score. *Nucleic Acids Res.* 2005;33(7):2302–9.
158. Siew N, Elofsson A, Rychlewski L, Fischer D. MaxSub: an automated measure for the assessment of protein structure prediction quality. *Bioinformatics.* 2000;16(9):776–85.
159. Antczak PLM, Ratajczak T, Lukasiak P, Blazewicz J. SphereGrinder - reference structure-based tool for quality assessment of protein structural models. In: *2015 IEEE International Conference on Bioinformatics and Biomedicine (BIBM).* Washington, DC: IEEE; 2015.
160. Abagyan RA, Totrov MM. Contact area difference (CAD): a robust measure to evaluate accuracy of protein models. *J Mol Biol.* 1997;268(3):678–85.
161. Olechnovič K, Kulberkytė E, Venclovas C. CAD-score: a new contact area difference-based function for evaluation of protein structural models. *Proteins.* 2013;81(1):149–62.
162. Davis IW, Murray LW, Richardson JS, Richardson DC. MOLPROBITY: structure validation and all-atom contact analysis for nucleic acids and their complexes. *Nucleic Acids Res.* 2004;32(Web Server issue):W615–9.
163. Vriend G. WHAT IF: a molecular modeling and drug design program. *J Mol Graph.* 1990;8(1):52–6.
164. Benkert P, Tosatto SCE, Schomburg D. QMEAN: a comprehensive scoring function for model quality assessment. *Proteins.* 2008;71(1):261–77.
165. Shen M-Y, Sali A. Statistical potential for assessment and prediction of protein structures. *Protein Sci.* 2006;15(11):2507–24.
166. Pawlowski M, Gajda MJ, Matlak R, Bujnicki JM. MetaMQAP: a meta-server for the quality assessment of protein models. *BMC Bioinform.* 2008;9(1):403.
167. Eramian D, Shen M-Y, Devos D, Melo F, Sali A, Marti-Renom MA. A composite score for predicting errors in protein structure models. *Protein Sci.* 2006;15(7):1653–66.

168. Elmezayen AD, Yelekçi K. Homology modeling and in silico design of novel and potential dual-acting inhibitors of human histone deacetylases HDAC5 and HDAC9 isozymes. *J Biomol Struct Dyn*. 2021;39(17):6396–414. https://doi.org/10.108007391102.2020.1798812
169. Al-Obaidi A, Elmezayen AD, Yelekçi K. Homology modeling of human GABA-AT and devise some novel and potent inhibitors via computer-aided drug design techniques. *J Biomol Struct Dyn*. 2021;39(11):4100–10.
170. Ji X, Wang W, Zheng Y, Hao J, Sun M. Homology modeling and molecular dynamics simulation studies of a marine alkaline protease. *Bioinform Biol Insights*. 2012;6:255–63.
171. Yarnitzky T, Levit A, Niv MY. Homology modeling of G-protein-coupled receptors with X-ray structures on the rise. *Curr Opin Drug Discov Devel*. 2010;13(3):317–25.
172. Vyas VK, Ukawala RD, Ghate M, Chintha C. Homology modeling a fast tool for drug discovery: current perspectives. *Indian J Pharm Sci*. 2012;74(1):1–17.
173. Petrey D, Honig B. GRASP2: visualization, surface properties, and electrostatics of macromolecular structures and sequences. *Methods Enzymol*. 2003;374:492–509.

7 Molecular Docking Guided Drug Design and Discovery

Feroz Khan, Palak Singh, and Aqib Sarfraz
Technology Dissemination & Computational Biology Division, CSIR-Central Institute of Medicinal & Aromatic Plants, Lucknow, India

INTRODUCTION

The computer-aided molecular modeling for drug designing includes pharmacogenomics (gene-based design of diagnostics and therapeutics) and drug molecule design and is based on the knowledge of molecular properties, e.g., molecular structure, functional groups present, molecular geometry, etc. Drug designing is an iterative process involving drug discovery, lead optimization, and chemical synthesis to maximize functional activity and minimize adverse effects. It does not include pharmacokinetics, dosage, or drug administration analysis (pharmacokinetics is the study of what the body does to a drug, and pharmacodynamics is the study of what a drug does to the body). Only one out of 10^{100} molecules possibly led to a drug molecule. Hence, drug design is an optimization problem. Understanding the functioning of drugs before getting into the molecular aspects is essential. Drugs can work in one of the following ways: (i) they interfere with the biological functions, (ii) block the interactions of viruses, bacteria, or parasites with our systems; and (iii) interfere with or enhance our biological functions.

A crucial aspect of drug designing is the understanding of methods by which the active site of a *receptor* (a protein molecule capable of selectively binding a drug, hormone, or neurotransmitter, thereby eliciting a physiological response) selectively restricts the binding of inappropriate structures. Any potential molecule capable of binding to a receptor is termed a ligand that must possess the molecular key to bind with a receptor. The molecular key consists of a specific combination of atoms with the correct size, shape, and charge composition to bind and interact with the receptor. Ligand binding relies on the complementary relationship between the ligand and the binding site. This is a steric complementarity. The physicochemical complementarity between ligand and binding pocket residues allows molecular interactions between the two. For ligand and receptor to interact, there needs to be a driving force that compels the ligand to bind to the receptor by replacing the water. Several non-covalent interactions like electrostatic complementarity, hydrophobic interactions, *van der* Waals, etc. ensure the appropriate positioning of the ligand within the binding pockets. The study of such physicochemical properties to demarcate the activities responsible for their affinities toward the target is utilized for drug designing.

COMPUTATIONAL DRUG DESIGNING AND DISCOVERY APPROACHES

There are the following basic approaches for drug designing and discovery:

i. **Ligand-Based Approach**: QSAR and Pharmacophore modeling methods.
ii. **Target/Structure-Based Approach**: Molecular docking—Sampling algorithms and Scoring functions.
iii. ***De novo* Designing Approach**: Fragment-based method (e.g., Incremental constructions (IC), Multiple Copy Simultaneous Search (MCSS), and LUDI sampling algorithms)

Ligand-Based Approaches

Such methodologies are used when the receptor is unknown or the structures are unavailable. Combinatorial chemistry can generate large libraries (a set of compounds) of molecules to be screened for biologically active compounds. The method allows many molecules to be synthesized much more rapidly and at a lower cost than traditional synthetic chemistry. However, the enormous number and range of compounds produced using combinatorial chemistry are not necessarily synthetically accessible drug-like molecules. Ligand-based approaches that try to identify characteristics/properties common to known ligands can potentially be used in the screening of new or improved drugs. There are two methods to perform ligand-based drug-like compounds:

i. Quantitative Structure–Activity Relationship (QSAR) method
ii. Pharmacophore Modeling Method

Quantitative Structure–Activity Relationship (QSAR)

A QSAR study tries to establish a relationship between the molecule's physicochemical properties also known as descriptors and their respective functions. It involves the statistical analysis of a set of properties or descriptors for a series of biologically active molecules, e.g., dimensionality (i.e., 1D, 2D, & 3D descriptors), physicochemical

DOI: 10.1201/9781003464679-7

(i.e., steric, electrostatic, hydrophobic), chemical (i.e., physicochemical parameters of amino acids), and other descriptors. Typical QSAR descriptors include physicochemical properties such as molecular weight, thermodynamic, and electronic properties. The compound's logP value is the logarithmic partition coefficient between n-octanol and water. It is a good indicator of the compound's hydrophilicity. Low hydrophilicity (or high logP values) causes poor absorption or permeation. It has been proven that compounds have a reasonable probability of being well absorbed, so their logP value must not be greater than 5.0. ClogP (now included in Bio-Loom software; http://www.biobyte.com/index.html) measure is another valid QSAR descriptor. Other statistical approaches such as least-squares optimization and principal component analysis have also been taken into account for such studies. Tripos CoMFA (Comparative Molecular Field Analysis) is another implementation of 3D QSAR where a charged probe atom is rolled over a grid surrounding a fitted compound set. The non-bonded energies were calculated as descriptors using Partial Least Squares analysis and correlated with their biological activities. The QSAR-based model helps in predicting the members of a series of proposed compounds that are likely to be active and further can be synthesized. The additional experimental data refine the model as new compounds are assayed. Such an approach has been used extensively in the pharmaceutical industry. The results show the pharmacophoric fields determining activity.

Pharmacophore Modeling Method

In numerous potential interactions between ligands and receptors, the specific interactions crucial for ligand recognition and binding by the receptor are termed the pharmacophore. There may be uncountable steric, electrostatic, and hydrophobic contacts based on the size of the binding or active site. However, some are more crucial than others. Usually, these are the interactions that directly factor into the structural integrity of a receptor or are involved in the mechanism of its action. A pharmacophore model can be derived from a set of known ligands or by using standard features for a series of active compounds against a target. These features include acceptors, donors, ring centroids, hydrophobes, etc. A 3D pharmacophore defines the complementarity between the groups or features of the set of ligands and conformations of the binding sites to generate a pharmacophore hypothesis.

When the protein is unknown, it is an efficient approach to determine the active conformation of the small molecules. Searching the databases for new hit compounds is possible with the generated hypotheses. The low energy active conformations of ligands are searched to identify pharmacophores. Chem3-DBS, the program, can detect a pharmacophore in a set of ligands and do a flexible search in the 3BDS. It is fast and relatively easy to use. The **PMapper** (https://docs.chemaxon.com/) perceives the pharmacophoric properties of atoms of given molecular structures. In order to search large compound libraries for structures exhibiting similar therapeutic activities, the calculated pharmacophoric features are assigned to corresponding atoms and stored along with the structure.

The conformational search can be pretty extensive due to higher degrees of freedom and can be approached using the following methods:

i. **Systematic Search Method:** It generates sterically allowed molecular conformations by systematically varying sets of specified torsion angles. C2 Conformers (https://www-jmg.ch.cam.ac.uk/cil/SGTL/cerius2.html) is an implementation of such a method.
ii. **Distance Geometry Methods:** They are suitable for dealing with molecular matching and flexibility problems by using random sample conformations. Rubicon uses this method (https://www.daylight.com/dayhtml/doc/rubicon/).
iii. **Clique-Detection Algorithm:** It searches standard sets of inter-feature distances within a group of active molecules. Suitable to tolerate the distance match accounting for discrete conformations and uncertainties in the pharmacophore. The method is used by DISCOtech (https://www.yumpu.com/en/document/view/15425061/discotech-tripos).

Target/Structure-based Drug Designing and Discovery

Target-based design methods are based on the 3D structure of the biological target (usually a protein), either bound or unbound to an inhibitor or substrate. Target-based design methods are also called structure-based or rational design methods.

Molecular Docking Method

The molecular docking method is among the most popular approaches for target-based methods. Docking involves the screening of a library of known molecules against the known target/receptor. It generates possible binding geometries which are then evaluated using a scoring function. In docking, there are three degrees of translational and rotational freedom possessed by the receptor and the ligand. Being a vast space to sample, most docking methods to date use a rigid body assumption, i.e., no conformations of the ligand or receptor are considered. This is not a practical assumption in most cases, but it makes the problem tractable. A docking method requires (i) a representative structure of the receptor and ligand, (ii) a method for generating potential binding geometries, and (iii) a way of scoring these geometries.

Sampling Algorithms for the Detection of Active Conformations

1. **Matching Algorithms (MA)**: The method utilizes molecular shape, which maps a ligand onto a protein's active site regarding shape features and chemical information. Pharmacophores are used to exemplify the ligand and the protein.

Considerations for the match require chemical parameters such as donors and hydrogen bond acceptors. Several programs have been developed utilizing this algorithm, including LibDock, SANDOCK, FLOG, and DOCK.

2. **Incremental Construction (IC)**: Methods that involve fragmenting the ligand and placing it incrementally into the active binding site are widely used. The breaking of rotatable bonds splits the ligands into multiple fragments. One fragment, often the largest or most functionally important, is chosen as an anchor and further docked into the active site first. The rest of the fragments are then added step by step. This approach generates various orientations to accommodate the ligand's flexibility within the active site. Molecular docking tools that utilize this strategy include DOCK 4.0, FlexX, Hammerhead, SLIDE, and eHiTS.
3. **Fragment-Based Method**: MCSS and LUDI are fragment-based techniques used for designing new ligands from *de novo* or modified ones to improve their binding affinity to a target protein.
   a. **MCSS Method**: Using the MCSS approach, 1,000–5,000 copies of functional groups are produced, randomly positioned at the target binding site, quenched molecular dynamics in the protein's force field, and simultaneously subjected to energy minimization. The interaction energies are utilized to identify a set of energetically favorable binding conformations for specific functional groups. By mapping binding sites with various functional groups, new compounds can be created by connecting these groups to align with the mapped site.
   b. **LUDI Method**: It primarily analyses potential hydrogen bonds and hydrophobic interactions between a ligand and protein. The core idea revolves around interaction sites—specific spatial points ideal for establishing hydrogen bonds or occupying hydrophobic pockets. These interaction sites are identified through database searches, such as the Cambridge Structural Database (CSD), or by applying predefined rules. Next, fragments are aligned to these sites and assessed based on distance criteria**.** The final step is the connection of some or all of the fitted fragments to a single molecule.
4. **Stochastic Methods**: It randomly alters the several conformations of the ligand population to search the conformational space.
   a. **Monte Carlo (MC) Methods**: The method produces ligand positions by rotation, rigid-body translation, or bond rotation. The connection of fitted fragments into a single molecule is the final step. The resulting conformation from this process is evaluated using an energy-based selection criterion. If the criterion is satisfied, the conformation is retained and adjusted to generate the next one. This process continues until the desired number of conformations is achieved. One key advantage of the Monte Carlo (MC) method is its ability to introduce substantial changes, enabling the ligand to overcome energy barriers on the potential energy surface—something that molecular dynamics simulations struggle to accomplish. Notable examples of tools utilizing this approach include AutoDock, ICM, QXP, and Affinity.
   b. **Genetic Algorithms (GA)**: Genetic algorithms (GA) belong to a well-established group of stochastic techniques inspired by Darwin's theory of evolution. In GA, the flexibility of a ligand is represented as binary strings, known as genes, which together form a "chromosome" that symbolizes the ligand's conformation. Two primary genetic operations, mutation and crossover, are employed. Mutation introduces random alterations in the genes, while crossover involves swapping genes between two chromosomes. These genetic operations lead to new ligand structures, which are then evaluated using a scoring function. The structures meeting certain criteria are carried forward to the next generation. Notable examples of GA-based tools include AutoDock, GOLD, DIVALI, and DARWIN.
5. **Molecular Dynamics (MD)**: Molecular dynamics (MD) simulations are extensively used in various areas of molecular modeling due to their effectiveness. In the context of docking, MD simulations offer a more accurate representation of the flexibility of ligands and proteins compared to other methods, as they allow the movement of each atom independently within the field of the surrounding atoms. However, one limitation of MD simulations is their gradual progression in small steps, which can make it difficult to overcome high-energy conformational barriers and result in insufficient sampling. Despite this, MD simulations are typically efficient in performing local optimizations. A common approach is to first use a random search to determine the ligand's conformation and then apply MD simulations for further refinement (Meng et al., 2011).

Scoring Functions

The objective of the scoring function is to differentiate between correct and incorrect poses within a reasonable computational timeframe, or to separate binders from non-binding substances. Also, they rely on several assumptions and simplifications when estimating the binding affinities between the protein and ligand, rather than directly calculating them (Beccari, 2015). Scoring functions can be classified into three categories: knowledge-based, empirical, and force-field-based (Meng et al., 2011).

- **Classical Force-Field-based Scoring Functions**: These functions evaluate the binding energies by adding up the non-bonded interactions, such as electrostatics and *van der* Waals. Extensions of scoring functions based on force fields consider the contributions of entropy, solvation, and hydrogen bonds. Examples: DOCK, GOLD and AutoDock.
- **Empirical Scoring Functions**: The binding energies are calculated using hydrophobic effects, ionic interactions, hydrogen bonding, and binding entropies. A final score is calculated by multiplying each element by a coefficient and summing them all up. Regression analysis applied to a test set of ligand–protein complexes with known binding affinities produces coefficients. Examples include LUDI, PLP, and ChemScore.
- **Knowledge-based Scoring Functions**: To assess the contact frequencies and distances between atoms of the ligand and protein, statistical analysis of ligand–protein complex crystal structures is employed. The analysis is based on the principle that more favorable interactions are encountered more frequently. From these frequency distributions, pairwise atom-type potentials are derived. The score is calculated within a defined cut-off by rewarding favorable interactions and penalizing repulsive ones between each atom in the ligand and protein (Beccari, 2015). **Examples include PMF, DrugScore, SMoG, and Bleep.**
- **Consensus Scoring**: A recent approach involves combining multiple scoring methods to evaluate docking conformations. A ligand pose or potential binder is deemed acceptable if it performs well across several scoring systems. Consensus scoring typically enhances enrichments (i.e., the proportion of true binders among high-scoring ligands) during virtual screening and improves the accuracy of predicted bound conformations and poses. However, the prediction of binding energies still requires refinement. For instance, CScore is an example that integrates the scoring functions of **DOCK, ChemScore, PMF, GOLD, and FlexX**.
- **MM-PB/SA and MM-GB/SA**: Conventional scoring methods often face challenges in predicting affinity, partly due to the inadequate consideration of solvation effects. One approach to address this is through physics-based scoring methods like MM-PB/SA and MM-GB/SA, where MM refers to molecular mechanics, PB and GB denote Poisson–Boltzmann and Generalized Born models, and SA stands for solvent-accessible surface area. A key step in refining binding affinity predictions is lead optimization or rescoring. Trials using MM-PB/SA or MM-GB/SA have yielded promising results in certain cases.

Types of Molecular Docking

- **Rigid Ligand and Rigid Receptor Docking**: The search space is limited to three translational and rotational degrees of freedom when the ligand and receptor are considered rigid. Ligand flexibility can be accounted for by using a pre-calculated set of conformations or by permitting some level of atom–atom overlap between the protein and ligand. For example, this approach was used in earlier versions of **DOCK, FLOG**, and the protein–protein docking program **FTDOCK.**
- **Flexible Ligand and Rigid Receptor Docking**: In systems that adhere to the induced fit model, it is crucial to account for the flexibility of both the ligand and the receptor, as both components undergo conformational changes to create an optimal energy fit. However, allowing the receptor to be flexible can significantly increase computational costs. Consequently, the conventional approach strikes a balance between accuracy and computational efficiency by treating the ligand as flexible while maintaining the receptor in a rigid state during docking. This method is the default in most docking software, including **AutoDock and FlexX**.
- **Flexible Ligand and Flexible Receptor Docking:** The inherent movement of proteins has been shown to be strongly linked to their interaction with ligands. Accounting for receptor flexibility remains a major challenge in docking studies. While molecular dynamics (MD) simulations could theoretically capture all the movements in a ligand–receptor complex, they suffer from insufficient sampling. Additionally, the high computational cost of MD makes it impractical for screening large chemical databases. Example:
  - **Ensemble Flexible Docking**: DOCK, FLexE
  - **Hybrid Flexible Docking**: IFREDA, QXP, Affinity, SLIDE
- **Local Move Monte Carlo (LMMC) Sampling for Flexible Receptor Docking**: Local move, also known as "window move," involves adjusting the driver torsion, which is the first of six torsions that must be adjusted in order for the remainder of the chain to stay in its original position while maintaining all bond lengths and bond angles (Meng et al., 2011).

Computer-Aided Molecular Design (CAMD) or Molecular Modeling

Molecular modeling, or CAMD, is an emerging technology that utilizes the knowledge of steric and electronic aspects of the receptor/ligand and enzyme/substrate interaction to identify pharmacophores. Some of the CAMD tools are as follows:

- **ChemDoodle**: It is a chemical sketcher with many features for working with chemical graphics. It can be applied to molecular mass calculations, elemental analysis, database searches for chemical structures, and other tasks. Additionally, it features a visualization tool named ChemDoodle 3D, which transforms 2D chemical structures into personalized 3D models (https://www.chemdoodle.com/).
- **Hypercube**: It is a widely used molecular modeling tool among chemists, known for its extensive capabilities and intuitive user interface. It offers a range of features for 3D chemical design, including protein simulations, molecular modeling, visualization, chemical structure computations, and various bioinformatics tasks (http://www.hyper.com/).
- **Avogadro**: The software allows for the import of chemical files and the generation of various computational chemistry packages. It is widely used across scientific disciplines, including computational chemistry, molecular modeling, bioinformatics, and materials science (https://avogadro.cc/).
- **ChemAxon**: It is a toolkit for representing chemical structures that can be highly beneficial for students, researchers, academicians, or chemical engineers working in the fields of chemistry or biology (https://www.chemaxon.com/products/chemical-structure-representation-toolkit).
- **BIOVIA Draw**: This molecular drawing software is primarily designed for chemists and professionals with expertise in the chemical industry. It provides researchers with a comprehensive set of tools to create and edit intricate molecules, chemical reactions, and biological sequences (https://www.3ds.com/products/biovia/draw).
- **MolView**: It is a chemistry modeling tool that offers a distinct interface compared to the other software previously mentioned. Its key distinguishing feature is that it is a free, online web application. Designed for simplicity, MolView is accessible even for those without prior experience in 3D modeling software. In addition to generating structural formulas, it also allows for protein visualization, modeling, and the simulation of protein assembly and chain representation (http://molview.org/).
- **ACD/ChemSketch**: A chemical drawing software available for free download, with an option to purchase a commercial license. However, the free version does not offer all the features available in the paid version (http://www.acdlabs.com/resources/freeware/chemsketch/).
- **VMD (Visual Molecular Dynamics)**: It is a software tool for visualizing, animating, and analyzing large biomolecular systems through 3D graphics and integrated scripting capabilities. It is supported by both Unix and Windows platforms (http://www.ks.uiuc.edu/Research/vmd/).
- **SwissPdbViewer**: A tool that offers an easy-to-use interface for analyzing multiple proteins at once. It enables the superimposition of proteins to determine structural alignments and compare active sites or other key regions. The intuitive graphical and menu-based interface simplifies the extraction of information such as amino acid mutations, hydrogen bonds, angles, and atomic distances (http://www.expasy.ch/spdbv/mainpage.htm).
- **Spartan**: It is used for Molecular Mechanics and QSAR modeling studies as part of molecular modeling analysis (http://www.wavefun.com/).
- **SCIGRESS**: Developed with experimental chemists in mind, SCIGRESS is a multiplatform software suite for molecular design, modeling, and dynamics. The researcher can quickly construct new structures using molecular builders and visualization tools. Without the burden of learning computational chemistry jargon, calculations of various chemical and physical characteristics and reaction modeling on many theory levels are achievable. In the fields of organic, pharmaceutical, and biochemistry, simulations' speed and ease of use enable on-the-spot testing of research concepts before moving on to a wet lab, saving time, money, and reagents lost on unsuccessful experiments (https://www.fqs.pl/en/chemistry/products/scigress).
    - **The CAChe Family of Software**: CAChe, Quantum CAChe, CAChe Worksystem, CAChe GroupServer, and CAChe Satellite (for Chemists). The Oxford Molecular Group (Parsippany, New Jersey (NJ), USA) was the developer of Quantum CAChe and Project Leader (earlier Scigress Explorer, now Scigress).
- **Other commonly used molecular modeling software**
    - Please explore more molecular modeling and simulation software list at RCSB PDB—Modeling and Simulation software website: https://www.rcsb.org/docs/additional-resources/modeling-and-simulation-software
    - Also, explore comparative features of different molecular modeling software at the Wikipedia molecular mechanics modeling website: https://en.wikipedia.org/wiki/Comparison_of_software_for_molecular_mechanics_modeling

De novo Drug Designing Method

De novo drug design is an approach to build a customized ligand for a given receptor. This approach involves the ligand optimization. Ligand optimization can be done by analyzing protein active site properties that could be a probable area of contact by the ligand. The analyzed active site properties are described as the negative images of proteins such as hydrogen bond donor, hydrogen bond acceptor, and a hydrophobic contact region.

There are two types of de novo design: (i) *inside-out* and (ii) *outside-in*.

In each of the methods, you need to follow these steps:

- Identify points on the receptor where a ligand can form favorable interactions.
- Choose ligand groups of atoms that are complementary to these sites.
- Connect them into a molecular framework in an exhaustive, unbiased way.

LIGAND DESIGN

The development of computer programs for ligand designing has been trendy. Researchers develop several programs. The basic approaches for ligand design methods are as follows:

Problem	Ligand Design Method	Tools
Ligands with conformational and structural variations	atom-by-atom ligand design	LEGEND, CONCEPTS, GROWMOL
Generate ligands that are chemically or synthetically plausible	fragment-addition ligand design	GROW, LUDI, SPROUT, MCSS
Ligands for ring structures or no receptor structure available	pharmacophore-pattern ligand design	CAVEAT, LINKOR, RCEPS

PREDICTIVE PHARMACOKINETICS AND TOXICOLOGY (ADMET)

Drug discovery needs the study of what the drug does on the body (pharmacodynamics), curing the disease without causing harmful side effects, and of what the body does on the drug (pharmacokinetics), namely ADMET (absorption, distribution, metabolism, excretion, and toxicity) of the drug. In medication design and development, the rule of three about activity-exposure toxicity presents the most significant challenge. In order to maximize the balance of qualities required to transform lead compounds into safe and effective medications for human patients, ADMET features are frequently employed in drug discovery. The importance of predicting the ADMET properties of compounds quite early in the drug discovery process is unquestionable. Pharmaceutical companies are under severe strain to declare the safety of their products devoid of side effects; failing to do so, the products have to be withdrawn from the market.

Step 1	Target identification and verification	Bioinformatics: Genomics, Proteomics, Transcriptomics, Metabolomics, etc.
Step 2	Hit identification	Compound libraries, Ultra HTS, Hit confirmation, *in silico* ADMET (empirical models)
Step 3	Lead identification	Combinatorial chemistry, Hit confirmation, Affinity/selectivity assays, ADMET tools, *in vitro*, and *in silico*
Step 4	Lead optimization	Structure-based medicinal chemistry, Disease models, Affinity/selectivity assays, *in-vitro*, *in-vivo*, and *in-silico* ADMET

IN SILICO ADMET STUDIES

Get an early assessment of compounds by calculating the predicted ADMET properties for collections of molecules such as synthesis candidates, vendor libraries, and screening collections. The elimination of the compounds with unfavorable ADMET properties and evaluating proposed structural refinements designed to improve these properties before synthesis (BIOVIA, Dassault Systems). Experimental ADMET tools may need to be more adequate as they only evaluate a fraction of compounds. Hence, in-*silico* (computational) models are devised to make predictions based on chemical structures to meet industry demands. *In-silico* models are built by calculating the compound descriptors to fit the model to the observed data. The computation need depends on the model's complexity, which is determined by the particular modeled process. The ADMET models can further be classified into two categories:

Empirical Models

Empirical models use statistical tools to explore linear or non-linear relationships between a particular ADMET property and structural descriptors. These models are called QSPR models using a regression method. The BIOVIA Discovery Studio 2023 software includes a QSAR module, which can be used for the following:

- Comprehensive and consistent data preparation:
  - Ligands: Remove duplicates and handle tautomers and ionization
  - Prepare response property (scaling and binning)

  - Split data into training and test sets with the appropriate methods
- Select from various chemical descriptors based on physicochemical, topological, electronic, geometric, fingerprint, and quantum mechanics.
- Develop statistical models, such as genetic functional analysis (GFA), partial least squares (PLS), multiple linear regression (MLR), and Bayesian.
- Use model applicability domains (MAD), cross-validation, automatic test set validation, and statistical metrics to analyze and evaluate QSAR/QSPR/QSTR models.
- Determine the changes of Matched Molecular Pairs (MMPs) and research activity cliffs.

Mechanistic Models

Mechanistic models are more complex and utilize quantum mechanics to calculate atomic interactions between small (ligand) and macro (receptor) molecules. Several enzymes or transporter proteins are involved in assessing ADMET properties. The mechanistic model requires the determination of 3D structures of small molecules (ligands) and macromolecules (receptor protein) to evaluate interactions. According to established SMARTS guidelines, ADMET descriptors include hepatotoxicity, blood–brain barrier penetration, aqueous solubility, CYP2D6 binding, plasma protein binding, and filter sets of small molecules (ligands) for undesired function groups (BIOVIA Discovery Studio, 2023—QSAR, ADMET & Toxicity modules) (https://www.3ds.com/products-services/biovia/products/molecular-modeling-simulation/biovia-discovery-studio/qsar-admet-and-predictive-toxicology/)

LIPINSKI'S RULE OF FIVE

A study by **Chris Lipinski** and co-workers at **Pfizer** involved a statistical analysis of ~2,200 drugs from the World Drug Index (WDI). They established a set of heuristics known as **Lipinski's Rule of Five** for oral bioavailability. This rule states that the absorption or permeation of a drug (that is not a substrate for a biological transporter) is likely to be impaired when:

- logP (an indicator of hydrophobicity) > 5
- Molecular weight > 500
- Number of hydrogen donor groups > 5
- Number of hydrogen acceptor groups > 10

EXAMPLE: BIOVIA DISCOVERY STUDIO SOFTWARE (FORMERLY ACCELRYS, USA)

The Discovery Studio (DS) 4.0 is a software suite for simulating small molecules and macromolecule systems. It can be used for Ligand design, Pharmacophore modeling, Structure-based design, Macromolecule design and validation, Macromolecule engineering, QSAR, ADME, predictive toxicity, and simulations. The simulation module includes molecular mechanics, molecular dynamics, and quantum mechanics to perform hybrid QM/MM calculations. DS includes implicit and explicit solvent and membrane models for MM-based simulations. It is developed and distributed by Dassault Systèmes BIOVIA, USA (formerly Accelrys) (https://www.3ds.com/products-services/biovia/products/molecular-modeling-simulation/biovia-discovery-studio/). DS software includes CHARMM, MODELLER, DELPHI, ZDOCK, DMol3, and more programs, including QSAR, ADMET, and Toxicity modules. The ADMET module can be used as a component module of Cerius2 or as an add-on to DS (https://discover.3ds.com/discovery-studio-visualizer-download). The DS ADMET module provides ADME models that enable researchers to compute and predict ADMET properties for chemical synthesis candidates and commercially available structures. Researchers can pre-screen small and virtual libraries with the DS ADMET module to predict pharmacokinetic and toxicity risk assessment properties in advance. Macromolecule engineering modules include tools for protein–protein docking, antibody design, and optimization, as well as tools for membrane-bound proteins, including GPCRs. QSAR modules include methods, e.g., multiple linear regression, partial least-squares, recursive partitioning, genetic function approximation, and 3D field-based QSAR. Besides, DS 4.0 visualizer software is free to use and has commonly used features. Other widely used software are Schrodinger molecular modeling suite, MOE, Cresset Forge suite, VLife MDS QSARPlus, GQSAR & Docking suite, Sybyl-X, Scigress, Gold (CCDC, UK), ChemBioOffice suite (CambridgeSoft, UK), LeadIT Flex suite, SeeSAR, etc.

PREDICTIVE TOXICOLOGY

Evaluate compounds' performance in experimental assays and animal models. Compute and validate assessments of chemicals' toxic and environmental effects solely from their molecular structure. For example, BIOVIA Discovery Studio TOPKAT® (TOxicity Prediction by Komputer Assisted Technology) module, which employs robust and cross-validated Quantitative Structure–Toxicity Relationship (QSTR) models for assessing various endpoints and utilizing the patented Optimal Predictive Space validation method to assist in interpreting the results. Details on extensible DS TopKat models are published in QSAR Model Report Format (QMRF) on the European Commission Joint Research Center (JRC)'s "QSAR Model Database" 2020 (https://data.jrc.ec.europa.eu/dataset/e4ef8d13-d743-4524-a6eb-80e18b58cba4). These models are related to Ames

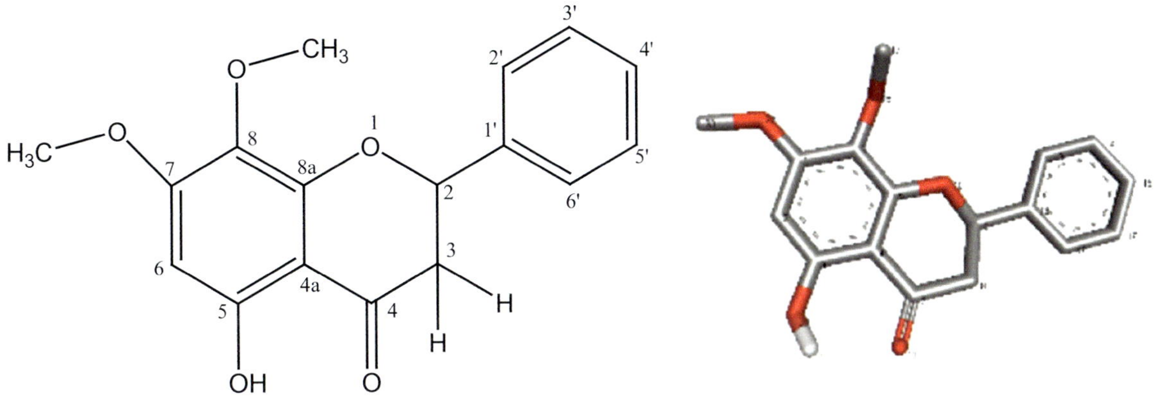

FIGURE 7.1 Chemical structure of ligand (e.g., Kaemferol) compound (2D & 3D).

mutagenicity, Rodent carcinogenicity (NTP and FDA data), Weight of evidence carcinogenicity, Carcinogenic potency TD50, Developmental toxicity potential, Rat oral LD50, Rat maximum tolerated dose, Rat inhalation toxicity LC50, Rat chronic LOAEL, Skin irritancy and sensitization, Eye irritancy, Aerobic biodegradability, Fathead minnow LC50, and Daphnia magna EC50.

EXAMPLE CASE STUDY (MOLECULAR DOCKING)

The docking between compound Kaempferol (Mol. formula: C17H16O5; Mol. mass: 300.31) (Figure 7.1) and Triacylglycerol acyl-hydrolase (synonym: Triacylglycerol lipase) (EC: 3.1.1.3) (PDB: 1ETH) was performed using the AutoDock Vina molecular docking software (http://vina.scripps.edu) through PyRx program, a virtual screening software for computational drug discovery. AutoDock Vina uses a sophisticated gradient optimization method in its local optimization procedure. Utilizing solely movable heavy atoms, the Root Mean Square Deviation (RMSD) values are computed about the optimal mode. Vina uses a united-atom scoring function. The structure of the ligand was identified and elucidated in the Analytical Chemistry Division, CSIR-CIMAP, Lucknow (www.cimap.res.in), and the crystal structure of Triacylglycerol acyl-hydrolase (PDB ID:1ETH) was retrieved from the RCSB-PDB protein database (www.rcsb.org). Hydrogen atoms and polarity were added to 1ETH. The predicted binding affinity was calculated in kcal/mol. The pancreatic triacylglycerol lipase (Gene name: PNLIP) of Sus scrofa (Pig) (PDB ID: 1ETH; UniProtKB: P00591) possesses two domains. Its N-terminal domain is an α/β hydrolase fold, which contains the catalytic sites of Ser153, Asp177, and His264. The docking calculation of the established PNLIP crystal structure and ligand was performed according to the iterated Local Search global optimizer and genetic algorithm (Trott & Olson, 2010). The docking complex with the lowest binding energy was used as the best conformational model for further receptor–ligand interaction analysis.

RESULTS OF BINDING AFFINITY DETECTED THROUGH MOLECULAR DOCKING & INTERPRETATION

Molecular docking analyzes the interaction mechanism between the receptor and ligand by directly observing the binding conformation and binding site of the receptor and ligand. Figure 7.2 showed that the ligand was located in the hydrophobic cavity (active site) of the Pancreatic Lipase (PNLIP), and the lowest binding energy on the ligand–receptor docking site was −9.4 kcal/mol. The central residues binding to ligand were Arg368, Gln369, Tyr370, Glu371, Leu41, Lys42, Glu64, Arg65, and Tyr404 (Figure 7.3). Moreover, three hydrogen bonds were found in the combination, the C5–OH of ligand formed a hydrogen bond with the hydrogen atom of Lys42, the second hydrogen bond existed between the hydrogen atoms of C7–OH and Tyr404, and the third was between the carbonyl oxygen atom of the Carbon ring and the hydrogen atom of Lys42; the distances were 4.98 Å, 2.85 Å, and 2.20 Å, respectively. It could also be seen from Figure 7.3 that other interactions (pi–pi T-shaped, pi-alkyl, pi-donor hydrogen bond, Van der Waals, et al.) also contributed to the formation of ligand–PNLIP complex; for example, ligand formed a pi–pi T-shaped interaction with amino acid residues Tyr370, the distances were 6.71 Å, and ligand formed pi–alkyl interactions with amino acid residues Lys42, Leu41, Arg368, with bond lengths of 5.06, 5.91, and 6.10 Å, respectively. However, according to the relationship between bond length and energy, hydrogen bonds contributed most to forming the ligand–PNLIP complex. These results demonstrated that hydrogen bonding was essential in the interaction between ligand and PNLIP. The production of hydrogen bonds increased the hydrophobicity of PNLIP and made the complex more stable; the ligand competed for the active site of PNLIP and blocked the substrate's channel into PNLIP, inhibiting PNLIP activity.

Inference: Hydrophobic forces and hydrogen bonding mainly drove the binding of ligands to PNLIP. Molecular docking simulation showed the binding mode of the ligand with PNLIP (PDB: 1ETH); this demonstrated that the ligand

S. No.	Docked Ligand Conformations (Binding conformation)	Binding Affinity (kcal/mol)	(Upper Bound)	RMSD (Lower Bound)
1	1eth_chainA_Orlistat (+ve Control/drug)	−9.4	0	0
2	1eth_chainA_ligand (Best fit conformation)	−9.4	0	0
3	1eth_chainA_ligand	−8.8	6.331	2.073
4	1eth_chainA_ligand	−8.4	6.169	1.55
5	1eth_chainA_ligand	−8.2	2.163	1.58
6	1eth_chainA_ligand	−7.9	6.551	2.6
7	1eth_chainA_ligand	−7.7	13.298	11.57
8	1eth_chainA_ligand	−7.4	4.454	1.92
9	1eth_chainA_ligand	−7.1	3.234	1.966
10	1eth_chainA_ligand	−6.7	8.119	4.596

was attached to the target enzyme's active site, preventing the substrate from binding to PNLIP, thereby suppressing the catalytic activity of PNLIP. This study may provide a theoretical basis for further studies of ligands as a natural anti-obesity supplement.

CONCLUSION

Computer-aided drug design is a practical and realistic way of helping medicinal chemists and drug discovery researchers. The QSAR and molecular docking approaches, including MD simulation studies, are essential in early drug design and discovery programs. QSAR modeling uses statistical and machine learning methods to predict active hits and is also used in lead optimization. A molecular docking study is essential in exploring the binding affinity of small molecules and their mechanism of action. Receptor flexibility, especially backbone flexibility, and movement

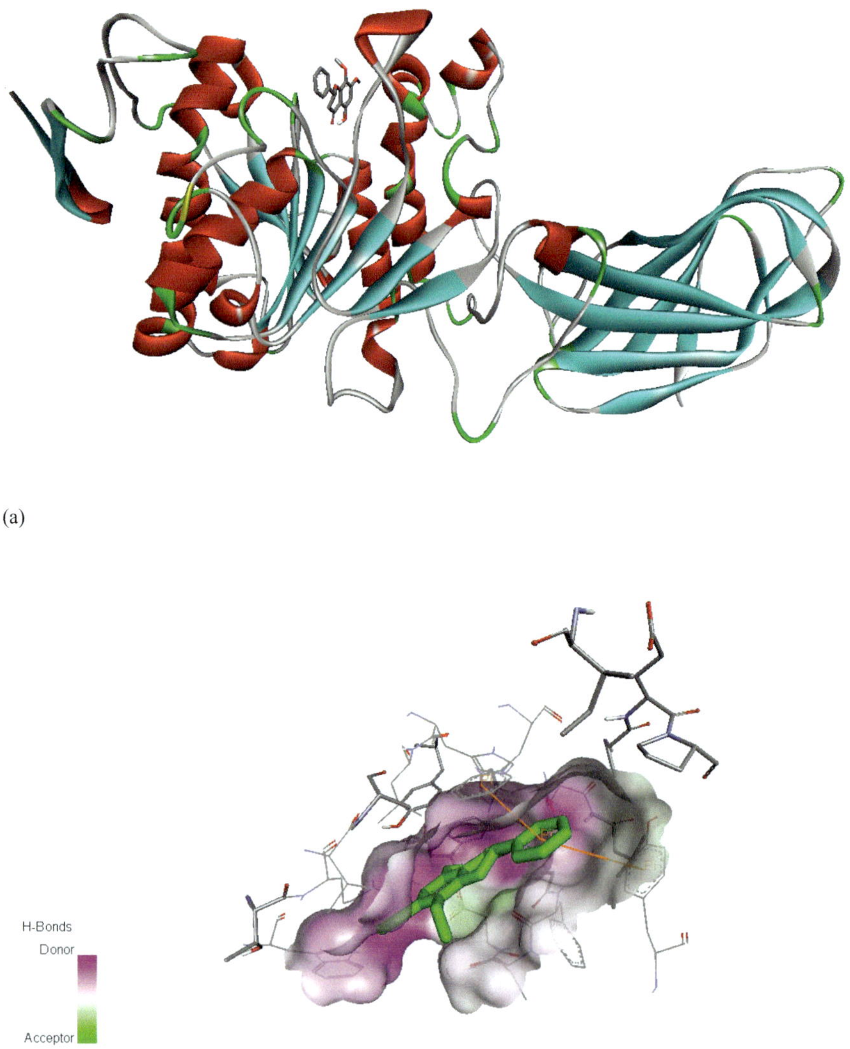

(a)

(b)

FIGURE 7.2 Representative docking result of PNLIP and the possible binding site of the ligand on PNLIP. (a) Ribbon form and (b) Visualization of the binding pocket and bound ligand.

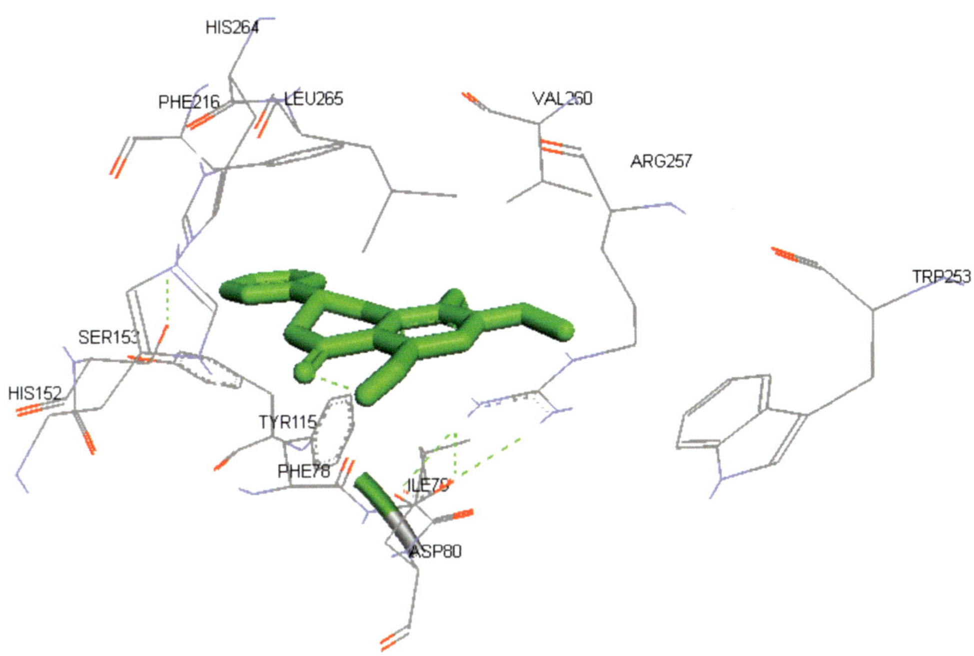

(a)

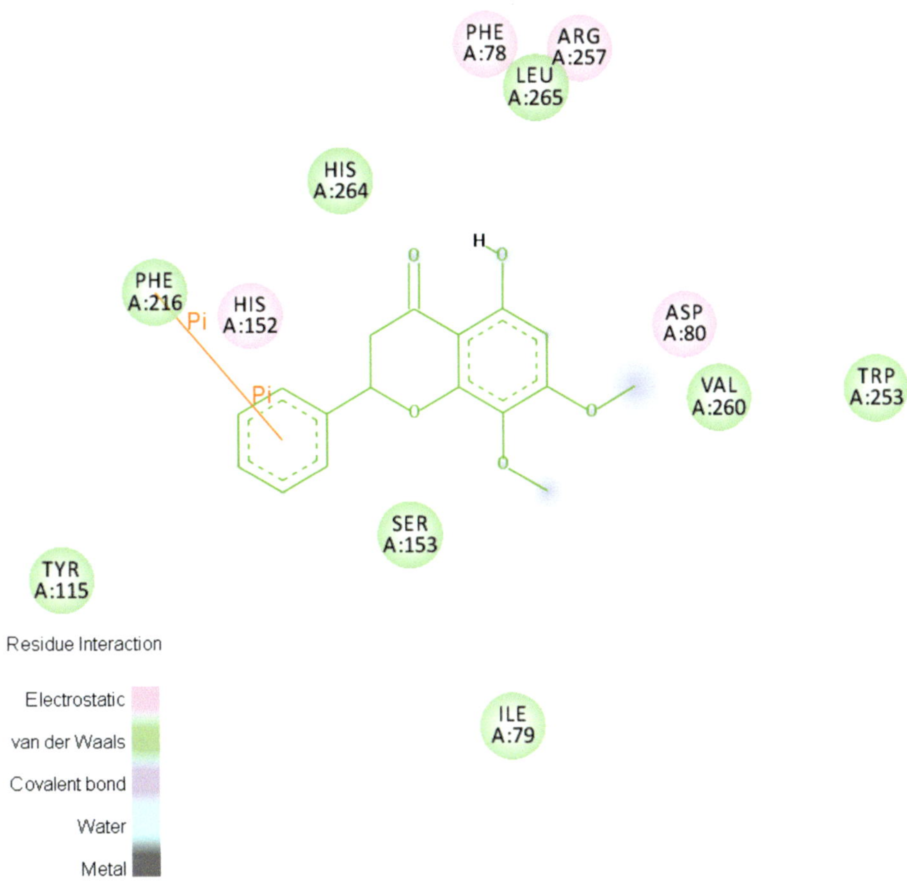

(b)

FIGURE 7.3 (a) Representative docking result of PNLIP and the best fit binding site of the ligand in PNLIP. (b) Corresponding amino acid residues of PNLIP interacted with ligands.

of several vital secondary elements of receptors involving ligand binding are still significant problems. Docking scoring is also an important criterion that still requires more improvement. Predicted leads may be checked through Lipinski's rule of five for oral bioavailability and ADMET compliance. Drugs must be synthesized and tested by computational techniques that provide a clear molecular rationale and a shot to the imagination.

BIBLIOGRAPHY

Beccari, A.R. Development and validation of in silico tools for efficient library design and data analysis in high throughput screening campaigns. Ph.D. thesis, (2015). https://core.ac.uk/download/pdf/41182622.pdf

Good, A.C., S.S. So, W.G. Richards. Structure-activity relationships from molecular similarity matrices. *Journal of Medicinal Chemistry*, 36 (1993) 433.

Haworth, I.S., A. Rodger, W.G. Richards. A molecular dynamics simulation of a polyamhe-induced conformational change of DNA. A possible mechanism for the B to Z transition. *Journal of Biomolecular Structure and Dynamics*, 10 (1992) 195.

Haworth, I.S., A. Rodger, W.G. Richards. A molecular mechanics study of spermine complexation to DNA: A new model for spermine-poly(dG-dC) binding. *Proceedings of the National Academy of Sciences, India Section B: Biological Sciences*, 244 (1991a) 107.

Haworth, L.S., A.H. Elcock, A. Rodger, W.G. Richards. A binding mode of A-[tris(l,lO-phenanthrohe)rutheni~m(II)]2^+ exhibiting preference for purine-3′,5′-pyrimidine sites of DNA. *Journal of Biomolecular Structure and Dynamics*, 9 (1991b) 553–569.

Haworth, L.S., A.H. Elcock, A. Rodger, W.G. Richards. Sequence selective binding to the DNA major groove: tris (1, lO phenanhline) metal complexes binding to poly(dGdC) and poly (dAdT). *Journal of Biomolecular Structure and Dynamics*, 9 (1991c) 23.

Haworth, L.S., C. Burt, F. Gago, C.A. Reynolds, W.G. Richards. A prototype bioreductive DNA groove binding ligand. *Anticancer Drug Design,* 6 (1991d) 59.

King, P.M., C.A. Reynolds, J.W. Andex, G.A. Worth, W.G. Richards. Free energy calculations of pharmaceutically essential properties. *Molecular Simulation*, 5 (1990) 265.

Li, S., J. Pan, X. Hu, Y. Zhang, D. Gong, G. Zhang. Kaemferol inhibits the activity of pancreatic lipase and its synergistic effect with orlistat. *Journal of Functional Foods*, 72 (2020), 104041.

Lister, S.G., C.A. Reynolds, W.G. Richards. Theoretical calculation of electrode potentials: Electron withdrawing compounds. *International Journal of Quantum Chemistry*, 41 (1992) 293.

Manikpuri, A.D., S. Joshi, P.V. Khadikar. Synthetic, spectral, antimicrobial and QSAR studies on novel mannich bases of glutarimides. *Journal of the Chilean Chemical Society*, 55 (2010) 283–292.

Meng, X.Y., H.X. Zhang, M. Mezei, M. Cui. Molecular docking: A powerful approach for structure-based drug discovery. *Current Computer-Aided Drug Design*, 7(2) (2011) 146–157.

Menziani, M.C., M. Cocchi, P.G. De Benedetti, R.G. Gilbert, W.G. Richards, M. Zamai, V.R. Caiolfa. A theoretical study of the structure of big endothelin. *Journal of Chemical Physics*, 88 (1991) 2687.

Meyer, A.Y., W.G. Richards. Similarity of molecular shape. *Journal of Computer-Aided Molecular Design*, 5 (1991) 427.

Mulholland, A.J., G.H. Grant, W.G. Richards. Computer modelling of enzyme catalysed reaction mechanisms. *Protein Engineering*, 6 (1993) 133.

Reynolds, C.A., P.M. King, W.G. Richards. Free energy calculations in molecular biophysics. *Molecular Physics*, 76 (1992) 251.

Richards, W.G., C.A. Reynolds. Bioreductive anticancer drugs. In D.L. Beveridge, R. Lavery (Eds.). *Theoretical Biochemistry and Molecular Biophysics*, Adenine Press, Schenectady (1990).

Trott, O., A.J. Olson. AutoDock Vina: Improving the speed and accuracy of docking with a new scoring function, efficient optimization and multithreading. *Journal of Computational Chemistry*, 31 (2010) 455–461.

8 Pharmacophore Modeling
A Cornerstone of Modern Drug Design

Asif Husain[1], *Neelima Shrivastava*[2], *and Shah Alam Khan*[3]
[1]Department of Pharmaceutical Chemistry, School of Pharmaceutical Education and Research, Jamia Hamdard, New Delhi, India
[2]Department of Pharmaceutical Chemistry, School of Pharmacy, JSS University, Noida, India
[3]Department of Pharmaceutical Chemistry, College of Pharmacy, National University of Science and Technology, Muscat, Sultanate of Oman

INTRODUCTION

The concept of the *pharmacophore* is a fundamental principle for drug design in medicinal chemistry. It was first introduced by Paul Ehrlich in the early 1900s. He suggested that certain structural features of molecules are responsible for their biological activity. In 1960, Schueler further refined the idea by proposing that a pharmacophore represents the spatial arrangement of certain chemical features that are essential for therapeutic effects. Later, between 1967 and 1971, Lemont B. Kier contributed significantly to developing and representing the pharmacophore concept, bringing deeper insight into its theoretical basis and practical applications in drug design [1].

The pharmacophore involves the study of the three-dimensional (3D) arrangement of molecular features—such as hydrogen bond donors (HBD) or acceptors (HBA), hydrophobic regions, aromatic rings, and charged groups—which are essential for interactions with biological targets. It serves as a foundational concept for understanding receptor–ligand interactions. Today, pharmacophores have become an indispensable tool in the early stages of drug discovery in pharmaceutical industries, particularly in virtual screening, novel drug design, and lead optimization processes [2].

The development of pharmacophore-based computational tools evolved in parallel with advances in structural databases and molecular modeling. One of the earliest software programs, MOLPAT, was introduced in 1974 [3]. However, due to the absence of comprehensive 3D structural databases, which are crucial for effective pharmacophore modeling and 3D searches, significant progress could not be made. During the 1970s, the Cambridge Structural Database (CSD) emerged as one of the earliest sources of 3D molecular data paving the way for further development [4].

However, the real momentum came between 1987 and 1989, with the advent of rapid 3D structure generation tools such as CONCORD, CORINA, AIMB, and WIZARD. These tools allowed for the efficient creation of 3D molecular models, which fueled the demand for advanced 3D searching software [5–7]. This period marked a turning point in computational chemistry and molecular modeling. As 3D databases and modeling tools gained popularity, pharmaceutical research institutions began to develop specialized software platforms tailored for pharmacophore-based screening. Notable examples include CAST-3D, developed by Chemical Abstracts Service; DOCK, created at the University of California, San Francisco; and CAVEAT, developed at the University of California, Berkeley. These tools enabled more precise ligand–receptor modeling and structure-based drug design, making pharmacophore modeling a vital asset in the pharmaceutical industry's computational toolkit [8–10].

WHAT IS PHARMACOPHORE?

A *pharmacophore* refers to the essential 3D arrangement of structural features within a molecule that is required to produce a specific physiological action. These features may include HBD orHBA, aromatic rings, hydrophobic centers, positive or negative ionizable groups, and other functional groups of a ligand that interact with a macromolecular biological target, such as a receptor or enzyme.

According to the modern definition, a pharmacophore is "the ensemble of steric and electronic features that is necessary to ensure optimal supramolecular interactions with a specific biological target and to trigger (or block) its biological response." This concept plays a vital role in rational drug design by helping researchers identify and optimize compounds that can effectively bind to the desired target with high specificity and activity [11,12]. Pharmacophoric features are the key molecular characteristics that enable a compound to interact effectively with a biological target. These typically include HBD and HBA, sites for electronic interactions (such as charged or polar groups), and hydrophobic interaction sites. These features play a central role in defining the molecule's ability to bind to its receptor and initiate or block a biological response [13,14].

DOI: 10.1201/9781003464679-8

The identification of pharmacophores is usually done with the help of structural biology techniques, particularly X-ray crystallography. High-resolution X-ray crystal structures of protein–ligand complexes, obtained from the protein data bank (PDB), allow researchers to visualize how ligands interact within the binding site of their target proteins. This structural insight helps in the accurate mapping and modeling of pharmacophoric features.

Interestingly, the term "pharmacophore" itself is derived from two Greek words; *Phoros* meaning "to bear," referring to molecular features that carry or express certain properties, and *Pharmacon* meaning "drug" or "medicine," referring to the properties responsible for biological activity [1]. The International Union of Pure and Applied Chemistry (IUPAC) defines a pharmacophore as: "An ensemble of steric and electronic features that is necessary to ensure the optimal supramolecular interactions with a specific biological target and to trigger or block its biological response" [15]. This modern definition highlights the importance of pharmacophores in the design of therapeutic molecules with desired biological effects.

PHARMACOPHORE MODEL OR QUERY

A pharmacophore model (also referred to as a pharmacophore query) represents a set of chemical features arranged in a specific 3D pattern that is essential for biological activity. Each feature—such as HBD, HBA, hydrophobic region, or aromatic ring—is typically represented as a sphere with a defined radius. This radius reflects the degree of permissible spatial variation from the ideal position, allowing for some flexibility during molecular matching [16]. These defined pharmacophoric features serve as a structural framework to interrogate compound libraries, especially during virtual screening, by identifying molecules that match the key interaction patterns [17,18]. Pharmacophore models can be constructed manually by an expert or automatically using computational algorithms [19]. Several software platforms support pharmacophore modeling and screening, including MOE (Molecular Operating Environment), Discovery Studio, Phase, and LigandScout [14,20,21].

ROLE OF PHARMACOPHORE MODELING IN VIRTUAL SCREENING

Pharmacophore modeling plays a main role in virtual screening, a computational technique used to identify potential drug candidates from large compound libraries (Figure 8.1). In this process, the pharmacophore model encodes the ideal 3D spatial arrangement of interaction features required for a molecule to exhibit a desired therapeutic effect. The screening algorithm then searches for compounds that conform to this model, helping prioritize molecules for synthesis and biological testing [22].

The feasibility and strategy of pharmacophore modeling depend on the availability of structural information:

- If neither the protein target nor any active ligand structures are known, pharmacophore modeling cannot be applied [23,24].
- If the protein structure is unknown but the structures of active ligands are available, ligand-based pharmacophore modeling is used. This approach extracts common features from known active molecules to define the pharmacophore [25,26].
- If the protein structure is available but no ligand structures are known, a structure-based or putative pharmacophore model can be generated using the binding site information to predict likely interaction points [27].

PHARMACOPHORE FINGERPRINTING: A RAPID SCREENING APPROACH

Pharmacophore fingerprints represent an important advancement in the field of pharmacophore modeling. This technique enables the rapid screening of large chemical libraries by converting pharmacophoric features into simplified binary or numerical representations. These fingerprints capture the spatial and functional characteristics of a molecule's pharmacophore, allowing for high-throughput comparisons and similarity assessments [28].

One notable method, PharmPrint, is a fast and efficient pharmacophore fingerprinting technique designed specifically to support virtual screening workflows [23]. The generated fingerprints can also serve as molecular descriptors in QSAR modeling. By applying statistical tools like partial least squares (PLS) regression, researchers can develop predictive models that relate molecular features to biological activity.

Another innovative application of pharmacophore fingerprinting is found in KRIPO (Key Representation of Interaction in POckets). KRIPO enables the comparison of binding site sub-pockets by encoding their pharmacophoric interaction patterns. This allows for the quantification of similarities between protein binding sites, facilitating the identification of potential cross-reactivity or the design of selective ligands [29].

APPLICATIONS OF PHARMACOPHORE MODELING

Pharmacophore modeling offers a wide range of applications across different stages of the modern drug discovery and development process (Figure 8.2) [30–32]. Some of the important applications of pharmacophore modeling are discussed below:

i. **Ligand–Receptor Interaction Analysis**: Pharmacophore models help in the identification and visualization of key interactions between a

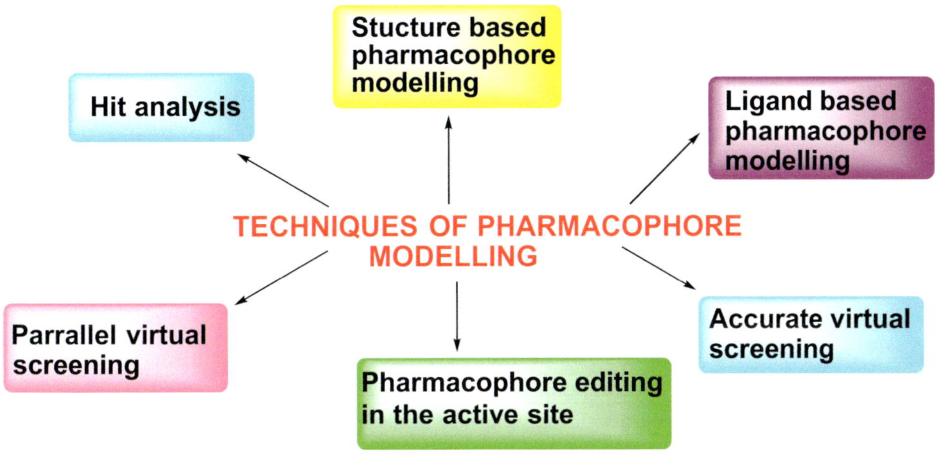

FIGURE 8.1 Different techniques used for pharmacophore modeling.

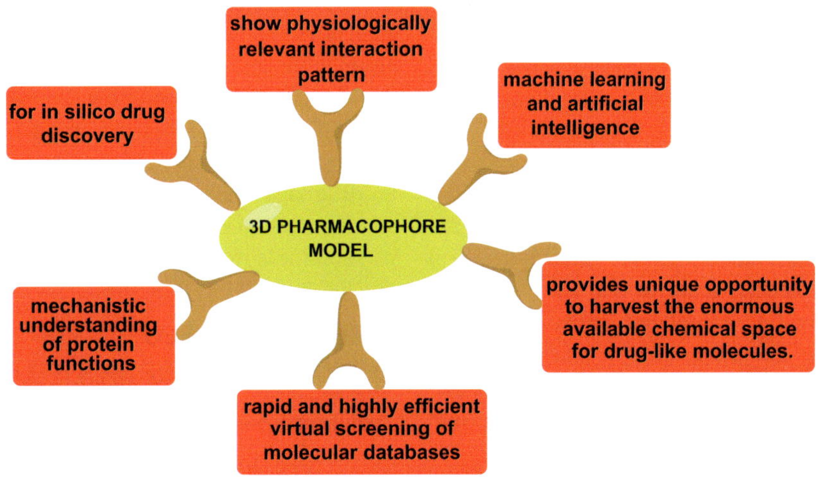

FIGURE 8.2 Various functions of 3D pharmacophore model [15,54].

ligand and its biological target. These interactions are then used to guide the identification and optimization of lead compounds.

ii. **Computer-Aided Drug Design (CADD)**: Pharmacophore-based approaches are integral to CADD, enabling virtual screening, hit-to-lead optimization, and structure–activity relationship (SAR) analysis.
iii. **Target Identification**: By comparing pharmacophoric features across known ligands and targets, researchers can predict possible targets for new or existing compounds.
iv. **ADME-Tox Prediction**: Pharmacophore models are employed to predict absorption, distribution, metabolism, excretion, and toxicity (ADME-Tox) profiles of drug candidates, improving early-stage decision-making.
v. **Off-Target Prediction**: By analyzing pharmacophoric similarities across different proteins, this approach helps predict potential off-target interactions, which is crucial for assessing safety and reducing side effects.
vi. **Integration with Machine Learning (ML)**: Emerging methodologies now combine pharmacophore mapping with ML algorithms, allowing for smarter, data-driven predictions and improved accuracy in virtual screening and activity prediction.
vii. **Pharmacophore Modeling with Molecular Dynamics (MD)**: When integrated with MD simulations, pharmacophore modeling helps capture the dynamic behavior of ligand–protein interactions, leading to more accurate drug design strategies.
viii. **Lead Optimization and Functional Group Identification**: Upon discovering a compound with promising biological activity, medicinal chemists use pharmacophore modeling to pinpoint critical functional groups and modify the structure for enhanced potency, selectivity, or pharmacokinetics.
ix. **Designing Multi-Target Drugs**: Pharmacophore approaches are also employed to design molecules capable of interacting with multiple targets, which is particularly useful in treating complex diseases like cancer, neurodegenerative disorders, or metabolic syndromes.

COMPUTER-AIDED DRUG DESIGN (CADD) AND PHARMACOPHORE MODELING

Traditionally, drug development has been a time-consuming and costly process, often limited to herbal medicine and natural compounds. However, with advances in synthetic chemistry and computational methods, the application of CADD has expanded to include a wide range of synthetic drug candidates. In the early days of drug development, the lack of rational design methods often resulted in compounds with low potency, poor target selectivity, and limited safety profiles. The emergence of rational drug design approaches and the eventual integration of computational tools led to the evolution of CADD—revolutionizing the way novel therapeutics are discovered and optimized.

CADD utilizes various computational strategies to generate and evaluate a broad spectrum of chemical derivatives based on core scaffolds. These derivatives are scored and ranked to identify those with improved potency and desired pharmacokinetic properties [33–36]. A major reason for drug failure in the late stages of development is poor ADME properties. To address this, CADD technologies now incorporate tools that predict these pharmacokinetic behaviors early in the development pipeline [37,38].

One of the core techniques in rational drug design within CADD is pharmacophore modeling. Pharmacophores help identify and define molecular features that are critical for biological activity, enabling compound screening at both the 2D and 3D levels. The integration of pharmacophore models with molecular docking simulations enhances virtual screening efficiency by identifying hits with the right spatial and electronic characteristics [39]. CADD aims to streamline and rationalize the drug discovery process, reducing both time and cost [40]. The primary goal is to identify hit compounds through high-throughput virtual screening, which are then validated using specific bioassays [41]. Following hit identification, the focus shifts to lead optimization, where compound potency and selectivity are enhanced [33,34].

By employing computational tools and vast chemical and biological data resources, CADD supports the investigation of molecular properties before synthesis and biological testing, thereby improving decision-making at early stages of discovery [42,43]. The application of CADD has led to significant advancements in research and development, particularly in areas such as virtual screening, molecular docking, target identification, ligand profiling, and ADMET prediction—with pharmacophore modeling playing a central role throughout [44,45].

The potential of pharmacophore modeling is further enhanced by integrating it with MD simulations, which capture the dynamic behavior of ligand–receptor interactions, improving prediction accuracy and model robustness [11,12]. Especially in times of urgent medical needs, such as during pandemics or health crises, CADD proves invaluable by rapidly extracting and analyzing bioactive compounds from extensive chemical libraries [46].

3D PHARMACOPHORE MODELS IN DRUG DISCOVERY

Pharmacophore modeling in 3D focuses on understanding the spatial arrangement of ligand functional groups in their bioactive conformations. This 3D interaction pattern reveals essential information about ligand binding and has become a fundamental aspect of modern drug design [47,48]. A typical 3D-QSAR pharmacophore model includes features such as aromatic rings, hydrogen bond acceptors or donors, and hydrophobic moieties arranged in a specific geometric configuration. For example, a common model might consist of one aromatic ring, two hydrophobic features, and one HBA, which collectively contributes to the ligand's biological activity [49,50].

Using the HypoGen algorithm, predictive pharmacophore models can be generated based on known inhibitors, such as UPPS (Undecaprenyl Pyrophosphate Synthase) inhibitors. These models enable virtual screening of novel compounds and prediction of their potential activity, making the process both efficient and scientifically grounded [49,51,52]. Pharmacophore modeling—particularly 3D approaches—has emerged as one of the most effective tools in drug discovery. Its widespread application in virtual screening, lead identification, and optimization highlights its usefulness in both academia and industry [53].

STRUCTURE-BASED PHARMACOPHORE MODELING

Structure-based pharmacophore (SBP) modeling is a powerful approach in drug discovery that focuses on the identification of critical interaction features within a receptor's binding pocket without relying on prior ligand information. Unlike ligand-based pharmacophore models, SBPs do not disclose the structure of the original molecule or the specific arrangement of its functional groups. Instead, they are derived from analyzing the interactions of one or more ligands within the receptor-binding site, enabling the identification of essential features for biological activity.

SBP modeling can be conducted using both apo (ligand-free) and holo (ligand-bound) protein structures. In the case of ligand-free proteins, the pharmacophore is derived solely from the structural information of the protein's active site. This approach overcomes several limitations associated with ligand-based pharmacophore modeling, such as challenges in ligand alignment, flexibility of ligands, and improper selection of training sets [55]. One of the major advantages of SBP modeling is that it does not require a template ligand in its bioactive conformation, a prerequisite in ligand-based pharmacophore modeling. This independence makes SBPs particularly valuable in situations where ligand information is limited or unavailable.

The construction of a SBP typically involves the following four steps (Figure 8.3):

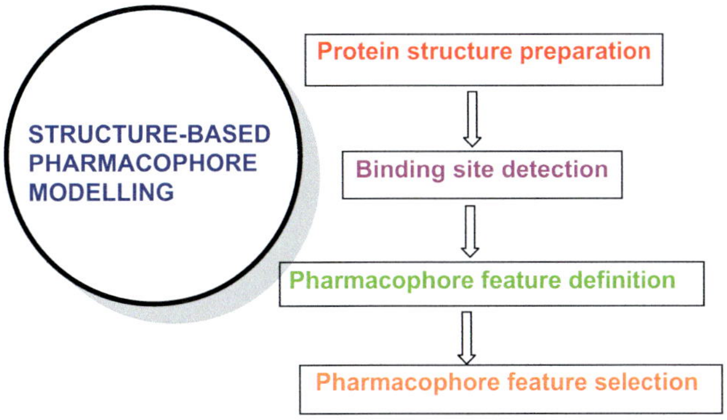

FIGURE 8.3 Steps in structure-based pharmacophore modeling construction.

1. Preparation of the protein structure (usually retrieved from the PDB),
2. Detection of the binding site, often using cavity-detection algorithms,
3. Definition of pharmacophoric features based on interaction points such as HBD/HBA, hydrophobic regions, and charged groups,
4. Selection of relevant features to build the pharmacophore model [56].

SBP models have become essential tools for hit identification and lead optimization, designing focused libraries for screening, and scaffold hopping, especially where ligand data are sparse or unavailable [12,57]. With the increasing availability of high-resolution 3D structures of biological macromolecules in the PDB, interest in SBP modeling has surged. These models facilitate the exploration of ligand–protein interactions, support structural chemogenomics, and aid in identifying new ligands for known proteins or novel targets for existing drugs [58,59].

Some of the advantages of SBP modeling include the following:

- The ability to identify novel scaffolds,
- Insight into the binding modes of ligands within the protein's structural framework,
- Structural optimization of ligand molecules,
- and a deeper understanding of ligand-binding site characteristics.

Such benefits enable the discovery of ligands for orphan receptors, exploration of binding site similarities across protein families, and proposal of new targets for known therapeutic agents [60].

Steindlt et al. used parallel pharmacophore-based virtual screening in order to assess bioactivity profiles of small molecules. They aimed to develop a fast, efficient, scalable system which could provide the rapid prediction of *in silico* activity. The tool used for SBP modeling was Ligand Scout and the high-performance database mining platform Catalyst. This model works on a set of 50 SBP models which targets 100 antiviral compounds. The end result showed that about 90% of all input molecules showed successful activity profiling. Thus, this pharmacophore-based parallel screening is a reliable and fast *in silico* method which helps in the prediction of potential biological targets of compound libraries [61].

Suma et al. developed an approach based on SBP modeling in order to design azaindole derivatives as DprE1 inhibitors for the treatment of tuberculosis. The compounds having better molecular features were subjected to Glide docking and Prime re-scoring calculations. Induced fit docking and molecular dynamics processes were used to search for *in silico* potential of compounds. One potential hit (**1**) was identified and selected from the ZINC database, having drug likeness, without any violations [62].

Kaur et al. designed a structure-based inhibitor in order to evaluate the CYP3A4 pharmacophore model. Human cytochrome P4503A4(CYP 3A4) is a xenobiotic-metabolizing enzyme which oxidizes many drugs. Thus, inhibition of CYP3A4 may result in drug–drug interactions and thus possible side effects. Based on the investigations on analogues of ritonavir, potential CYP3A4 inactivator and pharmaco-enhancer, a pharmacophore model was built. The flexible backbone, H-bond acceptor/donor group, and aromaticity of the side group in the CYP3A4 inhibitor were found to have a leading role in hydrophobic interactions at sites adjacent to the ligand. Compounds **2a** and **2b** were proposed to serve as a template for the synthesis of a second-generation inhibitor and for optimization of the pharmacophore model [63].

Saxena et al. studied novel inhibitors of *Mycobacterium tuberculosis* MurG: homology modeling, SBP, molecular docking, and molecular dynamics simulations. The quality of the model was assessed by PROCHECK, ProSA, and ERRAT software. The model was optimized by molecular dynamic simulation. A total of ten compounds were found to be in good fitness range, with a docking score and better interactions with the protein active site. For the design of Mtb MurG inhibitors, the most potent compound (**3**) was identified as a promising lead molecule [64].

Thangapandian et al. designed a new and dynamic SBP Model to identify Histone Deacetylase 8 (HDAC8) inhibitors. Inhibition of this enzyme will lead to a decrease in

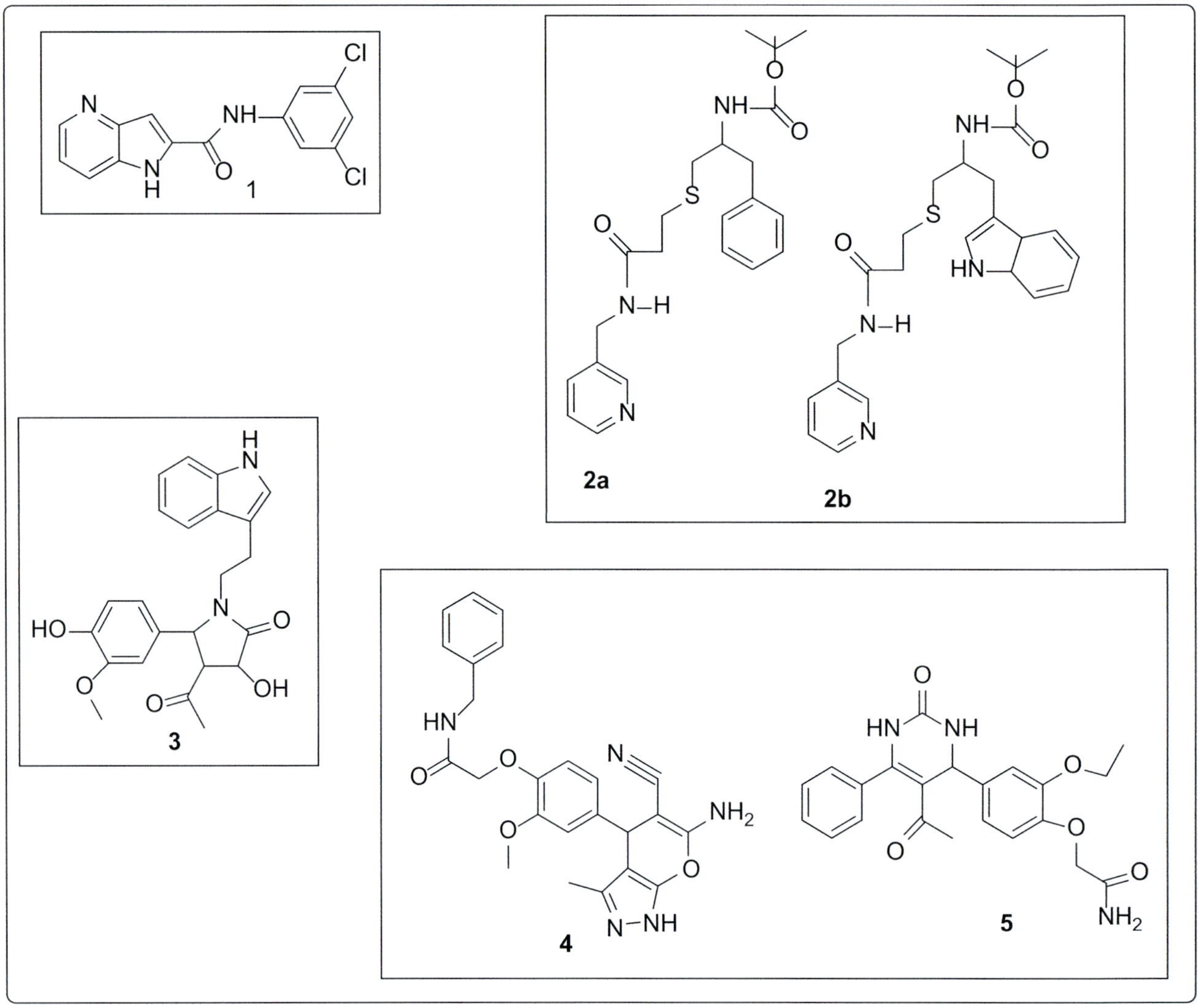

FIGURE 8.4 Chemical structure of some lead compounds identified using SBP modeling.

cancer levels. Thus, it acts as a better anticancer therapeutic strategy. Two new pharmacophore models were developed which contain six and five pharmacophoric features having HDAC8 inhibitor activity. The false positives were reduced by molecular docking. These two compounds (**4** and **5**) could be used to design future HDAC8 inhibitors [65] (Figure 8.4).

LIGAND-BASED PHARMACOPHORE MODELING

Ligand-based pharmacophore modeling remains one of the most widely used strategies in the early stages of drug discovery, particularly when the 3D structure of the target protein is unknown (Figure 8.5). In such scenarios, pharmacophore models are constructed based on the structural and functional features of known active ligands rather than receptor information. These models help identify new hit compounds by capturing the spatial arrangement of features responsible for biological activity.

Traditionally, ligand-based pharmacophore modeling relies on the alignment of template molecules to identify common pharmacophoric features. However, the dependency on molecular alignment can be limiting, especially when dealing with diverse chemical scaffolds. To address this, several advanced methods have been developed for 3D pharmacophore representation and matching that do not require molecular alignment. These alignment-free approaches allow for the rapid identification of identical or highly similar pharmacophores within large datasets [66,67]. Such methods are designed to differentiate pharmacophores that are preferentially associated with active compounds while excluding those associated with inactive ones. This capability significantly enhances the accuracy and efficiency of virtual screening efforts [68,69]. Richmond et al. developed an alignment-free 3D pharmacophore representation that enables rapid matching within chemical libraries. This approach identifies pharmacophore models that correlate strongly with biologically active compounds while filtering out models linked to inactive molecules [70].

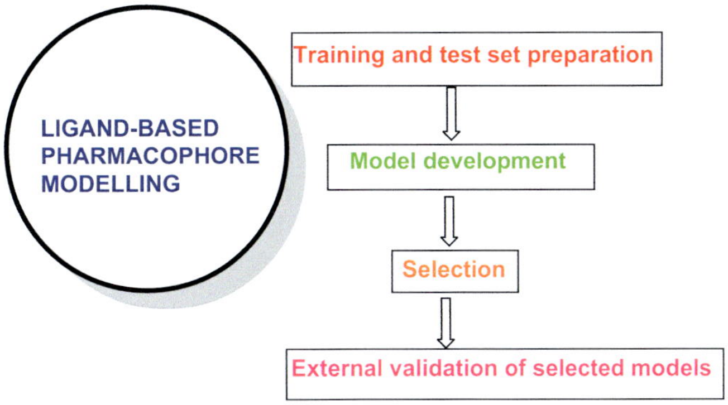

FIGURE 8.5 Steps in Ligand-based pharmacophore modeling construction.

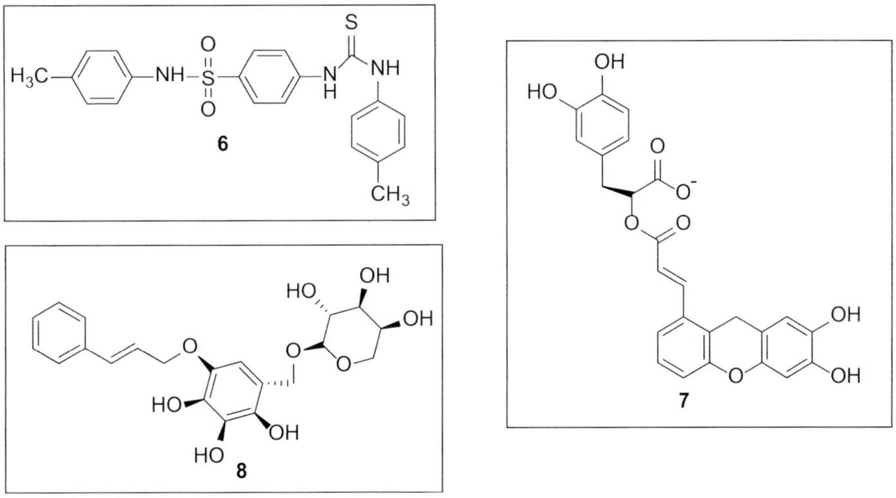

FIGURE 8.6 Chemical structure of hit compounds (6–8) identified using ligand-based pharmacophore modeling.

Ligand-based pharmacophore models are typically derived from a set of structurally diverse compounds known to exhibit activity against a specific biological target. This method has been widely adopted by researchers due to its reliability in hit identification and lead optimization.

Several computational tools and software platforms have been developed for ligand-based pharmacophore modeling including LigandScout, Discovery Studio, MOE, PHASE (Schrödinger), PharmaGist, etc. Each of these tools employs different algorithms for common pharmacophore identification, many of which are based on genetic optimization techniques or machine learning strategies [71,72]. These tools provide intuitive interfaces and powerful engines for pharmacophore model generation, refinement, and validation.

Khan et al. developed a ligand-based pharmacophore model to identify novel antiepileptic compounds. Currently available drugs for epilepsy are associated with many side effects, so there is a need to develop novel antiseizure drugs with better efficacy and tolerability. Forty Four new antiepileptic analogues are developed by ligand-based approach using the software Ligand Scout and J mol. Three features are contained in the developed pharmacophore: hydrophobic unit, hydrogen bonding domain, and electron donor. Orally bioavailable compounds are generated by this approach which gives inhibitory activity toward seizures. One of the compounds (**6**) is recommended for further investigation [73].

Dhanjal et al. applied the ligand-based pharmacophore modeling combined with molecular docking to discover novel acetylcholinesterase (AChE) inhibitors for treating Alzheimer's disease (AD). In this study, a pharmacophore was developed based on existing drugs. The generated pharmacophore was used to search for new AChE inhibitors with varying scaffolds by the use of high-throughput virtual screening and molecular docking studies. Among the screened compounds, **7** and **8** (Figure 8.6) displayed strong binding affinity to the catalytic site of AChE enzyme and were proposed as potential AChE inhibitors for the treatment of AD [74].

Expanding on pharmacophore applications to different targets, Zerroug et al. focused on the identification of CDK5 inhibitors, a crucial target in AD, using interaction studies, pharmacophore modeling, and molecular dynamics simulations. In total, five CDK5 anti-hits were developed. Docking-based Comparative Intermolecular Contacts Analysis (dbCICA) model was used to target AD. In order to develop new CDK inhibitors, 3D-QSAR, genetic algorithm, and pharmacophore modeling were used [75].

Jiang et al. identified new potential inhibitors of butyrylcholinesterase (BChE) with the help of various models such as virtual screening model, pharmacophore model, molecular docking model, and de-novo evolution model. After initial screening using Lipinski's rule of five and ADMET analysis, 35 compounds were narrowed down to 10 compounds based on -CDOCKER_ENERGY values obtained after molecular docking. This resulted in the identification of the best potential candidate (**9**; Figure 8.7) for further modification [76].

John et al. developed and validated 3D QSAR pharmacophore model for the discovery of potential human renin inhibitors aimed at blocking the conversion of angiotensinogen to angiotensin I. The results showed that the best pharmacophore model selected was made of one hydrophobic, one HBD, and two HBA properties having a high correlation value of 0.944. The binding characteristics and essential features were confirmed by molecular docking methods and density functional theory (DFT). Final hits, **10** and **11** emerging as promising candidates (Figure 8.8) showed good inhibitory properties and could be used to design and develop novel renin inhibitors [77].

Addressing hypercholesterolemia, Teli et al. studied atom-based 3D QSAR of N-iso-propyl-pyrrole-based derivatives as HMG-CoA-reductase inhibitors. Compound **12** through a five-point pharmacophore model was

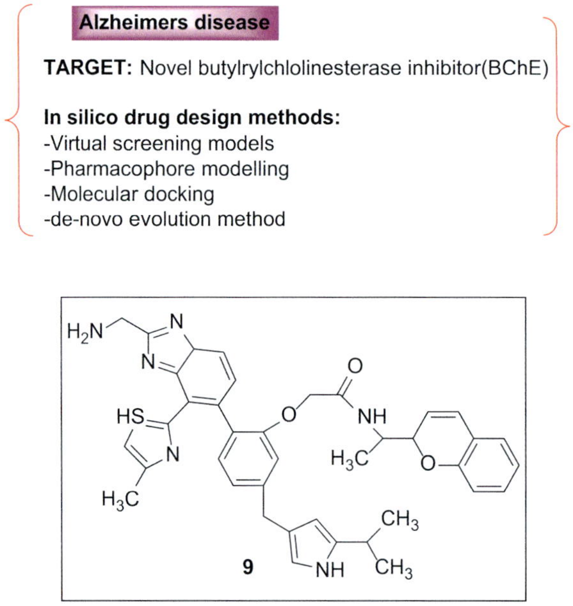

FIGURE 8.7 Chemical structure of compound 9 as a potential inhibitor of butyrylcholinesterase.

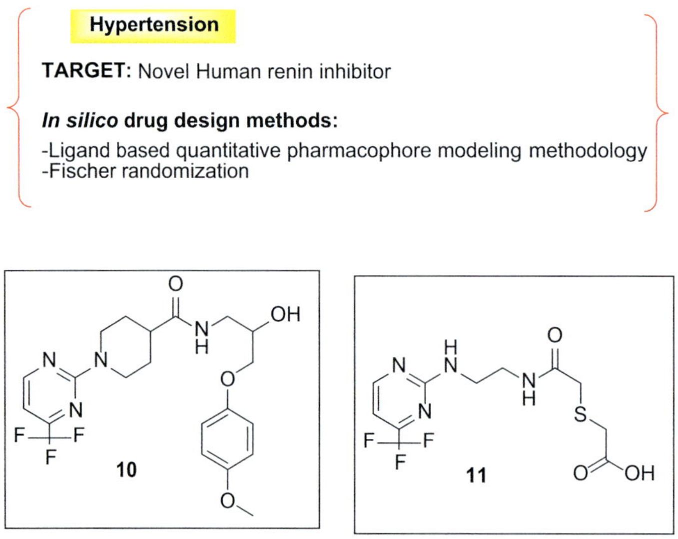

FIGURE 8.8 Chemical structure of compounds 10 and 11 as human renin inhibitors.

identified as a promising candidate for further development (Figure 8.9). The QSAR model suggests that electron-withdrawing groups are important for HMG-CoA reductase activity. However, electron-donating groups, H-bonding, and hydrophobic interactions also play an important role in the activity [50].

Hua et al. explored the potential of AChE inhibitors by the use of various techniques like pharmacophore modeling, virtual screening, and molecular docking studies. The study aimed to identify potential novel pharmacophores to target AD by inhibiting AChE. Using Discovery Studio 2.5.5 program, 3D pharmacophore models were constructed. The obtained hits (**13** and **14**; Figure 8.10) exhibited strong interactions with the important residues present on the active site of AChE [78].

Gao et al. have provided valuable insights on the use of pharmacophore-based drug discovery strategies. They discussed in detail about the fundamental approach strategies for both structure-based pharmacophore and ligand-based pharmacophore approaches along with their respective advantages and limitations. Their virtual screening efforts identified compounds **15** and **16** as promising leads (Figure 8.11). In general, the modern drug discovery process includes pharmacophore modeling and validation, virtual hits profiling, and identification of leads [79].

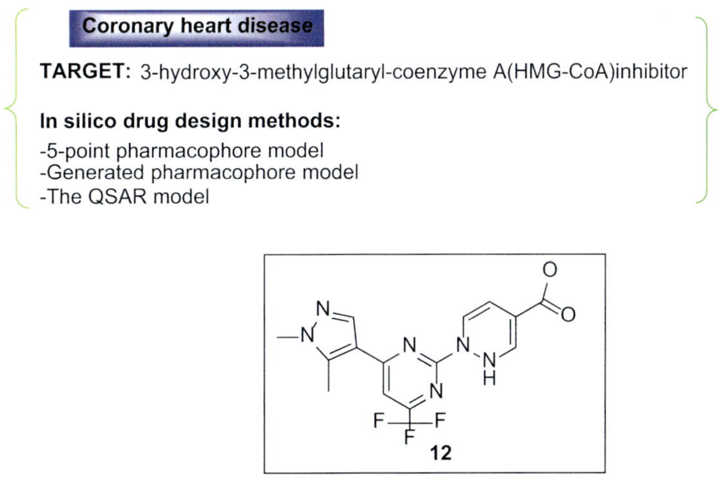

FIGURE 8.9 Chemical structure of compound 12 as an HMG-CoA-reductase inhibitor.

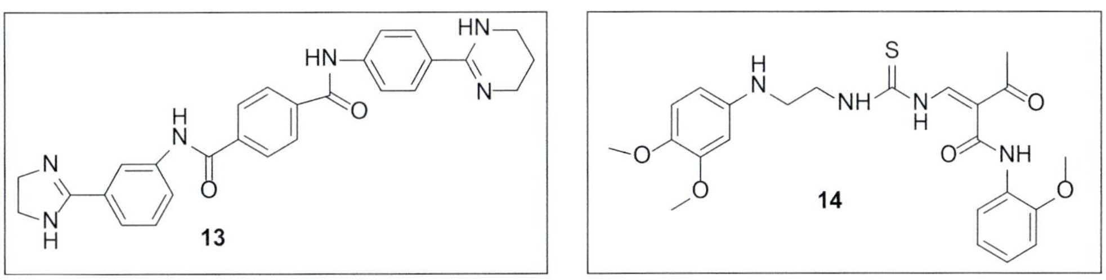

FIGURE 8.10 Chemical structure of compounds 13 &14 as AChE inhibitors.

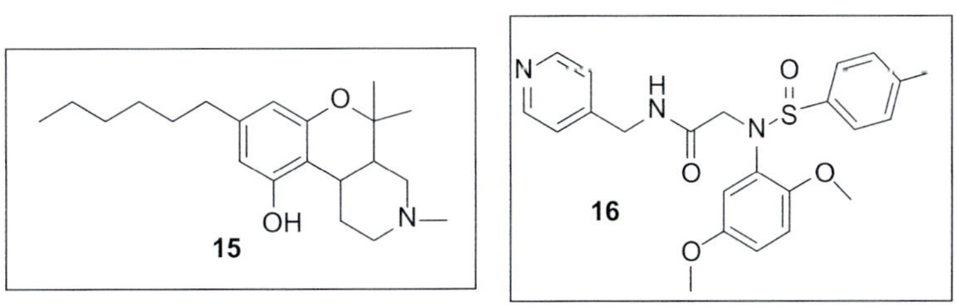

FIGURE 8.11 Chemical structure of compounds 15 &16 discovered with the help of pharmacophore-based drug design.

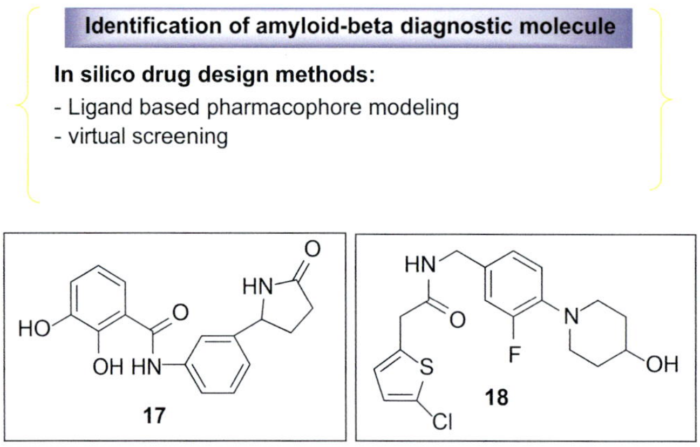

FIGURE 8.12 Chemical structure of compounds 17 &18 discovered with the help of ligand-based pharmacophore modeling amyloid-beta diagnostic compounds.

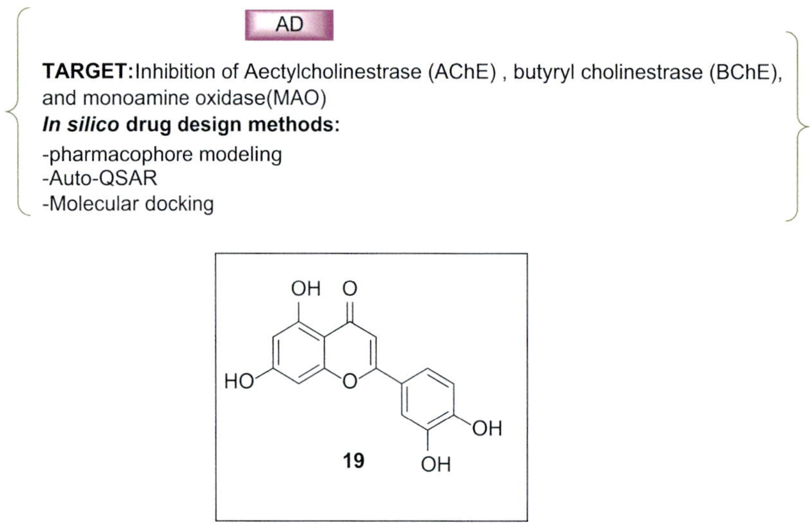

FIGURE 8.13 Chemical structure of luteolin (19) as AChE, BChE, and MAO inhibitor.

Kiran et al. developed pharmacophore-based models for the treatment of AD. One of the effective targets for treating AD is phosphorylated tau (P-tau) and this target is also involved with synaptic damage and neuronal dysfunction [80].

Marondedze et al. applied a new ligand-based pharmacophore model and virtual screening methods to identify the amyloid-beta diagnostic agents. Currently, there is an urgent need for novel diagnostic molecules having good physiological and pharmacokinetic properties. Compounds **17** and **18** were identified from a series of 166 amyloid-beta diagnostic compounds that effectively target amyloid-beta fibrils in AD (Figure 8.12) [81].

Ojo et al. studied the interactions of bioactive compounds present in traditional medicinal plants against AD through techniques such as pharmacophore modeling, QSAR, and molecular docking. Plant products like flavonoids have the capability to cross the blood–brain barrier and treat neurodegenerative diseases. In this study, nine different flavonoids obtained from three plants were studied for inhibitory role on AChE, BChE, and mono-amine oxidase (MAO) using different techniques. The generated pharmacophore model was shown to be effective in identifying bioactive compounds present in these plants. Among all the flavonoid compounds, luteolin (**19**) was found to be the most active (Figure 8.13) [82].

In another study, Kaserer et al. used pharmacophore models and pharmacophore-based virtual screening techniques to identify inhibitors of hydroxy steroid dehydrogenases, enzymes implicated in several pathophysiological conditions. The identified compound (**20**; Figure 8.14) target interactions associated with potentially harmful effects are also investigated [83].

Valasani et al. combined structure-based design, synthesis, pharmacophore modeling, virtual screening, and molecular docking studies to discover a novel Cyclophilin D inhibitor (**21**; Figure 8.15). Targeting cyclophilin-D, a peptidyl prolyl isomerase F and a mitochondrial membrane permeability pore regulator implicated in AD-related cell death, represents a promising therapeutic strategy [84].

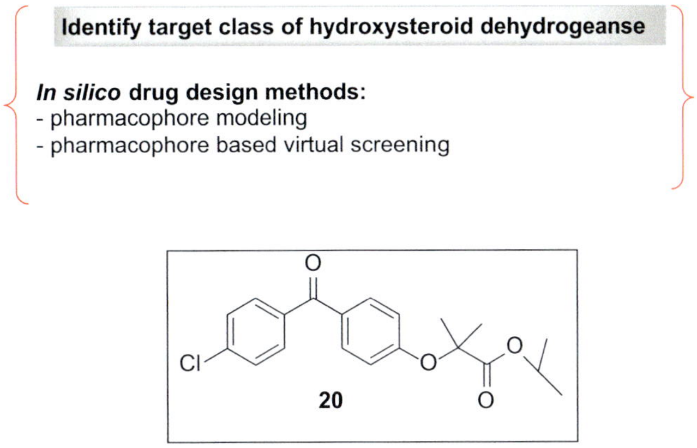

FIGURE 8.14 Chemical structure of compound 20 as hydroxy steroid dehydrogenases inhibitor.

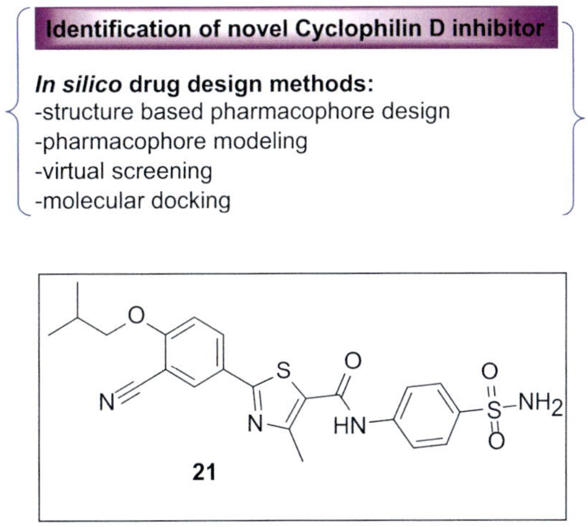

FIGURE 8.15 Chemical structure of compound 21 as Cyclophilin D inhibitor.

CONCLUSION

Pharmacophore modeling has emerged as a powerful and versatile tool in the early stages of drug discovery, facilitating the identification of novel therapeutic agents. This book chapter highlights the strategic application of both ligand-based and structure-based pharmacophore models in designing compounds with enhanced efficacy, selectivity, and pharmacokinetic properties. As computational tools continue to evolve, pharmacophore modeling will remain central to rational drug design, accelerating the path from molecular insight to clinical innovation.

REFERENCES

1. Güner, O. F., & Bowen, J. P. Setting the record straight: The origin of the pharmacophore concept. *Journal of Chemical Information and Modeling*, 2014, *54*, 1269–1283.
2. VanDrie, J. H. 3D database searching in drug discovery. *Network Science*, 1996, *1*, 225–266.
3. Allen, F. H., Davies J. E., Galloy, J. J., Johnson, O., Kennard, O., Macrae, C. F., Mitchell, E. M., Mitchell, G. F., Smith, J. M., & Watson, D. G. The development of versions 3 and 4 of the Cambridge structural database system. *Journal of Chemical Information & Computer Sciences*, 1991, *31*, 187–204.
4. Pearlman, R. S. Rapid generation of high-quality approximate 3D molecular structures. *Chemical Design Automation*, 1987, *2*, 1–7.
5. Hiller, C., & Gasteiger, J. Ein Automatisierter Molekülbaukasten. *Software-Entwicklung in der Chemie*, 1987, 1, 53–66.
6. Wipke, W. T., & Hahn, M. A. AIMB: Analogy and intelligence in model building. System description and performance characteristiocs. *Tetrahedron Computer Methodology*, 1988, *1*, 141.
7. Dolata, P. D., Leach, A. R., & Prout, K. Wizard: AI in conformational analysis. *Journal of Computer-Aided Molecular Design*, 1987, *1*, 73–85.
8. Fisanick, W., Cross, K. P., & Rusinko, A. III. Similarity searching on CAS registry substances. 1. Global molecular property and generic atom triangle geometric searching. *Journal of Chemical Information*, 1992, *32*, 664–674.
9. Kuntz, I. D. Structure-based strategies for drug design and discovery. *Science*, 1992, *257*, 1078–1082.
10. Bartlett, P. A., Shea, G. T., Telfer, S. J., & Waterman, S. C. O. T. T. CAVEAT: A program to facilitate the structure-derived design of biologically active molecules. *Journal of Molecular Recognition*, 1990, *78*, 182.
11. Guner, O. History and evolution of the pharmacophore concept in computer-aided drug design. *Current Topics in Medicinal Chemistry*, 2005, *2*, 1321–1332.

12. Sanders, M. P., McGuire, R., Roumen, L., de Esch, I. J., de Vlieg, J., Klomp, J. P., & de Graaf, C. From the protein's perspective: The benefits and challenges of protein structure-based pharmacophore modeling. *MedChemComm*, 2012, *3*(1), 28–38.
13. Lin, S.-K. S., & Hoffman J. M. R. HypoGen: An automated system for generating predictive 3D pharmacophore models. *Pharmacophore Perception, Development and Use in Drug Design*, 2000, 2, 171–189.
14. Yang, S. Y. Pharmacophore modeling and applications in drug discovery: Challenges and recent advances. *Drug Discovery Today*, 2010, *15*, 444–450.
15. Tyagi, R., Singh, A., Chaudhary, K. K., & Yadav, M. K. Pharmacophore modeling and its applications. *Journal of Bioinformatics*, 2022, *2022*, 269–289.
16. Geppert, T. D., & Lipsky, P. E. Antigen presentation at the inflammatory site. *Critical Reviews in Immunology*, 1989, *9*, 313–362.
17. Muegge, I. Pharmacophore features of potential drugs. *Chemistry: A European Journal*, 2002, *8*(9), 1976–1981.
18. Sutter, J., Li, J., Maynard, A., Goupil, A., Luu, T., & Nadassy, K., 2011. New features that improve the pharmacophore tools from Accelrys. *Current Computer-Aided Drug Design*, 2011, *7*(3), 173–180.
19. Wolber, G., Seidel, T., Bendix, F., & Langer, T. Molecule-pharmacophore super positioning and pattern matching in computational drug design. *Drug Discovery Today*, 2008, *13*, 23–29.
20. Leach, A. R., Gillet, V. J., Lewis, R. A., & Taylor, R. Three-dimensional pharmacophore methods in drug discovery. *Journal of Medicinal Chemistry*, 2009, *53*(2), 539–558.
21. Dragos, H. Pharmacophore-based virtual screening. *Chemoinformatics and Computational Chemical Biology. Methods in Molecular Biology*, 2011, *672*, 261–298.
22. Mannhold, R., Kubinyi, H., & Folkers, G. *Pharmacophores and Pharmacophore Searches*. Berlin: Wiley-Vch, 2006.
23. McGregor, M. J., & Muskal, S. M. Pharmacophore fingerprinting. 1. Application to QSAR and focused library design. *Journal of Chemical Information and Computer Sciences*, 1999, *39*, 569–574.
24. McGregor, M. J., & Muskal, S. M. Pharmacophore fingerprinting. 2. Application to primary library design. *Journal of Chemical Information and Computer Sciences*, 2000, *40*, 117–125.
25. Jones, G., Willett, P., & Glen, R. C. A genetic algorithm for flexible molecular overlay and pharmacophore elucidation. *Journal of Computer-Aided Molecular Design*, 1995, *9*, 532–549.
26. Hähnke, V., & Schneider, G. Pharmacophore alignment search tool: Influence of scoring systems on text-based similarity searching. *Journal of Computational Chemistry*, 2011, *32*, 1635–1647.
27. Böhm, H. J. The computer program LUDI: A new method for the de novo design of enzyme inhibitors. *Journal of Computer-Aided Molecular Design*, 1992, *6*, 61–78.
28. Mason, J. S., Morize, I., Menard, P. R., Cheney, D. L., Hulme, C., & Labaudiniere, R. F. New 4-point pharmacophore method for molecular similarity and diversity applications: Overview of the method and applications, including a novel approach to the design of combinatorial libraries containing privileged substructures. *Journal of Medicinal Chemistry*, 2014, *42*, 3251–3264.
29. Wood, D. J., Vlieg, J. D., Wagener, M., & Ritschel, T. Pharmacophore fingerprint-based approach to binding site subpocket similarity and its application to bioisostere replacement. *Journal of Chemical Information and Modeling*, 2012, *52*, 2031–2043.
30. Maji, S., Pattanayak, S. K., Sen, A., & Badavath, V. N. (2022). Pharmacophore modeling in drug design. In: Mithun, R., & Chukwuebuka, E. (eds) *Computer Aided Drug Design (CADD): From Ligand-Based methods to structure-based approaches*, 2022, pp. 157–179. Amsterdam, Netherlands: Elsevier.
31. Ghose, A. K., & Wendoloski, J. J. Pharmacophore modeling: Methods, experimental verification and applications. In: Hugo, K., Gerd, F., & Martin, Y. C. (eds) *3D QSAR in Drug Design: Ligand-Protein Interactions and Molecular Similarity,* 2002, pp. 253–271. Dordrecht: Springer Netherlands.
32. Pirhadi, S., Shiri, F., & Ghasemi, J. B. Methods and applications of structure-based pharmacophores in drug discovery. *Current Topics in Medicinal Chemistry*, 2013, *13*, 1036–1047.
33. Hopkins, A. L., Keserü, G. M., Leeson, P. D., Rees, D. C., & Reynolds, C. H. The role of ligand efficiency metrics in drug discovery. *Nature Reviews Drug Discovery*, 2014, *13*(2), 105–121.
34. Ballester, P. J., Mangold, M., Howard, N. I., Robinson, R. L. M., Abell, C., Blumberger, J., & Mitchell, J. B. (2012). Hierarchical virtual screening for the discovery of new molecular scaffolds in antibacterial hit identification. *Journal of the Royal Society Interface*, 2012, *9*(77), 3196–3207.
35. Boyd, M. R. The position of intellectual property rights in drug discovery and development from natural products. *Journal of Ethnopharmacology*, 1996, *51*(1–3), 17–27.
36. Thiel, K. A. Structure-aided drug design's next generation. *Nature Biotechnology*, 2004, *22*(5), 513–519.
37. Seidel, T., Bryant, S. D., Ibis, G., Poli, G., & Langer, T. 3D pharmacophore modeling techniques in computer-aided molecular design using ligandscout. *Tutorials in Chemoinformatics*, 2017, *2017*, 279–309.
38. Guner, O. F., & Bowen, J. P. Pharmacophore modeling for ADME. *Current Topics in Medicinal Chemistry*, 2013, *13*(11), 1327–1342.
39. Qing, X., Yin Lee, X., De Raeymaeker, J., Tame, J., Zhang, K., De Maeyer, M., & Voet, A. Pharmacophore modeling: Advances, limitations, and current utility in drug discovery. *Journal of Receptor, Ligand and Channel Research*, 2014, *2014*, 81–92.
40. Taft, C. A., Da Silva, V. B., & Da Silva, C. H. Current topics in computer-aided drug design. *Journal of Pharmaceutical Sciences*, 2008, *97*(3), 1089–1098.
41. Bajorath, J. Integration of virtual and high-throughput screening. *Nature Reviews Drug Discovery*, 2002, *1*(11), 882–894.
42. Song, C. M., Lim, S. J., & Tong, J. C. Recent advances in computer-aided drug design. *Briefings in Bioinformatics*, 2009, *10*, 579–591.
43. Anh Vu, L., Thi Cam Quyen, P., & Thuy Huong, N. In silico drug design: Prospective for drug lead discovery. *International Journal of Engineering and Science Invention (IJESI)*, 2015, *4*(10), 60–70.

44. Macalino, S. J. Y., Gosu, V., Hong, S., & Choi, S. Role of computer-aided drug design in modern drug discovery. *Archives of Pharmacal Research*, 2015, *38*(9), 1686–1701.
45. Prachayasittikul, V., Worachartcheewan, A., Shoombuatong, W., Songtawee, N., Simeon, S., Prachayasittikul, V., & Nantasenamat, C. Computer-aided drug design of bioactive natural products. *Current Topics in Medicinal Chemistry*, 2015, *15*(18), 1780–1800.
46. Voet, A., Banwell, E. F., Sahu, K. K., Heddle, J. G., & Zhang, K. Y. J. Protein interface pharmacophore mapping tools for small molecule protein: Protein interaction inhibitor discovery. *Current Topics in Medicinal Chemistry*, 2013, *3*, 989–1001.
47. Talele, T. T. Opportunities for tapping into three-dimensional chemical space through a quaternary carbon. *Journal of Medicinal Chemistry*, 2020, *63*(22), 13291–13315.
48. Qandeel, B. M., Mowafy, S., Abouzid, K., & Farag, N. A. Lead generation of UPPS inhibitors targeting MRSA: Using 3D-QSAR pharmacophore modeling, virtual screening, molecular docking, and molecular dynamic simulations. *Bioorganic & Medicinal Chemistry*, 2024, *18*(1), 14.
49. Sobhy, M. K., Mowafy, S., Lasheen, D. S., Farag, N. A., & Abouzid, K. A. 3D-QSAR pharmacophore modeling, virtual screening and docking studies for lead discovery of a novel scaffold for VEGFR 2 inhibitors: Design, synthesis and biological evaluation. *Bioorganic Chemistry*, 2019, *89*, 102988.
50. Teli, M. K., & Rajanikant, G. K. Pharmacophore generation and atom-based 3D-QSAR of N-iso-propyl pyrrole-based derivatives as HMG-CoA reductase inhibitors. *Bioorganic & Medicinal Chemistry Letters*, 2012, *2*, 1–10.
51. Vrontaki, E., Melagraki, G., Voskou, S., Phylactides, M., Mavromoustakos, T., Kleanthous, M., & Afantitis, A. (2017). Development of a predictive pharmacophore model and a 3D-QSAR study for an in-silico screening of new potent Bcr-Abl kinase inhibitors. *Mini-Reviews in Medicinal Chemistry*, 2017, *17*(3), 188–204.
52. Braga, R. C., & Andrade, C. H. (2013). Assessing the performance of 3D pharmacophore models in virtual screening: How good are they?. *Current Topics in Medicinal Chemistry*, 2013, *13*(9), 1127–1138.
53. Schaller, D., Šribar, D., Noonan, T., Deng, L., Nguyen, T. N., Pach, S., Machalz, D., Bermudez, M., & Wolber, G. Next generation 3D pharmacophore modeling. *Wiley Interdisciplinary Reviews: Computational Molecular Science*, 2020, *10*(4), 1468.
54. Mason, J. S., Good, A. C., & Martin, E. J. 3-D pharmacophores in drug discovery. *Current Pharmaceutical Design*, 2001, *7*(7), 567–597.
55. Ehrlich, P. (1909). Über den jetzigen stand der chemotherapie. *Berichte der deutschen chemischen Gesellschaft*, 1909, *42*(1), 17–47.
56. Kier, L. B. Molecular orbital calculation of preferred conformations of acetylcholine, muscarine, and muscarone. *Molecular Pharmacology*, 1967, *3*(5), 487–494.
57. Kier, L. B. The prediction of molecular conformation as a biologically significant property. *Medicinal Chemistry III*, 1973, *1973*, 509–520.
58. Leach, A. R., Gillet, V. J., Lewis, R. A., & Taylor, R. Three-dimensional pharmacophore methods in drug discovery. *Journal of Medicinal Chemistry*, 2010, *53*(2), 539–558.
59. Sun, H. Pharmacophore-based virtual screening. *Current Medicinal Chemistry*, 2008, *15*(10), 1018–1024.
60. Kubinyi, H. Success stories of computer-aided design. *Computer Applications in Pharmaceutical Research and Development*, 2006, *2006*, 377–424.
61. Steindl, T. M., Schuster, D., Wolber, G., Laggner, C., & Langer, T. High-throughput structure-based pharmacophore modeling as a basis for successful parallel virtual screening. *Journal of Computer-Aided Molecular Design*, 2006, *20*, 703–715.
62. Suma, K. B., Kumari, A., Shetty, D., Fernandes, E., Chethan, D. V., Jays, J., & Murahari, M. Structure based pharmacophore modeling approach for the design of azaindole derivatives as DprE1 inhibitors for tuberculosis. *Journal of Molecular Graphics and Modelling*, 2020, *101*, 107718.
63. Kaur, P., Chamberlin, A. R., Poulos, T. L., & Sevrioukova, I. F. Structure-based inhibitor design for evaluation of a CYP3A4 pharmacophore model. *Journal of Medicinal Chemistry*, 2016, *59*(9), 4210–4220.
64. Saxena, S., Abdullah, M., Sriram, D., & Guruprasad, L. Discovery of novel inhibitors of Mycobacterium tuberculosis MurG: Homology modeling, structure-based pharmacophore, molecular docking, and molecular dynamics simulations. *Journal of Biomolecular Structure and Dynamics*, 2018, *36*(12), 3184–3198.
65. Thangapandian, S., John, S., Lee, Y., Kim, S., & Lee, K. W. Dynamic structure-based pharmacophore model development: A new and effective addition in the histone deacetylase 8 (HDAC8) inhibitor discoveryInt. *Journal of Molecular Science*, 2011, *12*(12), 773–788.
66. Hinsberger, S., Hüsecken, K., Groh, M., Negri, M., Haupenthal, J., & Hartmann, R. W. Discovery of novel bacterial RNA polymerase inhibitors: Pharmacophore-based virtual screening and hit optimization. *Journal of Medicinal Chemistry*, 2013, *56*, 8332–8338.
67. Krautscheid, Y., Senning, C. J. Å., Sartori, S. B., Singewald, N., Schuster, D., & Stuppner, H. Pharmacophore modeling, virtual screening, and in vitro testing reveal haloperidol, eprazinone, and fenbutrazate as neurokinin receptors ligands. *Journal of Chemical Information*, 2014, *54*, 1747–1757.
68. Polishchuk, P. G., Samoylenko, G. V., Khristova, T. M., et al. Design, virtual screening, and synthesis of antagonists of $\alpha IIb\beta 3$ as antiplatelet agents. *Journal of Medicinal Chemistry*, 2015, *58*, 7681–7694.
69. Wolber, G., Dornhofer, A. A., & Langer, T. Efficient overlay of small organic molecules using 3D pharmacophores. *Journal of Computer-Aided Molecular Design*, 2006, *20*, 773–788.
70. Richmond, N. J., Abrams, C. A., Wolohan, P. R. N., Abrahamian, E., Willett, P., & Clark, R. D. GALAHAD: 1. Pharmacophore identification by hypermolecular alignment of ligands in 3D. *Journal of Computer-Aided Molecular Design*, 2006, *20*, 567–587.
71. Schneidman-Duhovny, D., Dror, O., Inbar, Y., Nussinov, R., & Wolfson, H. J. PharmaGist: A webserver for ligand-based pharmacophore detection. *Nucleic Acids Research*, 2008, *36*, W223–W228.
72. Patel, J. R., Joshi, H. V., Shah, U. A., & Patel, J. K. A review on computational software tools for drug design and discovery. *Indo Global Journal of Pharmaceutical Sciences*, 2022, *12*, 53–81.

73. Khan, H. N., Kulsoom, S., & Rashid, H. Ligand based pharmacophore model development for the identification of novel antiepileptic compound. *Epilepsy Research*, 2012, *98*(1), 62–71.
74. Dhanjal, J. K., Sharma, S., Grover, A., & Das, A. Use of ligand-based pharmacophore modeling and docking approach to find novel acetylcholinesterase inhibitors for treating Alzheimer's. *Biomedicine & Pharmacotherapy*, 2015, *71*, 146–152.
75. Zerroug, E., Belaidi, S., Chtita, S., Tuffaha, G., AbulQais, F., Kciuk, M., Dubey, A., & Taha, M. O. Structure-based approaches for the prediction of Alzheimer's disease inhibitors: Comparative interactions analysis, pharmacophore modeling and molecular dynamics simulations. *Chemistry Select*, 2024, *9*, 202303307.
76. Jiang, Y., & Gao, H. Pharmacophore-based drug design for the identification of novel butyrylcholinesterase inhibitors against Alzheimer's disease. *Phytomedicine*, 2019, *54*, 278–290.
77. John, S., Thangapandian, S., Arooj, M., Hong, J. C., Kim, K. D., & Lee, K. W. Development, evaluation and application of 3D QSAR Pharmacophore model in the discovery of potential human renin inhibitors. *BMC Bioinformatics*, 2011, *12*, 1–14.
78. Lu, S. H., Wu, J. W., Liu, H. L., Zhao, J. H., Liu, K. T., Chuang, C. K., Lin, H. Y., Tsai, W. B., & Ho, Y. The discovery of potential acetylcholinesterase inhibitors: A combination of pharmacophore modeling, virtual screening, and molecular docking studies. *Journal of Biomedical Science*, 2011, *18*, 1–13.
79. Gao, Q., Yang, L., & Zhu, Y. Pharmacophore based drug design approach as a practical process in drug discovery. *Current Computer-Aided Drug Design*, 2010, *6*(1), 37–49.
80. Pradeep Kiran, J. A., Reddy, A. P., & Reddy, P. H. Pharmacophore-based models for therapeutic drugs against phosphorylated tau in Alzheimer's disease. *Drug Discovery Today*, 2019, *24*(2), 616–623.
81. Marondedze, E. F., Govender, K. K., & Govender, P. P. Ligand-based pharmacophore modeling and virtual screening for the identification of amyloid-beta diagnostic molecules. *Journal of Molecular Graphics and Modelling*, 2020, *101*, 107711.
82. Ojo, O. A., Ojo, A. B., Okolie, C., Nwakama, M. A. C., Iyobhebhe, M., Evbuomwan, I. O., Nwonuma, C. O., Maimako, R. F., Adegboyega, A. E., Taiwo, O. A., & Alsharif, K. F. Deciphering the interactions of bioactive compounds in selected traditional medicinal plants against Alzheimer's diseases via pharmacophore modeling, auto-QSAR, and molecular docking approaches. *Molecules*, 2021, *26*(7), 1996.
83. Kaserer, T., Beck, K. R., Akram, M., Odermatt, A., & Schuster, D. Pharmacophore models and pharmacophore-based virtual screening: Concepts and applications exemplified on hydroxysteroid dehydrogenases. *Molecules*, 2015, *20*(12), 22799–22832.
84. Valasani, K. R., Vangavaragu, J. R., Day, V. W., & Yan, S. S. Structure based design, synthesis, pharmacophore modeling, virtual screening, and molecular docking studies for identification of novel cyclophilin D inhibitors. *Journal of Chemical Information and Modeling*, 2014, *54*(3), 902–912.

9 Quantitative Structure–Activity Relationship in Drug Design

Satyamshyam Vishwakarma[1], *Prashant Sahu*[2], *Pragyanshu Khare*[3], and *Mukesh Chandra Sharma*[4]
[1]Integrated Drug Discovery Research Laboratory, Department of Pharmaceutical Sciences, Dr. Harisingh Gour University (A Central University), Sagar, India
[2]Babulal Tarabai Institute of Pharmaceutical Science Narsinghpur Road, Sironja Sagar, India
[3]Department of Pharmacy, Birla Institute of Technology & Science, Pilani, India
[4]School of Pharmacy, Devi Ahilya University, Takshila Campus, Khandwa Road, Indore, India

INTRODUCTION

Over the last few decades, rapid developments in information and communication technology have revolutionized our ability to acquire, analyse, store, and exchange data. This transition has had a significant impact on scientific research in a variety of areas, particularly the discovery of novel and effective therapies. Researchers now have access to excellent resources for drug development because of large databases such as PubChem, which contain millions of investigated chemical compounds. Efficient screening of huge chemical databases against known active compounds is required to uncover potential novel therapeutic leads. In this context, "Quantitative Structure–Activity Relationship" modeling proves to be a useful tool. It aids in the exploration and exploitation of the relationships between biological functions and chemical structures, which allows the development of new therapeutic candidates [1].

The basic principle of Quantitative Structure–Activity Relationship (QSAR) states that biological activity is directly proportional to molecular structure. As a result, molecules with comparable structures will have similar bioactivities for proteins/receptors/enzymes, and structural alterations will be reflected in changes in bioactivities. The general description of a QSAR model will be (Figure 9.1)

According to the IUPAC (International Union of Pure and Applied Chemistry), "a three-dimensional (3D) quantitative structure–activity relationship is the analysis of the quantitative relationship between the biological activity of a set of compounds and their spatial properties using statistical methods" [2].

OBJECTIVES OF QSAR

In general, the following objectives describe most of the QSAR approaches:

- Quantitatively connect and concluded the changes in chemical structure along with changes in biological activities to understand those chemical characteristics which can be the predictors of biological activity.
- Optimize existing leads for better biological activity.
- Predicting the biological actions of experimental and sometimes unavailable drugs [3].

ASPECTS OF QSAR APPROACHES

The level of credibility attained through QSAR modeling is influenced by the nature of the property being predicted, the project's phase, and the simplicity and cost-effectiveness of synthesizing compounds and conducting subsequent tests. QSAR models frequently provide valuable predictions, yet they often fall short, even when supported by robust statistics derived from the internal data utilized during training. Despite these challenges, QSAR presents a significant alternative for several reasons:

- Conventional synthetic processes are costly and time-consuming.
- Biological experiments are costly due to time constraints, animal sacrifice, and the need for purified chemicals.
- Poor ADMET profiles can lead to costly and traumatic drug failures in later phases of research, including after commercialization. Combinatorial chemistry and HTS provide access to a large number of molecules, but prioritization and screening require estimates [3] (Table 9.1).

VARIOUS QSAR METHODOLOGIES

Based on Dimensions

QSAR approaches are generally categorized based on their structural illustration or the methodology used to derive the descriptor values:

DOI: 10.1201/9781003464679-9

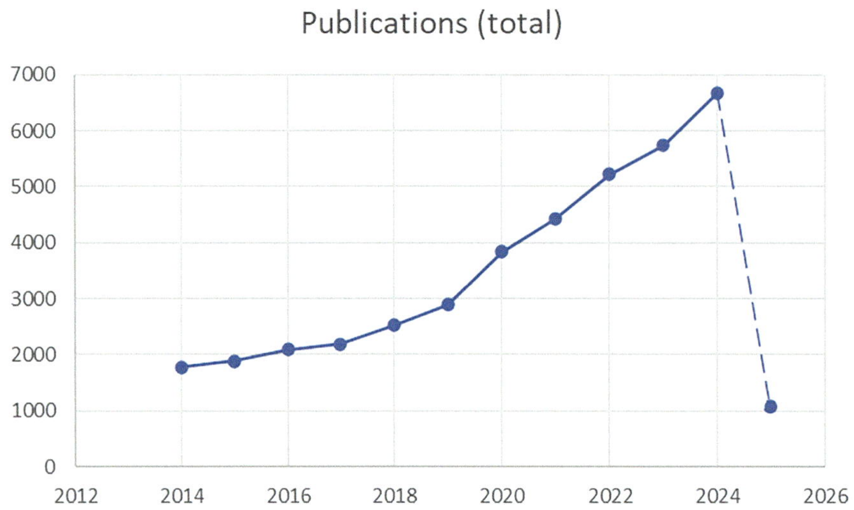

FIGURE 9.1 3D QSAR-related papers for a period from 2014 to 2025 (Dimension database as source database, 3D QSAR as keyword in full data, and articles as publication type).

TABLE 9.1
A Concise Overview of Previous QSAR Approaches

Contributions of Authors	Authors
Examined the effects of ionization and electron distribution, as well as steric access, on the potencies of various aminoacridines.	Albert [4]
Proposed a way to separate polar, steric, and resonance effects and introduced the first steric parameter, ES.	Taft [5]
Proposed the relationship between the biological actions of plant growth regulators and Hammett constants, as well as hydrophobicity.	Hansch and Muir [6]
Combined the hydrophobic constants with Hammett's electrical constants to produce the linear Hansch equation and its various extended variants.	Hansch and Fujita [7]
Developed the parabolic Hansch equation for expanded hydrophobicity ranges.	Hansch [8]
Created additive model that represents activity as a total of substituent contributions.	Free and Wilson [9]
The Free-Wilson equation was simplified to estimate the activity of the non-substituted component of the series. Additionally, the Fujita–Ban equation was proposed, utilizing the logarithm of activity to align the activity parameter with other free energy-related parameters.	Fujita and Ban [10]
Investigated drug transport via aqueous and lipoidal compartment systems, and further modified the parabolic equation of Hansch to produce a superior bilinear (non-linear) QSAR model.	Kubinyi [11]
Developed comparative quantitative structure–activity relationship (C-QSAR) and integrated it into the C-QSAR program.	Hansch and Gao [12]
Developed Hologram QSAR (HQSAR), which generates molecular holograms by transforming structures into all possible pieces, assigning integers, and hashing into a fingerprint. These holograms encode molecular, chemical, and topological information as QSAR descriptors using bin occupancies.	Heritage and Lowis, 1997 [13,14]
Created inverse QSAR to determine chemical descriptor values with the desired activity or attribute. Thus, it involves finding the best descriptor values for a particular activity and building a concentrated library of potential structures from them.	Cho and workers [15]
Binary QSAR was developed to handle binary activity measures derived from high-throughput screening (such as pass/fail or active/inactive), utilizing chemical descriptor vectors as input. Bayes' Theorem is applied to determine the distribution of probabilities for active and inactive elements.	Labute [16]

- 1DQSAR establishes relationships between chemical properties like pKa and log P and activity.
- 2DQSAR connects structural patterns like 2D pharmacophores and connection indices to activity, disregarding their 3D representation.
- 3DQSAR compares the non-covalent interaction fields around the molecules with the activity of molecules.
- 4DQSAR also includes a group of ligand configurations in 3DQSAR.
- 5DQSAR clearly represents several induced-fit models in 4DQSAR.
- 6DQSAR involved the alternative solvation models from 5DQSAR.

On the Basis of Chemometric Methods Involved

The method of correlation utilized to establish a link between structural properties and biological activity also allows for the categorization of QSAR techniques into two broad categories:

- Principal component analysis/regression (PCR), partial least squares (PLS), multiple linear regression (MLR), and linear regression (LR) are all examples of linear methods.
- ANNs, k-nearest neighbors (kNNs), and Bayesian neural networks are examples of non-linear methods.

GENERAL METHODOLOGY INVOLVED IN A QSAR STUDY

The fundamental method of developing a QSAR model is depicted in Figure 9.2b. The method begins with gathering the molecules of interest and computing all potential descriptors. Next, the significant descriptors are chosen, and a model is built. If the appropriate descriptors are chosen, the model will perform well when assessed on the Test Set. Typically, a bad model is built, a changed or new collection of descriptors is selected, and the procedure is repeated. Once a suitable model has been identified, it is used to create novel chemicals for the system of interest. Unfortunately, this is all that is commonly understood or seen when utilizing a QSAR software.

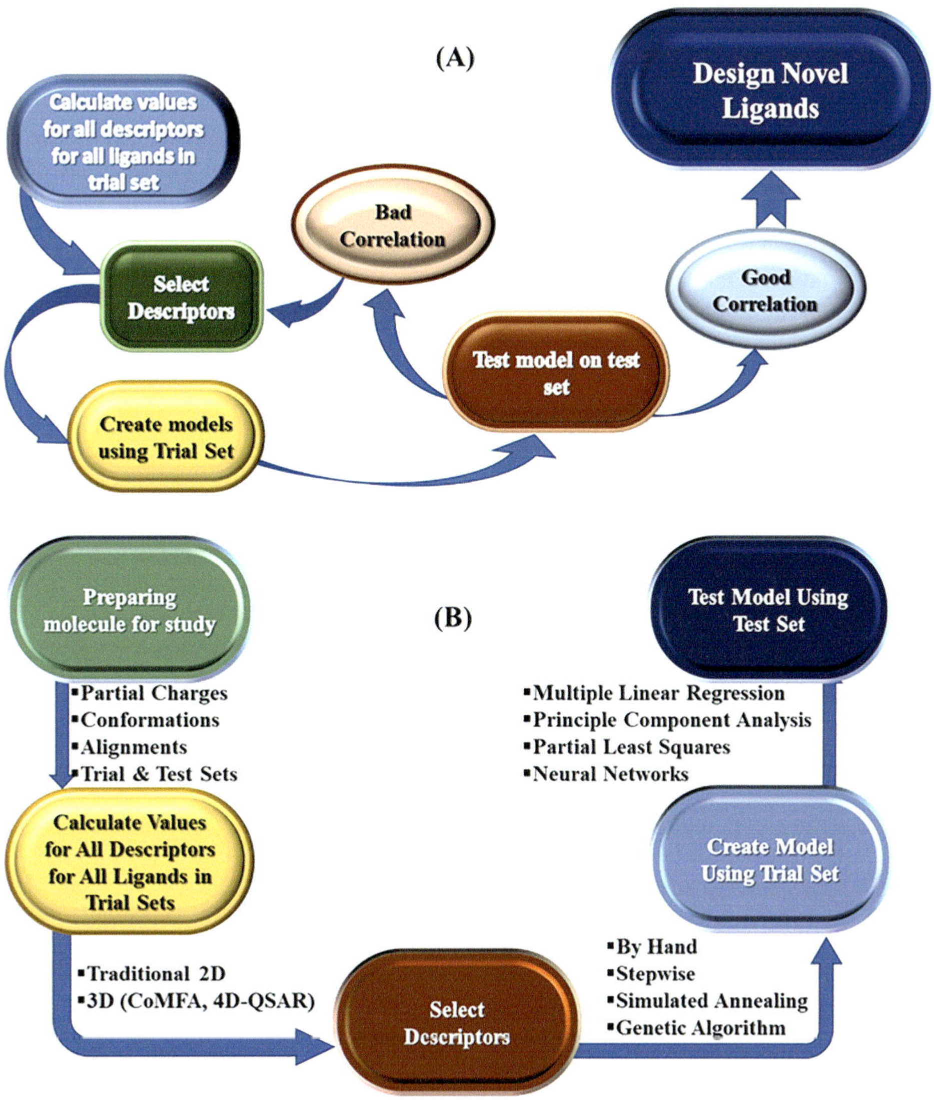

FIGURE 9.2 (a) Procedure for developing and applying a QSAR model. Molecules are gathered and descriptors computed. Before being tested on the test set, the model is constructed using the training set. If the model shows a poor association with experimental results, the descriptor set is updated or a new set of descriptors is chosen. Otherwise, the model is disassembled and new ligands are created utilizing the information from the descriptor set. (b) Alternative approaches to creating a QSAR model. There are various ways for calculating, selecting, and generating QSAR models, depending on the study type [3].

Each QSAR package has a distinct approach for calculating descriptors and a consistent method for picking descriptors and building models. The backgrounds of these strategies are frequently glossed over in user manuals. At each phase in a QSAR investigation (Figure 9.2b), several methodologies are used to complete the same objective, notably the selection of descriptors and model building. Preparing the molecules for the study appears simple, but the quantity of options for the various parameters can be overwhelming. During the initial setup of a QSAR investigation, there are no clear guidelines for implementing many of the parameters, simply "rules-of-thumb." The rule-of-thumb criteria include atomic partial charges, molecular conformers, molecule alignment, and test set and Training Set composition.

STATISTICAL TOOLS FOR VALIDATING AND DEVELOPING MODELS

The effectiveness of any QSAR model depends on the choice of molecular descriptors and the ability to generate the appropriate mathematical link between the descriptors and the model, the biological action in question. Since the beginning of QSAR, it has been clear that defining molecular descriptors is a crucial part of the process [17,18]. Recent software developments have made it possible to build a large number of chemical descriptors that can be used with QSAR approaches. This makes it more challenging to select appropriate descriptors to describe the activity data [18,19]. When selecting chemical properties that are essential for activity, linear QSAR algorithms often employ three primary statistical techniques:

i. Multivariable linear regression (MLR) Analysis
ii. Principal components (PCA) Analysis
iii. Partial least squares (PLS) Analysis

The most straightforward technique for measuring a molecular descriptor with a strong link to activity change is MLR analysis. The process of creating an MLR model may involve either backward or forward stepwise regression to determine the best model (i.e., systematically adding or eliminating molecular descriptors, respectively), depending on the statistical test. However, for large collections of descriptors, the MLR approach can be time-consuming, and the user needs to be careful when avoiding variable combinations with high internal correlation. However, this problem can be fixed by using statistical software, which in some cases enables the user to automate the MLR process. The PCA method was developed to overcome the problems with MLR analysis by distilling information from multiple, possibly unnecessary variables into a smaller number of uncorrelated variables [20]. Consequently, PCA is a useful technique for reducing the number of independent variables in QSAR models. For systems where there are more molecular descriptors than observations, this approach is quite helpful. However, it is frequently challenging to evaluate PCA data in order to pinpoint the specific structural or physicochemical traits that are crucial for activity. In order to maximize the correlation, PLS combines MLR and PCA approaches by extracting the dependent variable (such as biological activity) into new components [21].

Systems with multiple dependent variables benefit from PLS. Other approaches to variable selection are also available, including Bayesian techniques and evolutionary algorithms, which are employed in linear QSAR models. The link between molecular characteristics and activity is frequently non-linear in biological systems. The neural network is one of the most used non-linear regression methods for building QSAR models. The self-learning ability of neural networks is the cornerstone of neural network-supported QSAR modeling. In this approach, the network learns the connection between the molecular descriptors and the related biological activity using the training set of ligand molecules. Back-propagation can also be used by a "trained" neural network to predict the activity of a novel molecule based on its structural and physicochemical properties. Like other regression techniques, neural networks can generate models with limited predictive power and are prone to overfitting. It can also be time-consuming and subjective to determine the best architecture for the neural network [18]. The Bayesian regularized neural network approach is a neural network variation that may be applied to the modeling of non-linear QSAR data [22,23]. Because it can automatically adjust the neural architecture and solve the overfitting issue, this approach has advantages over a standard neural network [18]. Recent work has developed a modified Bayesian regularized artificial neural network (BRANN) with a Laplacian prior to increase the number of descriptors used in QSAR models by removing ineffective descriptors [24]. It is necessary to validate the first QSAR model after it has been created. To form a model for QSAR studies, two primary forms of validation are needed:

i. Internal validation of models
ii. External validation of models

The most popular internal validation method is leave-one-out cross-validation. With this method, one observation is kept as validation data and the rest of the data is used as the training set to calculate the QSAR model's coefficients. The action of the test compound is then predicted using the model built from the compounds in the training set. This procedure is repeated until each additional chemical has been employed as a test set once. The predictive strength of the model is then assessed by computing the cross-validated r2 or Q2 using the following formula:

$$Q^2 = 1 - \frac{\sum\left(y_{\text{pred}} - y_{\text{obs}}\right)^2}{\sum\left(y_{\text{obs}} - y_{\text{mean}}\right)^2}$$

TABLE 9.2
Some 3D-QSAR Approaches and Their Characteristics

3D-QSAR Methods	Intermolecular Modeling	Alignment Criteria	Chemometric Approach
CoMFA	Ligand-Based	Dependent	Linear
CoMSIA	Ligand-Based	Dependent	Linear
GERM	Ligand-Based	Dependent	Linear
AFMoC	Receptor-Based	Dependent	Linear
CoMMA	Ligand-Based	Independent	Linear
SoMFA	Ligand-Based	Independent	Linear
COMPASS	Ligand-Based	Independent	Non-Linear

However, the disadvantage of this approach is that the calculation time grows as the training set size squares. K-fold cross-validation is an additional variant of this technique. This approach creates the training set by excluding a subset of k molecules at a time, as opposed to excluding a single component. By reducing the size of the effect training set, this strategy can shorten the calculation time. Nevertheless, the k-fold approach depends on the value of k being chosen. Furthermore, neither of the cross-validation techniques successfully validates the model using all of the data at once. Predicting the activity of the test set molecules that weren't utilized for model construction is part of the external validation technique [25] (Table 9.2).

3D-QSAR METHODS CLASSIFICATION

CoMFA: In 1987, Cramer developed the first 3D techniques, the DYnamic Lattice-Oriented Molecular Modeling System (DYLOMMS). Vectors are extracted from the molecular interaction fields using PCA, and these are then linked to biological activity [26]. He swiftly modified it by combining the two existing methods, GRID and PLS, to produce the powerful 3D QSAR methodology known as CoMFA [27]. Currently, CoMFA is a prototype of the 3D-QSAR approach [28]. Tripos Inc.'s Sybyl Software [29] uses a typical CoMFA approach that consists of the following consecutive phases:

- Each molecule's bioactive conformations are identified.
- Using either automated or human techniques, all of the molecules are aligned or superimposed according to the purported way of receptor interaction.
- Molecules have been superimposed at a 2 Å spacing in the middle of a lattice grid.
- The program compares the steric and electrostatic fields calculated surrounding the molecules in three dimensions using different probe groups positioned at each lattice junction.
- The results are expressed using correlation equations that contain the number of latent variable terms, each of which represents a linear combination of the initial independent lattice descriptors.
- The PLS technique connects the interaction energy or field values via the biological activity data by identifying and extracting the quantitative influence of specific chemical properties of molecules on their biological activity.

The following describes a number of characteristics that greatly affect the created CoMFA model's overall performance; Some of the characteristics are also applicable to other QSAR modeling methods too:

INFORMATION ON BIOLOGICAL ACTIVITIES OF MOLECULES

"GIGO" (garbage in, garbage out) principle is well known among various computational techniques. To create a good 3D-QSAR model, one also needs to use correct activity data. Despite the fact that 3D-QSAR techniques can be used on diverse datasets, some factors must be taken into account in order to preserve biological data accuracy [30]:

- It is more crucial for compounds to be part of a congeneric sequence in the case of classical QSAR.
- The mechanisms of action and binding modes of compounds should be identical or comparable.
- Compounds' biological activities should be correlated with their binding affinities, and the biological reactions they elicit should be quantifiable.
- It is optimal to collect biological data for molecules using standard laboratory methods (radioligand, activator, cofactor, pH, buffer, etc.) from just one object (cells, organisms, tissues, and/or from proteins).
- The activity data for each component must be expressed in identical units of measurement (binding, functional, IC50, Ki). The Ki value is preferred over the IC50 data since it is independent of the substrate concentration.
- The spectrum of biological function covered should be as wide as possible while keeping an identical mode of action. A uniform distribution of information with more than three logarithm units

is optimal, and the activity range should ideally be much larger than the data standard deviations.

- If at all possible, biological data should be equally distributed at its mean and uniformly distributed across its range of variation. Otherwise, to remove this skewness, the data might be log converted and reported as log (1/*C*), where *C* is the molar concentration of the drug that generates a typical reaction. Interestingly, the change in free energy is correlated with the 1/log of the compound's concentration.

CHOOSING COMPOUNDS AND OPTIMIZING SERIES

Optimizing the current leads through structural changes to increase their activity and lessen or eliminate their negative effects is one of the main uses of QSAR. When choosing substituents to modify compounds, there are numerous considerations to be made; some of the more significant ones are listed as follows [31,32]:

- To reduce collinearity across the variables, the chemicals or substitutes chosen should be believable departures from the ones that are currently in use.
- In order to maximize dissimilarity and orthogonality, the selected compounds or substitutes should possess characteristics that behave independently of one another.
- The selection process should be carried out in a way that maps the substituent (descriptor) space with the fewest possible compounds.
- The practicality and accessibility of the chosen compounds should also be taken into account.

OPTIMIZING THE MOLECULAR STRUCTURE IN THREE DIMENSIONS

One of the most important issues in 3D-QSAR is how to generate and represent the original molecular structure for analysis. The problem can be resolved using experimental and computational approaches. Databases contain a large number of well-resolved experimentally determined crystal structures like the Cambridge Structural Database [33] and the Protein Data Bank [34], among others. The fact that the crystal structures offer some structural information about the versatile molecule is one advantage. Molecular modeling methods, however, are particularly useful for compounds that are either absent or difficult to make under normal conditions. The 3D structures can be calculated using three methods:

- Use existing 3D structures from fragment libraries or create your own using a 3D computational graphics interface;
- quantitatively through the application of mathematical methods such as quantum or molecular mechanics and distance geometry; and
- using automated techniques frequently employed in the construction of databases with 3D structures.

Using theoretical calculation techniques, the geometries of 3D structures of molecules are adjusted by minimizing their conformal energies. Typical methods for structural optimization include the following:

- The molecular mechanics techniques, which are quick, reasonably accurate, and applicable to very big molecules like enzymes because they often do not specifically take electronic motion into account.
- Quantum mechanics or ab-initio techniques, whichconsider 3D electronic distribution surrounding the nucleus, are incredibly precise, laborious, computationally demanding, and unable to handle big molecules.
- Semi-empirical approaches, which require a lot of approximations, similar to molecular mechanics, but are r eally quantum mechanical in nature.

Molecular mechanics techniques are typically used to optimize the molecular shape, while semi-empirical or, less frequently, ab initio approaches are used to determine the atomic charges.

MOLECULE CONFORMATIONAL ANALYSIS

It is commonly known that every chemical with the existence of single or other bonds at any one time exists in a variety of so-called conformers of molecules. Bigger and adaptable molecules exist in various conformations under physiological conditions, but tiny molecules may only have one lowest energy conformation. Consequently, it becomes essential to incorporate different molecular conformations in a 3D-QSAR analysis [35,36]. Depending on the kind of molecules being studied, any of the conformational search techniques can be utilized:

- The systematic search approach, also known as grid search, creates every possible conformation by methodically changing each molecule's torsion angle by a certain amount while constantly maintaining bond angles and lengths.
- Creation of a set of conformations by a random search approach allows repeatedly and arbitrarily altering the internal (bond lengths, bond angles, and torsion/dihedral angles) or Cartesian (*x*, *y*, *z*) points of the starting molecular structure.
- Randomly altering a molecule's structure, the Monte Carlo approach mimics its dynamic behavior, creates conformations, calculates and compares their energy with that of the previous conformation, and if it is unique, then accept the model.

- The molecular dynamics technique, which simulates time-dependent movements and modifications to conformation in a molecular system using the second law of Newton's (force=mass acceleration), creates a so-called trajectory that shows how the positions and velocity of molecules in a molecular system vary over time.
- Simulated annealing, which breaks through massive energy barriers by heating the molecular system under study to high temperatures using molecular dynamics. To get low-energy conformations consistent with the Boltzmann distribution, the system gets cooler gradually over time.

The distance geometry technique generates a random set of coordinates by selecting random distances inside each pair of upper and lower bounds, so generating restrictions in a distance matrix that are subsequently used to generate energetically feasible conformations of a group of molecules.

Biological evolution serves as the foundation for genetic and evolutionary algorithms, which begin by generating a population of potential solutions to the issue. Over time, the solutions with the highest fitness scores experience crossings and mutations, passing on their positive traits to subsequent generations, leading to improved solutions in the form of new conformers.

IDENTIFYING THE BIOLOGICAL ACTIVITY OF MOLECULES

The term "bioactive conformation" describes the molecule's state when it is bound to the receptor. Both intrinsic forces between atoms and extrinsic forces between a molecule and its environment have a significant impact on its bioactive conformation. The identification of bioactive conformations is essential to the dependability of any 3D-QSAR technique [3,25,35,36]. Both theoretical and experimental methods can be used to obtain the molecules' bioactive conformations. The following are some experimental techniques for creating bioactive conformations: X-ray crystallography: This method is the sole way to obtain an exact 3D structure of the macromolecules. Although X-ray crystallography-generated drug–receptor complexes offer information that is reasonably precise, this technique has a number of drawbacks, including the following:

- The protein must crystallize, and the crystallizing media's composition typically differs from physiological circumstances.
- Since data collection typically takes a lengthy period, the approach creates a time-averaged structure; also, crystal packing frequently distorts the structures.
- It is often impossible for substrates or other molecules that are important for health to diffuse into existing crystals because the crystals are unstable and the active sites are blocked.
- Determining the locations of hydrogen atoms is difficult.
- Mistakes during precisely identifying the ligand's structure.

SPECTROSCOPY USING NMR

Using this technology, the 3D structural information is gathered in solution. The favored choice is when a molecule cannot be experimentally crystallized, such as in the case of receptors bound to the membrane or ligands that have not yet been isolated due to stability, resolution, or other difficulties. This strategy's essential elements are as follows:

- Since protein crystallization is not necessary, packing pressures in the crystal environment have no effect on the protein's conformation.
- It is possible to modify the solution parameters (temperature, pH, ionic strength, substrate, etc.) to correspond with the physical conditions. Additionally, the solvent has a significant impact on the outcomes.
- It is possible to get important data about the dynamic characteristics of molecular mobility.
- Saves time, but only works with tiny molecules.
- Hydrogen atom positions are resolvable.
- An overestimation of hydrogen bonding events could result from using apolar solvents.
- The structure derived via NMR may differ from that derived from experimental techniques, and it frequently does not depict the receptor-bound conformation.

Protein/homology modeling is a knowledge-based technique that theoretically provides 3D structural information. It compares the main sequence of a new protein with all sequences of structurally known proteins that are kept in a database such as PDB. The unknown protein's structure is predicted using templates derived from a database of proteins that are homologous to it. This strategy, however, is only applicable to target proteins whose structures can be determined. Additionally, the degree of sequence similarity between the protein to be modeled (the target) and the protein with a known structure (the template) determines the method's quality and applicability.

ALIGNMENT OF MOLECULES

One of the most important problems with most alignment-based 3D-QSAR methods is that their results are highly sensitive to the way the bioactive conformation of each molecule is positioned on top of the others [25,35,36]. When every molecule in a dataset shares a rigid core structure, aligning molecules is made easy with the least-squares fitting method. However, when there is structural variability in the sample, aligning very flexible molecules becomes difficult and time-consuming. Some techniques that have been proposed for overlapping the molecules as exactly as feasible include the following: Superimposition based on atom overlap: This method pairs molecules based on their atoms. It is the most widely utilized method and is also referred to as the pharmacophore approach since it offers the best matching for the preselected atom locations. It works well for identifying changes between similar molecules, but it cannot be applied to molecules with different structural types, where it is difficult to select related atoms. Binding site-based superimposition: To accomplish molecular alignment, this method entails superimposing the receptor residues or active sites that interact with the ligands. This approach is believed to be more practical despite conformational analysis problems caused by the additional degrees of freedom. The superimposition based on fields/pseudofields: This method assesses the similarity between molecules by comparing their calculated interaction energy fields. Using electrostatic similarity and molecule surface similarity indexes, the researchers have also used molecular alignment. Pharmacophore-based superimposition: This method uses a fake pharmacophore as a useful shared target template. Conformation guides each molecule to adopt the shape necessary for its sub-molecular properties to match a known pharmacophore or the one produced during the conformation analysis. superimposition according to different conformers: This method is particularly useful when the correct binding mode is uncertain or when ligands have a significant degree of conformational adaptability and can bind to a receptor in several ways. For example, the 3D-QSAR method COMPASS consistently selects the best biologically active geometry and optimal alignment from a set of initial postures.

CALCULATING THE ENERGY FIELD OF MOLECULAR INTERACTIONS

The stacked set of molecules is positioned in the center of a grid box or lattice to ascertain the relationship of energies among the ligands and different probe atoms placed at each lattice intersection [35–37]. When utilizing the CoMFA methodology to determine the interaction energies, the following aspects need to be taken into account:

- Generally, the grid spacing is 2 Å. The grid spacing has an inverse relationship with the level of calculation precision. As the grid spacing is reduced to 1 Å or less, the calculations grow increasingly intricate and require a significant amount of processing time and disk storage space. As previously indicated, the smaller grid spacing (0.5 Å) is usually employed when producing interaction energy fields for a reference (most active) molecule during molecular superimposition based on fields.
- The grid box is typically 3–4 Å larger than the union surface of the overlay molecules. The long-range nature of the electrostatic/Coulombic interactions may necessitate a larger grid box. Due to the inherent correlation between the electrostatic energies of neighboring lattice points, a grid box of similar size can be used for steric/van der Waals interactions.
- The statistics, particularly the number of components in the final CoMFA model, are often greatly affected by the position of the grid box. Usually, the initial models are made in several locations to find the optimal grid position. Two solutions have been proposed to reduce the instability. The first suggests that the set of molecules that are superimposed ought to be rotated so that none of the grid edges are parallel to them. The second approach proposes to replace the field value at a lattice point with the average of the field values at the vertices of a cube centered on the grid point, whose side length is two-thirds of the grid spacing.
- In CoMFA, the interaction energy is calculated using probes. A small molecule like water or a chemical fragment like a methyl group could serve as the probe. The electrostatic energies are calculated using an H+ probe, and the steric energies are included using an sp3 hybridized carbon atom with an effective radius of 1.53 Å and a charge of +1.0. Each probe is positioned alternately at each point where the lattice intersects, and the interaction energies between the probe and each of the compounds are calculated using different molecular force fields.
- By combining bond lengths, bond angles, dihedral angles, interatomic distances, coordinates, and other properties, a force field is a mathematical formula that empirically fits the potential energy surface. The primary intermolecular forces that are encountered in drug–receptor interactions include hydrophobic, steric/van der Waals, hydrogen bonding, and electrostatic/coulombic forces. Hydrophobic interactions frequently contribute to the intensity of binding, whereas electrostatic and hydrogen bonding interactions result in ligand–receptor specificity. The most often employed fields in CoMFA are steric and electrostatic fields, which are mostly enthalpic in nature. However, the entropic effects, which appear as hydrophobic interactions, are sometimes also considered in the CoMFA analysis. The uniqueness of the study and the validity of the underlying theory are the primary determinants of the type of field that should be developed and included in a CoMFA model.
- In CoMFA, electrostatic interactions are determined using Coulomb's equation, whereas van der Waals interactions are approximated using the standard [6–12] Lennard-Jones function. The potential energy at lattice sites close to the van der Waals surface changes significantly due to the steep slope of the Lennard-Jones potentials.

This implies that a small difference in the mutual shift or conformational changes between the compounds could result in extremely large variances in energy levels. Additionally, the Lennard-Jones and Coulombic potentials show singularities, or uncomfortably high values, at the atomic locations. Therefore, to avoid all of these problems in CoMFA, the cut-off values for steric and electrostatic energy (± 30 kcal/mol) are set.

PRE-TREATMENT AND DATA SCALING

In 3D-QSAR, the raw data are typically pre-treated to reduce redundancy prior to the actual chemometric analysis [35,36]. The standard deviation cut-off is a popular reduction technique that removes all energy columns with low standard deviations from the data because they take longer to calculate and don't materially affect the results. Similarly, there are a number of variable selection techniques that may be applied to lessen collinearity between the descriptors, reducing data overfitting and enhancing the model's prediction performance. Additionally, depending on where the lattice point is, cut-offs (± 30 kcal/mol) are used in CoMFA to modify the steric and electrostatic values. The data are often scaled after pretreatment, assigning the same weight to each description and combining them on a single platform for a statistically significant analysis. Scaling significantly improves the signal-to-noise ratio and allows one to rate the relative importance of various factors. Several scaling techniques can be effectively used with 3D-QSAR methodologies. For example, autoscaling divides each column by its standard deviation to scale the variables to zero mean and a unit standard deviation; block-adjusted scaling, which is particularly useful when other variables are included in the analysis along with the energy values; and block-scaling, which gives each category of variables the same weight by dividing the initial autoscaling weights of descriptors in one class by the square root of the number of descriptors in that class (CoMFA standard scaling). This scaling allows other variables to be assigned a weight that is similar to the total variables. Sometimes the column means are subtracted from the total data to apply centering to the pretreatment data. No coefficient values or comparative weights of the descriptors are changed, although the number of significant components from PLS may be one less than from the data without centering. The method aims to improve interpretability and numerical stability.

DEVELOPING AND VALIDATING MODELS

It is assumed that the descriptor (interaction energies and additional variables, if necessary) has a linear connection with the biological activities of the molecules after pretreatment and scaling [35–38]. Because there are substantially more independent (x) variables in CoMFA than there are chemicals in the dataset, the fitting process cannot be completed using standard linear regression analysis. Therefore, from a range of possible solutions, a stable and ideal QSAR model is chosen using thePLS technique. PCR, MLR, PCA, and other methods are also used to model linear connections. When the relationship between the dependent (y) and independent (x) variables is often non-linear or unpredictable, non-linear chemometric techniques—like neural networks—are employed; these techniques make no assumptions regarding the relationship between the variables during training and model development. Most of those chemometric methods used in QSAR modeling are discussed in the sections that follow. The most important criterion for assessing the quality of a QSAR model is its ability to accurately forecast the actions of molecules that were not part of the model's creation (external prediction) as well as those that are included in the training set (internal prediction) [38]. The internal predictive capabilities of the model can be evaluated through cross-validation using techniques like leave-one-out and leave-group-out, while the external predictivity of the model can be evaluated using an additional set of molecules (the test set) that weren't utilized in the model's creation. The statistical confidence and robustness of the created models are further assessed using randomization (y-scrambling), bootstrapping analysis, and Fisher's statistics. Each one of these cross-validation methods has been explained in the following sections.

DISPLAY OF RESULTS AFTER STUDIES

CoMFA generates an equation at each grid point that connects the interaction energy field contribution to biological activity. For easy visual understanding, results are usually presented as contour plots of coefficients (or the scalar product of coefficients and standard deviation). These plots show important spatial areas around molecules where specific structural changes significantly modify the activity [29,35,36]. Both the positive and negative contouring are frequently shown for every interaction energy field. The steric field contours are displayed in green (positive contours, more bulk favored) and yellow (negative contours, less bulk favored), whereas the electrostatic field contours are displayed in red (positive contours, electronegative substituents favored) and blue (negative contours, electropositive substituents favored). In addition to contour plots, CoMFA provides two other plot styles from PLS models: score plots and loading/weight plots. While score graphs between biological activity (Y scores) and latent variables (X-scores) demonstrate the relationship between the activity and the structures, plots of latent variables (X-scores) reveal the similarity/dissimilarity between the structures of molecules and their clustering propensities.

THE LIMITATIONS AND DRAWBACKS OF COMFA

Despite having several benefits over traditional QSAR and performing well in a range of real-world applications, CoMFA has a number of drawbacks and flaws, which are listed below [26,30,35,36]:

- Since the method's introduction in 1988, numerous CoMFA method applications in diverse fields have been described [39]. The first set of data analyzed is the binding affinity of the steroid dataset [40] for human testosterone-binding globulins (TBG) and corticosteroid-binding globulins (CBG). Enzymes that are highly sensitive to bioactive conformation, different ligand binding modalities, alignment requirements, and component count have a large number of successful CoMFA technique projects.
- An excessive number of tunable parameters, such as probe atom type, step size, lattice positioning, and overall orientation.
- Uncertainty in the variables and compound selection.
- Variable selection processes in fragmented contour maps.
- Hydrophobicity is not accurately measured.
- Cut-off limit application.
- A low signal-to-noise ratio caused by a lot of irrelevant field variables.
- Potential energy functions with flaws.
- A number of real-world PLS issues.
- Only applicable to data obtained in vitro.

COMSIA

The limitations of CoMFA were addressed by developing CoMSIA. Utilizing molecular similarity indices produced from modified SEAL similarity fields as descriptors, CoMSIA concurrently considers steric, electrostatic, hydrophobic, and hydrogen bonding information. These indices are indirectly calculated by comparing the similarity of each molecule in the dataset to a common probe atom which is positioned at the intersections of a surrounding grid or lattice and has a radius of 1 Å, charge of +1, and hydrophobicity of +1. For the purpose of determining similarity at each grid point, the alignment dataset takes into account the mutual distances between the probe atom and the atoms of the molecules. To calculate the molecular features and describe this distance dependency, functions of the Gaussian type are employed. Because the underlying Gaussian-type functional forms are "smooth" and devoid of singularities, their slopes are not steeper compared to those of the Coulombic and Lennard-Jones potentials in CoMFA, eliminating the necessity to set arbitrary cut-off limits. In most cases, these functions nevertheless produce acceptable results, even when atoms overlap. Despite sharing many of the identical issues as CoMFA, CoMSIA does offer some distinct benefits, such as the following:

- The interactions between grid-based probe atoms are protected from abrupt changes by employing the similarity indices distribution, which follows the normal distribution.
- Hydrophobic and hydrogen bonding fields, including acceptors and donors of hydrogen bonds, as well as steric and electrostatic potential fields, are all part of the similarity probe selection process.
- The effect of the solvent entropic terms can be included in a hydrophobic probe as well.
- Typical CoMFA contours indicate the spatial positions of aligned molecules that will interact positively or negatively with a possible receptor environment. Conversely, CoMSIA contours reveal the areas of the region occupied by the ligands that "like" or "dislike" a group having a particular physicochemical characteristic. By examining the correlation between the required attributes and a probable ligand form, it is easier to verify if all the features required for function are found in the structures being considered.

A few examples of recent applications of CoMSIA include the development of 3D-QSAR models for thiazolidin-4-one derivatives as anti-HIV-1 agents [39], proteasome inhibitors [37], urease inhibitors [38], and aldose reductase inhibitors [40] based on boron-containing dipeptides [37]. Tripos Inc. provides Sybyl users with access to both CoMFA and CoMSIA [29].

MSA

With the goal of merging conformational analysis with the conventional Hansch approach, MSA is a ligand-based 3D QSAR formalism. The quantitative characterization, representation, and manipulation of molecule shape are addressed in order to construct a QSAR model [41]. An intramolecular conformational analysis is performed on all molecules in the dataset initially, with a fixed valence geometry. The torsion angles of all the other types are scanned at 30° intervals, except for the amide N (C=O) torsion, which is scanned at 180° intervals. To estimate the confoundational energies, a molecular mechanics force-field with a constant valence geometry is utilized, which includes dispersion/steric, electrostatic, and, if applicable, hydrogen bonding components. Rigid fixed valence geometry energy minimizations are based on the systematic identification and recording of all apparent intramolecular energy minima for each molecule. The "active" conformation of each analog can be either rigorously minimized or apparently minimized in the following steps. Each analog's active conformation is determined using the LBA-LCS (loss in biological activity–loss in conformational stability) method. The idea behind it is that active analogs share stable, low-energy intramolecular conformer states, while inactive analogs have a more unstable, high-energy state. We select the mutant shape as the reference structure based on the common and different volume combinations that result from different compound alignments or active conformations. Each chemical in the database has its potential active conformation matched with the shape reference and checked in pairs. The next step is to calculate a number of descriptors that compare the morphologies of different molecules. As a function of conformation and relative intermolecular geometry, the common overlap steric volume (COSV) between molecular pairs is an important shape component. Actually, it determines the amount of steric space shared by two molecules in a particular intermolecular bond. One of the other descriptors has length dimensions but is not a cumulative measure of molecular distances, and the other has area dimensions but is not a physical measure of common atomic surface areas between two molecules. Both of these alternative mathematical representations of COSV are arbitrary definitions that can be helpful in developing empirical QSARs [41]. You can make a 3D MSA by combining these shape-by-shape variables with electronic and non-shape therapeutic descriptors, as the ones in the Hansch equation (Es, QSAR model). The MLR technique, along with other chemometric methods like PLS and GA, correlates the shape similarity descriptors and non-shape factors with the biological activities of the molecules. Two options to visually represent the MSA results are to show the most active analog in its active conformation or to superimpose shape descriptors onto the molecular structure of the most active molecule.

GRID

As an alternative to the CoMFA approach, the first program developed for medicinal chemists was GRID. Through molecular field analysis, it calculates the interaction energy fields and identifies the energetically beneficial binding sites on known-structure molecules. There are two key differences between this method and CoMFA, despite the fact that both compute explicit non-bonded or non-covalent interactions between a macromolecule and a probe, which is a small chemical group with user-defined properties, at sample positions on a lattice around the macromolecule. Initially, the interaction energies are computed using a 6-4 potential function, which is more refined compared to the 6–12 version of the Lennard-Jones type in CoMFA. The software does more than just compute the usual electrostatic and steric potentials; it also uses a "DRY probe" to find the hydrophobic potential and a hydrogen bond donor and accelerator to find the hydrogen bonding potential. The addition of a water probe allowed for the computation of hydrophobic interactions [3]. Since it is both electrically neutral and capable of donating and accepting hydrogen bonds, the water probe is anticipated to provide energies that encompass both hydrophobic and steric bonding interactions, in addition to those represented by log P due to its molecular surface area. It is common practice to use the methyl, amine NH_2, carboxylate, and hydroxyl groups as standalone probes in addition to the water and DRY groups. On the grid, you can see the protein structure and, for each probe, the contour surfaces at various energy levels. In most cases, the contours' positive energy levels depict the shape of the target, while the contours' negative energy levels indicate the best places for the ligand to bind. Some recent applications of the GRID method include ligand-binding site classification and comparison [42], predictive pharmacophore model construction [43], energetically favorable binding site identification and surface property characterization [44], and rationally designing potent inhibitors of influenza virus sialidase [43]. Many 3D QSAR and QSAR analyses in CoMFA, GOLPE, and SIMCA use GRID maps as input descriptors. A statistical approach can be used to quantitatively connect these contact energies to the biological activity. In order to run GRID, one needs Molecular Discovery Ltd. [45].

HASL

One approach to approximating molecular morphologies within an active site is the inverse grid-based Hypothetical Active Site Lattice (HASL) method. The first step is to transform the Cartesian coordinates of a layered set of molecules (x, y, z) into a 3D grid consisting of the regularly spaced points that are as follows:

- Positioned perpendicular.
- Divided by a specific distance known as the resolution, which establishes how many grid points a molecule is represented by.
- All spread out inside the atoms' van der Waals radii that make up the molecule.

Molecular lattice, a resultant point structure, represents the receptor active site map. The total lattice dimensions are determined by the resolution used and the size of the molecules. One common way to build the HASL is to take an energy-minimized, similarly shaped conformation from a reference molecule and utilize it as a starting point. After that, the reference molecule in its chosen conformation is placed over a regular grid with a resolution that is determined. The grid lies at the origin of a Cartesian coordinate system. All "occupied" grid points inside the van der Waals radii of the molecule's atoms make up the molecular lattice. The electronic properties of the occupying atoms are faithfully captured by assigning the lattice points a "HASL-type" value that is dictated by the electron density of the atoms, which is the fourth dimension of the molecular lattice. The values of +1, −1, and 0 for electron-rich (like O and N), electron-poor (like C in C=O), and neutral atoms/substituents, respectively, are used to indicate H-bond acceptors, donors, and lipophilic atom types. Such designs based on internal atoms allow for the superimposition of molecules with similar atomic properties in 3D space, allowing for the greatest potential spatial and physicochemical complementarity within that space. These structures appear to be completely different from one another, but they actually have an electronic "sense" that is equivalent to one another. Hydrophobicity is just one of several user-selected physicochemical properties that can serve as the fourth dimension, similar to electron density. The next step is to select a different molecule and subject it to the identical procedure; this will generate a 4D lattice that can be compared to the stationary reference molecule's and then aligned with it. Using a sequence of translational and rotational movements, an intermediate lattice is constructed at each stage to systematically seek for the optimal alignment of the molecules according to their lattices. Until an ideal match is discovered, this procedure is carried out again. The similarity between the two molecules is determined using a fitting function that takes into account the degree of correlation or the number of shared spots between the two lattices. Once both molecular lattices are aligned as best they can, the remaining lattice points of the fit molecule are added to the reference molecule to build a larger reference lattice or a new composite construct that incorporates the information from both molecules. This fitting and merging method continues to incorporate every molecule of the training set into the growing HASL, until a reference lattice is created that contains every point from all of the molecular lattices. To determine the relative contributions of different lattice sites to a molecule's experimental activity value, it is necessary to first distribute the value evenly across all of the lattice points. Averaging the partial biological activity values over many-molecule-shared lattice positions is the initial step. Next, the partial bioactivity values are adjusted iteratively until they match the experimental activity data from the complete training set. As a standardized model, this iteratively revised HASL is then used to predict the activities of untested substances. Predicting a chemical's bioactivity is as simple as tallying up its partial activity values at sites that are comparable to the composite reference graph. One useful application of the HASL method is to examine the antimalarial activity of artemisinin analogs in vitro [3]. The HASL module is part of Sybyl Software [29].

GERM

For 3D-QSAR and the practical creation of 3D models of macromolecular binding sites where the target receptor's structure cannot be established via crystallography or homology modeling, the Genetically Evolved Receptor Models (GERM) technique is utilized. The primary condition for GERM is a structure–activity series that has a reasonable alignment of realistic conformers. The procedure involves enclosing the stacked collection of molecules with an atomic shell (like the initial layer of atoms in the active site) and assigning these atoms specific atomic kinds (aliphatic H, polar H, etc.) so they conform to the types of atoms often present in proteins. An even distribution of aliphatic carbon atoms around a sphere surrounding the training set of aligned ligands maximizes the van der Waals contact between the model carbon atoms and the substrate molecules. After their positions have been determined, the carbons can be occupied by any type of atom, even none at all. There is a real-world problem with increasing the number of shell atoms and the kind of atoms they include; there are just too many combinations for a systematic way to find the best model. Consequently, genetic algorithms (GA) are employed to resolve this exceedingly intricate search issue. Following the docking of each ligand from the training set into a receptor active site model generated by GA, the intermolecular non-bonded interaction energies (including electrostatic and van der Waals components) are computed using a CHARMm molecular mechanics force field. The last step is to relate these calculated interaction energies to the biological activities of the molecules. One advantage of this approach is that it displays the model as a 3D depiction of the spatial properties of the receptor. A limitation of the GERM approach is that it limits consideration of ligand conformations to a single orientation in the training set and a single orientation in the binding site. Due to its reliance on computing interaction energies with the hypothetical receptor, this approach has all of the shortcomings of other approaches, including the alignment issue. If all of the molecules in the set bind without drastically altering the binding site, then GERM could be an effective technique. Several sweeteners' bioactivities have been linked to their calculated intermolecular energies using this method. Using the method in tandem with de novo ligand-design techniques or searching 3D-structural databases to find new leads is not out of the question [3].

COMBINE

The Comparative Binding Energy Analysis (COMBINE) method has been created to utilize the structural data from ligand–macromolecule complexes in a 3D-QSAR paradigm. The approach is based on the notion that the free energy of binding can be linked to a subset of energy components that are obtained from the structures of ligands and receptors in both their bound and unbound states [3,46]. Each compound is given an equal number of fragments once the ligands are broken up. "Dummy" fragments are provided to ligands that do not have a certain fragment. The non-bonded (van der Waals and electrostatic) interaction energies between each receptor residue and each ligand fragment are computed using a force field from molecular mechanics. Every pair of residues or fragments' energy is also calculated for the complexes as well as for unbound ligands and receptors. The electrostatic interactions are calculated using a distance-dependent dielectric constant, while the non-bonded interactions are not subject to cut-off limits. After eliminating the unnecessary descriptors from the data using the variable selection tool in the GOLPE application, the PLS technique is utilized to link the interaction energy values to the biological activities of the molecules. Like all other interaction energy-based 3D-QSAR approaches, COMBINE suffers from fundamental flaws in computing these energies. The method's predictive power can also be increased by making improvements in a number of areas, including the description of the electrostatic term, adding suitable descriptors for solvation and entropic effects, and optimizing certain aspects of the methodology, such as the choice of ligand fragment definitions and the details of the variable selection protocol. Recently, 3D-QSAR models have been built using the COMBINE methodology to determine amino acid residues in haloalkane dehalogenase LinB that modify its substrate specificity [47], estimate the binding affinity of non-peptide inhibitors of HIV-1 protease [48], and evaluate the selectivity and specificity of Ras proteins [49].

CoMMA

Comparative Molecular Moment Analysis (CoMMA), one of the unique alignment-independent 3D-QSAR methods, calculates molecular similarity descriptors using spatial moments of molecular mass (shape) and charge distributions up to and including second order as well as related characteristics. For every molecular structure, two Cartesian reference frames are then created. One frame consists of the fundamental inertial axes as calculated with respect to the center of mass. With respect to the molecular "center-of-dipole," the principal quadrupolar axes provide an alternate frame of reference for neutral molecular species. The dipolar, quadrupolar, and displacement descriptors are then calculated using the principal inertial axes translated so that their origin is superposed on the center-of-dipole. Interestingly, these descriptors are obtained after translation to the center of mass and center of dipole of each molecule, preserving the alignment independence of the system. Finally, those molecular moment descriptors are correlated against the biological activities of drugs using the PLS technique. Published results show that CoMMA descriptors are less sensitive to molecular conformations than CoMFA field parameters. The authors claim that the CoMMA descriptors could help solve issues like large-scale screening and molecular diversity. The method has been used to build robust 3D-QSAR models and produce combinatorial QSAR of ambergris fragrance compounds in order to comprehend the structure–activity relationships of the benchmark steroid dataset [3,50].

CoMSA

Comparative Molecular Surface Analysis (CoMSA), a non-grid 3D-QSAR technique, uses the molecular surface to determine which regions of the compounds require comparison using the mean electrostatic potentials [3,51]. The procedure entails optimizing the geometry of the molecules in the dataset and giving them partial atomic charges. Following the extraction of the signals to the Cartesian coordinates of points randomly sampled at the molecules' van der Waals surface, the 3D surface of the molecules is transformed into 2D topographical maps using Kohonen's self-organizing maps (SOM, a type of neural network). Furthermore, the 2D topographical maps are projected onto the partial atomic charges of the atomic molecule representations. Each grid cell is explained by a mean value of the potential equal to the corresponding points found in that cell. At the surface locations, the molecule electrostatic potentials (MEPs) are calculated. The calculated mean electrostatic potential values are converted into vectors by comparing the molecules' corresponding topographical maps, and the vectors for each molecule in the series are then stacked into a matrix. The resulting comparative matrix of the mean electrostatic potentials (converted into vectors) is then used to generate a 3D-QSAR model using the PLS approach. In contrast to CoMFA and related methods, CoMSA computes average property values (MEPs) for a particular area of the molecular surface as opposed to a discrete set of locations in order to compare the molecular characteristics. Recently, a receptor-dependent CoMSA model was developed and applied to investigate sulforaphane compounds as quinone reductase activators using multipose molecular docking and iterative variable elimination PLS (IVE-PLS) [3,52].

AFMoC

Adaptation of Fields for Molecular Comparison (AFMoC), often known as a "reverse" CoMFA (=AFMoC) approach, is a 3D-QSAR method that use fields derived from the protein environment instead of the superimposed ligands as in CoMFA [3,53]. The receptor binding site is initially filled with a grid that is regularly spaced. The potential fields are then created by mapping the knowledge-based pair-potentials connecting protein atoms and ligand atom probes onto the grid intersections. Based on these potential fields, interaction fields are generated by multiplying the nearby grid values by the distance-dependent atom-type properties of actual ligands bound into the active site. The PLS approach, which gives each field value a unique weighting factor, is then used to connect these atom-type specific interaction fields with the compounds' binding affinities. Lastly, contribution maps are used to graphically illustrate the results, and the generated 3D-QSAR equation is used to estimate the binding affinities of novel ligands. This approach's distinguishing characteristics include the following:

- By combining a protein-based CoMFA technique with a customized scoring function, the requirement to use entire ligand training sets is removed.
- The steady progression from generally acceptable knowledge-based potentials toward protein-specific pair-potentials reflects the amount and degree of structural variability incorporated in the ligand training data.
- The methodology is expected to incorporate the entropic effects resulting from (de-)solvation in addition to the enthalpic contribution, since structural information from empirically determined complexes has been converted into statistical pair potentials.
- PLS results are easier to interpret when atom-type specific interaction fields, which are mutually orthogonal, are used.

CoRIA

Comparative Residue Interaction Analysis (CoRIA) is a 3D-QSAR technique that uses descriptors that describe the thermodynamic processes involved in ligand binding to analyze both the quantitative and qualitative facets of the ligand–receptor recognition process. The initial CoRIA methodology was limited to calculating a non-bonded (vander Waals and Coulombic) interaction energies between the ligand and each of the active site residues of the receptor that interact with the ligand [3,54,55,56]. Using the genetic version of the PLS technique (G/PLS), these energies were then correlated with the biological activities of molecules through with other physicochemical variables describing the thermodynamics of binding, such as lipophilicity, molar refractivity, surface area, molecular volume, Jurs descriptors, strain energy, etc. In order to address the problems with peptide QSAR, this approach was then modified and expanded to produce two new CoRIA variants: reverse-CoRIA (rCoRIA) and mixed-CoRIA (mCoRIA). These methods separate the peptide (ligand) into individual amino acids and determine the interaction energies (vander Waals, Coulombic, and hydrophobic interactions) for each amino acid in the peptide with the receptor as a whole (rCoRIA) as well as the individual active site residues in the receptor (mCoRIA), in conjunction with other thermodynamic descriptors (such as free energy of solvation, entropy loss on binding, strain energy, and solvent assessable surface area). The G/PLS chemometric technique is then applied to link the biological activity with these independent variables [57]. CoRIA approaches use the wealth of information present in ligand–receptor complexes to extract crucial details regarding the kind and character of important interactions at the ligand and receptor levels. The design of novel molecules and receptors can then make direct use of this information. The techniques can forecast alterations in the corresponding ligand and the receptor as long as the structures of certain ligand–receptor complexes are known. The method has been effectively applied to examine the interactions between inhibitors and Cyclooxygenase-2 [54]. It is difficult to apply these methods to small organic compounds because, unlike peptides, there is no commonly accepted or rational method for disassembling small molecules. The methods can be improved by solving entire ligand–protein complexes, doing extensive conformational sampling with molecular dynamics, and adding other important interactions such hydrogen bonds [3].

REFERENCES

1. Tropsha A. Best practices for QSAR model development, validation, and exploitation. *Mol Inform.* 2010;29(6–7): 476–88. Available from: https://doi.org/10.1002/minf.201000061.
2. Wermuth CG, Ganellin CR, Lindberg P, Mitscher LA. Glossary of terms used in medicinal chemistry (IUPAC Recommendations 1998). *Pure Appl Chem.* 1998;70(5):1129–43. Available from: https://doi.org/10.1351/pac199870051129.
3. Verma J, Khedkar VM, Coutinho EC. 3D-QSAR in drug design—A review. *Curr Top Med Chem.* 2010;10(1):95–115. Available from: https://doi.org/10.2174/156802610790232260.
4. Albert A, Goldacre R, Phillips J. 455. The strength of heterocyclic bases. *J Chem Soc.* 1948;1948:2240–2249. Available from: https://doi.org/10.1039/jr9480002240
5. Taft RW. Polar and steric substituent constants for aliphatic and o-Benzoate groups from rates of esterification and hydrolysis of esters. *J Am Chem Soc.* 1952;74:3120–8.
6. Hansch C, Maloney PP, Fujita T, Muir RM. Correlation of biological activity of phenoxyacetic acids with Hammett substituent constants and partition coefficients. *Nature.* 1962;194(4824): 178–80. Available from: https://doi.org/10.1038/194178b0.
7. Hansch C, Fujita T. p-σ-π analysis. A method for the correlation of biological activity and chemical structure. *J Am Chem Soc.* 1964;86(8):1616–26. Available from: https://doi.org/10.1021/ja01062a035.
8. Hansch C. Quantitative approach to biochemical structure-activity relationships. *Acc Chem Res.* 1969;2(8):232–9. Available from: https://doi.org/10.1021/ar50020a002.
9. Free SM Jr, Wilson JW. A mathematical contribution to structure-activity studies. *J Med Chem.* 1964;7(4):395–9. Available from: https://doi.org/10.1021/jm00334a001.
10. Fujita T, Ban T. Structure-activity study of phenethylamines as substrates of biosynthetic enzymes of sympathetic transmitters. *J Med Chem.* 1971;14(2):148–52. Available from: https://doi.org/10.1021/jm00284a016.
11. Kubinyi H. Quantitative structure-activity relationships. IV. Nonlinear dependence of biological activity on hydrophobic character: A new model. *Arzneimittelforschung.* 1976;26(11):1991–7.
12. Hansch C, Gao H. Comparative QSAR: Radical reactions of benzene derivatives in chemistry and biology. *Chem Rev.* 1997;97:2995–3060.
13. Hurst T, Heritage T. HQSAR - A highly predictive QSAR technique based on molecular holograms. In: *The 213th ACS National Meeting.* San Francisco, CA; 1997.
14. Naumann T, Lowis D. HQSAR: A new, highly predictive QSAR technique. In: *Proceedings of the 1st International Electronic Conference on Synthetic Organic Chemistry.* Basel, Switzerland: MDPI; 1997.
15. Cho SJ, Zheng W, Tropsha A. Rational combinatorial library design. 2. Rational design of targeted combinatorial peptide libraries using chemical similarity probe and the inverse QSAR approaches. *J Chem Inf Comput Sci.* 1998;38(2):259–68. Available from: https://doi.org/10.1021/ci9700945.
16. Labute P. Binary QSAR: A new method for the determination of quantitative structure activity relationships. *Pac Symp Biocomput.* 1999;1999:444–55. Available from: https://doi.org/10.1142/9789814447300_0044.
17. Akamatsu M. Current state and perspectives of 3D-QSAR. *Curr Top Med Chem.* 2002;2(12):1381–94. Available from: https://doi.org/10.2174/1568026023392887.
18. Winkler DA. Overview of quantitative structure-activity relationships (QSAR). In: Rapley R, Harbron S (Eds.), *Molecular Analysis and Genome Discovery.* Chichester, UK: John Wiley & Sons, Ltd; 2005, pp. 347–67.
19. Topliss JG, Edwards RP. Chance factors in studies of quantitative structure-activity relationships. *J Med Chem.* 1979;22(10):1238–44. Available from: https://doi.org/10.1021/jm00196a017
20. Kubinyi H. QSAR and 3D QSAR in drug design Part 1: Methodology. *Drug Discov Today.* 1997;2(11):457–67. Available from: https://doi.org/10.1016/s1359-6446(97)01079-9.

21. Geladi P, Kowalski BR. Partial least-squares regression: A tutorial. *Anal Chim Acta.* 1986;185:1–17. Available from: https://doi.org/10.1016/0003-2670(86)80028-9.
22. Caballero J, Fernández M. Linear and nonlinear modeling of antifungal activity of some heterocyclic ring derivatives using multiple linear regression and Bayesian-regularized neural networks. *J Mol Model.* 2006;12(2):168–81. Available from: https://doi.org/10.1007/s00894-005-0014-x.
23. Winkler DA, Burden FR. Modelling blood-brain barrier partitioning using Bayesian neural nets. *J Mol Graph Model.* 2004;22(6):499–505. Available from: https://doi.org/10.1016/j.jmgm.2004.03.010.
24. Burden FR, Winkler DA. An optimal self-pruning neural network and nonlinear descriptor selection in QSAR. *QSAR Comb Sci.* 2009;28(10):1092–7. Available from: https://doi.org/10.1002/qsar.200810202.
25. Cronin MTD (Eds.) *Predicting Chemical Toxicity and Fate.* Boca Raton, FL: CRC Press; 2004. Available from: https://doi.org/10.1201/9780203642627
26. Wise M, Cramer RD, Smith D, Exman I. Progress in three dimensional drug design: The use of real time colour graphics and computer postulation of bioactive molecules in DYLOMMS. In: Dearden J (Eds.), *Quantitative Approaches to Drug Design.* Amsterdam; Elsevier; 1983, pp. 145–6.
27. Cramer I, Patterson RD, Bunce DE. Comparative molecular field analysis (CoMFA). 1. effect of shape on binding of steroids to carrier proteins. *J Am Chem Soc.* 1988;110:5959.
28. Podlogar BL, Ferguson DM. QSAR and CoMFA: A perspective on the practical application to drug discovery. *Drug Des Discov.* 2000;17(1):4–12.
29. Tripos Associates Inc. *Sybyl, Version 7.1.* St. Louis, MO: Tripos Associates Inc.; 2005.
30. Oprea T. 3D QSAR Modeling in Drug Design. In: Anwar T, Kumar P, Khan AU (Eds.), *Computational Medicinal Chemistry for Drug Discovery.* Boca Raton, FL: CRC Press; 2003.
31. Tokarski JS, Hopfinger AJ. Three-dimensional molecular shape analysis-quantitative structure-activity relationship of a series of cholecystokinin-A receptor antagonists. *J Med Chem.* 1994;37(21):3639–54. Available from: https://doi.org/10.1021/jm00047a021.
32. Cho WJ, Kim EK, Park MJ, Choi SU, Lee CO, Cheon SH, et al. Synthesis and comparative molecular field analysis (CoMFA) of antitumor 3-arylisoquinoline derivatives. *Bioorg Med Chem.* 1998;6(12):2449–58. Available from: https://doi.org/10.1016/s0968-0896(98)80019-9.
33. The Cambridge Structural Database. Cam.ac.uk. [cited 2025 May 5]. Available from: https://www.ccdc.cam.ac.uk/solutions/software/csd/
34. RCSB Protein Data Bank. RCSB PDB: Homepage. Rcsb.org. [cited 2025 May 5]. Available from: https://www.rcsb.org/
35. Zhao X, Chen M, Huang B, Ji H, Yuan M. Comparative molecular field analysis (CoMFA) and comparative molecular similarity indices analysis (CoMSIA) studies on α(1A)-adrenergic receptor antagonists based on pharmacophore molecular alignment. *Int J Mol Sci.* 2011;12(10):7022–37. Available from: https://doi.org/10.3390/ijms12107022
36. Putz MV, Duda-Seiman C, Duda-Seiman D, Putz A-M, Alexandrescu I, Mernea M, et al. Chemical structure-biological activity models for pharmacophores 3D-interactions. *Int J Mol Sci.* 2016;17(7):1087. Available from: https://doi.org/10.3390/ijms17071087.
37. Zhu Y-Q, Lei M, Lu A-J, Zhao X, Yin X-J, Gao Q-Z. 3D-QSAR studies of boron-containing dipeptides as proteasome inhibitors with CoMFA and CoMSIA methods. *Eur J Med Chem.* 2009;44(4):1486–99. Available from: https://doi.org/10.1016/j.ejmech.2008.07.019.
38. Ul-Haq Z, Wadood A, Uddin R. CoMFA and CoMSIA 3D-QSAR analysis on hydroxamic acid derivatives as urease inhibitors. *J Enzyme Inhib Med Chem.* 2009;24(1):272–8. Available from: https://doi.org/10.1080/14756360802166665.
39. Murugesan V, Prabhakar YS, Katti SB. CoMFA and CoMSIA studies on thiazolidin-4-one as anti-HIV-1 agents. *J Mol Graph Model.* 2009;27(6):735–43. Available from: https://doi.org/10.1016/j.jmgm.2008.11.006.
40. Liu H-Y, Liu S-S, Qin L-T, Mo L-Y. CoMFA and CoMSIA analysis of 2,4-thiazolidinediones derivatives as aldose reductase inhibitors. *J Mol Model.* 2009;15(7):837–45. Available from: https://doi.org/10.1007/s00894-008-0439-0.
41. Hopfinger AJ. A QSAR investigation of dihydrofolate reductase inhibition by Baker triazines based upon molecular shape analysis. *J Am Chem Soc.* 1980;102(24):7196–206. Available from: https://doi.org/10.1021/ja00544a005.
42. Hoppe C, Steinbeck C, Wohlfahrt G. Classification and comparison of ligand-binding sites derived from grid-mapped knowledge-based potentials. *J Mol Graph Model.* 2006;24(5):328–40. Available from: https://doi.org/10.1016/j.jmgm.2005.09.013.
43. Ortuso F, Langer T, Alcaro S. GBPM: GRID-based pharmacophore model: Concept and application studies to protein-protein recognition. *Bioinformatics.* 2006;22(12):1449–55. Available from: https://doi.org/10.1093/bioinformatics/btl115.
44. Caron G, Nurisso A, Ermondi G. How to extend the use of grid-based interaction energy maps from chemistry to biotopics. *ChemMedChem.* 2009;4(1):29–36. Available from: https://doi.org/10.1002/cmdc.200800259
45. GRID 2021. Moldiscovery.com. [cited 2025 May 5]. Available from: https://www.moldiscovery.com/software/grid/
46. Lushington GH, Guo J-X, Wang JL. Whither combine? New opportunities for receptor-based QSAR. *Curr Med Chem.* 2007;14(17):1863–77. Available from: https://doi.org/10.2174/092986707781058814.
47. Kmunícek J, Hynková K, Jedlicka T, Nagata Y, Negri A, Gago F, et al. Quantitative analysis of substrate specificity of haloalkane dehalogenase LinB from Sphingomonas paucimobilis UT26. *Biochemistry.* 2005;44(9):3390–401. Available from: https://doi.org/10.1021/bi047912o.
48. Nakamura S, Nakanishi I, Kitaura K. Binding affinity prediction of non-peptide inhibitors of HIV-1 protease using COMBINE model introduced from peptide inhibitors. *Bioorg Med Chem Lett.* 2006;16(24):6334–7. Available from: https://doi.org/10.1016/j.bmcl.2006.09.006.
49. Tomić S, Bertosa B, Wang T, Wade RC. COMBINE analysis of the specificity of binding of Ras proteins to their effectors. *Proteins.* 2007;67(2):435–47. Available from: https://doi.org/10.1002/prot.21321.
50. Kovatcheva A, Golbraikh A, Oloff S, Xiao Y-D, Zheng W, Wolschann P, et al. Combinatorial QSAR of ambergris fragrance compounds. *J Chem Inf Comput Sci.* 2004;44(2):582–95. Available from: https://doi.org/10.1021/ci034203t.
51. Polanski J, Walczak B. The comparative molecular surface analysis (COMSA): A novel tool for molecular design. *Comput Chem.* 2000;24(5):615–25. Available from: https://doi.org/10.1016/s0097-8485(00)00064-4.

52. Magdziarz T, Mazur P, Polanski J. Receptor independent and receptor dependent CoMSA modeling with IVE-PLS: Application to CBG benchmark steroids and reductase activators. *J Mol Model.* 2009;15(1):41–51. Available from: https://doi.org/10.1007/s00894-008-0373-1.
53. Gohlke H, Klebe G. DrugScore meets CoMFA: Adaptation of fields for molecular comparison (AFMoC) or how to tailor knowledge-based pair-potentials to a particular protein. *J Med Chem.* 2002;45(19):4153–70. Available from: https://doi.org/10.1021/jm020808p.
54. Datar PA, Khedkar SA, Malde AK, Coutinho EC. Comparative residue interaction analysis (CoRIA): A 3D-QSAR approach to explore the binding contributions of active site residues with ligands. *J Comput Aided Mol Des.* 2006;20(6):343–60. Available from: https://doi.org/10.1007/s10822-006-9051-5.
55. Dhaked DK, Verma J, Saran A, Coutinho EC. Exploring the binding of HIV-1 integrase inhibitors by comparative residue interaction analysis (CoRIA). *J Mol Model.* 2009;15(3):233–45. Available from: https://doi.org/10.1007/s00894-008-0399-4.
56. Khedkar SA, Malde AK, Coutinho EC. Design of inhibitors of the MurF enzyme of Streptococcus pneumoniae using docking, 3D-QSAR, and de Novo design. *J Chem Inf Model.* 2007;47(5):1839–46. Available from: https://doi.org/10.1021/ci600568u.
57. Verma J, Khedkar VM, Prabhu AS, Khedkar SA, Malde AK, Coutinho EC. A comprehensive analysis of the thermodynamic events involved in ligand-receptor binding using CoRIA and its variants. *J Comput Aided Mol Des.* 2008;22(2):91–104. Available from: https://doi.org/10.1007/s10822-008-9172-0.

10 Computational Approaches for Drug Target Identification

Shankar Gupta[1], *Abhimanyu Shome*[2], *Moumita Saha*[1], *and Vivek Asati*[1]
[1]Department of Pharmaceutical Chemistry, ISF College of Pharmacy, Moga, India
[2]Department of Pharmacy, Sanaka Educational Trust Group of Institutions, Durgapur, India

INTRODUCTION

The drug research and development pipeline encompasses several critical steps: target identification and validation, hit to lead molecule generation, lead molecule optimization [1] and characterization, drug formulation and delivery, pharmacokinetics and drug disposition, preclinical drug candidate identification, and bioanalytical testing and clinical trials [2]. Over the past few decades, computational drug discovery has emerged as a vital tool, offering reduced risks, time savings, cost-effectiveness, and efficient resource utilization compared to traditional experimental approaches [3]. This transformation has been facilitated by advancements in computational power and *in silico* methods, which work alongside experimental techniques to focus research efforts and guide *in vivo* validation. Traditional drug development incurs significant costs, averaging around $2.558 billion over 10–15 years, with a success rate of approximately 13% [4,5]. The high attrition rate during clinical development, especially in phases II and III, is often attributed to unexpected clinical side effects and cross-reactivity, mainly revolving around drug targets such as disease candidate proteins or genes, biological pathways, disease-associated microRNAs, biomarkers, crucial nodes of biological networks, or molecular functions [6,7].

In drug development, challenges often arise due to inadequate knowledge about drug targets, unpredictable pharmacokinetic responses upon target interaction, or off-target effects [8]. This challenge stems from the methods and population data used to identify targets, particularly for complex diseases with multiple genetic factors, creating a bottleneck in drug development [9]. The initial stage of drug development involves identifying and validating drug targets for further analysis, which is crucial for modulating targets to improve the observed disease state and achieve the desired biological response [10]. Experimental approaches for drug target identification typically involve characterizing proteins of interest and validating them through techniques like gene knockouts, animal studies, and mutagenesis [11]. However, these methods face difficulties in predicting off-target effects, which are often unpredictable [12]. Off-target effects have led to the pursuit of highly selective drug-like molecules, following the magic-bullet paradigm. With the advent of the post-genomic era and the abundance of biological data, drug discovery has evolved to incorporate various biological datasets [13,14]. This has prompted the adoption of *in silico* methods to design and redesign drug-like molecules with desired bioactivity profiles and predict and validate drug targets more efficiently [15].

With the rising prevalence of drug-resistant strains, the efficacy of common drugs is increasingly under threat, making it crucial to explore new avenues for drug discovery [16]. Computational methods have revolutionized the process by providing rational and systematic approaches to combinatorial drug discovery [17]. These *in silico* methods have significantly impacted the identification of new targets for existing drugs and the prediction of side effects and therapeutic indicators for approved drugs [15]. By streamlining processes and reducing the time required for drug development, computational approaches have played a vital role in accelerating the availability of drugs in the market. The effectiveness of a drug candidate depends on its potency and selectivity, with the selection of drug targets prioritizing their essential role in disease outcomes [18]. For genetic diseases, identifying disease-associated genes is crucial, while for infectious diseases, understanding the interaction between the host and pathogen is essential [19]. Computational drug discovery focuses on identifying unique host targets and eliminating pathogen protein targets that may lead to adverse drug reactions [20]. Leveraging analytical platforms and omics databases, computational approaches have become integral in the drug discovery pipeline, facilitating the identification of novel drug targets, potential off-targets, drug leads, and repurposable drug candidates through essential chemogenomic relationships [21].

The current trajectory of drug discovery is centered around a deep understanding of disease mechanisms, leading to the identification of targets and the discovery of potential lead compounds [22]. Moreover, there's a growing emphasis on meeting increasingly stringent safety regulations. To navigate this complex landscape efficiently, the scientist advocates for the integration of information technologies with chemoinformatics. However, managing big and diverse datasets requires rigorous filtering and thorough analysis to extract meaningful insights [23].

DOI: 10.1201/9781003464679-10

Thus, the implementation of such a strategy is crucial for advancing drug discovery in a systematic and efficient manner. Simultaneously, biomedical scientists must prioritize efficient collaboration and decision-making processes [24]. This necessitates the meaningful aggregation, mining, and analysis of vast amounts of complex, multifaceted data. In this environment, the dependable identification and validation of targets, in conjunction with drug discovery methods, are crucial for advancing computer-aided drug discovery effectively [3]. Furthermore, the integration of new network-based computational models and systems biology approaches plays a pivotal role, leveraging omics databases and optimizing combinational regimens in drug development.

In the realm of drug discovery, advanced computational models, including network-based and machine learning methods, hold promise for enhancing our understanding of drug target interactions and the molecular mechanisms underlying diseases. By leveraging computational approaches, biological data can be translated into actionable insights and treatment interventions against diseases more rapidly. This method offers a comprehensive view of the disease within the biological system, shedding light on intricate processes and molecular networks that may be challenging to explore experimentally [25]. By revealing these patterns, predictive models can be designed to pinpoint disease biomarkers and potential drug targets [26]. For complex diseases characterized by their ability to disrupt biological functions and pathways, computational methods offer a pathway to comprehend the regulatory mechanisms through gene regulatory network analysis [27]. Moreover, the development of an integrative computational framework, combining biological processes with functional datasets such as protein–protein interactions between disease-causing pathogens and hosts, facilitates the identification of drug targets and the exploration of drugs suitable for repositioning or repurposing against infections [28].

In the realm of pharmacogenomics and pharmacomicrobiomics, computational techniques have played a crucial role in predicting drug metabolism by identifying inhibitors and substrates of specific enzymes involved in metabolism [29]. This advancement has facilitated a comprehensive evaluation of absorption, distribution, metabolism, excretion, and toxicity (ADMET) properties through interactive lead optimization, thereby reducing the likelihood of drug failure [30]. For instance, *in silico* models have been developed for predicting cytochrome metabolism. While some review studies on *in silico* approaches to drug design exist, they predominantly focus on protein-associated methods for predicting candidate drug targets and drug-like molecules [31].

In this chapter, we delve into various informatics-based approaches employed in drug target identification. It includes cheminformatics-based methods, which utilize computational techniques to predict potential drug targets based on chemical properties and structural features of molecules. Ligand-based target prediction, another approach discussed, focuses on analyzing similarities between ligands to infer potential targets. Additionally, we explore chemogenomics databases, which provide valuable resources for understanding the relationships between drugs, targets, and diseases. Moving forward, we delve into the integration of omics data, particularly emphasizing functional genomics and the revolutionary CRISPR screening technology. By combining genomic data with CRISPR-based experimental techniques, researchers gain insights into gene function and identify potential drug targets with high precision. Furthermore, we highlight the importance of database and literature mining in target identification, showcasing the role of target-specific databases and text mining methods in extracting valuable information from scientific literature. As we progress, we explore the application of machine learning and artificial intelligence in target prediction, including network pharmacology, systems biology-based approaches, the BLAST method, and deep learning techniques. Finally, we conclude by discussing future perspectives in the field, emphasizing the ongoing advancements and the potential of these informatics-based approaches to revolutionize drug discovery and development. The broad classification of approaches for drug target identification is shown in Figure 10.1. In Table 10.1, the discussion of various tools classification with different web server tools is classified in different modes with URL.

INFORMATICS-BASED APPROACHES

Cheminformatics-Based Target Prediction

Chemoinformatic tools offer immense potential for advancing *in silico* drug design and discovery by integrating information across various levels to improve the reliability of data outcomes [32]. These tools encompass a range of techniques such as chemical structure similarity searching, data mining, machine learning, panel docking, and bioactivity spectra-based algorithms, which have been consistently and successfully applied in drug research [33]. For instance, the ligand-based interaction fingerprint approach utilizes physics-based docking and sampling methods to predict potential targets for small-molecule drugs [34]. Similarly, the protein–ligand interaction fingerprints method summarizes interactions between ligands and proteins using a fingerprint scheme. These methods contribute significantly to the computational modeling of drug–target interactions, thereby accelerating the drug discovery process [35,36] (Figure 10.2).

Ligand-Based Target Prediction

Target prediction based on single structures relies on ligand-based methods, which operate on the principle of chemical similarity. These methods exploit the notion that structurally similar compounds often exhibit similar biological activities, albeit with potential variations in their interactions with protein targets [37,38]. By leveraging

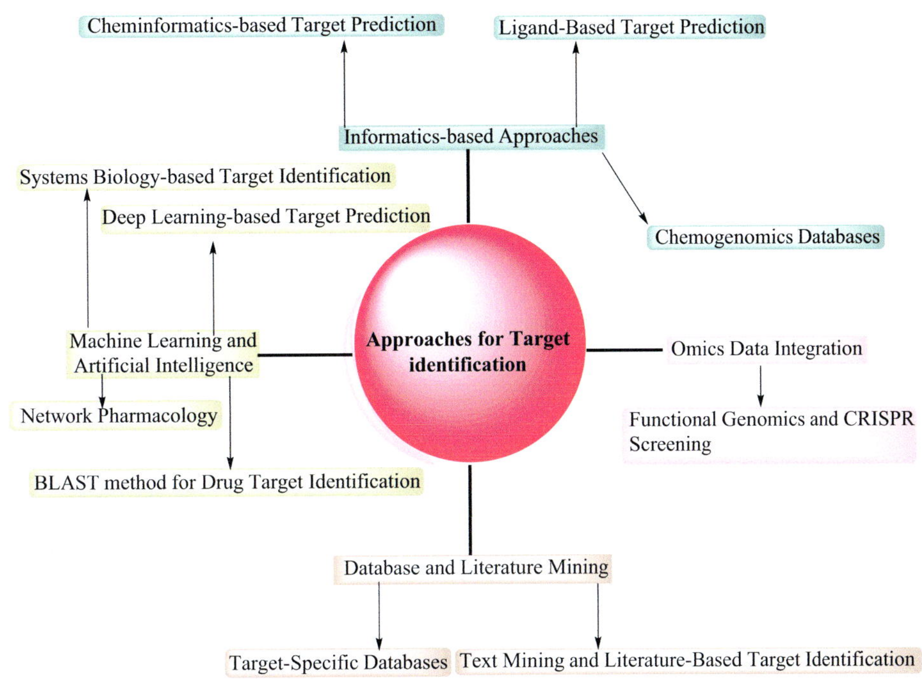

FIGURE 10.1 Broad classification of different approaches of drug target identification.

TABLE 10.1
Different Tools with Their URL

Tools	Mode	URL
RDKit	Cheminformatics-based Target Prediction	https://www.rdkit.org/
CDK (Chemistry Development Kit)	Cheminformatics-based Target Prediction	https://cdk.github.io/
KNIME	Cheminformatics-based Target Prediction	www.knime.com
scikit-learn:	Cheminformatics-based Target Prediction	https://scikit-learn.org/stable/
ChEMBL	Cheminformatics-based Target Prediction	https://www.ebi.ac.uk/chembl
PubChem:	Cheminformatics-based Target Prediction	pubchem.ncbi.nlm.nih.gov
OpenEye	Cheminformatics-based Target Prediction	https://www.eyesopen.com/
SwissTargetPrediction	Ligand-Based Target Prediction	https://www.swisstargetprediction.ch/
LigTMap	Ligand-Based Target Prediction	https://cbbio.cis.um.edu.mo/LigTMap/
Molinspiration	Ligand-Based Target Prediction	https://www.molinspiration.com/
Way2Drug	Ligand-Based Target Prediction	https://www.way2drug.com/passonline/
TargetHunter	Ligand-Based Target Prediction	https://www.cbligand.org/TargetHunter/login.php
D3Similarity	Ligand-Based Target Prediction	https://www.d3pharma.com/D3Targets-2019-nCoV/D3Similarity/index.php
STITCH	Chemogenomics Database	http://stitch.embl.de/
DrugBank:	Chemogenomics Database	https://go.drugbank.com
BindingDB database	Chemogenomics Database	https://www.bindingdb.org
Omics Integrator	Omics Data Integration	https://fraenkel-nsf.csbi.mit.edu/omicsintegrator/
MixOmics	Omics Data Integration	http://mixomics.org/
MeV (MultiExperiment Viewer)	Omics Data Integration	en.bio-soft.net
Galaxy	Omics Data Integration	https://usegalaxy.org/
Gene Ontology	Functional Genomics Tools	http://geneontology.org/
Kyoto Encyclopedia of Genes and Genomes (KEGG)	Functional Genomics Tools	https://www.genome.jp/kegg/pathway.html
BioGRID	Functional Genomics Tools	https://thebiogrid.org/

(*Continued*)

TABLE 10.1 (*Continued*)
Different Tools with Their URL

Tools	Mode	URL
CHOPCHOP	Functional Genomics Tools	https://chopchop.cbu.uib.no/
CRISPRdesign	Functional Genomics Tools	
Guide Design Tool (MIT)	Functional Genomics Tools	
MAGeCK (Model-based Analysis of Genome-wide CRISPR/Cas9 Knockout)	Functional Genomics Tools	https://bitbucket.org/liulab/mageck/src/master
CAST (CRISPR Activation Screening Tool)	Functional Genomics Tools	http://crispr-era.stanford.edu/

FIGURE 10.2 Graphical representation of cheminformatics-based target prediction.

prior knowledge of bioactive ligands and protein structures, ligand-based approaches swiftly analyze vast chemical libraries to identify molecules with substructural similarities that are likely to interact with similar protein targets [39]. However, these methods are limited in cases where no ligands for a specific protein are available, rendering it impossible to train them on ligand-based information. A crucial consideration in this process is how to effectively represent a molecular structure to a computer, with various molecular descriptors being compared to assess their correlation with bioactivity. Notably, circular fingerprints have demonstrated strong correlation with bioactivity across different definitions, making them a promising choice for target prediction endeavors [40,41].

The methods focusing on individual molecules primarily involve similarity searching tasks, where molecules with similar structures to the one being analyzed are identified from bioactivity databases like ChEMBL or BindingDB [42]. These databases provide experimentally measured activities that serve as a reference for predicting the bioactivity of the molecule in question. While this method offers rapid predictions, its major limitation lies in its inability to effectively extrapolate findings in practice. This is because it only compares entire molecules to each other, potentially overlooking novel combinations of molecular fragments [43,44]. For instance, if a database contains molecules with fragments A–B and fragments C–D as active against a specific protein target, a new molecule with the combination A–C (or A–D, B–C, B–D) might not be recognized as similar to the molecules in the database using similarity searching alone [45]. However, machine learning models, as discussed later, are likely to have higher success rates in capturing such novel feature combinations in previously unseen molecular structures [46].

Another method for target prediction is based on data mining, which has become increasingly important in bioactivity prediction of chemical compounds. Unlike chemical similarity methods that rely on identifying potential targets solely based on ligand similarities of molecular substructures, data mining methods perform pattern recognition across vast chemogenomic space [47,48]. These methods excel at clustering compounds that exhibit similarities in multidimensional space, making them preferable and credible for identifying and correlating relationships among large numbers of compounds that would otherwise be challenging to connect based solely on substructural similarities. Machine learning techniques such as Support Vector Machines (SVMs) and Linear Regression Models (LRMs) are the driving force behind these approaches. Over the last decade, a significant number of methods employing data mining approaches for mapping the chemogenomic space have been published, highlighting the growing importance of this field in drug discovery [49,50].

The Similarity Ensemble Approach (SEA) stands out as a target prediction method rooted in the distribution of features within each bioactivity class of molecules [51,52]. It utilizes a BLAST-derived algorithm to construct minimal spanning trees, taking into account chemical similarities [50,53]. SEA evaluates target similarity by ranking them based on the chemical similarity of their ligands [54]. In a noteworthy study, researchers applied the SEA method to analyze over 3,000 FDA-approved drugs across hundreds of targets, uncovering 23 new instances of drug–target interactions [55]. Five of these interactions were subsequently validated *in vitro*, demonstrating potent affinities of less than 100 nM [56]. Furthermore, the SEA method was utilized to explore potential off-target inhibition by commercially available drugs against the enzyme protein

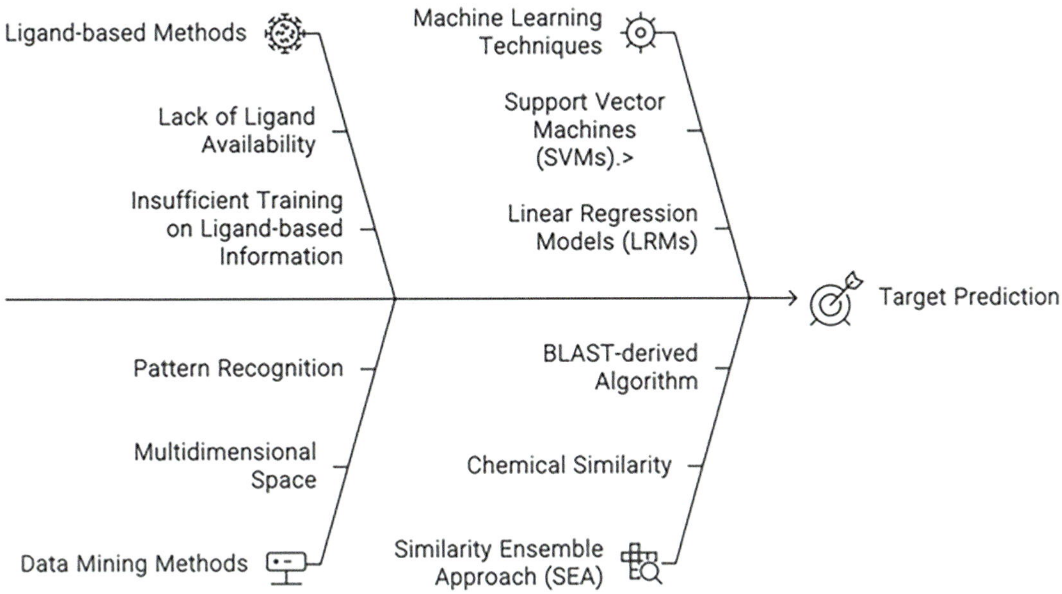

FIGURE 10.3 Different approaches for target protection.

farnesyltransferase (PFTase). This analysis identified loratadine and miconazole as ligands for PFTase, a finding that was experimentally confirmed [57] (Figure 10.3).

Chemogenomic Databases

Chemogenomic data generally refer to the activity data of chemical compounds on an array of protein targets and represents an important source of information for building *in silico* target prediction models, the emergence of systematic screening programs aimed at accelerating the identification of potential drug candidates [3,58]. One such approach involves screening targeted chemical libraries against specific protein families [59]. For instance, in the pursuit of new cancer treatments, researchers may screen libraries containing known kinase inhibitors to identify promising compounds, kicking off a medicinal chemistry program [60]. Similar strategies have been employed with libraries focused on G protein-coupled receptors (GPCRs) and protein–protein interaction inhibitors [61]. Additionally, more generalized chemical libraries have been established, comprising selective small molecules capable of modulating various protein targets across the human proteome, potentially affecting different biological processes [62,63]. With increasing access to these large chemical libraries, novel approaches like chemogenomics, proteochemometrics, and polypharmacology have emerged, enabling the exploration of extensive protein–ligand interactions and the prediction of a single ligand's effects on multiple targets [64]. Researchers have also begun investigating associations between drug–target interactions and gene–disease relationships through druggable genome studies [65]. Furthermore, network pharmacology approaches have facilitated the integration of diverse data sources, allowing for a comprehensive understanding of a drug's effects on multiple protein targets and biological processes [66]. Numerous studies have reported valuable insights into drug target clinical outcomes by combining chemogenomics, network analysis, and disease research [67]. While many chemical libraries used in these studies have been developed by industrial companies, some are available for public screening programs, such as the Mechanism Interrogation PlatE (MIPE) library [68]. Despite the resurgence of phenotypic screening in drug discovery, existing chemical libraries may not always be optimized for such studies. Advances in cell-based phenotypic screening technologies, including induced pluripotent stem cell (iPS) technologies, CRISPR-Cas gene editing tools, and imaging assays, have fueled new avenues of research [69,70].

In the initial stage of the process, sub-scaffolds at level 2 were selected from a dataset of bioactive molecules as mentioned in Figure 10.4. This step involved analyzing the dataset to identify suitable sub-scaffolds capturing selectivity while avoiding overly specific ones [71,72]. Additionally, scaffolds linked to more than 6 targets were pruned to reduce promiscuity, resulting in a refined subset of bioactive molecules with level 2 sub-scaffolds. Moving on to the second stage, attention shifted to the protein space, where first level protein classes were selected from the ChEMBL database [73,74]. These protein classes were then connected to their associated protein targets, laying the groundwork for further analysis [75]. In stage 3, the objective was to establish a set of 5,000 molecules covering as many as possible of the 850 UniprotInter nodes [76]. Leveraging hierarchical clustering and Pareto multiobjective optimization, a subset of 5,000 molecules was curated based on

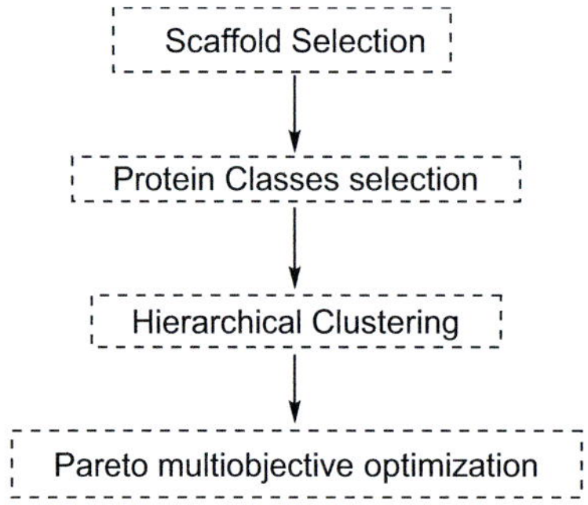

FIGURE 10.4 Flow diagram of the chemogenomic database with different stages.

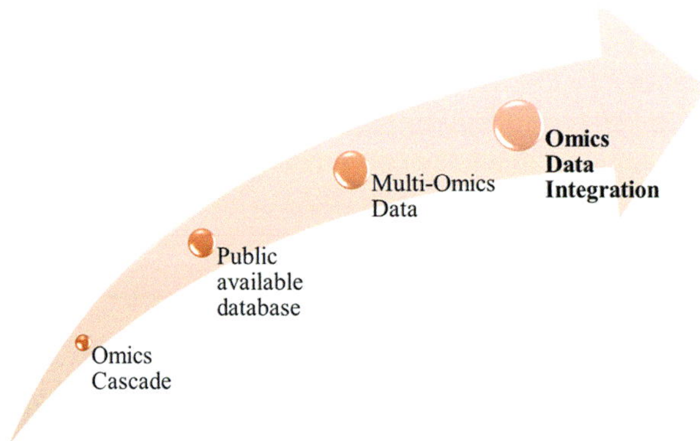

FIGURE 10.5 Workflow of omics data integration.

criteria satisfaction. Subsequently, in stage 4, the library was refined to ensure maximum coverage of the chemogenomic space, incorporating bioactive compounds for missing proteins [73,77]. This culminated in the creation of a final library comprising 5,100 molecules with enhanced coverage [78]. Finally, in stage 5, a thorough analysis of the library's characteristics and potential applications was conducted [79]. This involved assessing the diversity of scaffolds, evaluating the average number of active compounds per protein, and examining overlaps with Cell painting data and ChEMBL bioactivity data [80]. Overall, these stages represent a comprehensive approach to the development and evaluation of a chemogenomic library for potential use in phenotypic screening and drug discovery endeavors [81].

OMICS DATA INTEGRATION

The term "omics" encompasses a broad approach to understanding biological systems, focusing on the comprehensive assessment of various features [82]. This approach involves analyzing different aspects such as genomics, metagenomics, transcriptomics, metabolomics, and proteomics, among others, in a global and typically high-throughput manner [83]. While traditional methods concentrated on single genes or proteins, we now recognize that complex diseases often arise from multiple factors [84]. Thanks to advancements in technology like high-throughput mass spectrometry and sequencing platforms, we can now efficiently profile samples on a large scale [85]. This has allowed us to delve into information-rich "omic datasets," leading to a more holistic understanding of biological systems, which we refer to as the "omics cascade" [86]. By studying these cascades, we can gain insights into disease progression and potentially identify early diagnostic indicators, revolutionizing approaches to public health screening and personalized treatments [87]. Multi-omics data, covering aspects like the genome, proteome, transcriptome, metabolome, and epigenome, offer a comprehensive view of biological processes and can aid in unraveling the mechanisms underlying various conditions [88]. Furthermore, publicly available databases provide valuable multi-omics datasets, facilitating further research and analysis in this field [89] (Figure 10.5).

Functional Genomics and CRISPR Screening

Functional genomics involves studying the function and interactions of genes within a biological system [90]. One powerful tool in functional genomics research is CRISPR screening, which utilizes the CRISPR-Cas9 system to systematically perturb genes and observe resulting changes in phenotype [91]. CRISPR screening enables researchers to investigate gene function on a large scale, uncovering novel gene interactions, identifying potential therapeutic targets, and elucidating the genetic basis of diseases [92].

In the workflow for CRISPR screening, several key steps are involved to effectively identify genes associated with specific phenotypic traits. Firstly, the process begins with the design of CRISPR Guide RNAs (gRNAs), which are tailored to target specific genomic regions of interest, such as protein-coding genes or regulatory elements within the genome [93]. These gRNAs serve as guides for the Cas9 enzyme, directing it to the precise location for gene editing [94]. Next, the CRISPR components, including the Cas9 protein and gRNA, are delivered into the target cells using various methods such as viral vectors, lipid nanoparticles, or electroporation [95]. Once inside the cells, the Cas9 enzyme, guided by the gRNA, induces double-stranded breaks at the predetermined genomic loci [96]. This triggers the cellular DNA repair mechanisms, resulting in gene knockout, knock-in, or other modifications based on the experimental design [97]. Following genome editing, the next step involves either applying selective pressure or conducting a screening process to identify cells with desired phenotypic changes resulting from the CRISPR-mediated gene editing [98]. This step may entail cell culture assays, high-throughput sequencing, or other screening methods to identify cells with the desired traits [99]. Subsequently, the data obtained from the screening experiment undergo thorough analysis to identify genes that influence the observed phenotype [100]. This analysis may involve bioinformatics techniques, statistical modeling, and validation experiments to confirm candidate gene hits [101]. Finally, the identified candidate genes undergo functional validation through additional experiments, such as loss-of-function or gain-of-function studies, to confirm their roles in the observed phenotype and further elucidate their biological functions [102]. Through this systematic workflow, CRISPR screening offers a powerful tool for uncovering the functional significance of genes and understanding their roles in various biological processes [103].

CRISPR screening has revolutionized functional genomics research by enabling systematic interrogation of gene function at a genome-wide scale. It has applications across various fields, including cancer biology, drug discovery, developmental biology, and neurobiology, among others [104]. By combining the power of CRISPR technology with functional genomics approaches, researchers can gain deeper insights into the complex mechanisms underlying biological processes and diseases, paving the way for new therapeutic strategies and personalized medicine [105].

DATABASE AND LITERATURE MINING

Database and literature mining involve the systematic extraction of information from various sources, including scientific articles, databases, and other repositories. This process is essential for collecting, organizing, and analyzing vast amounts of data to generate insights and support scientific research. In the first step, the identification of relevant databases and resources is done. Researchers determine the databases, repositories, and literature sources that contain relevant information for their research topic. This may include academic databases like PubMed [106], Web of Science [107], Scopus [108], and specialized databases specific to their field of study.

Once the relevant sources are identified, researchers formulate search queries using appropriate keywords, Boolean operators, and search filters to retrieve relevant data. Search queries are refined as needed to optimize the retrieval of relevant information from the selected databases and literature sources. Researchers then execute search queries across the selected databases and literature sources to retrieve relevant articles, datasets, and other resources. Advanced search features and filters provided by the databases are used to narrow down search results based on specific criteria such as publication date, study type, and experimental techniques. After retrieving data, researchers extract relevant information from articles and databases, including text, tables, figures, and metadata. Extracted data are annotated with relevant metadata such as publication details, author information, study methodology, and experimental results to facilitate further analysis [109,110].

Data from multiple sources and databases are integrated to create a comprehensive dataset for analysis. Data formats are normalized, and inconsistencies or discrepancies are resolved to ensure data compatibility and coherence. Text mining and natural language processing (NLP) techniques are then applied to extract structured information from unstructured text data, such as articles, abstracts, and full-text documents [111,112]. This may involve entity recognition, relationship extraction, sentiment analysis, and topic modeling to identify key concepts, entities, and relationships within the text. The analyzed data are then subjected to statistical methods, machine learning algorithms, and data visualization techniques to uncover patterns, trends, and relationships. Visualizations, charts, graphs, and summary statistics are generated to communicate findings effectively [113,114]. Quality assessment and validation are conducted to assess the quality, reliability, and relevance of extracted data and findings. Any limitations or biases in the data and analysis process are addressed [115].

Finally, researchers interpret the analyzed data in the context of the research question or objectives to derive meaningful insights, hypotheses, or conclusions. Findings are synthesized with existing knowledge and literature to generate new knowledge and contribute to the scientific understanding of the topic. The entire process is documented and reported in a clear and transparent manner through publications, presentations, or other channels [116,117] (Figure 10.6).

TARGET-SPECIFIC DATABASES

Target-specific databases play a crucial role in biomedical research by providing curated collections of data focused on specific molecular targets or biological processes [118,119]. These databases aggregate information from various sources, including experimental studies, literature, and computational predictions, to offer comprehensive resources for researchers studying particular genes, proteins, pathways, or diseases. One notable example of

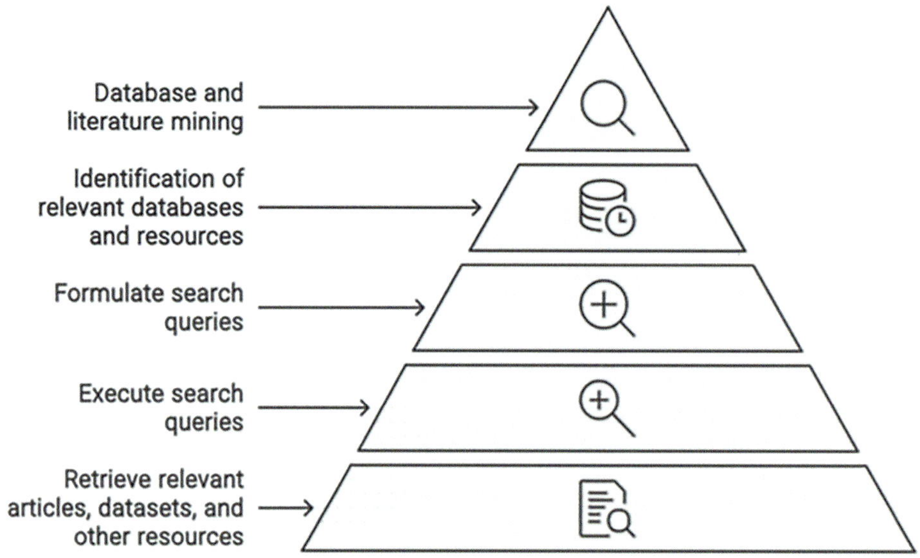

FIGURE 10.6 Visual representation of database and literature mining.

target-specific databases is the Human Protein Atlas (HPA), which catalogs information about the expression, localization, and function of proteins in human tissues and cells [120]. HPA integrates data from immunohistochemistry, RNA sequencing, mass spectrometry, and antibody-based profiling to create detailed maps of protein expression patterns across different tissues and cell types. Researchers can use HPA to explore the role of specific proteins in health and disease, identify potential biomarkers, and prioritize targets for further investigation [121].

Another example is the DrugBank database [122], which focuses on pharmaceutical targets and drug-related information. DrugBank provides comprehensive data on drug targets, including protein structures, biochemical pathways, pharmacological properties, and drug interactions. Researchers can use DrugBank to explore the mechanisms of action of existing drugs, identify potential drug targets for novel therapeutics, and predict drug–drug interactions or adverse effects. Additionally, databases such as the Cancer Genome Atlas (TCGA) and the Genotype-Tissue Expression (GTEx) project offer valuable resources for studying the genetic basis of cancer and normal tissue gene expression, respectively [123]. TCGA provides genomic, transcriptomic, and clinical data from thousands of cancer patients across multiple cancer types, enabling researchers to identify genetic alterations associated with cancer development and progression. GTEx, on the other hand, offers insights into gene expression patterns across different tissues and cell types in healthy individuals, providing a reference for understanding normal gene expression profiles and regulatory mechanisms [124,125].

Furthermore, databases focused on specific molecular pathways or biological processes, such as the Kyoto Encyclopedia of Genes and Genomes (KEGG) and Reactome, offer curated collections of pathway maps, gene annotations, and functional annotations [126,127]. These resources help researchers elucidate the molecular mechanisms underlying various biological processes, identify key signaling pathways involved in disease pathogenesis, and prioritize targets for drug discovery and therapeutic intervention. Overall, target-specific databases serve as invaluable resources for biomedical research, providing researchers with access to curated data and analytical tools to accelerate discoveries, advance our understanding of disease mechanisms, and facilitate the development of novel diagnostics and therapeutics. These databases play a critical role in driving translational research efforts aimed at improving human health and tackling complex diseases [128].

Text Mining and Literature-Based Target Identification

Text mining and literature-based target identification are powerful approaches in drug discovery and biomedical research, leveraging computational methods to extract valuable information from vast amounts of scientific literature. These techniques enable researchers to identify potential drug targets, elucidate molecular mechanisms, and prioritize candidates for further experimental validation [129].

1. **Data Collection:**
   - Retrieve scientific literature from databases such as PubMed, Web of Science, and Scopus [130].
   - Gather unstructured text sources including scientific articles and patents related to the research topic.

2. **Text Mining:**
   - Apply text mining techniques to extract relevant information from the collected literature [131].
   - Utilize named entity recognition (NER) to identify and extract mentions of biological entities such as genes, proteins, diseases, and drugs [132].
   - Employ NLP techniques to standardize and annotate the extracted entities, linking them to established identifiers and ontologies [133].
3. **Entity Annotation and Standardization:**
   - Normalize and annotate the extracted entities using standardized vocabularies, ontologies, and identifiers [134].
   - Connect the identified genes, proteins, diseases, and drugs to established databases and ontologies for further analysis.
4. **Literature-Based Target Identification:**
   - Systematically analyze the curated scientific literature to identify potential drug targets.
   - Conduct comprehensive literature reviews and data curation to compile information on genes, proteins, and pathways associated with disease processes or treatment outcomes.
   - Evaluate the evidence from experimental studies, genetic associations, and functional annotations to prioritize potential drug targets.
5. **Integration and Analysis:**
   - Integrate information obtained from text mining and literature analysis with other data sources such as genomic databases and pathway databases.
   - Analyze the integrated datasets to identify key genes, proteins, and pathways implicated in disease pathogenesis or therapeutic responses [135].
6. **Validation and Further Exploration:**
   - Validate the identified drug targets through experimental studies, such as functional genomics or biochemical assays [136].
   - Explore the functional roles of the identified targets in disease mechanisms and treatment responses through additional experiments and analyses.
7. **Insights and Decision-Making:**
   - Interpret the findings from text mining and literature-based target identification to generate insights into potential drug targets.
   - Make informed decisions regarding the selection and prioritization of drug targets for further development and validation [137].
8. **Iterative Process:**
   - Iterate the workflow as needed based on new findings, feedback, and emerging literature to continually refine and improve target identification strategies [138].

Overall, text mining and literature-based target identification are iterative processes that require interdisciplinary expertise in bioinformatics, computational biology, and domain-specific knowledge. By leveraging the vast amount of information available in scientific literature, these approaches facilitate the discovery of novel drug targets, elucidation of disease mechanisms, and development of targeted therapeutics for various diseases and conditions.

MACHINE LEARNING AND ARTIFICIAL INTELLIGENCE

Network Pharmacology

Network pharmacology is an interdisciplinary approach that combines principles from pharmacology, systems biology, and network science to understand how drugs interact with biological systems. The basic idea is to view the body as a complex network of molecules, proteins, and pathways, where changes in one component can affect the entire system and lead to various physiological responses and diseases. To achieve this understanding, network pharmacology integrates diverse biological datasets, including genomics, transcriptomics, proteomics, metabolomics, and pharmacology, to construct detailed network models. These models are then analyzed using advanced computational algorithms and network analysis techniques to identify important nodes and functional modules involved in drug action and disease processes [139,140]. One key aspect of network pharmacology is predicting drug–target interactions, which involves leveraging information about drug structures, target protein sequences, and network topology to forecast potential interactions and understand how drugs exert their effects. Additionally, network pharmacology explores the concept of poly-pharmacology, where drugs can interact with multiple targets and pathways simultaneously, and it provides insights into disease mechanisms by mapping disease-associated genes, proteins, and pathways onto molecular networks. This approach not only aids in identifying potential drug targets and biomarkers but also facilitates drug repurposing efforts and rational drug design. Furthermore, network-based drug discovery integrates data from drug–protein and protein–disease networks, utilizing computational frameworks to identify new drug targets and understand disease mechanisms comprehensively [141]. With the emergence of omics technologies and computational tools, network pharmacology has become an indispensable tool in drug discovery and systems biology research, enabling researchers to analyze vast biological datasets, reconstruct biological networks, and advance our understanding of complex biological systems [142].

This multifaceted approach encompasses several key components. Firstly, network pharmacology involves the comprehensive integration of diverse biological datasets spanning genomics, transcriptomics, proteomics, metabolomics, and pharmacology. These datasets are amalgamated to construct intricate network models that elucidate the complex interactions between drugs, target proteins, signaling pathways, and diseases [140].

Once constructed, these molecular interaction networks undergo rigorous analysis utilizing advanced computational algorithms and network analysis techniques. Through graph theory and network analysis, network pharmacology uncovers the topology, connectivity, and dynamics of these networks, identifying crucial nodes (such as hub proteins or drug targets) and functional modules that play pivotal roles in drug action and disease processes. Another pivotal aspect of network pharmacology is the prediction of drug–target interactions. Leveraging information about drug chemical structures, target protein sequences, and network topology, computational methods are employed to forecast potential drug–target interactions and assess their biological significance. By doing so, network pharmacology aids in identifying promising drug candidates and understanding their mechanisms of action. Moreover, network pharmacology sheds light on the concept of poly-pharmacology, whereby drugs exert their therapeutic effects by modulating multiple targets and pathways simultaneously. By examining drug–target interaction networks and drug–protein interaction profiles, network pharmacology uncovers multi-target drugs and delineates their intricate mechanisms of action within biological systems. Furthermore, network pharmacology provides a systems-level understanding of disease mechanisms by mapping disease-associated genes, proteins, and pathways onto molecular interaction networks. This facilitates the identification of key drivers of disease, potential biomarkers, and therapeutic targets for drug intervention, thus paving the way for personalized medicine approaches [143].

Lastly, network pharmacology serves as a valuable tool for drug repurposing efforts and rational drug design. By leveraging the comprehensive network models of drug–target interactions and disease pathways, network pharmacology identifies new indications for existing drugs based on their poly-pharmacological profiles and guides the development of novel therapeutics that target specific nodes or pathways within the molecular network associated with disease [144].

Systems Biology-Based Target Identification

Systems biology-based target identification represents a sophisticated strategy in biomedical research, harnessing computational modeling and analysis of biological systems to pinpoint potential drug targets and unravel intricate disease mechanisms. This approach integrates diverse types of biological data, including genomics, transcriptomics, proteomics, metabolomics, and interactomics, offering a comprehensive view of cellular functions and their dysregulation in disease states [145,146].

The process begins with the integration of heterogeneous omics data obtained from various experimental techniques such as high-throughput sequencing and mass spectrometry. These datasets provide detailed insights into the molecular landscape of cells and tissues under different conditions, forming the foundation for systems biology analyses. Once the data are integrated, researchers construct molecular interaction networks, including protein–protein interaction networks, gene regulatory networks, and metabolic networks [147]. These networks capture the complex web of molecular interactions within biological systems, enabling researchers to model and analyze the underlying biological processes. Computational algorithms and statistical methods are then employed to analyze these molecular networks and identify key nodes that drive disease-related pathways. By assessing network centrality metrics, performing clustering analyses, and conducting pathway enrichment studies, researchers can prioritize potential drug targets based on their functional relevance and network properties [148]. Moreover, systems biology incorporates mathematical modeling techniques, such as ordinary differential equations (ODEs) and Boolean networks, to simulate and predict the dynamic behavior of biological systems. These models integrate experimental data with computational simulations to generate testable hypotheses and predict the effects of perturbations on system behavior. Once candidate drug targets are identified through systems biology approaches, they undergo experimental validation using *in vitro* and *in vivo* assays. Functional genomics screens, biochemical assays, and animal models are employed to assess the effects of target modulation on cellular phenotypes and disease outcomes [149]. Following validation, promising drug targets are pursued in drug discovery pipelines, where small molecule inhibitors, biologics, or gene therapies are developed to modulate target activity and treat diseases. Rational drug design strategies, structure-based drug discovery, and high-throughput screening are employed to identify lead compounds and optimize their efficacy, selectivity, and safety profiles [150]. Further validated drug candidates undergo preclinical and clinical testing to evaluate their safety and efficacy in human patients. Clinical trials are conducted to assess drug pharmacokinetics, pharmacodynamics, and therapeutic outcomes, leading to regulatory approval and clinical use of new drugs for treating diseases. In summary, systems biology-based target identification offers a holistic and data-driven approach to drug discovery, enabling researchers to uncover novel therapeutic targets, elucidate disease mechanisms, and develop precision medicine strategies tailored to individual patients. Through the integration of computational modeling with experimental validation, this approach accelerates the translation of basic research findings into clinically relevant therapies, ultimately improving patient outcomes and advancing human health.

BLAST Method for Drug Target Identification

The Basic Local Alignment Search Tool (BLAST) method serves as a cornerstone in drug target identification, enabling researchers to identify potential drug targets by comparing protein or nucleotide sequences against a vast database of known sequences. This widely used bioinformatics tool offers a rapid and efficient means to uncover similarities between query sequences and sequences of interest, facilitating the identification of homologous proteins or genes with known functions or associations with diseases [151,152]. The BLAST method begins with the submission of a query sequence, typically a protein or nucleotide sequence associated with a particular biological process or disease. This query sequence is then compared against a comprehensive database containing sequences from diverse organisms, including bacteria, viruses, plants, and animals, as well as sequences from genomic and protein databases. It employs a heuristic algorithm to identify regions of local similarity between the query sequence and sequences in the database. The algorithm identifies short, contiguous stretches of identical or similar residues (known as "hits") between the query sequence and database sequences, using a scoring system based on sequence identity, alignment length, and the presence of gaps or mismatches [153].

The results of a BLAST search are presented as a list of matches ranked according to their similarity to the query sequence. Each match includes information about the sequence identity, alignment score, E-value (a measure of the statistical significance of the alignment) [154], and alignment length, allowing researchers to assess the quality and significance of the matches. In the context of drug target identification, BLAST searches are often used to identify proteins or genes that share sequence similarity with known drug targets or proteins implicated in disease pathways. By identifying homologous proteins or genes, researchers can infer potential functions, interactions, and associations of the query sequence based on the known characteristics of the matched sequences. Furthermore, BLAST searches can aid in the identification of conserved domains, motifs, or functional residues within proteins, providing insights into their structural and functional properties. This information is invaluable for predicting the effects of genetic variations, mutations, or post-translational modifications on protein function and for designing targeted therapeutic interventions. Overall, the BLAST method serves as a powerful and versatile tool for drug target identification, offering researchers a systematic approach to explore sequence similarities, functional relationships, and evolutionary conservation across diverse biological systems. By leveraging the wealth of sequence data available in public databases, BLAST enables researchers to uncover novel drug targets, validate existing targets, and accelerate the drug discovery process in various fields of biomedical research [155].

Deep Learning-Based Target Prediction

Deep learning-based target prediction stands at the forefront of modern drug discovery efforts, revolutionizing the identification of potential drug targets through the utilization of sophisticated artificial intelligence techniques [156]. This approach involves the deployment of deep learning algorithms, which are a subset of machine learning methods, to scrutinize complex biological data and forecast interactions between drug compounds and target proteins. By leveraging vast amounts of biological information and computational power, deep learning-based target prediction offers a transformative avenue for accelerating drug discovery processes. At the outset, deep learning-based target prediction entails the acquisition and preprocessing of diverse datasets encompassing details about drug compounds, target proteins, and their known interactions. These datasets often encompass a wide array of information such as chemical structures, molecular fingerprints, protein sequences, structural characteristics, gene expression profiles, and biological annotations. Before analysis, the data undergoes preprocessing steps to standardize formats, eliminate noise, and normalize features to ensure compatibility for subsequent model training. The core of deep learning-based target prediction lies in the construction and training of neural network architectures tailored to the specific characteristics of the input data and the task at hand. These architectures, which may include convolutional neural networks (CNNs) [157], recurrent neural networks (RNNs) [158], or graph neural networks (GNNs) [159], are adept at discerning intricate patterns and relationships within the data. During training, the model learns to optimize its parameters using labeled data, where known drug–target interactions guide the iterative adjustment of model parameters through optimization algorithms such as stochastic gradient descent (SGD) or Adam optimization [160].

One of the critical aspects of deep learning-based target prediction is the representation of features extracted from the input data. For drug compounds, molecular descriptors, chemical fingerprints, and structural properties are translated into numerical vectors that encapsulate the chemical space. Conversely, for target proteins, amino acid sequences, structural motifs, and functional domains are encoded into feature vectors representing the protein space. These representations enable the model to discern meaningful patterns underlying drug–target interactions [161].

Following model training, the performance of the deep learning model is evaluated using validation datasets and metrics such as accuracy, precision, recall, and area under the receiver operating characteristic curve (AUC-ROC) [162,163]. Cross-validation techniques and external validation using independent datasets are often employed to assess the generalizability and robustness of the model's predictions. Furthermore, interpretability and visualization techniques elucidate the molecular determinants driving drug–target interactions, enhancing our understanding of the underlying mechanisms [164]. The applications of deep learning-based target prediction are diverse and far-reaching. These models play a crucial role in large-scale virtual screening endeavors, where millions of drug compounds are screened against a library of target proteins to identify promising candidates for experimental validation. Moreover, these models are integrated into drug discovery pipelines and computational platforms to facilitate decision-making processes, optimize lead compounds, and expedite the development of novel therapeutics for various diseases and conditions. Continual refinement and improvement are integral to deep learning-based target prediction. Models are continually updated with new data and insights to enhance their accuracy, robustness, and generalization performance. Techniques such as transfer learning, ensemble learning, and active learning enable the incorporation of new knowledge and adaptation to emerging challenges, ensuring that deep learning-based target prediction remains at the forefront of innovation in drug discovery. Through iterative refinement and adaptation, these models evolve to address the evolving needs of the field and drive advancements in therapeutic development [165].

CONCLUSION AND FUTURE PERSPECTIVE

Challenges in drug development relate to the limited knowledge available about a new target, unpredictable pharmacokinetic behavior upon engagement of that target, and off-target side effects. This bottleneck in drug development is driven by the strategies and population data we employ to identify targets, particularly for disorders influenced by multiple genetic risk factors. First is target identification and validation, key to controlling disease states or pathological activities with desired biological responses. Chemoinformatics, Chemogenomic methods, and BLAST methods for Drug Target Identification are different approaches used to tackle the challenges.

Chemoinformatic tools are extensively used in current drug discovery, bringing together data at different levels to increase reliability. Chemoinformatic tools and technologies, such as chemical structure similarity searching, data mining, and machine learning, have advanced drug research to a new stage. Ligand-based methods have been able to better handle chemical similarity than any of the structure-based methods, and they are fast in analyzing large sets of molecules for protein interaction with similar properties. But their limitations come about when there are no ligands for a certain protein. These fingerprints are also circular and have been shown to have high correlation with bioactivity, making them very interesting tools for target prediction.

With data mining, the datasets may be filed in some databases and with machine learning models to cluster compounds according to multidimensional similarities, which helps delineate relationships between numerous compounds. The Similarity Ensemble Approach (SEA) ranks targets on ligand chemical similarity, successfully pinpointing novel drug–target interactions and off-targets due to its fast execution. Chemogenomic information linking the biological and pharmacological space of chemical compounds to protein targets greatly aids in silico target prediction. Targeting of screening programs specific to protein families leads to accelerated drug discovery. Greater precision and efficiency in drug development have been supported by the advances that cell-based phenotypic screening technologies continue to deliver.

Recently, the advent of systems-wide omics strategies together with high-throughput gene editing makes us enter a new stage in our efforts to understand biological systems as one comprehensive landscape. These tools have dramatically increased our capacity to probe complex biological systems, providing novel perspectives on the mechanisms of disease and gene function, as well as potential targets for therapeutic development.

Integrating omics data types, such as genomics, proteomics, transcriptomics, and metabolomics, offers a global view of biological processes that can help researchers disentangle complex molecular networks and identify key drivers of normal health states versus disease. The most notable of these is CRISPR screening, which has now become a transformative technology in functional genomics. It has proven to be a powerful tool for the systematic perturbation of genes at a genome-wide level, revealing new and unexpected levels of gene interactions and promising drug targets. CRISPR screening, covering the workflow from guide RNA design to functional validation of candidate genes, provides a powerful tool for uncovering gene function and its associated cellular phenotypes. In the future, multi-omics and CRISPR technologies will benefit from combinatorial uses in biomedical research. Despite this, it will surely transform our approach to complex diseases and provide us with the accurate diagnostics and personalized treatments, which at no stage of an illness is nothing more than a ticket into hell. We believe that these emerging and maturing technologies will have a major impact on cancer biology, neurobiology, and developmental biology, among others, in the years to come.

Overall, future experiments could refine CRISPR screens by increasing the ability to discern real hits from off-target effects, improve data analysis for multi-omics datasets, and put in place methods that control against false positive identification and assess synergism between hits arising within each dataset or across different omics classes (RNA-seq versus proteome) as well as juxtapose these approaches with emerging technologies such as single-cell sequencing techniques or organoid models. Furthermore, the increasing public accessibility of multi-omics databases should drive more synergy in research initiatives and uncover new aspects.

The construction of assignment-specific databases such as Human Protein Atlas and DrugBank has revolutionized the investigation of protein expression profiles, drug–target interactions, and aetiology. On the other hand, text-mining and literature-based target identification have become essential tools in the discovery of potential new drug targets and understanding complex disease biology. In summary, the future of database and literature mining in biomedical research is bright and transformative. Advances in artificial intelligence and machine learning are expected to lead to more effective and accurate text mining algorithms that can be used for the extraction of fine-grained information from scientific literature. Combining multi-omics data with literature-derived knowledge will be crucial for a better and broader picture of the functioning of biological systems and pathogenesis. Finally, the establishment of common controlled vocabularies and advanced data harmonization methodologies will contribute to making different databases interoperable, streamlining integrated analysis across disparate research fields.

But important hurdles remain, including how to deal with the growing amount of scientific data and ensuring that the information extracted is high-quality. Overcoming these challenges will require continuous dialogue between bioinformaticians, data scientists, and domain experts to refine mining methods and develop sound validation approaches. With the maturation of these technologies, one would anticipate them taking on a more central role in speeding drug discovery and uncovering new targets for therapy.

Also, the new landscape of drug target identification in the emerging era has significantly enhanced the entire process from discovery by combining network pharmacology, systems biology, and deep learning strategies. These increasingly complex methods have transformed our understanding of complex biology and disease processes, allowing us to gain insights into drug–target interactions and molecular pathways. Looking ahead, the combination of these methods with new technologies offers an opportunity to heighten both drug discovery and personalized medicine.

Hybrid models incorporating network pharmacology, systems biology, and deep learning are likely to appear, making more accurate predictions about drug–target interactions as well as underlying disease mechanisms. In addition, real-time biological data from wearable devices could be integrated with high-throughput experiments to allow for personalized longitudinal modeling of drug response profiles and disease courses. However, despite these advances that hold promise for clinical exceptionalism, considerable challenges remain in validating and translating such computational predictions into clinically actionable therapies. This will be followed by efforts to narrow the gap between *in silico* predictions and *in vivo* success, possibly via improved organ-on-chip technologies or even the development of entirely digital clinical trials. Interpretability of deep learning models is another important issue that has to be addressed before the regulatory and clinical acceptance of AI-driven drug discovery, but it needs a separate article.

With the maturation of these technologies, a paradigm shift to highly accurate, effective, and affordable drug development processes is likely. Over time, this evolution will yield cracked codes that can be translated into novel therapeutic targets, repurposing them in existing drugs for new indications and the rise of highly targeted combination therapies designed.

REFERENCE

1. Dambach DM, Misner D, Brock M, Fullerton A, Proctor W, Maher J et al. Safety lead optimization and candidate identification: integrating new technologies into decision-making. *Chemical Research in Toxicology* 2016; 29(4): 452–472.
2. Rogge M, Taft DR. *Preclinical Drug Development. Drugs and the Pharmaceutical Sciences*. 2010; 187: CRC Press, Boca Raton, FL.
3. Kiriiri GK, Njogu PM, Mwangi AN. Exploring different approaches to improve the success of drug discovery and development projects: a review. *Future Journal of Pharmaceutical Sciences* 2020; 6: 1–12.
4. Ram RN, Gadaleta D, Allen TEH. The role of 'big data'and 'in silico' new approach methodologies (NAMs) in ending animal use–a commentary on progress. *Computational Toxicology* 2022; 23: 100232.
5. Juárez-López D, Schcolnik-Cabrera A. Drug repurposing: considerations to surpass while re-directing old compounds for new treatments. *Archives of Medical Research* 2021; 52(3): 243–251.
6. Keerthana N, Koteeswaran K. Target identification and validation in research. *World Journal of Biology Pharmacy and Health Sciences* 2024; 17(3): 107–117.
7. Rinaldi S, Moroni E, Rozza R, Magistrato A. Frontiers and challenges of computing ncRNAs biogenesis, function and modulation. *Journal of Chemical Theory and Computation* 2024.
8. Manzari MT, Shamay Y, Kiguchi H, Rosen N, Scaltriti M, Heller DA. Targeted drug delivery strategies for precision medicines. *Nature Reviews Materials* 2021; 6(4): 351–370.
9. Dugger SA, Platt A, Goldstein DB. Drug development in the era of precision medicine. *Nature Reviews Drug discovery* 2018; 17(3): 183–196.
10. Santos R, Ursu O, Gaulton A, Bento AP, Donadi RS, Bologa CG et al. A comprehensive map of molecular drug targets. *Nature Reviews Drug discovery* 2017; 16(1): 19–34.
11. West DB, Iakougova O, Olsson C, Ross D, Ohmen J, Chatterjee A. Mouse genetics/genomics: an effective approach for drug target discovery and validation. *Medicinal Research Reviews* 2000; 20(3): 216–230.
12. Urban L, Maciejewski M, Lounkine E, Whitebread S, Jenkins JL, Hamon J et al. Translation of off-target effects: prediction of ADRs by integrated experimental and computational approach. *Toxicology Research* 2014; 3(6): 433–444.
13. Sadri A. Is target-based drug discovery efficient? Discovery and "off-target" mechanisms of all drugs. *Journal of Medicinal Chemistry* 2023; 66(18): 12651–12677.
14. Chanda SK, Caldwell JS. Fulfilling the promise: drug discovery in the post-genomic era. *Drug Discovery Today* 2003; 8(4): 168–174.
15. Agamah FE, Mazandu GK, Hassan R, Bope CD, Thomford NE, Ghansah A et al. Computational/in silico methods in drug target and lead prediction. *Briefings in Bioinformatics* 2020; 21(5): 1663–1675.
16. Harikumar G, Krishanan K. The growing menace of drug resistant pathogens and recent strategies to overcome drug resistance: a review. *Journal of King Saud University-Science* 2022; 34(4): 101979.
17. Thomford NE, Senthebane DA, Rowe A, Munro D, Seele P, Maroyi A et al. Natural products for drug discovery in the 21st century: innovations for novel drug discovery. *International Journal of Molecular Sciences* 2018; 19(6): 1578.
18. Schaduangrat N, Lampa S, Simeon S, Gleeson MP, Spjuth O, Nantasenamat C. Towards reproducible computational drug discovery. *Journal of Cheminformatics* 2020; 12: 1–30.
19. Wang Y, Cornett A, King FJ, Mao Y, Nigsch F, Paris CG et al. Evidence-based and quantitative prioritization of tool compounds in phenotypic drug discovery. *Cell Chemical Biology* 2016; 23(7): 862–874.

20. Barman RK, Mukhopadhyay A, Maulik U, Das S. Identification of infectious disease-associated host genes using machine learning techniques. *BMC Bioinformatics* 2019; 20: 1–12.
21. Sliwoski G, Kothiwale S, Meiler J, Lowe EW. Computational methods in drug discovery. *Pharmacological Reviews* 2014; 66(1): 334–395.
22. Matthews H, Hanison J, Nirmalan N. "Omics"-informed drug and biomarker discovery: opportunities, challenges and future perspectives. *Proteomes* 2016; 4(3): 28.
23. Rao M, McDuffie E, Sachs C. Artificial intelligence/machine learning-driven small molecule repurposing via off-target prediction and transcriptomics. *Toxics* 2023; 11(10): 875.
24. Verweij J, Griffin G, Guchelaar HJ, Hollak CEM, Leufkens HGM, Moons KGM et al. Royal Netherlands Academy of Arts and Science (KNAW)© Some rights reserved This publication is subject to rights of use as laid down in the [Naamsvermelding 3.0 Nederland] Creative Commons Licence. The full text of the licence can be consulted at https://www.creativecommons.org/licenses, 2021.
25. Tsiliki G, Kossida S, Friesen N, Rüping S, Tzagarakis M, Karacapilidis N. A data mining based approach for collaborative analysis of biomedical data. *International Journal on Artificial Intelligence Tools* 2014; 23(04): 1460010.
26. Niazi SK, Mariam Z. Computer-aided drug design and drug discovery: a prospective analysis. *Pharmaceuticals* 2023; 17(1): 22.
27. Cho D-Y, Kim Y-A, Przytycka TM. Chapter 5: Network biology approach to complex diseases. *PLoS Computational Biology* 2012; 8(12): e1002820.
28. Frank R, Hargreaves R. Clinical biomarkers in drug discovery and development. *Nature Reviews Drug Discovery* 2003; 2(7): 566–580.
29. Delgado FM, Gómez-Vela F. Computational methods for gene regulatory networks reconstruction and analysis: a review. *Artificial Intelligence in Medicine* 2019; 95: 133–145.
30. Agamah FE. *Large–Scale Data–Driven Network Analysis of Human–Plasmodium Falciparum Interactome: Extracting Essential Targets and Processes for Malaria Drug Discovery.* Thesis Presented for the Degree of MSc Med Human Genetics. 2020.
31. Singh S, Sarma DK, Verma V, Nagpal R, Kumar M. Unveiling the future of metabolic medicine: omics technologies driving personalized solutions for precision treatment of metabolic disorders. *Biochemical and Biophysical Research Communications* 2023; 682: 1–20.
32. Speck-Planche A, Dias Soeiro Cordeiro MN. Review of current chemoinformatic tools for modeling important aspects of CYPsmediated drug metabolism. Integrating metabolism data with other biological profiles to enhance drug discovery. *Current Drug Metabolism* 2014; 15(4): 429–440.
33. Mai H, Le TC, Chen D, Winkler DA, Caruso RA. Machine learning for electrocatalyst and photocatalyst design and discovery. *Chemical Reviews* 2022; 122(16): 13478–13515.
34. Cao R, Wang Y. Predicting molecular targets for small-molecule drugs with a ligand-based interaction fingerprint approach. *ChemMedChem* 2016; 11(12): 1352–1361.
35. Leelananda SP, Lindert S. Computational methods in drug discovery. *Beilstein Journal of Organic Chemistry* 2016; 12(1): 2694–2718.
36. Wilson GL, Lill MA. Integrating structure-based and ligand-based approaches for computational drug design. *Future Medicinal Chemistry* 2011; 3(6): 735–750.
37. Nandi S, Bhaduri S, Das D, Ghosh P, Mandal M, Mitra P. Deciphering the lexicon of protein targets: a review on multifaceted drug discovery in the era of artificial intelligence. *Molecular Pharmaceutics* 2024; 21(4): 1563–1590.
38. Surade S, Blundell TL. Structural biology and drug discovery of difficult targets: the limits of ligandability. *Chemistry & Biology* 2012; 19(1): 42–50.
39. Wu H, Liu J, Zhang R, Lu Y, Cui G, Cui Z et al. A review of deep learning methods for ligand based drug virtual screening. *Fundamental Research* 2024; 4(4): 715–737.
40. Li X, Yang Q, Xu L, Dong W, Luo G, Wang W et al. DrugMGR: a deep bioactive molecule binding method to identify compounds targeting proteins. *Bioinformatics* 2024; 40(4): btae176.
41. Martinez-Mayorga K, Rosas-Jiménez JG, Gonzalez-Ponce K, López-López E, Neme A, Medina-Franco JL. The pursuit of accurate predictive models of the bioactivity of small molecules. *Chemical Science* 2024; 15(6): 1938–1952.
42. Adriaensen S, Rakotoarison H, Müller S, Hutter F. Efficient bayesian learning curve extrapolation using prior-data fitted networks. *Advances in Neural Information Processing Systems* 2024; 872: 36.
43. Chan BW, Lynch NB, Tran W, Joyce JM, Savage GP, Meutermans W et al. Fragment-based drug discovery for disorders of the central nervous system: designing better drugs piece by piece. *Frontiers in Chemistry* 2024; 12: 1379518.
44. Utgés JS, MacGowan SA, Ives CM, Barton GJ. Classification of likely functional class for ligand binding sites identified from fragment screening. *Communications Biology* 2024; 7(1): 320.
45. Kamal H, Zafar MM, Parvaiz A, Razzaq A, Elhindi KM, Ercisli S et al. Gossypium hirsutum calmodulin-like protein (CML 11) interaction with geminivirus encoded protein using bioinformatics and molecular techniques. *International Journal of Biological Macromolecules* 2024; 269: 132095.
46. Ding Y, Qiang B, Chen Q, Liu Y, Zhang L, Liu Z. Exploring chemical reaction space with machine learning models: representation and feature perspective. *Journal of Chemical Information and Modeling* 2024; 64(8): 2955–2970.
47. Li Z, Huang R, Xia M, Patterson TA, Hong H. Fingerprinting interactions between proteins and ligands for facilitating machine learning in drug discovery. *Biomolecules* 2024; 14(1): 72.
48. Shi W, Yang H, Xie L, Yin X-X, Zhang Y. A review of machine learning-based methods for predicting drug–target interactions. *Health Information Science and Systems* 2024; 12(1): 1–16.
49. Charest N, Lowe CN, Ramsland C, Meyer B, Samano V, Williams AJ. Improving predictions of compound amenability for liquid chromatography–mass spectrometry to enhance non-targeted analysis. *Analytical and Bioanalytical Chemistry* 2024; 416(10): 1–15.
50. Pan S, Xia L, Xu L, Li Z. SubMDTA: drug target affinity prediction based on substructure extraction and multi-scale features. *BMC Bioinformatics* 2023; 24(1): 334.
51. Manochkumar J, Ramamoorthy S. Artificial intelligence in the 21st century: the treasure hunt for systematic mining of natural products. *Current Science* 2024; 126(1): 00113891.
52. Naranjo-Castañeda C, Coello-Coello CA, Juaristia E. Application of artificial intelligence and machine learning methods in drug discovery and development. *Arkivoc* 2024; 1: 202412195.

53. Bank NC, Singh V, McCourt B, Burberry A, Roberts KD, Grubb B et al. Antigenic operon fragmentation and diversification mechanism in Bacteroidota impacts gut metagenomics and pathobionts in Crohn's disease microlesions. *Gut Microbes* 2024; 16(1): 2350150.
54. Boldini D, Ballabio D, Consonni V, Todeschini R, Grisoni F, Sieber SA. Effectiveness of molecular fingerprints for exploring the chemical space of natural products. *Journal of Cheminformatics* 2024; 16(1): 35.
55. Jain P, Jain SK, Jain M. Harnessing drug repurposing for exploration of new diseases: an insight to strategies and case studies. *Current Molecular Medicine* 2021; 21(2): 111–132.
56. Gleeson MP, Hersey A, Montanari D, Overington J. Probing the links between in vitro potency, ADMET and physicochemical parameters. *Nature Reviews Drug Discovery* 2011; 10(3): 197–208.
57. Koutsoukas A, Simms B, Kirchmair J, Bond PJ, Whitmore AV, Zimmer S et al. From in silico target prediction to multi-target drug design: current databases, methods and applications. *Journal of Proteomics* 2011; 74(12): 2554–2574.
58. Wang L, Ma C, Wipf P, Liu H, Su W, Xie X-Q. TargetHunter: an in silico target identification tool for predicting therapeutic potential of small organic molecules based on chemogenomic database. *The AAPS Journal* 2013; 15: 395–406.
59. Breinbauer R, Vetter IR, Waldmann H. From protein domains to drug candidates—natural products as guiding principles in the design and synthesis of compound libraries. *Angewandte Chemie International Edition* 2002; 41(16): 2878–2890.
60. Bridges AJ. Chemical inhibitors of protein kinases. *Chemical Reviews* 2001; 101(8): 2541–2572.
61. Nubbemeyer B, Pepanian A, Paul George AA, Imhof D. Strategies towards targeting Gαi/s proteins: scanning of protein-protein interaction sites to overcome inaccessibility. *ChemMedChem* 2021; 16(11): 1697–1716.
62. Burslem GM, Crews CM. Small-molecule modulation of protein homeostasis. *Chemical Reviews* 2017; 117(17): 11269–11301.
63. Doddareddy MR, van Westen GJ, van der Horst E, Peironcely JE, Corthals F, Ijzerman AP et al. Chemogenomics: looking at biology through the lens of chemistry. *Statistical Analysis and Data Mining: The ASA Data Science Journal* 2009; 2(3): 149–160.
64. Chaudhari R, Tan Z, Huang B, Zhang S. Computational polypharmacology: a new paradigm for drug discovery. *Expert Opinion on Drug Discovery* 2017; 12(3): 279–291.
65. Li P, Fu Y, Wang Y. Network based approach to drug discovery: a mini review. *Mini Reviews in Medicinal Chemistry* 2015; 15(8): 687–695.
66. Noor F, Tahir ul Qamar M, Ashfaq UA, Albutti A, Alwashmi AS, Aljasir MA. Network pharmacology approach for medicinal plants: review and assessment. *Pharmaceuticals* 2022; 15(5): 572.
67. Wang L, Ma S, Hu Z, McGuire TF, Xie X-Q. Chemogenomics systems pharmacology mapping of potential drug targets for treatment of traumatic brain injury. *Journal of Neurotrauma* 2019; 36(4): 565–575.
68. Jones LH, Bunnage ME. Applications of chemogenomic library screening in drug discovery. *Nature Reviews Drug Discovery* 2017; 16(4): 285–296.
69. Chaterji S, Ahn EH, Kim D-H. CRISPR genome engineering for human pluripotent stem cell research. *Theranostics* 2017; 7(18): 4445.
70. Wagner BK. The resurgence of phenotypic screening in drug discovery and development. *Expert Opinion on Drug Discovery* 2016; 11(2): 121–125.
71. Manelfi C, Gemei M, Talarico C, Cerchia C, Fava A, Lunghini F et al. "Molecular Anatomy": a new multi-dimensional hierarchical scaffold analysis tool. *Journal of Cheminformatics* 2021; 13: 1–19.
72. Schuffenhauer A. Computational methods for scaffold hopping. *Wiley Interdisciplinary Reviews: Computational Molecular Science* 2012; 2(6): 842–867.
73. Diéguez-Santana K, Casanola-Martin GM, Torres R, Rasulev B, Green JR, González-Díaz H. Machine learning study of metabolic networks vs ChEMBL data of antibacterial compounds. *Molecular Pharmaceutics* 2022; 19(7): 2151–2163.
74. Zhang Y, Fonslow BR, Shan B, Baek M-C, Yates III JR. Protein analysis by shotgun/bottom-up proteomics. *Chemical Reviews* 2013; 113(4): 2343–2394.
75. Dafniet B, Cerisier N, Boezio B, Clary A, Ducrot P, Dorval T et al. Development of a chemogenomics library for phenotypic screening. *Journal of Cheminformatics* 2021; 13(1): 91.
76. Xie Y, Xu Z, Ma J, Mei Q. How much space has been explored? measuring the chemical space covered by databases and machine-generated molecules. arXiv preprint arXiv:2112.12542 2021.
77. Athanasiadis P, Ravikumar B, Elliott RJ, Dawson JC, Carragher NO, Clemons PA et al. Chemogenomic library design strategies for precision oncology, applied to phenotypic profiling of glioblastoma patient cells. *iScience* 2023; 26(7): 107209.
78. Richard AM, Judson RS, Houck KA, Grulke CM, Volarath P, Thillainadarajah I et al. ToxCast chemical landscape: paving the road to 21st century toxicology. *Chemical Research in Toxicology* 2016; 29(8): 1225–1251.
79. Grant MJ, Booth A. A typology of reviews: an analysis of 14 review types and associated methodologies. *Health Information & Libraries Journal* 2009; 26(2): 91–108.
80. Fredin Haslum J, Lardeau C-H, Karlsson J, Turkki R, Leuchowius K-J, Smith K et al. Cell Painting-based bioactivity prediction boosts high-throughput screening hit-rates and compound diversity. *Nature Communications* 2024; 15(1): 3470.
81. Cruz-Monteagudo M, Schürer S, Tejera E, Perez-Castillo Y, Medina-Franco JL, Sanchez-Rodriguez A et al. Systemic QSAR and phenotypic virtual screening: chasing butterflies in drug discovery. *Drug Discovery Today* 2017; 22(7): 994–1007.
82. Misra BB, Langefeld C, Olivier M, Cox LA. Integrated omics: tools, advances and future approaches. *Journal of Molecular Endocrinology* 2019; 62(1): R21–R45.
83. Aguiar-Pulido V, Huang W, Suarez-Ulloa V, Cickovski T, Mathee K, Narasimhan G. Metagenomics, metatranscriptomics, and metabolomics approaches for microbiome analysis: supplementary issue: bioinformatics methods and applications for big metagenomics data. *Evolutionary Bioinformatics* 2016; 12: S36436.
84. Loscalzo J, Kohane I, Barabasi AL. Human disease classification in the postgenomic era: a complex systems approach to human pathobiology. *Molecular Systems Biology* 2007; 3(1): 124.
85. Kempa EE, Hollywood KA, Smith CA, Barran PE. High throughput screening of complex biological samples with mass spectrometry–from bulk measurements to single cell analysis. *Analyst* 2019; 144(3): 872–891.

86. Srivastav S, Lavanya, Avasthi A, Kaur N, Gaertner W, Kesheri M et al., Integromics: Tracking the multi-omic expanse in theragnostics. In: Kesheri M, Kanchan S, Salisbury TB, Sinha RP (eds) *Microbial Omics in Environment and Health*. 2024, Springer, Singapore. pp. 135–173.
87. Hampel H, O'Bryant SE, Durrleman S, Younesi E, Rojkova K, Escott-Price V et al. A precision medicine initiative for Alzheimer's disease: the road ahead to biomarker-guided integrative disease modeling. *Climacteric* 2017; 20(2): 107–118.
88. Sun YV, Hu Y-J. Integrative analysis of multi-omics data for discovery and functional studies of complex human diseases. *Advances in Genetics* 2016; 93: 147–190.
89. Krassowski M, Das V, Sahu SK, Misra BB. State of the field in multi-omics research: from computational needs to data mining and sharing. *Frontiers in Genetics* 2020; 11: 610798.
90. Bunnik EM, Le Roch KG. An introduction to functional genomics and systems biology. *Advances in Wound Care* 2013; 2(9): 490–498.
91. Shen H, McHale CM, Smith MT, Zhang L. Functional genomic screening approaches in mechanistic toxicology and potential future applications of CRISPR-Cas9. *Mutation Research/Reviews in Mutation Research* 2015; 764: 31–42.
92. Kampmann M. CRISPR-based functional genomics for neurological disease. *Nature Reviews Neurology* 2020; 16(9): 465–480.
93. Awwad DA. Beyond classic editing: innovative CRISPR approaches for functional studies of long non-coding RNA. *Biology Methods and Protocols* 2019; 4(1): bpz017.
94. Svitashev S, Young JK, Schwartz C, Gao H, Falco SC, Cigan AM. Targeted mutagenesis, precise gene editing, and site-specific gene insertion in maize using Cas9 and guide RNA. *Plant Physiology* 2015; 169(2): 931–945.
95. Yip BH. Recent advances in CRISPR/Cas9 delivery strategies. *Biomolecules* 2020; 10(6): 839.
96. Chen X, Tasca F, Wang Q, Liu J, Janssen JM, Brescia MD et al. Expanding the editable genome and CRISPR–Cas9 versatility using DNA cutting-free gene targeting based on in trans paired nicking. *Nucleic Acids Research* 2020; 48(2): 974–995.
97. Lau C-H, Tin C, Suh Y. CRISPR-based strategies for targeted transgene knock-in and gene correction. *Faculty Reviews* 2020; 9: 20.
98. Zhang F. Development of CRISPR-Cas systems for genome editing and beyond. *Quarterly Reviews of Biophysics* 2019; 52: e6.
99. An WF, Tolliday N. Cell-based assays for high-throughput screening. *Molecular Biotechnology* 2010; 45: 180–186.
100. Fey PD, Endres JL, Yajjala VK, Widhelm TJ, Boissy RJ, Bose JL et al. A genetic resource for rapid and comprehensive phenotype screening of nonessential Staphylococcus aureus genes. *MBio* 2013; 4(1): e00537–e00512. https://doi.org/10.1128/mbio.00537-12.
101. Zhu M, Zhao S. Candidate gene identification approach: progress and challenges. *International Journal of Biological Sciences* 2007; 3(7): 420.
102. Naxerova K, Di Stefano B, Makofske JL, Watson EV, de Kort MA, Martin TD et al. Integrated loss-and gain-of-function screens define a core network governing human embryonic stem cell behavior. *Genes & Development* 2021; 35(21–22): 1527–1547.
103. Bock C, Datlinger P, Chardon F, Coelho MA, Dong MB, Lawson KA et al. High-content CRISPR screening. *Nature Reviews Methods Primers* 2022; 2(1): 1–23.
104. O'Loughlin TA, Gilbert LA. Functional genomics for cancer research: applications in vivo and in vitro. *Annual Review of Cancer Biology* 2019; 3(1): 345–363.
105. Pasipoularides A. Genomic translational research: paving the way to individualized cardiac functional analyses and personalized cardiology. *International Journal of Cardiology* 2017; 230: 384–401.
106. White J. PubMed 2.0. *Medical Reference Services Quarterly* 2020; 39(4): 382–387.
107. Qiu J, Lv H. An overview of knowledge management research viewed through the web of science (1993–2012). *Aslib Journal of Information Management* 2014; 66(4): 424–442.
108. Burnham JF. Scopus database: a review. *Biomedical Digital Libraries* 2006; 3: 1–8.
109. Nasar Z, Jaffry SW, Malik MK. Information extraction from scientific articles: a survey. *Scientometrics* 2018; 117(3): 1931–1990.
110. Saier T, Färber M. unarXive: a large scholarly data set with publications' full-text, annotated in-text citations, and links to metadata. *Scientometrics* 2020; 125(3): 3085–3108.
111. Salloum SA, Al-Emran M, Monem AA, Shaalan K. Using text mining techniques for extracting information from research articles. *Intelligent Natural Language Processing: Trends and Applications* 2018; 18: 373–397.
112. Feldman R, Sanger J, *The Text Mining Handbook: Advanced Approaches in Analyzing Unstructured Data*. 2006: Cambridge University Press, Cambridge.
113. Ratner B, *Statistical and Machine-Learning Data Mining: Techniques for Better Predictive Modeling and Analysis of Big Data*. 2017: Chapman and Hall/CRC, Boca Raton, FL.
114. Wang Q, Chen Z, Wang Y, Qu H. Applying machine learning advances to data visualization: a survey on ML4VIS. arXiv preprint arXiv:2012.00467 2020: 1–18.
115. Hayden JA, Côté P, Bombardier C. Evaluation of the quality of prognosis studies in systematic reviews. *Annals of Internal Medicine* 2006; 144(6): 427–437.
116. Suri H, Clarke D. Advancements in research synthesis methods: from a methodologically inclusive perspective. *Review of Educational Research* 2009; 79(1): 395–430.
117. Hiebl MR. Sample selection in systematic literature reviews of management research. *Organizational Research Methods* 2023; 26(2): 229–261.
118. Yadav S, Bhardwaj A, Mathur P. Databases in drug design. *Computational Biology in Drug Discovery and Repurposing* 2024; 10: 185–200.
119. Wooller SK, Benstead-Hume G, Chen X, Ali Y, Pearl FM. Bioinformatics in translational drug discovery. *Bioscience Reports* 2017; 37(4): 20160180.
120. Raghav PK, Mann Z, Pandey PK, Mohanty S, Systems biology resources and their applications to understand the cancer. In: Chakraborti S, Ray BK, Roychowdhury S. (eds) *Handbook of Oxidative Stress in Cancer: Mechanistic Aspects*. 2021, Springer, Singapore. pp. 1–35.
121. Thul PJ, Lindskog C. The human protein atlas: a spatial map of the human proteome. *Protein Science* 2018; 27(1): 233–244.
122. Knox C, Wilson M, Klinger CM, Franklin M, Oler E, Wilson A et al. DrugBank 6.0: the DrugBank knowledgebase for 2024. *Nucleic Acids Research* 2024; 52(D1): D1265–D1275.

123. Tan H, Kim P, Sun P, Zhou X. miRactDB characterizes miRNA–gene relation switch between normal and cancer tissues across pan-cancer. *Briefings in Bioinformatics* 2021; 22(3): bbaa089.
124. Chen B, Butte A. Leveraging big data to transform target selection and drug discovery. *Clinical Pharmacology & Therapeutics* 2016; 99(3): 285–297.
125. Rintala TJ, Ghosh A, Fortino V. Network approaches for modeling the effect of drugs and diseases. *Briefings in Bioinformatics* 2022; 23(4): bbac229.
126. Ghulam A, Lei X, Guo M, Bian C. Comprehensive analysis of features and annotations of pathway databases. *Current Bioinformatics* 2020; 15(8): 803–820.
127. Ooi HS, Schneider G, Lim T-T, Chan Y-L, Eisenhaber B, Eisenhaber F. Biomolecular pathway databases. *Data Mining Techniques for the Life Sciences* 2010; 407: 129–144.
128. Moffat JG, Vincent F, Lee JA, Eder J, Prunotto M. Opportunities and challenges in phenotypic drug discovery: an industry perspective. *Nature Reviews Drug Discovery* 2017; 16(8): 531–543.
129. Zhu F, Shen B. Biomarker discovery with text mining and literature based discovery. *Bioinformatics for Diagnosis, Prognosis and Treatment of Complex Diseases* 2013; 5: 57–78.
130. Baykoucheva S. Selecting a database for drug literature retrieval: a comparison of MEDLINE, Scopus, and Web of Science. *Science & Technology Libraries* 2010; 29(4): 276–288.
131. Jo T. Text mining. *Studies in Big Data* 2019; 4(1): 45.
132. Perera N, Dehmer M, Emmert-Streib F. Named entity recognition and relation detection for biomedical information extraction. *Frontiers in Cell and Developmental Biology* 2020; 8: 673.
133. Liu K, Hogan WR, Crowley RS. Natural language processing methods and systems for biomedical ontology learning. *Journal of Biomedical Informatics* 2011; 44(1): 163–179.
134. Jimeno A, Jimenez-Ruiz E, Lee V, Gaudan S, Berlanga R, Rebholz-Schuhmann D. Assessment of disease named entity recognition on a corpus of annotated sentences. *BMC Bioinformatics* 2008; 9(Suppl 3): S3.
135. Lapatas V, Stefanidakis M, Jimenez RC, Via A, Schneider MV. Data integration in biological research: an overview. *Journal of Biological Research-Thessaloniki* 2015; 22: 1–16.
136. Wang S, Sim TB, Kim Y-S, Chang Y-T. Tools for target identification and validation. *Current Opinion in Chemical Biology* 2004; 8(4): 371–377.
137. Yang Y, Adelstein SJ, Kassis AI. Target discovery from data mining approaches. *Drug Discovery Today* 2012; 17: S16–S23.
138. Stoudt S, Vásquez VN, Martinez CC. Principles for data analysis workflows. *PLoS Computational Biology* 2021; 17(3): e1008770.
139. Chandran U, Mehendale N, Patil S, Chaguturu R, Patwardhan B. Network pharmacology. *Innovative Approaches in Drug Discovery* 2017; 14: 127.
140. Zhang G-B, Li Q-Y, Chen Q-L, Su S-B. Network pharmacology: a new approach for Chinese herbal medicine research. *Evidence-Based Complementary and Alternative Medicine* 2013; 2013(1): 621423.
141. Jin Q, Lu J, Gao R, Xu J, Pan X, Wang L. Systematically deciphering the pharmacological mechanism of Fructus Aurantii via network pharmacology. *Evidence-Based Complementary and Alternative Medicine* 2021; 2021(1): 6236135.
142. Li S. Network pharmacology evaluation method guidance-draft. *World Journal of Traditional Chinese Medicine* 2021; 7(1): 146–154.
143. Xin W, Zi-Yi W, Zheng J-H, Shao L. TCM network pharmacology: a new trend towards combining computational, experimental and clinical approaches. *Chinese Journal of Natural Medicines* 2021; 19(1): 1–11.
144. Bennett DA, Yu L, De Jager PL. Building a pipeline to discover and validate novel therapeutic targets and lead compounds for Alzheimer's disease. *Biochemical Pharmacology* 2014; 88(4): 617–630.
145. Hampel H, Nisticò R, Seyfried NT, Levey AI, Modeste E, Lemercier P et al. Omics sciences for systems biology in Alzheimer's disease: state-of-the-art of the evidence. *Ageing Research Reviews* 2021; 69: 101346.
146. Veenstra TD. Omics in systems biology: current progress and future outlook. *Proteomics* 2021; 21(3–4): 2000235.
147. Gao Y, Xu T, Zhao Y-X, Ling-Hu T, Liu S-B, Tian J-S et al. A novel network pharmacology strategy to decode metabolic biomarkers and targets interactions for depression. *Frontiers in Psychiatry* 2020; 11: 667.
148. Zhao L, Zhang H, Li N, Chen J, Xu H, Wang Y et al. Network pharmacology, a promising approach to reveal the pharmacology mechanism of Chinese medicine formula. *Journal of Ethnopharmacology* 2023; 309: 116306.
149. Nogales C, Mamdouh ZM, List M, Kiel C, Casas AI, Schmidt HH. Network pharmacology: curing causal mechanisms instead of treating symptoms. *Trends in Pharmacological Sciences* 2022; 43(2): 136–150.
150. Hopkins AL. Network pharmacology: the next paradigm in drug discovery. *Nature Chemical Biology* 2008; 4(11): 682–690.
151. Altschul SF, Gish W, Miller W, Myers EW, Lipman DJ. Basic local alignment search tool. *Journal of Molecular Biology* 1990; 215(3): 403–410.
152. Lobo I. Basic local alignment search tool (BLAST). *Nature Education* 2008; 1(1): 215.
153. Mount DW. Using the basic local alignment search tool (BLAST). *Cold Spring Harbor Protocols* 2007; 2007(7): 17.
154. Kerfeld CA, Scott KM. Using BLAST to teach "E-value-tionary" concepts. *PLoS Biology* 2011; 9(2): e1001014.
155. Taylor BE, Reynolds AJ, Etz KE, MacCalla NM, Cotter PA, DeRuyter TL et al. BUILDing BLaST: promoting rural students' biomedical research careers using a culturally responsive, one health approach. *BMC Proceedings* 2017; 11(Suppl 12): 13.
156. Kurzweil R. What Is Artificial Intelligence Anyway? As the techniques of computing grow more sophisticated, machines are beginning to appear intelligent—but can they actually think? *American Scientist* 1985; 73(3): 258–264.
157. O'Shea K, Nash R. An introduction to convolutional neural networks. arXiv preprint arXiv:1511.08458 2015.
158. Grossberg S. Recurrent neural networks. *Scholarpedia* 2013; 8(2): 1888.
159. Wu Z, Pan S, Chen F, Long G, Zhang C, Philip SY. A comprehensive survey on graph neural networks. *IEEE Transactions on Neural Networks and Learning Systems* 2020; 32(1): 4–24.
160. Bottou L, Stochastic gradient descent tricks. In: Montavon G, Orr GB, Müller KR. (eds) *Neural Networks: Tricks of the Trade:* Second Edition. 2012, Springer, Berlin. pp. 421–436.

161. Ding H, Takigawa I, Mamitsuka H, Zhu S. Similarity-based machine learning methods for predicting drug–target interactions: a brief review. *Briefings in Bioinformatics* 2014; 15(5): 734–747.
162. Narkhede S. Understanding auc-roc curve. *Towards Data Science* 2018; 26(1): 220–227.
163. Tafvizi A, Avci B, Sundararajan M. Attributing AUC-ROC to analyze binary classifier performance. arXiv preprint arXiv:2205.11781 2022.
164. Chiang AP, Butte AJ. Data-driven methods to discover molecular determinants of serious adverse drug events. *Clinical Pharmacology & Therapeutics* 2009; 85(3): 259–268.
165. Nunez S, Venhorst J, Kruse CG. Target–drug interactions: first principles and their application to drug discovery. *Drug Discovery Today* 2012; 17(1–2): 10–22.

11 Protein–Ligand Interaction and Molecular Dynamics Simulation

Rinki Prasad Bhagat and Shovanlal Gayen
Laboratory of Drug Design and Discovery, Department of Pharmaceutical Technology, Jadavpur University, Kolkata, India

INTRODUCTION

Proteins are the catalysts of biological systems, coordinating a broad range of cellular functions, including signal transduction inside the cell or between cells [1]. The interaction between proteins and ligands (small molecules) is essential for many of these activities, and it needs to be analysed for therapeutic intervention during drug design and discovery [2,3]. Conventional experimental methods, including differential scanning fluorimetry, electrospray ionization mass spectrometry, surface plasmon resonance, fluorescence polarization assay, fluorescence resonance energy transfer, isothermal titration calorimetry, enzyme-linked immunosorbent assay, X-ray crystallography, nuclear magnetic resonance spectroscopy, etc., offer valuable insights into protein–ligand interactions [4–9]. However, these methodologies often provide snapshots of static structures, which limits our potential to understand the dynamic nature of these interactions. In addition, these methods are time-consuming and costly and demand a high amount of purified proteins [10]. Researchers are looking for alternative, inexpensive methods to understand protein–ligand interactions and protein–ligand complex conformations to speed up the drug design and discovery process. Thus, a wide range of computational techniques have been evolved and established to understand such interactions and protein conformations (bound/unbound) for the development of inhibitors/drug molecules [11,12]. Molecular docking is one such computer-based method that is used to find exact ligands for different protein targets and generate 3D models of protein–ligand complexes highlighting different important interactions [13]. On the other hand, molecular dynamics (MD) simulation is another computational tool that has improved our ability to investigate the dynamic behaviour of protein–ligand complexes at the atomic level. By computationally solving Newton's equations of motion, MD simulations allow for the dynamic, time-dependent modelling of protein–ligand complexes, capturing the intricate details showing transitory interactions that are difficult to witness experimentally [14,15].

In this chapter, an overview of molecular docking and MD simulation in drug design and discovery is presented. The basic concepts, applications, and various software used in molecular docking and MD simulations are also discussed, which will help computational/medicinal chemists in the drug design and discovery process.

PROTEIN–LIGAND INTERACTION

The interaction between a protein and a ligand is regulated by some non-covalent forces, such as hydrophobic interactions, hydrogen bonding, electrostatic interactions, van der Waals forces, etc. These interactions determine the selectivity and affinity of the binding process between the protein and ligand. Predicting and explaining these interactions requires an understanding of the structural and chemical characteristics of both the ligand and the protein [16–18]. With the help of computational methods such as molecular docking, we can understand the nature and type of interactions between important amino acid residues of the protein and the ligand.

MOLECULAR DOCKING ANALYSIS

Molecular docking is used in drug design and discovery to forecast the binding mechanism and affinity of ligands to a target protein. Docking algorithms rank potential ligands according to their binding affinity by simulating the interaction between the ligand and the protein binding site. It is an important part of computational virtual screening that examines vast databases to reveal potential binding candidates. Furthermore, it also helps in structure-based drug design by directing the optimization of lead compounds to increase binding affinity and selectivity [19]. Although molecular docking offers insightful information on protein–ligand interactions, the accuracy of results is contingent upon the quality of the protein structure and the scoring function employed. Nowadays, docking algorithms and scoring methodologies are constantly improving, increasing the reliability and utility of molecular docking in drug discovery [20]. Some common docking programs with different scoring mechanisms and algorithms are shown in Table 11.1.

Molecular Docking Algorithms

Several molecular docking approaches and algorithms are available, each based on a specific set of parameters, to achieve protein–ligand docking with optimal performance.

DOI: 10.1201/9781003464679-11

TABLE 11.1
Summary of Some Docking Software Along with Its Features

Software	Ligand Sampling	Scoring Function	Features	Year of Introduction	Reference and Source
GOLD	Stochastic	Force field	It is a genetic algorithm-based docking approach, which enables the overlapping of atoms between ligand and protein.	1995	[21], https://www.ccdc.cam.ac.uk/solutions/software/gold/
Surflex-Dock	Systematic	Empirical	This is a protocol-based docking mechanism that may be developed automatically.	1998	[22], https://www.biopharmics.com/downloads/
LigandFit	Stochastic	Empirical	Monte Carlo-generated ligand conformations are docked into an active site according to its shape, followed by CHARMM minimization.	2003	[23], http://accelrys.com/
Glide	Stochastic	Empirical	It is an exhaustive-based docking approach with enhanced precision, standard precision, and high-throughput virtual screening scoring modes provided by Schrodinger.	2006	[24], https://newsite.schrodinger.com/platform/products/glide/
PLANTS (Protein-Ligand ANT System)	Stochastic	Empirical	It is based on ant colony optimization, a successful swarm intelligence technique.	2006	[25], https://github.com/discoverdata/parallel-PLANTS
AutoDock	Stochastic	Force field	It is a set of automatic docking tools and has a wide range of applications.	2009	[26], http://autodock.scripps.edu/
AutoDock Vina	Stochastic	Empirical	It is one of the most popular and quick open-source docking engines.	2010	[27], http://vina.scripps.edu/
LeDock	—	Knowledge	It involves annealing ligand positions while considering bond formation, utilizing function scoring information. It is free for academics and is managed by Lephar Research.	2014	[28], http://lephar.com/
rDock	Stochastic	Empirical	It is a blend of stochastic and deterministic search techniques used for low-energy ligand posing. It is maintained by rDock development.	2012	[29], http://rdock.sourceforge.net/
FRED	Systematic	Gaussian	It employs a Gaussian function for docking, resulting in a smooth and searchable energy surface. It has a broad tolerance for mistakes in protein atom location and is a rapid exhaustive docking approach.	2011	[30], https://www.eyesopen.com/oedocking
FlexAID	Stochastic	Empirical	It can use proteins and nucleic acids as docking targets and small compounds and peptides as ligands. It enables both comprehensive ligand flexibility and selective side-chain flexibility.	2015	[31], http://biophys.umontreal.ca/nrg/resources.html
MOLS 2.0	Systematic	Force field (AMBER)	It selects a relatively small but accurate sample of the multidimensional search space.	2016	[32], https://sourceforge.net/projects/mols2-0/

The important aspects of molecular docking algorithms, ligand sampling and scoring functions, are illustrated in Figure 11.1 [33,34].

Molecule Flexibility

Depending on how proteins/ligands are treated for flexibility, molecular docking can be classified into three categories: rigid, flexible, and semiflexible. There is also another method, soft docking [35], which can explain minor conformational changes.

- **Rigid Docking**: In this method, protein and ligand molecules are handled as inflexible structures. Only translation and rotation are taken into account, not conformational degrees of freedom during the docking process. Hence, the locations of the molecules shift without changing their shape in three dimensions so that they best match with other compounds [36].
- **Flexible Docking**: In flexible docking, the flexibility of both protein and ligand is maintained

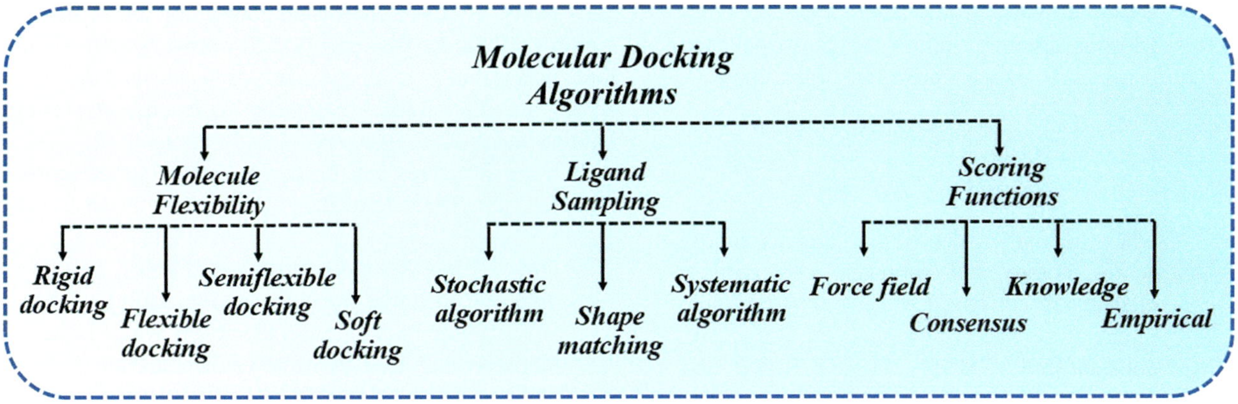

FIGURE 11.1 Different algorithms of molecular docking.

while calculating the energy required for various ligand conformations to fit into proteins. Although it takes a bit longer, this approach is more dependable in some cases since it can assess a wide range of potential conformations [37].

- **Semiflexible Docking**: This method is asymmetrical and considers one of the molecules, often the smaller ligand, to be flexible while the receptor is rigid. The idea behind semiflexible docking is to maintain the rigidity of the protein structure while permitting rotatable links to provide flexibility to the ligand structure. Compared to rigid docking, it yields more precise findings [38].
- **Soft Docking:** It involves the softening of van der Waals interactions between atoms, which indirectly enables some overlap and flexibility in the structures of both ligand and receptor molecules. To ensure a realistic soft docking procedure, it is necessary to make sure that the ligand and the protein can be rotated to resemble their native states. This method is easier to use and more effective than an explicit sampling of various receptor conformations [35,39].

Ligand Sampling

All docking programs must use protein–ligand sampling techniques to generate acceptable ligand poses. The sampling algorithm takes a protein target and produces potential ligand conformations around the selected protein binding site. Three general categories of ligand sampling methods exist, namely systematic search algorithms, stochastic algorithms, and shape-matching algorithms [40,41,42].

- **Systematic Search Algorithms**: They are mostly utilized in flexible ligand docking, which investigates all of the ligand's degrees of freedom. Three categories of systematic search algorithms exist: exhaustive search, fragmentation search, and conformational ensemble [34,43].
  a. **Exhaustive Search Algorithm**: It is a brute force algorithm that meticulously lists every solution to a given issue and then evaluates it to verify whether it is accurate or not. It is usually applied for issues with a compact and specific search space, where it is practical to test all possible solutions [44]. Some examples of software that use this algorithm are Glide [24] and FRED [30].
  b. **Fragmentation Search Algorithm**: In this process, the ligand breaks into tiny fragments, which are then progressively added to the binding site by covalently combining with one another. Some examples of software that use fragmentation search include DOCK [45] and FlexX [46].
  c. **Conformational Ensemble Algorithm**: In this process, an ensemble of pre-generated ligand conformations is docked together into the binding site of the protein. PhDOCK [47], FLOG [48], and MS-DOCK [49] are some examples of software that use conformational ensemble docking algorithms.
- **Stochastic Algorithms**: These algorithms investigate the conformational space of ligands by creating unique conformations with a random change that populates an extensive portion of the energy landscape. These random changes will be evaluated using a probabilistic criterion to determine acceptance or rejection. Since these algorithms rely on the generation of random ligand conformations, they cannot guarantee an exact outcome. Consequently, the docking process in these algorithms is iterative [50].

  Stochastic algorithms are classified into four types: Monte Carlo (MC), Tabu search, swarm optimization (SO), and evolutionary algorithms (EAs) [20,34].
  a. **Monte Carlo**: MC is a mathematical method for projecting potential outcomes of an unknown event. Using this technique, computer programs examine previous data and forecast a variety of future results depending on the

selection made. In this method, the probability of accepting a random change is calculated using the Boltzmann probability function:

$$P \sim \exp\left[\frac{-(E_1 - E_0)}{k_B T}\right]$$

where E_0 and E_1 denote the ligand's energy scores before and following the random change, respectively; T is the system's absolute temperature; and k_B represents the Boltzmann constant [50]. ICM [51], MCDOCK [52], and QXP [53] are some examples of docking programs that employ MC techniques.

b. **Evolutionary Algorithms**: EAs estimate the free energy of binding similarly to force field-based approaches, but they do not require a lot of processing power. A weighted total of factors, including the quantity of hydrogen bonds and hydrophobic/hydrophilic interactions, is evaluated to arrive at this estimate. These factors can be computed more quickly since they are less complicated than force field parameters. Examples of software that use EAs are PSI-DOCK [54], MolDock, [55] DARWIN [56], and AutoDock [26].
c. **Swarm Optimization**: SO is a computer technique that solves problems by repeatedly attempting to enhance a potential solution concerning a predetermined quality of statistics [57]. It uses a concept of swarm intelligence to identify the best solution in a search domain, and it is used in several docking programs, such as SODOCK [58] and PLANTS [25].
d. **Tabu Search**: It is an optimization technique that uses the so-called Tabu list to guide a local search process, avoiding local optimization and rejecting advances to previously explored sites in the search space [59]. An example of software using Tabu search is PSI-DOCK [54].

- **Shape-Matching Algorithms**: It is the process of altering a shape and comparing its similarity to another using a similarity measure. This method is one of the most basic sampling algorithms, is faster than other algorithms, and uses a sampling concept that requires ligands to be structurally similar to protein binding sites. This approach is frequently employed in the first phases of the docking process or as a prelude to more complex ligand sampling techniques. A few examples of software that use shape matching are LigandFit [23], FLOG [48], and MS-Dock [49].

Scoring Functions

Scoring protein–ligand interactions is essential for identifying, investigating, and optimizing ligands that bind to proteins. It is therefore a crucial part of drug design. It evaluates the quality of a protein–ligand interaction by estimating binding affinity or free energy. Most scoring methods use force fields from molecular mechanics to calculate the pose's energy and rely on minimizing the free energy involved in binding and protein–ligand complex formation [60].

Scoring functions can be divided into four categories: force field, empirical, knowledge-based, and consensus [34].

- **Force Field Scoring Function (FFSF)**: In this function, affinities are calculated by adding up the electrostatic interactions and van der Waals forces between each atom in the two molecules that make up the complex [61]. DOCK [45] force field is one of the examples of force fields used in docking.
- **Empirical Scoring Function**: It is easier to compute and faster than FFSF. In addition, by combining many parametric functions, it can replicate experimental data, including binding energies and/or conformations [31]. A few examples of software that use empirical scoring functions are GlideScore [24], SCORE [62], X-SCORE [63], and LigScore [64].
- **Knowledge-Based Scoring Function**: It is a new and promising method for grading protein–ligand complexes with known 3D structures based on their binding affinities. It does not bind energies; instead, they are developed to replicate experimental structures [65]. DrugScore [66] and SMoG [67] are examples of knowledge-based scoring functions.
- **Consensus Scoring Function**: It is a technique for increasing the hit rates generated by docking databases involving 3D structures into proteins. It increases the likelihood of uncovering "true" ligands by balancing the flaws in single scores by combining data from several scores [68]. An example of software that use this function is X-SCORE [63], a combination of FlexX, GOLD, and PMF DOCK.

Steps of Molecular Docking

A brief description of the steps [69] of molecular docking in general is presented in the following section (Figure 11.2).

- **Preparation of Ligand and Protein**: This involves obtaining the 3D structure of the ligand and the protein. Protein structures can be developed computationally (artificial intelligence/homology modelling) or can be obtained from databases like Protein Data Bank (PDB).
- **Grid Generation**: A 3D grid is constructed in or around the receptor to identify the possible binding site. This grid makes it easier to sample ligand conformations and orientations inside the binding site effectively.
- **Search Algorithm**: Search algorithms (systematic search, stochastic, and shape matching) are

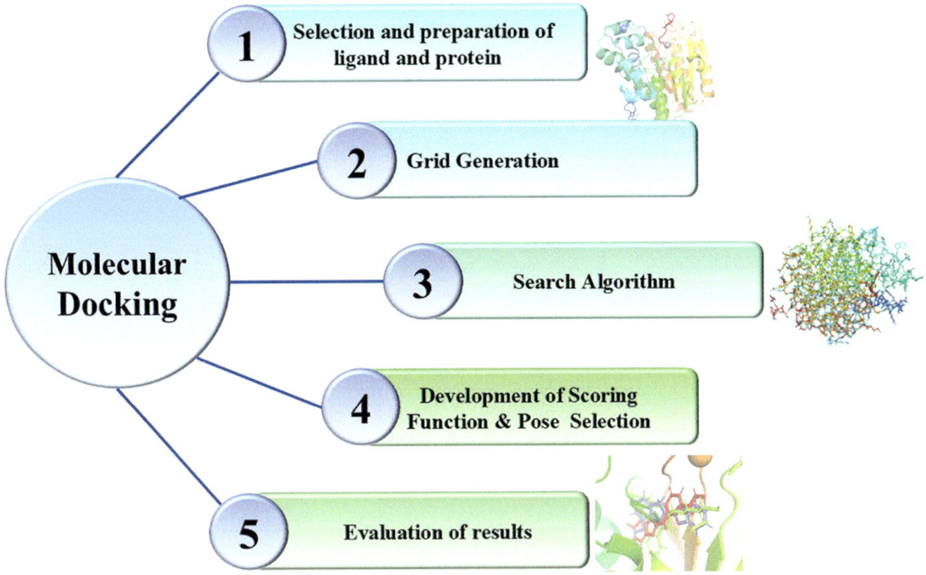

FIGURE 11.2 Steps of molecular docking.

employed to investigate the ligand's conformational space within the binding site. The aim of these algorithms is to maximize the binding affinity by determining the ideal ligand posture.

- **Development of Scoring Function**: A scoring function determines the fitness of protein–ligand interactions. The aim of scoring functions is to forecast the binding affinity between the ligand and the target protein.
- **Pose Selection**: Once a collection of ligand postures is created, the most energetically favourable positions are chosen using a selection criterion. This criterion is frequently based on the scoring function related to the highest predicted binding affinity.
- **Analysis of Results**: The selected ligand poses are examined to determine probable interactions and modes of binding with the protein. Visualization tools can be used to get insights into the structural basis of protein–ligand interactions.

Applications of Molecular Docking

Docking is a commonly used method for discovering protein–ligand interactions and plays an essential role in the drug design and discovery process [70,71], such as in the following methods:

- **Hit Identification and Lead Optimization**: When docking and scoring functions are coupled, inhibitors can be identified *in silico* by recognizing compounds that have a high probability of binding to a target protein.
- **Binding Site Identification**: Docking can be utilized to forecast the position in which the ligand will interact with the protein.
- **Bioremediation**: It aids in determining the active site for accommodating the contaminant molecules, based on the residue's stearic hindrance and the structure of the active site pocket.
- **Drug–DNA Interaction**: Docking can be used to predict the binding mode of a drug with the protein or DNA.
- **Other Uses**: Some other uses of docking are the prediction of the stability and interactions of protein–ligand complexes, protein–peptide interactions, adverse drug responses, drug repurposing, polypharmacology, target profiling, etc. [72,73].

MOLECULAR DYNAMICS SIMULATION

MD simulation plays a crucial role in understanding the complex interactions between proteins and ligands, which is important in various biological processes. Researchers can simulate the dynamic behaviour of protein–ligand complexes using various software (Table 11.2) at the atomic level, capturing the changes in their structure, binding routes, and binding affinities with time [74,75]. These simulations may disclose the locations of all atoms at the femtosecond (10^{-15}s, each) temporal resolution [76]. In addition to displaying structural variations in response to environmental factors such as pH, temperature, and residue mutations, MD simulations may also demonstrate the dynamic process of protein or peptide misfolding and aggregation [77]. MD simulation in conjunction with binding free energy calculations is an effective way to improve the enrichment factor of virtual screening as it increases the accuracy of the prediction of binding ability between the ligand and protein [75]. The binding free energy between the ligand and protein can be calculated using molecular mechanics/Poisson–Boltzmann surface area (MM-PBSA)

TABLE 11.2
List of Some MD Software Along with Its Features

S. No.	Software	Force Field	Features	Reference and Source
1	GROMACS	GROMOS 43A1-S3 MBER99SB ILDN	It delivers extremely high performance as compared to all other programs.	[82], http://www.gromacs.org/
2	AMBER	AMBER ff99SBildn LIPID14	It gives parameters for very small organic compounds, allowing the simulations of drugs and small molecule ligands in combination with biomolecules.	[83], http://ambermd.org/
3	DESMOND	AMBER CHARMM OPLS MMFF	It provides support for methods that are often used to achieve rapid and precise MD.	[84], https://www.deshawresearch.com/downloads/download_desmond.cgi/
4	NAMD	CHARMM AMBER	It is a free software for MD code developed for highly efficient simulation of huge biomolecular systems.	[85], http://www.ks.uiuc.edu/Research/namd/
5	CHARMM	CHARMM	It is utilized in macromolecular simulations, including MD, energy minimization, and Monte Carlo simulations.	[86], http://yuri.harvard.edu/
6	MOLDY	CHARMM	It makes it possible to use short-range potentials to simulate millions of atoms.	[87], http://www.tchpc.tcd.ie/projects/completed/moldy
7	LAMMPS	CHARMM, AMBER, OPLS, GROMACS	It has possibilities for soft and solid-state materials, as well as coarse-grained systems.	[88], http://lammps.sandia.gov/
8	TINKER	AMBER, CHARMM, OPLS, MMFF	It is an all-inclusive program for MD and mechanics that includes specific features for biopolymers.	[89], http://dasher.wustl.edu/tinker/
9	ACEMD	CHARM, AMBER, OPLS	It is the fastest MD engine available for a single workstation worldwide.	[90], https://www.acellera.com/acemd
10	ABALONE	AMBER, OPLS	It allows for long-term simulations.	[91], http://www.biomolecular-modeling.com/Abalone/Molecular-purchase.html

and molecular mechanics/generalized Born surface area (MM-GBSA) [78]. Much research has been done to find the interaction between the ligand and protein using MD simulations in combination with binding free energy calculations [79–81].

Steps of MD

The methods [92] involved in MD are defined below, and a workflow of the basic steps is illustrated in Figure 11.3.

- **Molecular Setup**: The initial geometries are downloaded from the PDB, and coordinate files and topology files are generated. The starting configuration of the system is described by giving the locations, velocities, and charges of the atoms or molecules.
- **Force Field Selection**: A suitable force field is selected to represent the interactions between the atoms or molecules in the system. In force fields, one may find parameters related to bond stretching, angle bending, torsional rotations, and non-bonded interactions such as electrostatic interactions, van der Waals forces, etc.
- **Solvation and Neutralization of the Simulation Box**: In the simulation box, molecules from various species (solutes and solvents) are inserted based on their concentrations. Coordinate files and topology files are assembled and neutralized by adding counter-ions. Periodic boundary conditions are also applied to each face of the simulation box.
- **Energy Minimization**: Energy is minimized to ease the initial configuration and get rid of any steric clashes or unfavourable interactions.
- **Equilibration**: After energy minimization, the next step is NPT and NVT. NVT guarantees that during the simulation, there will be a constant number of atoms (N), constant volume (V), and constant temperature (T). On the other hand, NPT guarantees that during the simulation, a constant number of atoms (N), constant pressure (P), and constant temperature (T) will be maintained.
- **Production**: The next step is to run the production MD simulation for the preferred time period, during which the system's dynamics are sampled

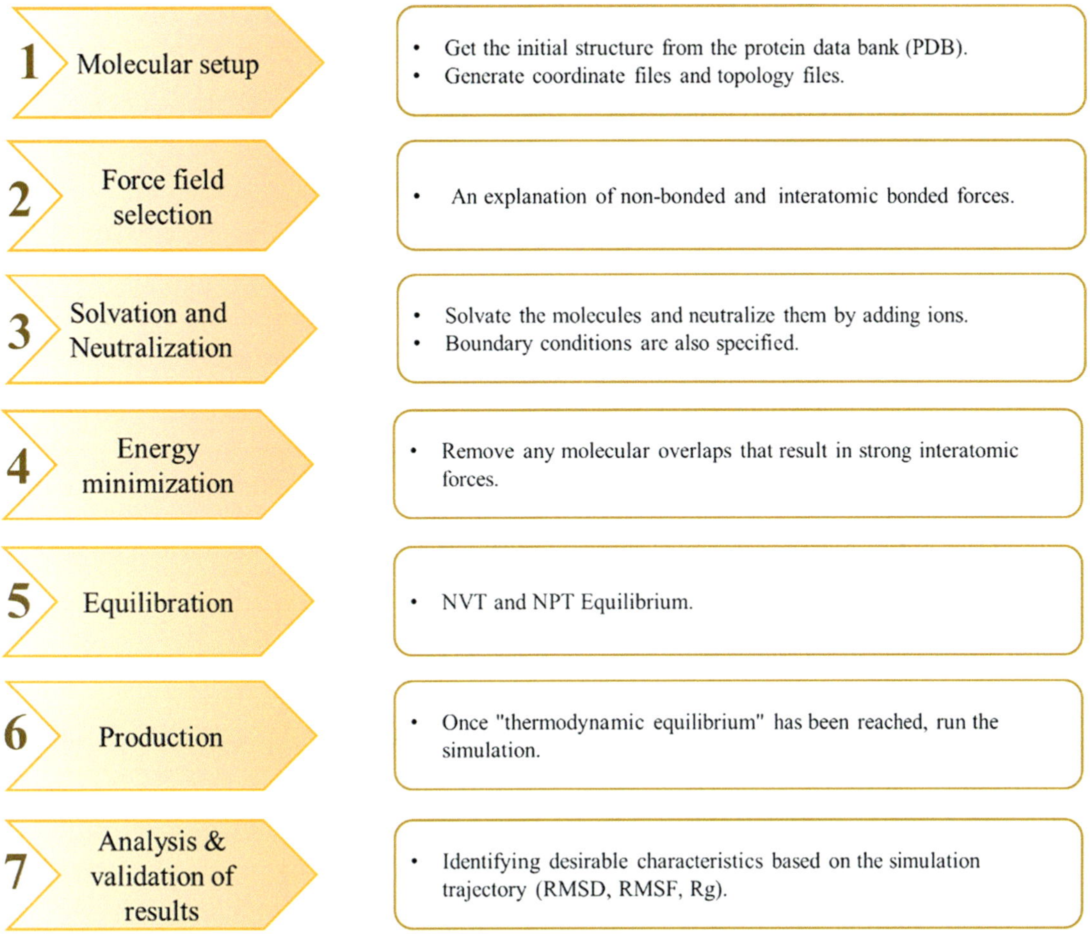

FIGURE 11.3 Steps of molecular dynamics simulation.

continuously. This stage produces trajectory data that may be examined to learn more about the dynamics, thermodynamic characteristics, and structural alterations of the system.

- **Analysis**: In this step, MD trajectory data is evaluated to extract system-relevant information, including structural fluctuations, radial distribution functions, diffusion coefficients, energy landscapes, etc.

Trajectory Analysis of MD

a. **Root Mean Square Deviation (RMSD)**: It is a statistical parameter used to quantify the change in a protein's backbone/side chains between different structural conformations.
b. **Radius of Gyration (Rg)**: It describes how atoms are arranged along a protein's axis. It is the length that indicates the separation between the point at which it is spinning and the point at which the energy transmission has the utmost effect.
c. **Root Mean Square Fluctuation (RMSF)**: It is a statistical parameter comparable to RMSD. Instead of demonstrating positional changes across whole structures over a period, RMSF measures distinct residue flexibility, or the degree to which a specific residue moves (fluctuates) throughout a simulation.
d. **Binding Free Energy**: It is calculated using force field energy projections of variations in electrostatic energies and van der Waals interactions.
e. **Free Energy Landscape**: It is used to analyse the protein's conformational variations. It provides precise information on amino acid interactions, including protein folding and binding.
f. **Principal Component Analysis**: It is a promising multivariate method that simplifies and reduces huge, complex data sets to characterize protein conformations.

Application of Molecular Dynamics Simulation

MD simulation has evolved as an essential tool in drug discovery, providing unparalleled knowledge on the dynamic behaviour of biomolecules at the atomic level. It offers useful insights that complement experimental methods to better comprehend drug–target interactions, ligand binding processes, and structure–activity relationships by computationally modelling the interactions and movements of atoms over time. We highlight here some of the applications of MD simulations:

- **Determining the Movement and Structure of Biomolecules**: Studying, analysing, and mimicking the flexibility, motions, and interactions between various proteins is one of the most popular uses of MD simulation in biomolecules. It also displays the dynamic behavioural characteristics of salt ions and water molecules, which are often crucial for ligand binding and protein function.
- **ADME/T Predictions**: MD simulations provide information about drug absorption, distribution, metabolism, excretion, and toxicity (ADME/T) features by simulating drug interactions with biological membranes, transporters, and metabolic enzymes. This facilitates the optimization of pharmacokinetic profiles and early detection of safety issues.
- **Evaluation of Accuracy and Refinement of Simulated Structures**: MD may be used to evaluate the accuracy of previously modelled structures or even to improve the structures developed scientifically in the laboratory or by molecular modelling approaches. For example, a computational MD-simulated annealing strategy refines experimentally discovered X-ray crystal structures, fitting the model to the experimental information even more precisely while preserving structural stability.
- **Providing Information about Temporal-Scale Molecular Interactions**: MD simulation provides atomic-level insights into the dynamic development of biomolecular systems. This characteristic, combined with the temporal scale for MD simulations, allows forecasting several potential cellular interactions and behaviours, depending on which changes to the current structures can be made and observed.
- **Protein Folding**: This methodology involves utilizing MD simulations to compute the energy of both folded and unfolded states at varying temperatures. This yields two crucial values (ΔH and ΔC_p), which are utilized to determine the protein's conformational stability (ΔG).
- **Drug Resistance Mechanisms**: MD simulations uncover the structural basis of drug resistance by simulating the dynamics of drug–target complexes in the existence of mutations or resistance-imparting events. This contributes to the development of more potent therapeutics and methods for overcoming resistance.
- **Identification of Allosteric Sites**: MD simulations aid in identifying allosteric sites and conformational changes, allowing for the design and development of allosteric modulators with therapeutic potential.

COMBINED MD SIMULATIONS AND MOLECULAR DOCKING

In drug discovery, the combination of molecular docking and MD simulations is considered an important approach, providing a deeper understanding of protein–ligand interactions and improved accuracy. While molecular docking predicts the binding mechanism and affinity of small molecules inside the binding region of a target protein, MD simulations provide dynamic information on the flexibility of proteins and the impact of solvents, compute precise energies, optimize final complex structures, and reveal changes in conformations over time. This combined strategy accelerates the drug discovery process, improves the understanding of drug–target interactions, and assists in rational drug design. In addition, it is an effective way to address the drawbacks of both approaches since these methods can counterbalance each other's shortcomings [93].

SUMMARY

This chapter aims to provide a general introduction to the idea of protein–ligand interaction through the use of molecular docking and MD simulations, along with their use in modern computational drug design. The first half of the chapter addressed the specifics of molecular docking, while the second part discussed MD simulations and their applications. This chapter offers a general introduction and a fundamental understanding of the rationale behind the application of MD simulation and molecular docking. This information will help acquire a basic knowledge of different parameters in the drug discovery process.

REFERENCES

1. Cala, O.; Guillière, F.; Krimm, I. NMR-based analysis of protein–ligand interactions. *Anal. Bioanal. Chem.* 2014, 406, 943–56.
2. Harding, S.E.; Chowdhry, B.Z. editors. *Protein-Ligand Interactions, Structure, and Spectroscopy: A Practical Approach*. Oxford University Press, Oxford, 2001, 243.
3. Sliwoski, G.; Kothiwale, S.; Meiler, J.; Lowe, E.W. Computational methods in drug discovery. *Pharmacol. Rev.* 2014, 66, 334–95.
4. Aguirre, C.; Cala, O.; Krimm, I. Overview of probing protein-ligand interactions using NMR. Curr. *Protoc. Protein Sci.* 2015, 81, 17–8.
5. Renaud, J.P.; Delsuc. M.A. Biophysical techniques for ligand screening and drug design. *Curr. Opin. Pharmacol.* 2009, 9, 622–8.
6. Niesen, F.H.; Berglun, H.; Vedadi, M. The use of differential scanning fluorimetry to detect ligand interactions that promote protein stability. *Nat. Protoc.* 2007, 2, 2212–21.
7. Jonker, N.; Kool, J.; Irth, H.; Niessen, W.M. Recent developments in protein-ligand affinity mass spectrometry. *Anal. Bioanal. Chem.* 2011, 399, 2669–81.
8. Kay, L.E.; NMR studies of protein structure and dynamics – a look backwards and forwards. *J. Magn. Reson.* 2011, 213, 492–4.

9. Joachimiak, A. High-throughput crystallography for structural genomics. *Curr. Opin. Struct. Biol.* 2009, 19, 573–84.
10. Colwell, L.J. Statistical and machine learning approaches to predicting protein–ligand interactions. *Curr. Opin. Struct. Biol.* 2018, 49, 123–8.
11. Du, X.; Li, Y.; Xia, Y.L.; Ai, S.M.; Liang, J.; Sang, P.; Ji, X.L.; Liu, S.Q. Insights into protein–ligand interactions: mechanisms, models, and methods. *Int. J. Mol. Sci.* 2016, 17, 144.
12. Leach, A.R. *Molecular Modelling: Principles and Applications*. Pearson Education, London, 2001.
13. Sethi, A.; Joshi, K.; Sasikala, K.; Alvala, M. Molecular docking in modern drug discovery: principles and recent applications. *Drug Disc. Develop. New Adv.* 2019, 2, 1–21.
14. Shukla, R.; Tripathi, T. Molecular dynamics simulation of protein and protein–ligand complexes. In: Singh, D.B. (eds) *Computer-Aided Drug Design*. Springer, Singapore, 2020, 133–61.
15. Hernández-Rodríguez, M.; Rosales-Hernández, M.C.; Mendieta-Wejebe, J.E.; Martínez-Archundia, M.; Basurto, J.C. Current tools and methods in molecular dynamics (MD) simulations for drug design. *Curr. Med. Chem.* 2016, 23, 3909–24.
16. Chen, K.; Kurgan, L. Investigation of atomic level patterns in protein—small ligand interactions. *PLoS One*. 2009, 4, e4473.
17. Bronowska, A.K. Thermodynamics of ligand-protein interactions: implications for molecular design. In: Moreno-Pirajan, J.C. (eds) *Thermodynamics-Interaction Studies-Solids, Liquids and Gases*. IntechOpen, New York, 2011, 2,
18. Salahudeen, M.S.; Nishtala, P.S. An overview of pharmacodynamic modelling, ligand-binding approach and its application in clinical practice. *Saudi Pharm. J.* 2017, 25, 165–75.
19. Ferreira, L.G.; Dos Santos, R.N.; Oliva, G.; Andricopulo, A.D. Molecular docking and structure-based drug design strategies. *Molecules*. 2015, 20, 13384–421.
20. Crampon, K.; Giorkallos, A.; Deldossi, M.; Baud, S.; Steffenel, L.A. Machine-learning methods for ligand–protein molecular docking. *Drug. Discov. Today*. 2022, 27, 151–64.
21. Cole, J.C.; Nissink, J.W.; Taylor, R. Protein-ligand docking and virtual screening with GOLD. *Virtual Screen. Drug Discov.* 2005, 1, 379–415.
22. Spitzer, R.; Jain, A.N. Surflex-dock: docking benchmarks and real-world application. *J. Comput. Aided Mol. Des.* 2012, 26, 687–99.
23. Venkatachalam, C.M.; Jiang, X.; Oldfield, T.; Waldman, M. LigandFit: a novel method for the shape-directed rapid docking of ligands to protein active sites. *J. Mol. Graph. Model.* 2003, 21, 289–307.
24. Friesner, R.A.; Banks, J.L.; Murphy, R.B.; Halgren, T.A.; Klicic, J.J.; Mainz, D.T.; et al. Glide: a new approach for rapid, accurate docking and scoring. 1. Method and assessment of docking accuracy. *J. Med. Chem.* 2004, 47, 1739–49.
25. Korb, O.; Stützle, T.; Exner, T.E. PLANTS: Application of ant colony optimization to structure-based drug design. In: *International Workshop on Ant Colony Optimization and Swarm Intelligence*. Springer, Berlin Heidelberg, 2006, 247–258.
26. Morris, G.M.; Goodsell, D.S.; Huey, R.; Hart, W.E.; Halliday, S.; Belew, R.; Olson, A.J. *AutoDock. Automated Docking of Flexible Ligands to Receptor-User Guide*. Molecular Graphics Laboratory, La Jolla, CA, 2001.
27. Trott, O.; Olson, A.J. AutoDock Vina: improving the speed and accuracy of docking with a new scoring function, efficient optimization, and multithreading. *J. Comput. Chem.* 2010, 31, 455–61.
28. Liu, N.; Xu, Z. Using LeDock as a docking tool for computational drug design. *IOP Conf. Ser.: Earth Environ.* 2019, 218, 012143.
29. Li, L.; Chen, R.; Weng, Z. RDOCK: refinement of rigid-body protein docking predictions. *Proteins: Struct. Funct. Bioinform.* 2003, 53, 693–707.
30. Mcgann, M.R.; Almond, H.R.; Nicholls, A.; Grant, J.A.; Brown, F.K. Gaussian docking functions. *Biopolym. Orig. Res. Biomole.* 2003, 68, 76–90.
31. Gaudreault, F.; Najmanovich, R.J. FlexAID: revisiting docking on non-native-complex structures. *J. Chem. Inf. Model.* 2015, 55, 1323–36.
32. Paul, D.S.; Gautham, N.M. MOLS 2.0: software package for peptide modeling and protein–ligand docking. *J. Mol. Model.* 2016, 22, 1–9.
33. Maden, S.F.; Sezer, S.; Acuner, S.E. Fundamentals of molecular docking and comparative analysis of protein–small-molecule docking approaches. In: Istifli, E.S. (eds) *Molecular Docking-Recent Advances*. IntechOpen, New York, 2022.
34. Huang, S.Y.; Zou, X. Advances and challenges in protein-ligand docking. *Int. J. Mol. Sci.* 2010, 11, 3016–34.
35. Jiang, F.; Kim, S.H. "Soft docking": matching of molecular surface cubes. *J. Mol. Biol.* 1991, 219, 79–102.
36. Teodoro, M.L.; Phillips, G.N.; Kavraki, L.E. Molecular docking: a problem with thousands of degrees of freedom. In: *Proceedings 2001 ICRA. IEEE International Conference on Robotics and Automation (Cat. No. 01CH37164)*. IEEE, New York, 2001, 960–5.
37. Wang, J.; Kollman, P.A.; Kuntz, I.D. Flexible ligand docking: a multistep strategy approach. *Proteins: Struct. Funct. Bioinform.* 1999, 36, 1–9.
38. Halperin, I.; Ma, B.; Wolfson, H.; Nussinov, R. Principles of docking: an overview of search algorithms and a guide to scoring functions. *Proteins: Struct. Funct. Bioinform.* 2002, 47, 409–43.
39. Ferrari, A.M.; Wei, B.Q.; Costantino, L.; Shoichet, B.K. Soft docking and multiple receptor conformations in virtual screening. *J. Med. Chem.* 2004, 47, 5076–84.
40. Amin, S.A.; Gayen, S.; Das, S.; Jha, T. Other modeling approaches: pharmacophore mapping, molecular docking, and molecular dynamic simulation studies. In: Jha, T. (eds) *Modeling Inhibitors of Matrix Metalloproteinases*. CRC Press, Boca Raton, FL, 2023, 54–86.
41. Sousa, S.F.; Fernandes, P.A.; Ramos, M.J. Protein–ligand docking: current status and future challenges. *Proteins: Struct. Funct. Bioinform.* 2006, 65, 15–26.
42. Torres, P.H.; Sodero, A.C.; Jofily, P.; Silva-Jr, F.P. Key topics in molecular docking for drug design. *Int. J. Mol. Sci.* 2019, 20, 4574.
43. Guedes, I.A.; de Magalhães, C.S.; Dardenne, L.E. Receptor–ligand molecular docking. *Biophys. Rev.* 2014, 6, 75 87.
44. Dauzhenka, T.; Kundrotas, P.J.; Vakser, I.A. Computational feasibility of an exhaustive search of side-chain conformations in protein-protein docking. *J. Comput. Chem.* 2018, 39, 2012–21.
45. Ewing, T.J.; Makino, S.; Skillman, A.G.; Kuntz, I.D. DOCK 4.0: search strategies for automated molecular docking of flexible molecule databases. *J. Comput. Aided. Mol. Des.* 2001, 15, 411–28.

46. Rarey, M.; Kramer, B.; Lengauer, T.; Klebe, G. A fast flexible docking method using an incremental construction algorithm. *J. Mol. Biol.* 1996, 261, 470–89.
47. Joseph-McCarthy, D.; Thomas, IV, B.E.; Belmarsh, M.; Moustakas, D.; Alvarez, J.C. Pharmacophore-based molecular docking to account for ligand flexibility. *Proteins: Struct. Funct. Bioinform.* 2003, 51, 172–88.
48. Miller, M.D.; Kearsley, S.K.; Underwood, D.J.; Sheridan, R.P. FLOG: a system to select 'quasi-flexible'ligands complementary to a receptor of known three-dimensional structure. *J. Comput.-Aided Mol. Des.* 1994, 8, 153–74.
49. Sauton, N.; Lagorce, D.; Villoutreix, B.; Miteva, M. MS-DOCK: accurate multiple conformation generator and rigid docking protocol for multi-step virtual ligand screening. *BMC Bioinform.* 2008, 9, 184.
50. Metropolis, N.; Ulam, S. The Monte Carlo method. *J. Am. Stat. Assoc.* 1949, 44, 335–41.
51. Abagyan, R.; Totrov, M.; Kuznetzov, D. ICM – a new method for protein modeling and design: applications to docking and structure prediction from the distorted native conformation. *J. Comput. Chem.* 1994, 15, 488–506.
52. Liu, M.; Wang, S. MCDOCK: a Monte Carlo simulation approach to the molecular docking problem. *J. Comput. Aided Mol. Des.* 1999, 13, 435–51.
53. Mcmartin, C.; Bohacek, R.S. QXP: powerful, rapid computer algorithms for structure-based drug design. *J. Comput.-Aided Mol. Des.* 1997, 11, 333–44.
54. Pei, J.; Wang, Q.; Liu, Z.; Li, Q.; Yang, K.L.; Lai, L. PSI-DOCK: towards highly efficient and accurate flexible ligand docking. *Proteins.* 2006, 62, 934–46.
55. Thomsen, R.; Christensen, M.H. MolDock: a new technique for high-accuracy molecular docking. *J. Med. Chem.* 2006, 49, 3315–21.
56. Taylor, J.S.; Burnett, R.M. DARWIN: a program for docking flexible molecules. *Proteins: Struct. Funct. Bioinform.* 2000, 41, 173–91.
57. Bonyadi, M.R.; Michalewicz, Z. Particle swarm optimization for single objective continuous space problems: a review. *Evol. Comput.* 2017, 25, 1–54.
58. Chen, H.M.; Liu, B.F.; Huang, H.L.; Hwang, S.F.; Ho, S.Y. SODOCK: swarm optimization for highly flexible protein–ligand docking. *J. Comput. Chem.* 2007, 28, 612–23.
59. Froger, A.; Gendreau, M.; Mendoza, J.E.; Pinson, É.; Rousseau, L.M. Maintenance scheduling in the electricity industry: a literature review. *Eur. J. Oper. Res.* 2016, 251, 695–706.
60. Sotriffer, C. Scoring functions for protein–ligand interactions. *Protein-Ligand Interact.* 2012, 237–63.
61. Kitchen, D.B.; Decornez, H.; Furr, J.R.; Bajorath, J. Docking and scoring in virtual screening for drug discovery: methods and applications. *Nat. Rev. Drug Discov.* 2004, 3, 935–49.
62. Wang, R.; Liu, L.; Lai, L.; Tang, Y. SCORE: a new empirical method for estimating the binding affinity of a protein-ligand complex. *J. Mol. Model.* 1998, 4, 379–94.
63. Wang, R.; Lai, L.; Wang, S. Further development and validation of empirical scoring functions for structure-based binding affinity prediction. *J. Comput. Aided Mol. Des.* 2002, 16, 11–26.
64. Krammer, A.; Kirchhoff, P.D.; Jiang, X.; Venkatachalam, C.M.; Waldman, M. LigScore: a novel scoring function for predicting binding affinities. *J. Mol. Graph. Model.* 2005, 23, 395–407.
65. Muegge, I. A knowledge-based scoring function for protein-ligand interactions: probing the reference state. *Perspect. Drug Discov. Design.* 2000, 20, 99–114.
66. Krüger, D.M.; Ignacio Garzón, J.; Chacón, P.; Gohlke, H. DrugScore-PPI knowledge-based potentials used as scoring and objective function in protein-protein docking. *PLoS One.* 2014, 9, e89466.
67. DeWhitte, R.S.; Shakhnovich, E.I. SMoG: de novo design method based on simple, fast, and accurate free energy estimates. 1. Methodology and supporting evidence. *J. Am. Chem. Soc.* 1996, 118, 11733–44.
68. Charifson, P.S.; Corkery, J.J.; Murcko, M.A.; Walters, W.P. Consensus scoring: a method for obtaining improved hit rates from docking databases of three-dimensional structures into proteins. *J. Med. Chem.* 1999, 42, 5100–9.
69. Chaudhary, K.K.; Mishra, N. A review on molecular docking: novel tool for drug discovery. *JSM Chem. Databases.* 2016, 3, 1029.
70. Singh, S.; Singh, V.K. Molecular dynamics simulation: methods and application. In: Singh, D., Tripathi, T. (eds) *Frontiers in Protein Structure, Function, and Dynamics.* Springer, Singapore, 2020, 213–38.
71. Gschwend, D.A.; Good, A.C.; Kuntz, I.D. Molecular docking towards drug discovery. *J. Mole. Recogn. Interdisc. J.* 1996, 9, 175–86.
72. Kaur, T.; Madgulkar, A.; Bhalekar, M.; Asgaonkar, K. Molecular docking in formulation and development. *Curr. Drug Discov. Technol.* 2019, 16, 30–9.
73. Pinzi, L.; Rastelli, G. Molecular docking: shifting paradigms in drug discovery. *Int. J. Mol. Sci.* 2019, 20, 4331.
74. Badar, M.S.; Shamsi, S.; Ahmed, J.; Alam, M.A. Molecular dynamics simulations: concept, methods, and applications. In: Rezaei, N. (eds) *Transdisciplinarity.* Springer, Cham, 2022, 131–51.
75. Alonso, H.; Bliznyuk, A.A.; Gready, J.E. Combining docking and molecular dynamic simulations in drug design. *Med. Res. Rev.* 2006, 26, 531–68.
76. Hollingsworth, S.A.; Dror, R.O. Molecular dynamics simulation for all. *Neuron.* 2018, 99, 1129–43.
77. Liu, X.; Shi, D.; Zhou, S.; Liu, H.; Liu, H.; Yao, X. Molecular dynamics simulations and novel drug discovery. *Expert Opin. Drug Discov.* 2018, 13, 23–37.
78. Genheden, S.; Ryde, U. The MM/PBSA and MM/GBSA methods to estimate ligand-binding affinities. *Expert Opin. Drug Discov.* 2015, 10, 449–61.
79. Amin, S.A.; Bhattacharya, P.; Basak, S.; Gayen, S.; Nandy, A.; Saha, A. Pharmacoinformatics study of Piperolactam A from Piper betle root as new lead for non steroidal anti fertility drug development. *Comput. Biol. Chem.* 2017, 67, 213–24.
80. Sardar, S.; Jyotisha, S.; Amin, S.A.; Khatun, S.; Qureshi, I.A.; Patil, U.K; Jha, T.; Gayen, S. Identification of structural fingerprints among natural inhibitors of HDAC1 to accelerate nature-inspired drug discovery in cancer epigenetics. *J. Biomol. Struct. Dyn.* 2023, 19, 1–5.
81. Amin, S.A.; Kumar, J.; Khatun, S.; Das, S.; Qureshi, I.A.; Jha, T.; Gayen, S. Binary quantitative activity-activity relationship (QAAR) studies to explore selective HDAC8 inhibitors: in light of mathematical models, DFT-based calculation and molecular dynamic simulation studies. *J. Mol. Struct.* 2022, 1260, 132833.
82. Van Gunsteren, W.F.; Berendsen, H.J. Computer simulation of molecular dynamics: methodology, applications, and perspectives in chemistry. *Angew. Chem. Int. Ed.* 1990, 29, 992–1023.
83. Weiner, P.K.; Kollman, P.A. AMBER: assisted model building with energy refinement. A general program for modeling molecules and their interactions. *J. Comput. Chem.* 1981, 2, 287–303.

84. Bowers, K.J.; Chow, E.; Xu, H.; Dror, R.O.; Eastwood, M.P.; Gregersen, B.A.; Klepeis, J.L.; Kolossvary, I.; Moraes, A.; Sacerdoti, F.D.; Salmon, J.K. Scalable algorithms for molecular dynamics simulations on commodity clusters. In *Proceedings of the 2006 ACM/IEEE Conference on Supercomputing*. IEEE, Tampa, FL, 2006, 11, 84.
85. Nelson, M.T.; Humphrey, W.; Gursoy, A.; Dalke, A.; Kalé, L.V.; Skeel, R.D.; Schulten, K. NAMD: a parallel, object-oriented molecular dynamics program. *Int. J. High Perform. Comput. Appl.* 1996, 10, 251–68.
86. Brooks, B.R.; Bruccoleri, R.E.; Olafson, B.D.; States, D.J.; Swaminathan, S.A.; Karplus, M. CHARMM: a program for macromolecular energy, minimization, and dynamics calculations. *J. Comput. Chem.* 1983, 4, 187–217.
87. Refson, K. Moldy: a portable molecular dynamics simulation program for serial and parallel computers. *Comput. Phys. Commun.* 2000, 126, 310–29.
88. Plimpton, S. Fast parallel algorithms for short-range molecular dynamics. *J. Comput. Phys.* 1995, 117, 1–9.
89. Ponder, J.W. *TINKER: Software Tools for Molecular Design*. Washington University School of Medicine, Saint Louis, MO, 2004, 116.
90. Harvey, M.J.; Giupponi, G.; Fabritiis, G.D. ACEMD: accelerating biomolecular dynamics in the microsecond time scale. *J. Chem. Theory Comput.* 2009, 5, 1632–9.
91. Spitznagel, B.; Pritchett, P.R.; Messina, T.C.; Goadrich, M.; Rodriguez, J. An undergraduate laboratory activity on molecular dynamics simulations. *Biochem. Mol. Biol. Educ.* 2016, 44, 130–9.
92. Lindahl, E.R. Molecular dynamics simulations. *Methods Mol. Biol.* 2008, 443, 3–23.
93. Okimoto, N.; Futatsugi, N.; Fuji, H.; Suenaga, A.; Morimoto, G.; Yanai, R.; Ohno, Y.; Narumi, T.; Taiji, M. High-performance drug discovery: computational screening by combining docking and molecular dynamics simulations. *PLoS Comput. Biol.* 2009, 5, e1000528.

12 Density Functional Theory and Its Applications

Satyam Yadav[1], Ayushi Gupta[2], Ajay Kumar[2], Neha Patel[3], and Alok Kumar Pandey[4]

[1] Department of Pharmaceutical Chemistry, Adina College of Pharmacy, Sagar, India
[2]School of Pharmaceutical Sciences, Department of Pharmaceutical Chemistry, C.S.J.M. University, Kanpur, India
[3]Department of Pharmacy, Friedrich-Alexander-University Erlangen-Nuremberg (FAU), Erlangen, Germany
[4]Department of BioSciences and Bioengineering, Indian Institute of Technology Guwahati, Kamrup, North Guwahati, India

INTRODUCTION

A lot of people still think that the idea of rational pharmaceutical design is attractive. It might help us learn more about the chemicals that make up life. It has a lot of potential to aid with disease treatment issues and come up with innovative ideas in healthcare. The pharmaceutical industry is always striving for faster and more accurate ways to uncover new drugs because the need for better drug discovery methods is always increasing. Because of this, people have never had bigger hopes for computer-aided molecular design. Hence, scientists have had to utilize a lot of different computer tools to look at biological systems and bioactive chemicals.

Biological macromolecules, including enzymes, receptors, immunoglobulins, and nucleic acids, are still exceedingly complicated and massive, which makes it hard to study them, even with these advances. The huge devices developed nonetheless make it impossible to execute appropriate quantum chemistry calculations since they need so much processing power. So, computational chemists who want to study biological systems using quantum mechanical (QM) methods have two big problems: (i) choosing the right part of the biological system to model at the QM level and (ii) finding a computational method that works quickly and gives reliable results [1].

One of the fundamental goals of rational drug design is to determine how much energy it takes for small molecules to bind to proteins. This is critical for developing medications that work [2]. QM approaches are being used more and more in computer-aided drug design (CADD). This is partly because computers have gotten better and first-principles methods are more accurate [3,4]. Some people believe that employing QM methods at every stage of CADD is becoming a reality [4]. Because of this new interest, there are now dedicated QM-based approaches for molecular docking, scoring, lead optimization, and finding out how reactions happen [5].

The need for technology that can provide a lot of molecular information in a short time is much higher now as people are looking for better ways to undertake medicinal chemistry and drug development. Hartree–Fock (HF) methods are useful, but they do not fully explain a lot of chemical properties since they do not consider how electrons interact with one another [6–9]. Post-HF methods such as Møller–Plesset perturbation theory (MPn) [10] and configuration interaction (CI) [11] get around this difficulty by taking electron correlation into account, but they are far more expensive to compute.

In this case, density functional theory (DFT) is a valuable and effective option. DFT is a very important tool in quantum chemistry because it achieves the right degree of accuracy without needing too much computational power [12].

In both quantum chemistry and condensed matter physics, DFT is a powerful approach that uses computers to look at the electronic structures and characteristics of atoms, molecules, and materials [13]. The best part is that it shifts the focus from wavefunctions to electron density. This makes the math easier while still being correct, especially for systems with a lot of electrons, where it is hard to solve the many-body equation of Schrödinger [14].

To use quantum computing approaches such as DFT to predict molecular structure, one should have a good grasp of quantum mechanics, be familiar with quantum chemistry tools, and know exactly which algorithms to use. These technologies keep making it easier for us to create accurate, high-resolution simulations of how molecules act as the field grows. This gives researchers a lot of fresh ways to come up with new ideas [15].

It seems that quantum computers would also operate well with traditional DFT and ab initio approaches. These new technologies are going to alter molecular modeling by making it possible to simulate quantum systems more quickly and precisely. When it comes to huge molecular assemblies

DOI: 10.1201/9781003464679-12

or sophisticated chemical interactions, quantum computers could solve hard electrical structure issues faster than conventional computers. This should make it much easier for us to precisely anticipate molecular properties, including bond dissociation energies, reaction pathways, and spectroscopic properties [16].

This chapter discusses the underlying theories behind DFT and how they can be applied to design new drugs. It is important to discuss how DFT interacts with AI and bioinformatics tools to help the research and development of new drugs.

THEORETICAL FOUNDATIONS OF DFT

Hohenberg–Kohn Theorems

People have long hypothesized that electron density may be a basic quantity in calculations of electronic structure, but it was not until Hohenberg and Kohn's groundbreaking work in 1964 that there was solid proof. They came up with two basic theorems that showed why electron density should be the main variable in QM calculations [17]. These theorems showed in a formal way that the electron density of a many-electron system uniquely determines its ground-state features. Because of this, the theorems made a clear association between electron density, external potential, Hamiltonian, and wavefunction. This made electron density the most important quantity in DFT.

Kohn–Sham Approach

Even though Hohenberg and Kohn laid the groundwork for this work, early attempts to calculate electronic structure based only on electron density and not on wavefunctions were not very successful. The main problem was that it was hard to adequately represent electronic kinetic energy as a function of electron density alone.

Kohn and Sham (1965) came up with a clever way to solve this problem by creating a fake system of non-interacting electrons that has the same electron density as the real system that interacts with them [17]. The Kohn–Sham formulation is a method that breaks down the difficult many-body issue into a set of equations for single electrons. Their "divide-and-conquer" technique lets them use the known kinetic energy of non-interacting electrons while using an approximate functional to deal with the complicated interactions between electrons, such as exchange and correlation. This method connects pure density-based formulas with wavefunction-based quantum mechanics.

DFT IN DRUG DESIGN AND MEDICINAL CHEMISTRY

DFT has become one of the most useful and commonly used QM approaches in drug design and medicinal chemistry. Calculating electron density (ρ) is a key part of DFT because it is the main input for figuring out molecular characteristics [18].

The ground-state energy is one of the most important factors that can be calculated from electron density. In DFT, it is written as a function of ρ. The following generic equation [19] shows this:

$$E_0[\rho] = V[\rho] + T[\rho] + \mathrm{EXC}[\rho]$$

where the functional $E_0[\rho]$ stands for the total ground-state energy, which is made up of three different parts:

$V[\rho]$: It is the electrostatic forces between electrons and nuclei and how they push each other away.
$T[\rho]$: It is the system's kinetic energy, which includes both the kinetic and potential energy of the non-interacting electron reference system.
$\mathrm{EXC}[\rho]$: It is the part of the exchange and correlation energy that $V[\rho]$ and $T[\rho]$ do not directly show.

We can figure out $V[\rho]$ and $T[\rho]$ with a lot of confidence, but the exchange–correlation term $\mathrm{EXC}[\rho]$ is still only an estimate. This built-in uncertainty is the main reason why DFT is called an approximation quantum approach. Several approximation methods have been developed to deal with this, such as local density approximation (LDA), generalized gradient approximation (GGA), and hybrid functionals. Each functional uses one of these approximations, and there is no one-size-fits-all way to improve the accuracy of DFT right now. Because of this, people often check the predictive reliability of DFT on a case-by-case basis [20].

DFT FUNCTIONALS AND B3LYP

Among the most well-known and successful DFT approximations is the B3LYP functional. In practice, assuming that exchange–correlation energy depends solely on local electron density often proves inadequate, especially in systems with varying electron density distributions. GGA functionals attempt to improve this by including gradients of electron density, leading to better representations of exchange and correlation energies [21].

The exchange–correlation functional is generally expressed as follows:

$$\mathrm{EXC}[\rho] = \mathrm{EX}[\rho] + \mathrm{EC}[\rho]$$

B3LYP is a hybrid functional, which blends exact HF exchange with DFT approximations. Specifically, it combines:

- Becke's three-parameter exchange functional (B3) [22]
- Lee–Yang–Parr correlation functional (LYP) [23].

These components are calibrated using empirical parameters that are fit to experimental data. B3LYP was first introduced by Stevens et al. [24] and has since become the most popular and extensively used hybrid functional in quantum chemistry [25].

Despite the approximate nature of its components, B3LYP delivers an optimal balance between computational cost and accuracy, making it a preferred choice for predicting molecular geometries, vibrational frequencies, reaction mechanisms, and binding energies in drug design applications.

ACCURACY OF DFT IN THE STUDY OF DRUG PROPERTIES

The accuracy of DFT in predicting a wide range of molecular properties has been the subject of extensive validation across numerous benchmark studies. In the context of drug discovery and design, reliable prediction of molecular descriptors such as geometry, dipole moment, ionization potential, binding affinity, and reaction energetics is crucial.

Among various DFT approaches, hybrid functionals—which combine HF exact exchange with DFT correlation functionals—have consistently outperformed LDA and GGA methods in terms of predictive accuracy.

The B3LYP functional, in particular, has emerged as the most widely adopted and validated hybrid functional in medicinal chemistry and pharmaceutical research, owing to its balance between computational cost and accuracy [26]. It has demonstrated exceptional reliability in reproducing experimental observables for a broad class of organic and bioactive molecules. A summary of the expected accuracy of B3LYP is presented in Table 12.1.

DFT, particularly with hybrid functionals like B3LYP, has demonstrated considerable success in accurately predicting a wide range of molecular properties essential for drug discovery and design. These include geometrical optimization, electronic properties, binding affinities, and reaction energetics. While DFT is not without limitations, it remains one of the most efficient and widely used quantum chemical approaches in modern computational medicinal chemistry.

The accuracy of B3LYP in predicting specific drug-relevant properties has been validated across numerous studies. Table 12.1 summarizes the expected performance of B3LYP for various molecular properties, along with a recommendation on whether DFT is suitable for their reliable calculation.

QM approaches such as ab initio methods and DFT represent some of the most direct and theoretically rigorous strategies for exploring and designing new drug candidates [27]. While DFT excels in its capacity to estimate electronic structure and molecular properties, it also plays a crucial indirect role in medicinal chemistry and drug development. Several key areas where DFT-based calculations contribute include:

a. Structure, conformation, and electronic property evaluation, which often correlate with biological activity. This includes molecular orbital analysis [28], electron density distributions, dipole moments [29], and electrostatic potential surfaces [30–33].
b. Descriptor generation for quantitative structure–activity relationship (QSAR) modeling, where DFT-calculated properties (such as highest occupied molecular orbital/lowest unoccupied molecular orbital (HOMO–LUMO) gaps, polarizability, chemical hardness, and electrostatic fields) are utilized in QSAR models to predict pharmacological behavior and optimize lead compounds [29–36].

The incorporation of DFT into drug design workflows thus not only enables mechanistic insights but also enhances the predictive power of AI-driven and data-intensive bioinformatics pipelines.

CHEMICAL BONDING PROPERTIES OF DRUG MOLECULES

The interaction between a putative drug molecule and its target is crucial in drug design. Potential interactions between a drug and target include covalent bonds, ionic interactions, ion–dipole interactions, dipole–dipole interactions, hydrogen bonding, charge transfer, and

TABLE 12.1
Expected Accuracy of DFT for Describing Key Properties of Drug Molecules Using the B3LYP Functional

S. No.	Properties	Expected Accuracy	Recommend DFT for Calculation
1.	Geometry	Bond lengths: 0.005Å Bond angles: 0.2°	Yes
2.	Relative conformational energies	1 kcal/mol	Yes
3.	Ionization energies and electron affinities	0.2 eV	Yes
4.	Covalent, ionic, and hydrogen bonding binding energies	1–2 kcal/mol	Yes
5.	Metal–ligand bond strength	4–5 kcal/mol	Yes
6.	Transition barriers	1 kcal/mol	Yes
7.	Transition metal reaction pathways	5 kcal/mol	Yes
8.	Atomization energies	2.2 kcal/mol	No
9.	Ion–dipole, charge transfer, and hydrophobic interactions	1–2 kcal/mol	No

TABLE 12.2
Chemical Bonds and Average Bond Energies (kcal/mol)

S. No.	Bond Types	Interaction Energy
1	Covalent	346
2	Ionic	450
3	Ion–dipole	171
4	Dipole–dipole	5
5	Hydrogen	37
6	Charge transfer	17
7	Hydrophobic	4.2

hydrophobic interactions. The typical strength of these interactions is presented in Table 12.2 [37]. The ability of DFT to predict the strengths of these interactions is highly significant when determining the success of DFT in CADD. DFT has been successful at predicting covalent, ionic, and hydrogen bonds but not weaker bonds such as ion–dipole, dipole–dipole, charge transfer, and hydrophobic interactions [37] (Table 12.3).

APPLICATIONS OF DFT IN DRUG DESIGN

DFT plays a pivotal role in drug design by enabling the evaluation of various molecular properties with high accuracy and computational efficiency. Its ability to provide insights into the electronic structure, interaction energies, and molecular conformations makes DFT a powerful tool for both direct and indirect applications in medicinal chemistry. Key applications of DFT in drug discovery are summarized below:

MODELING OF DRUG–RECEPTOR INTERACTIONS

DFT is instrumental in studying the interaction between drugs and their biological receptors. These interactions are often ligand-gated processes, where a drug (ligand) binds to a specific receptor on the cell membrane or a target site, initiating a biological response. DFT helps model the electronic and structural aspects of such interactions. For instance, Kee et al. (2009) investigated aromatic interactions involved in the binding of ligands to HMG-CoA reductase. Statin drugs, known for their cholesterol-lowering effects, act as competitive inhibitors of this enzyme. Using a combination

TABLE 12.3
Density Functional Theory Calculating Software

S.no	Software	Description	Applications	References
1	Gaussian	Based on the fundamental law of quantum mechanics and hybrid method; quicker than numerical models; provides accurate results that match the experimental data; computations can be carried out in the gaseous or liquid state.	Predicts the energies, molecular structure, and vibrational frequencies of complex molecular systems.	[38,39]
2	ORCA	Provides easy-to-learn input structure and high accessibility of quantum chemical approach; uses a variety of functionals: generalized gradient approximations (GGA), local density approximations (LDA), and hybrid functionals are used for the exchange–correlation functional.	Features virtually all modern electronic structure methods (DFT, many-body perturbation and coupled cluster theories and multireference and semi-empirical method); predicts catalytic mechanism, modeling molecular properties and drug–receptor interaction.	[40,41]
3.	Q-chem	Free and open source; supports functionals like meta-GGA, LDA hybrid, and range-separated hybrid.	Predicts molecular structure, reactivities, vibrational, electronic, and NMR spectra; to obtain the properties for the excited state, time-dependent DFT is performed.	[42]
4.	ADF (Amterdam density functional)	Uses Kohn–Sham approach; calculations are analyzed with its graphical user interface (GUI).	Predicts total energy and electron density; heavy metal and transition metals are accurately modeled.	[43]
5	CP2K	Versatile open source; aims at providing a broad range of models and stimulation methodologies, suitable for large and condensed phase systems; able to exploit the most advanced computer hardware.	Predicts material properties and rationally designs new compounds and experimental compounds; protein–ligand interaction.	[44,45]
6	PSI4	Provides efficient code for establishing an electronic structure model; flexible user input built on a python scripted language.	Allows rapid development of an electronic structure model with production-level efficiency.	[46]
7	Quantum ESPRESSO	Open-source distribution of computer codes; plane wave pseudopotential approach.	Predicts solid-state drug delivery and provides molecular interaction.	[47,48]

(Continued)

TABLE 12.3 (*Continued*)
Density Functional Theory Calculating Software

S.no	Software	Description	Applications	References
8	BIGDFT	Fast, precise, and flexible; uses linear DFT scaling algorithms and hybrid quantum mechanism and molecular mechanism.	Predicts drug–target binding and their energy; assesses lipophilicity and solubility; interprets spectroscopical parameters.	[49]
9	WEIN2K	Based on the full potential linearized augmented plane wave (FP-LAPW) method; explains ground-state properties of many-electron systems on the basis of electron density.	Solves crystal properties on an atomic scale; helps understand the stability of the drug molecule; surface and interface studies.	[50]
10	Octopus	Open-source software package; specialized in time-dependent DFT; uses real space grid to solve Kohn–Sham equation.	Ideal for stimulating optical absorption spectra, photoluminescence, and electronic excitation state properties; explores charge transfer in the molecule.	[51,52]
11	Molpro	Contains a number of utilities to support the analysis and manipulation of results, molecular orbitals, densities, etc. These include EXTRAPOLATE, which automatically extrapolates the last energy calculation to the basis set limit according to one of the standard formulae. MERGE and MATROP utilities perform matrix manipulations of objects holding molecular orbitals, densities, and properties.		

of DFT methods (local, hybrid, and gradient-corrected) alongside MP2 calculations and *in silico* mutation at the tyrosine 479 residue, the study revealed how ligand-binding energetics are influenced, guiding the design of novel statin drugs [20,44,53].

Modeling the Coordination Chemistry of Ligands Bound to Metal Ions

Understanding the structural, electronic, and thermodynamic properties of metal–ligand complexes is essential in designing metallodrugs or studying metalloenzymes. Xu et al. (2008) explored the vibrational frequencies of CO bound to iron in various protein environments using B3LYP functional and the 6–31G basis set (except for Fe). The investigation focused on Fe–CO bonding and how factors such as proximal histidine hydrogen bonding, heme distortion, and steric effects influence bonding. DFT calculations revealed that vibrational frequencies of Fe–CO complexes are highly sensitive to protein surroundings, affecting back-bonding interactions and electronic distribution [44,54].

Examination of Interactions in Polypeptide Structures

DFT has been employed to study non-covalent interactions and secondary structures in proteins. Improta et al. (2004) focused on dispersion interactions within an α-helix using an alanine dipeptide model. Multiple DFT functionals, including PBEPBE, PBELYP, BLYP, B3LYP, HCTH407, MPWPW91, MPWIPW91, and PBE0, were tested. The study found that most functionals overestimated residue energies locally, whereas PBE0 showed results consistent with MP2 and CCSD(T), making it suitable for modeling protein conformational energies [44,55].

DFT-Based Conformational Analysis with QSAR

DFT contributes to structure–activity relationship studies by providing reliable electronic descriptors and optimized conformations. Zhang et al. (2009) investigated 25 cyclic imide derivatives as protoporphyrinogen oxidase (PPO) inhibitors. PPO is a key enzyme in the biosynthesis of heme and chlorophyll. The geometries of the compounds were optimized using B3LYP/6–31G(d,p). Multiple conformations were analyzed depending on side-chain flexibility. DFT-derived conformations were validated through potential energy surface scans, molecular docking, and dynamics simulations, demonstrating how DFT enhances QSAR modeling by integrating QM accuracy with bioactivity prediction [35,44].

Modeling of Energy Interaction between Nucleotide Base Pairs

DFT has been used to model interaction energies between nucleotide base pairs, which is critical for understanding DNA stability and drug–DNA binding. Hesselmann et al. (2011) employed MP2, CCSD(T), and DFT combined with symmetry-adapted perturbation theory (SAPT) to calculate interaction energies between purine and pyrimidine bases. The study demonstrated a high level of agreement among the methods, validating DFT (when enhanced with SAPT) as a reliable and cost-effective approach for modeling nucleic acid interactions [44,56].

Modeling Dispersion Interactions in Protein Complexes

Modeling dispersion forces is essential for understanding protein–protein and protein–ligand interactions in drug design. Yun et al. (2024) studied the binding between tumor suppressor protein p53 and MDM2, a potential cancer drug target. Using MP2 and DFT, the interaction energy at five residues of the p53 active site was analyzed. MP2 results confirmed the significance of van der Waals (vdW) interactions. However, B3LYP failed to capture these dispersion effects, producing repulsive energies. This limitation underlines the need to use dispersion-corrected DFT functionals or a broader range of methods for accurate modeling in such systems [44,57].

Role of DFT in COVID-19 Drug Research

During the COVID-19 pandemic, DFT was extensively employed to analyze isolated drug molecules. These studies provided crucial insights into electronic properties that influence drug reactivity and binding preferences, even in the absence of full protein–ligand complex modeling. Geometry optimization using DFT yields the lowest energy conformation of molecules, critical for understanding bioactivity. In addition, frontier molecular orbital (FMO) analysis—focusing on HOMO and LUMO—is widely used. The HOMO–LUMO energy gap serves as an indicator of molecular reactivity and stability, where a larger gap corresponds to lower reactivity, higher stability, and stronger interaction potential—important attributes in the design of antiviral agents [58].

REFERENCES

1. Hoffmann M, Rychlewski J. Density functional theory (DFT) and drug design. *Reviews of Modern Quantum Chemistry: A Celebration of the Contributions of Robert G Parr.* 2022;2:1767–803.
2. Jorgensen WL. The many roles of computation in drug discovery. *Science.* 2004;303:1813–8.
3. Cavalli A, Carloni P, Recanatini M. Target-related applications of first principles quantum chemical methods in drug design. *Chemical Reviews.* 2006;106:3497–519.
4. Raha K, Peters MB, Wang B, Yu N, WollaCott AM, Westerhoff LM, Merz KM. The role of quantum mechanics in structure-based drug design. *Drug Discovery Today.* 2007;12:725–31.
5. Pasquini S, Mugnaini C, Tintori C, Botta M, Trejos A, Arvela RK, Larhed M, Witvrouw M, Michiels M, Christ F, Debyser Z, Corelli F. Investigations on the 4quinolone-3 carboxylic acid motif. 1. Synthesis-activity relationship of a class of human immunodeficiency virus type 1 integrase inhibitors. *Journal of Medicinal Chemistry.* 2008;51:5125–9.
6. Hartree DR. The wave mechanics of an atom with a non-coulomb central field. Part I. Theory and methods. *Mathematical Proceedings of the Cambridge Philosophical Society.* 1928;24(1): 89–110.
7. Hartree DR. The wave mechanics of an atom with a non-coulomb central field. Part II. Some results and discussion. *Mathematical Proceedings of the Cambridge Philosophical Society.* 1928;24(1):111–32.
8. Douglas R. The wave mechanics of an atom with a non-Coulomb central field. Part III. Term values and intensities in series in optical spectra. *Proceedings of the Cambridge Philosophical Society.* 1928;24:426–37.
9. Fock VZ. Näherungsmethode zur lösung des quantenmecha nischen mehrkörperproblems. *Physik.* 1930;61(1–2):126–48.
10. Møller C, Plesset MS. Note on an approximation treatment for ma ny-electron systems. *Physical Review.* 1934;46(7):618–22.
11. Slater JC. (1929) The theory of complex spectra. *Physical Review.* 34(10): 1293–1322.
12. Condon EU. The theory of complex spectra. *Physical Review.* 1930;36(7):1121–33.
13. Orio M, Pantazis DA, Neese F. Density functional theory. *Photosynthesis Research.* 2009;102:443–53.
14. Engel E. *Density Functional Theory.* 2011; Springer, Berlin.
15. Ahmed L, Rasulev B, Turabekova M, Leszczynska D, Leszczyn Ski J. Receptor-and ligand-based study of fullerene analogues: Comprehensive computational approach including quantum-chemical, qsar and molecular docking simulations. *Organic & Biomolecular Chemistry.* 2013;11(35):5798–808.
16. Kumar G, Yadav S, Mukherjee A, Hassija V, Guizani M. Recent advances in quantum computing for drug discovery and development. *IEEE Access.* 2024 May 11;12:64491–509.
17. Kohn W, Sham L. Density functional theory. In *Conference Proceedings-Italian Physical Society* (Vol. 49, pp. 561–572). 1996; Editrice Compositori, Italy.
18. Hohenberg W, Kohn P. Inhomogeneous electron gas. *Physical Review.* 1964;136:B864–71.
19. Kohn W, Sham LJ. Self-consistent equations including exchange and correlation effects. *Physical Review.* 1965;140(4A):A1133.
20. Tariq K, Shaheen I, Shaheena R, Khalil A. Role of DFT in drug design: A mini review. *Drug Diethylstilbestrol.* 2022;11:216.
21. Becke, AD. Density-functional thermochemistry. III. The role of exact exchange. *The Journal of Physical Chemistry.* 1993;98:1372–7.
22. Becke AD. Density-functional thermochemistry. III. The role of exact exchange. *The Journal of Physical Chemistry.* 1993;98:5648–52.
23. Lee C, Yang W, Parr RG. Development of the Colle-Salvetti correlation-energy formula into a functional of the electron density. *Physical Review B.* 1988;37:785–90.
24. Stevens PJ, Devlin JF, Chabalowski CF, Frisch MJ. Ab initio calculation of vibrational absorption and circular dichroism spectra using density functional force fields. *The Journal of Physical Chemistry.* 1994;98:11623.
25. Raghavachari K. Perspective on "Density functional thermochemistry. III. The role of exact exchange. *Theoretical Chemistry Accounts.* 2000;103:361–3.
26. Utkov H, Livengood M, Cafiero M. Using density functional theory methods for mod eling induction and dispersion interactions in ligand–protein complexes. *Annual Reports in Computational Chemistry.* 2010;6:96–112. doi:10.1016/S1574-1400(10)06007-X
27. Villar JAFP, Lima FTD, Veber CL, Oliveira ARM, Calgarotto AK, Marangoni S, da Silva SL. Synthesis and evaluation of nitrostyrene derivative compounds, new snake venom phospholipase A2 inhibitors. *Toxicon.* 2008;51(8):1467–78. doi:10.1016/j.toxicon.2008.03.023
28. Parameswari AR, Kumaradhas P. Exploring the conformation, charge density distri bution and the electrostatic properties of galanthamine molecule in the active site of AChE using DFT and AIM theory. *International Journal of Quantum Chemistry.* 2013;113(8):1200–8. doi:10.1002/qua.24251

29. Silva JRA, Lameira J, Alves CN. Insights for design of Trypanosoma cruzi GAPDH inhibitors: A QM/MM MD study of 1,3-bisphospo-d-glyceric acid analogs. *International Journal of Quantum Chemistry*. 2012;112(20):3398–402. doi:10.1002/qua.24253
30. Negri M, Recanatini M, Hartmann RW. Computational investigation of the binding mode of bis(hydroxylphenyl) arenes in 17β-HSD1: Molecular dynamics simulations, MM PBSA free energy calculations, and molecular electrostatic potential maps. *Journal of Computer-Aided Molecular Design*. 2011;25(9):795–811. doi:10.1007/s10822-011-9464-7
31. Arooj M, Thangapandian S, John S, Hwang S, Park JK, Lee KW. Computational studies of novel chymase inhibitors against cardiovascular and allergic diseases: Mechanism and inhibition. *Chemical Biology & Drug Design*. 2012;80(6):862–75. doi:10.1111/cbdd.12006
32. Muzet N, Guillot B, Jelsch C, Howardt E, Lecomte C. Electrostatic complementarity in an aldose reductase complex from ultra-high-resolution crystallography and first-princi ples calculations. *Proceedings of the National Academy of Sciences of the United States of America*. 2003;100(15):8747. doi:10.1073/pnas.1432955100
33. Van Damme S, Bultinck P. Conceptual DFT properties-based 3D QSAR: Analysis of inhibitors of the nicotine metabolizing CYP2A6 enzyme. *Journal of Computational Chemistry*. 2009;30(12):1749–57. doi:10.1002/jcc.21177
34. Wan J, Zhang LI, Yang G. Quantitative structure-activity relationships for phenyl triazolinones of protoporphyrinogen oxidase inhibitors: A density functional theory study. *Journal of Computational Chemistry*. 2004;25(15):1827–32. doi:10.1002/jcc.20122
35. Zhang L, Hao G-F, Tan Y, Xi Z, Huang M-Z, Yang G-F. Bioactive conformation analysis of cyclic imides as protoporphyrinogen oxidase inhibitor by combining DFT calcula tions, QSAR and molecular dynamic simulations. *Bioorganic & Medicinal Chemistry*. 2009;17(14):4935–42. doi:10.1016/j.bmc.2009.06.003
36. Puzyn T, Leszczynski J, Cronin MT (eds). *Recent Advances in QSAR Studies, In Series: Modern Techniques and Applications*, vol 8. 2010; Springer, Netherlands. doi:10.1007/978-1-40209783-6
37. Nogrady T, Weaver DF. *Medicinal Chemistry: A Molecular and Biochemical Approach*, 3rd Ed. 2005; Oxford University Press, New York.
38. Te Velde GT, Bickelhaupt FM, Baerends EJ, Fonseca Guerra C, van Gisbergen SJ, Snijders JG, Ziegler T. Chemistry with ADF. *Journal of Computational Chemistry*. 2001 Jul 15;22(9):931–67.
39. Hutter J, Iannuzzi M, Schiffmann F, VandeVondele J. Cp2k: Atomistic simulations of condensed matter systems. *Wiley Interdisciplinary Reviews: Computational Molecular Science*. 2014 Jan;4(1):15–25.
40. Lehtola S, Steigemann C, Oliveira MJ, Marques MA. Recent developments in libxc—A comprehensive library of functionals for density functional theory. *SoftwareX*. 2018 Jan 1;7:1–5.
41. Salahub DR, Fournier R, Młynarski P, Papai I, St-Amant A, Ushio J. Gaussian-based density functional methodology, software, and applications. In Labanowski JK, Andzelm JW (eds) *Density Functional Methods in Chemistry* (pp. 77–100). 1991; Springer, New York.
42. Tomberg A. *Gaussian 09w Tutorial. An Introduction to Computational Chemistry Using G09W and Avogadro Software*. 2013:1–36.
43. Neese F, Wennmohs F, Becker U, Riplinger C. The ORCA quantum chemistry program package. *The Journal of Chemical Physics*. 2020 Jun 14;152(22):224108.
44. Neese F. *ORCA - An ab Initio, Density Functional and Semiempirical Program Package Version*. 2009; Max Planck Institute for Bioinorganic Chemistry Stiftstr, Germany.
45. Epifanovsky E, Gilbert AT, Feng X. Software for the frontiers of quantum chemistry: An overview of developments in the Q-Chem 5 package. *The Journal of Chemical Physics*. 2021;155:084801.
46. Turney JM, Simmonett AC, Parrish RM, Hohenstein EG, Evangelista FA, Fermann JT, Mintz BJ, Burns LA, Wilke JJ, Abrams ML, Russ NJ. Psi4: An open-source ab initio electronic structure program. *Wiley Interdisciplinary Reviews: Computational Molecular Science*. 2012 Jul;2(4):556–65.
47. Giannozzi P, Baseggio O, Bonfà P, Brunato D, Car R, Carnimeo I, Cavazzoni C, De Gironcoli S, Delugas P, Ferrari Ruffino F, Ferretti A. Quantum ESPRESSO toward the exascale. *The Journal of Chemical Physics*. 2020 Apr 21;152(15):154105.
48. Rademaker L. A practical introduction to density functional theory. arXiv preprint arXiv:2011.09888. 2020 Nov 19.
49. Dawson W, Degomme A, Stella M, Nakajima T, Ratcliff LE, Genovese L. Density functional theory calculations of large systems: Interplay between fragments, observables, and computational complexity. *Wiley Interdisciplinary Reviews: Computational Molecular Science*. 2022 May;12(3):e1574.
50. Blaha P, Schwarz K, Madsen GK, Kvasnicka D, Luitz J. *Wien2k. An Augmented Plane Wave+ Local Orbitals Program for Calculating Crystal Properties*. 2001; Vienna University Technology, Vienna, Austria.
51. Andrade X, Strubbe D, De Giovannini U, Larsen AH, Oliveira MJ, Alberdi-Rodriguez J, Varas A, Theophilou I, Helbig N, Verstraete MJ, Stella L. Real-space grids and the Octopus code as tools for the development of new simulation approaches for electronic systems. *Physical Chemistry Chemical Physics*. 2015;17(47):31371–96.
52. Marques MA, Castro A, Bertsch GF, Rubio A. Octopus: A first-principles tool for excited electron–ion dynamics. *Computer Physics Communications*. 2003 Mar 1;151(1):60–78.
53. Kee EA, Livengood MC, Carter EE, McKenna M, Cafiero M. Aromatic interactions in the binding of ligands to HMGCoA reductase. *The Journal of Physical Chemistry B*. 2009 Nov 5;113(44):14810–5.
54. Xu C, Ibrahim M, Spiro TG. DFT analysis of axial and equatorial effects on heme−CO vibrational modes: Applications to CooA and H−NOX heme sensor proteins. *Biochemistry*. 2008 Feb 26;47(8):2379–87.
55. Improta R, Barone V. Assessing the reliability of density functional methods in the conformational study of polypeptides: The treatment of intraresidue nonbonding interactions. *Journal of Computational Chemistry*. 2004 Aug;25(11):1333–41.
56. Hesselmann A. Comparison of intermolecular interaction energies from SAPT and DFT including empirical dispersion contributions. *The Journal of Physical Chemistry A*. 2011 Oct 20;115(41):11321–30.
57. Zhu Y, Alqahtani S, Hu X. An assessment of dispersion-corrected DFT methods for modeling nonbonded interactions in protein kinase inhibitor complexes. *Molecules*. 2024 Jan 6;29(2):304.
58. Ye N, Yang Z, Liu Y. Applications of density functional theory in COVID-19 drug modeling. *Drug Discovery Today*. 2022 May 1;27(5):1411–9.

13 Drug Repurposing

Ravichandran Veerasamy[1], *Raghuraman Seenivasan*[2], *Prabha Thangavelu*[3], *and Harish Rajak*[4]
[1]Department of Pharmaceutical Sciences, Faculty of Pharmacy, AIMST University, Semeling, Kedah, Malaysia
[2]Department of Pharmaceutical Sciences, Department of Pharmaceutical Chemistry, Shri Venkateshwara College of Pharmacy, Ariyur, Pondicherry, India
[3]Department of Pharmaceutical Sciences, Department of Pharmaceutical Chemistry, Nandha College of Pharmacy, Erode, India
[4]Department of Pharmaceutical Sciences, Guru Ghasidas University, Bilaspur, India

INTRODUCTION

The process of finding new therapeutic applications for drugs that are currently in use or discontinued is known as drug repurposing, often called drug repositioning, redirecting, reprofiling, re-tasking, rescuing, and therapeutic switching. Using the well-established pharmacokinetic and pharmacodynamic profiles of well-known medications, this novel strategy drastically reduces the time, expense, and risk involved in the conventional drug development process. According to DiMasi et al., traditional drug development is an expensive and time-consuming process that can take up to 10 years and billions of dollars to get a new medication from discovery to market [1]. On the other hand, repurposing already approved drugs can reduce the timeframe and costs, offering a speedier path to meeting unmet medical needs (Table 13.1).

A number of historical instances demonstrate the effectiveness and promise of pharmacological repurposing. Originally created to treat angina, sildenafil was later repurposed to treat erectile dysfunction and sold under the brand name Viagra. Despite its notorious past as a teratogen, thalidomide was resurrected as a treatment for erythema nodosum leprosum and multiple myeloma [2].

Several different and constantly developing mechanisms underlie drug repurposing. Researchers can now find new uses for existing pharmaceuticals by analysing large datasets with the help of advances in computational methods such as machine learning (ML) and molecular docking [3]. Phenotypic screening is the study of drugs in cellular or animal models to monitor changes in sickness, which can reveal unexpected therapeutic benefits [4]. Moreover, real-world data (RWD) from insurance claims, patient registries, and electronic health records helps find out the off-label usage and unexpected advantages of drugs [5].

The COVID-19 pandemic has emphasized the importance and potential of drug repurposing. Repurposing initiatives during the COVID-19 pandemic revealed that the corticosteroid dexamethasone and remdesivir, the antiviral drug for Ebola, can be used to treat severe COVID-19 patients [6,7]. Although drug repurposing has significant potential, it has numerous obstacles, such as legal constraints, intellectual property concerns, and potential for unintended adverse effects. Nevertheless, the ongoing collaboration between the pharmaceutical industry and academics and the technical advancements achieved make the future of drug repurposing quite promising [8].

Drug repurposing is a beneficial method in drug development and discovery, and it is a more practical and cost-effective approach to address various medical requirements and enhance patient outcomes. There are two main strategies in drug repurposing: on-target and off-target strategies (Figure 13.1). When an approved drug acts on the

TABLE 13.1
Difference between Conventional Drug Development and Drug Repurposing

	Conventional Drug Development	Drug Repositioning
Average cost to bring a new drug to market	USD 1.24 billion	Lower cost
Time required	10–17 years	3–12 years
Failure rates in later stages	High	Low
Increasing focus on	Drugs to treat chronic and complex diseases	Drugs for rapidly emerging and re-emerging infectious diseases

DOI: 10.1201/9781003464679-13

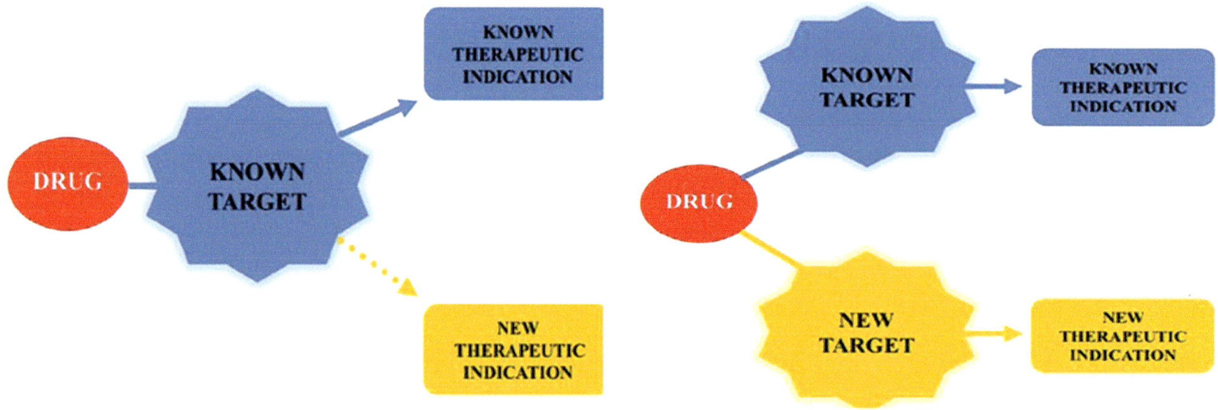

FIGURE 13.1 On-target and off-target strategies of drug repurposing. (Adopted from Ref. [9]).

same target as the original but produces a new therapeutic action, it is called an on-target strategy. However, if the approved drug produces a new therapeutic action by acting on a new target, then it is called an off-target strategy. This chapter covers the basic principles and procedures of drug repurposing, its obstacles, restrictions, and potential future directions along with examples.

CHALLENGES IN DRUG DISCOVERY AND DEVELOPMENT

The various stages of the drug discovery and development process, such as drug design, preclinical research, clinical trials, regulatory approval, and post-market surveillance, make it a lengthy, expensive, and difficult process.

Expensive and Protracted Timeliness

The substantial cost and long time required to bring a new drug to the market are among the primary difficulties in the drug discovery and development process. The entire process may take 10–15 years and require billions of dollars. According to DiMasi et al., the average cost for developing a new drug is nearly $2.6 billion [1]. The driving force behind these excessive costs is extensive preclinical and clinical research, as well as regulatory compliance, which is the basic necessity in new drug development.

High Rates of Failure

The attrition rate is a significant factor in the failure of numerous candidates at different stages of drug development. According to Hay et al., approximately 12% of drugs that commence clinical trials ultimately receive approval [10]. The factors for failures are ineffectiveness, unacceptable toxicity, or unanticipated undesirable consequences that arise during human trials. This high failure rate increases the risk and total cost of developing novel drugs.

Difficult Regulatory Needs

Another significant problem in drug development is that new drug approvals need compliance with rigorous regulations established by regulatory agencies, including the European Medicines Agency (EMA) and the US Food and Drug Administration (FDA). Although these constraints are intended to guarantee safety and efficacy, they can also prolong the development process and increase costs. Some of these regulations need the completion of numerous clinical research phases, accumulation of extensive documentation, and maintenance of close communication with authorities [11].

Technological and Scientific Difficulties

The drug discovery process has numerous scientific challenges. State-of-the-art scientific methodologies and technologies are necessary to successfully complete challenging tasks in drug development, such as the selection of appropriate therapeutic targets and the development of selective and efficient compounds. Moreover, the complex biology and the need for personalized methods make the development of drugs for complex disorders, such as cancer, Alzheimer's disease, and uncommon genetic defects, difficult [12].

Risks in the Market and Commerce

However, medications are subject to substantial market and commercial risks even after they have been approved, such as potential market saturation, competition from complementary therapies, and reimbursement issues. To face these market risks, pharmaceutical companies should implement comprehensive marketing and educational initiatives to guarantee that healthcare professionals and consumers are informed about the most recent alternative treatments.

Safety and Ethical Issues

Ethics is indispensable when it pertains to the development of novel medications. Conducting clinical trials in an ethically acceptable manner, obtaining informed consent, and ensuring patient safety are essential components. Transparency and the ethical implications of clinical trial designs, particularly in relation to vulnerable populations, are also receiving increased attention [13].

After-Market Monitoring

Even after a medicine has been approved, obstacles persist. It is critical to monitor its pharmaceutical safety and efficacy after it is released to the market. Increased use of the medication may disclose side effects that are not observed in clinical trials. This requires robust pharmacovigilance frameworks and the ability to respond quickly to emergent safety data [14].

The process of developing new drugs is extremely difficult due to high costs, lengthy development times, high failure rates, complex regulatory requirements, scientific and technological challenges, market and commercial risks, ethical concerns, and the need for ongoing post-market surveillance. A number of significant concerns and difficulties confronting the pharmaceutical industry have necessitated the need for alternative drug development methodologies. Alternative strategies should be used to overcome these constraints and improve the effectiveness, economy, and patient-centredness of the drug discovery process. Creative problem-solving strategies, cross-sector collaboration, and ongoing improvements in scientific research and regulatory processes are all required to manage these challenges in drug discovery and development.

SIGNIFICANCE OF FINDING NEW USES FOR EXISTING DRUGS

Existing drugs already have established safety profiles, lower development costs, faster timeframes to market, and fewer failure risks, which can reduce the risk and cost in the drug development process when these drugs are found to have new uses.

Time and Cost-Effectiveness

The development of a new drug is a time-consuming and costly process. However, in drug repurposing, the well-established safety and pharmacokinetic profiles of already approved drugs can significantly reduce the costs and timelines [1]. This proficiency is essential for meeting urgent medical needs, as evidenced during the COVID-19 pandemic, when dexamethasone and remdesivir were promptly repurposed to treat severe cases [6,7].

Increased Rates of Success

The failure rate of clinical trials is high for novel drug candidates, often as a result of unforeseen safety concerns. Repurposed drugs are more likely to be successful in novel indications as they already have established safety and toxicity profiles [15]. These increased rates of success may result in a more efficient utilization of resources in the development of new drugs.

Addressing Unmet Medical Needs

Repurposing drugs can lead to the development of novel therapeutic alternatives for diseases that now have no or very few therapeutic drugs. For example, the antidiabetic drug metformin has shown additional potential in treating a variety of malignancies [16]. Sildenafil is another example of repurposed drugs, which is now commonly used to treat erectile dysfunction but was originally intended to treat angina.

Enabling Personalized Healthcare

Drug repurposing is a critical component of personalized care. Doctors can better adapt their therapy to specific patients if they understand the molecular mechanisms and genetic conditions under which these drugs function. This technique enhances the therapeutic outcomes while minimizing the side effects.

Impact on Public Health

Drug repurposing can have a significant impact on public health, particularly in low- and middle-income countries where the cost of therapeutic development is a major problem. Repurposing low-cost, off-patent medications can assist healthcare professionals to expand treatment options and improve patient outcomes in general [8].

Accelerating Responses to Health Crises

The COVID-19 pandemic serves as an excellent example of how developing health issues should be addressed rapidly. Drug repurposing revealed the strategic importance of this method for the rapid identification and deployment of successful treatments in worldwide health emergencies [6,7].

Potential for Combination Therapies

Repurposed drugs can be combined with the existing prescriptions to improve treatment outcomes or address numerous aspects of a complex ailment. This approach could lead to more comprehensive treatment plans.

Extended Patent Life

Pharmaceutical companies can continue to profit from their products for long by repurposing old pharmaceuticals and extending their patent life even after their original indications have expired.

Targeting Multiple Diseases

Some modern drugs may have broad-spectrum activity that can increase their utility in treating a wide range of illnesses or diseases.

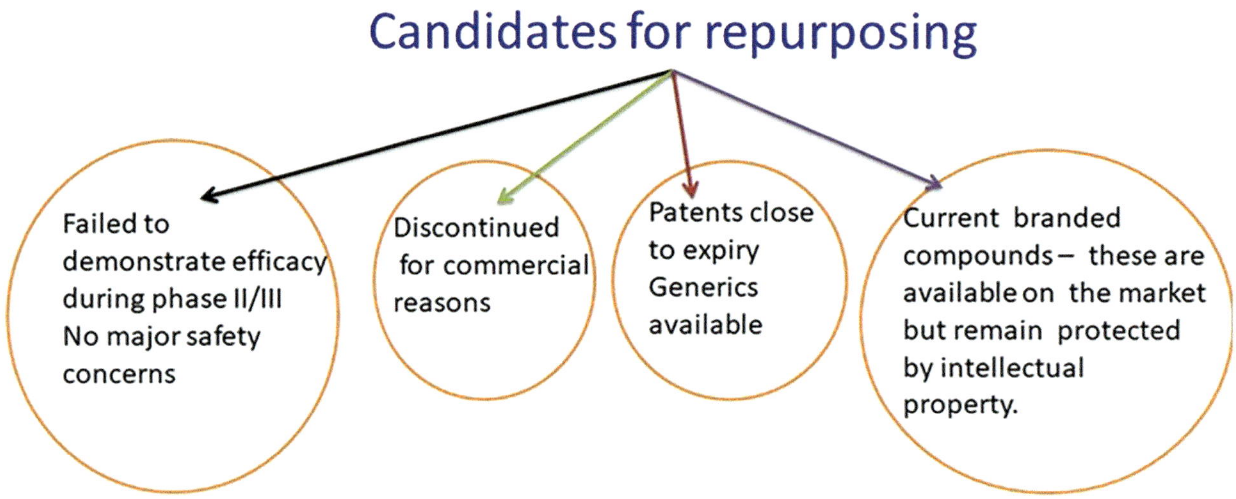

FIGURE 13.2 Candidate for drug repurposing.

Drug Rescue

Drug repurposing can "rescue" potentially helpful substances that failed in their initial clinical trials for one application but showed potential for another health problem, which reduces drug development costs (Figure 13.2).

KEY STEPS INVOLVED IN DRUG REPURPOSING

Identification and approval of repurposed drugs involve a few key steps, which include drug identification, mechanistic studies, preclinical validation, clinical trials, regulatory approval, and post-market surveillance.

Candidate Drug Identification

The first stage in repurposing drugs is to identify the possible candidates (Figure 13.2) for new therapeutic applications. Pushpakom et al. define "literature mining" as the practice of exploring scientific literature for anecdotal stories of good effects or proof of detrimental effects in various disease scenarios [8]. Data mining is the process of discovering medications with unexpected benefits or patterns of use in various situations by analysing big datasets from clinical trial databases, electronic health records, and pharmacovigilance data using bioinformatics tools [5]. Computing techniques, including molecular docking, virtual screening, and ML, employ computer algorithms to anticipate interactions between known medications and novel targets [3].

Mechanistic Studies

After viable candidates have been discovered, a mechanistic study is conducted to better understand how these medications work in the new indication. This includes:

***In Vitro* Research**: The medicine is examined in cell-based studies to assess its effects on important biological targets and pathways. "*In vivo* research" refers to the use of animal models to investigate the pharmacodynamics, pharmacokinetics, and efficacy of a drug in a living being [4].

Preclinical Validation

Preclinical validation is essential to confirm the drug's safety and efficacy in the new indication before proceeding further with clinical studies. This activity contains: animal models and dose–response studies are used to assess the range of effective doses as well as potential toxicity. Toxicology research involves assessing the drug's safety profile while taking into account potential adverse effects and long-term toxicity [12].

Clinical Trials

Clinical investigations can commence once preclinical validation is fruitful. The standard procedure for these types of investigations consists of three stages:

Phase I: Testing the drug in a limited number of healthy people to see how it affects their body and how long it takes for the drug to take effect.

Phase II: Testing the medication's effectiveness and dosage in a larger population of people with the target condition.

Phase III: Evaluating the medication's safety and effectiveness in a large group of people, typically in comparison with either a control group or the usual course of treatment [10].

Regulatory Approval

Upon completion of fruitful clinical research, applications for regulatory approval are sent to organizations such as the FDA in the USA and the EMA. A significant amount of data is required to assess the medication's effectiveness, safety, and manufacturing quality [11].

Post-Market Surveillance

The drug is released to the market upon approval; however, ongoing monitoring is necessary to guarantee its safety and efficacy for a broader range of patients. The following are included in post-market surveillance:

Pharmacovigilance: It is the ongoing surveillance of adverse effects and long-term safety.

Real-World Evidence: In order to confirm the medication's efficacy and identify any potential drawbacks or benefits, information is collected in real-world settings [14].

APPROACHES AND METHODS USED IN DRUG REPURPOSING

Drug repurposing finds novel therapeutic applications for already approved medications using a variety of techniques. These methods can be broadly divided into the following: database mining, clinical trial data analysis, literature mining, experimental analysis, chemical similarity and structural biology, omics approaches, and *in silico* studies. Every technique offers distinct benefits and insights that enhance the overall efficacy of medication repurposing endeavours. All these approaches are classified into four strategies: signature-based strategies, target-based strategies, drug-based strategies, and literature-based strategies.

In Silico Methodologies

Using sophisticated algorithms and big datasets, computational methods employ cheminformatics and bioinformatics tools to forecast novel applications for currently approved medications. Computational methods are cost-effective and can screen large datasets quickly, but require validation through experimental methods.

Molecular docking and virtual screening: These methods simulate drug–target interactions in order to forecast biological activity and binding affinity. Drugs that can bind to novel targets with high specificity are found using molecular docking [3]. Because of its capacity to bind to and inhibit histamine receptors expressed on specific cancer cells, cimetidine, which was initially developed as an antiulcer drug, was later identified as a potential anticancer agent through molecular docking studies [17]. In order to identify currently available antiviral medications that have the potential to halt the replication of SARS-CoV-2, such as favipiravir and remdesivir, virtual screening methods were implemented during the COVID-19 pandemic [18].

Network Pharmacology: This method investigates the relationships between medications, targets, and diseases using network analysis. Using these relationships, researchers may be able to uncover potential drug repurposing opportunities based on shared pathways and molecular networks. For example, thalidomide was originally intended to be a sedative, but researchers discovered through network pharmacology that it might modify immune responses and prevent angiogenesis. As a result, it was repurposed and approved to treat multiple myeloma [19]. According to network pharmacology analysis, metformin, a common antidiabetic drug, interacts with biochemical pathways involved in cancer progression, which instigated the investigation of metformin as an anticancer agent [16].

Artificial Intelligence (AI) and Machine Learning (ML): AI and ML are types of algorithms that investigate huge amounts of biomedical data for patterns and connections that could find new ways to use already approved drugs. Chen et al. reported that these methods can handle data from a lot of different places, such as genetic data, electronic health records, and clinical studies [20]. According to AI models, the anaesthetic drug ketamine can quickly reduce sadness [21]. This finding has led to its new use for people who do not respond to other treatments for depression. ML methods have shown that the antipsychotic drug pimozide can affect the survival and function of neurons, and this suggests that it might be useful in treating amyotrophic lateral sclerosis [22].

Experimental Methodologies

Experimental methodologies are direct biological testing methods in a laboratory environment to authenticate the possible use of novel drugs.

Phenotypic Screening: This method involves the use of animal or cell-based models to observe and analyse changes in observable characteristics. This method is employed to assess new therapeutic capabilities of already approved drugs. According to Swinney and Anthony, phenotypic screening is a valuable method for identifying undesirable effects since it does not rely on understanding the drug's mechanism of action [4]. Phenotypic screening conducted by Caly et al. revealed that the antiparasitic drug ivermectin exhibited potent antiviral effects against several viruses, including dengue, Zika, and SARS-CoV-2 [23]. In another phenotypic screening, it was found that minoxidil stimulated hair growth, and as a result, it was repurposed as a treatment for androgenetic alopecia, which was actually developed for the purpose of treating hypertension [24].

High-Throughput Screening (HTS): HTS is a technology that quickly analyses thousands of compounds using biological tests to identify their effects.

This technique allows for the rapid generation of data pertaining to the effectiveness and harmful effects of drugs across a range of disease models [25]. Using this technique, disulfiram, a medicine used for chronic alcoholism, was found to exhibit potent antibacterial properties against drug-resistant bacterial strains [26]. Another HTS study reported that the antimalarial drug chloroquine and hydroxychloroquine have anticancer properties via inhibiting autophagy in cancer cells [27].

Omics Technologies: During drug therapy, proteomics, metabolomics, and genomes can provide comprehensive profiles of biological systems. By aiding in the identification of molecular targets and pathways affected by medications, these instruments help focus repurposing efforts [28]. Tamoxifen, a breast cancer medication, has been shown to have an impact on oestrogen receptor signalling in malignancies other than breast cancer, such as pancreatic and liver cancer, which was identified using omics technologies. The immunosuppressant rapamycin, which is administered to transplant patients, has been shown through proteomics studies to modify ageing pathways. As a result, rapamycin is being investigated as a potential treatment for age-related disorders [29].

Clinical Trial Data Analysis

Clinical methods employ strategies from clinical trials and RWD to investigate novel therapeutic applications for currently available pharmaceuticals.

RWD Analysis: Unexpected therapeutic advantages and off-label uses of medications can be revealed by RWD from insurance claims, electronic health records, and patient registries. This data analysis assists in the identification of repurposing prospects by examining actual patient outcomes [5]. An analysis of RWD suggests that statins, which are primarily used to regulate cholesterol, may also reduce the risk of fractures. This finding infers that statins may be employed to manage osteoporosis [30]. In another RWD analysis, the use of beta-blockers in cancer therapy was investigated when being treated for cardiovascular disorders, which revealed the association of beta-blockers with decreased cancer progression and mortality [31].

Retrospective Clinical Trial Studies: The efficacy of the treatment in new indications is established through the analysis of current trial data. Researchers can identify secondary effects of a drug that may need additional investigation by re-examining data from completed trials [32]. A retrospective review of clinical trial data of aspirin has shown that it is effective in reducing the incidence of colorectal cancer; that is why, it is recommended for cancer prevention in high-risk populations [33]. The antiepileptic drug valproate has been repurposed for the treatment of bipolar illness as a result of retrospective clinical trials that demonstrated its mood-stabilizing effects [34].

Adaptive Clinical Trials: Clinical trial protocols can be adjusted in response to interim results through adaptive trial designs. This adaptability enables the efficient assessment of repurposed drugs in novel indications, thus decreasing the time and expense of clinical validation [35]. During the COVID-19 pandemic, the anti-Ebola drug remdesivir was promptly assessed and repurposed to treat COVID-19 patients as a result of adaptive trial designs [6]. Another repurposed drug identified during the COVID-19 pandemic was dexamethasone, a corticosteroid, which was extensively employed during the pandemic after the RECOVERY adaptive trial demonstrated that it significantly reduced mortality in severe COVID-19 cases [7].

Literature Mining

Literature mining, a text mining approach also known as bibliome mining, is a technique that involves extracting relevant data from vast amounts of scientific literature in order to discover potential therapeutic applications for already approved drugs. This process examines scientific texts in the literature and is supported by natural language processing (NLP), ML, and data mining approaches. Some of the primary objectives include identifying novel therapeutic targets, finding the links between drugs and disorders, and exploring the previously unknown mechanisms of action.

Text Preprocessing:

- The text is lemmatized, tokenized, and stemmed in order to make it ready for analysis.
- Named entity recognition (NER) is used to locate and categorize textual entities, such as drugs, illnesses, and genes.

Relationship Extraction:

- Syntactic analysis and co-occurrence analysis are employed to determine the relationships between objects.
- ML algorithms can be trained to recognize patterns and infer significant relationships using text.

Information Integration:

- The validation of conclusions and the provision of a comprehensive perspective are achieved by combining data collected from multiple sources.

- Databases such as PubMed, MEDLINE, and other specialized repositories (Table 13.2) are frequently employed to collect information [36,37].

Database Mining

Database mining is the systematic extraction and study of information from various scientific databases to uncover the potential new applications for already approved drugs, and it is an important strategy in drug repurposing. This strategy employs large-scale data integration, bioinformatics tools, and ML algorithms to reveal new drug–disease relationships, mechanisms of action, and therapeutic applications. Database mining is used to repurpose drugs by leveraging the quantity of data in various chemical and biological databases. These databases provide information about drug properties, biological targets, disease pathways, clinical trial findings, and patient records. Researchers can use data mining to develop theories for novel therapeutic applications of drugs.

Data Integration:

- It merges information from several sources to produce an extensive dataset.
- Often used databases include ClinicalTrials.gov, PubChem, ChEMBL, DrugBank, and a number of omics databases (Table 13.2) (e.g. proteomics, genomes).

Pattern Recognition:

- It involves identifying patterns and correlations in data using ML algorithms and bioinformatics tools, such as network analysis, categorization, and clustering.

Hypothesis Generation:

- It develops testable hypotheses for novel drug–disease connections using mining data.
- It validates these hypotheses through clinical and experimental research [8,38].

Chemical Similarity and Structural Biology

Structural biology and chemical similarities are used to find new uses for already approved drugs. The idea behind chemical similarity is that biological processes are often shared by compounds with similar structures. This method looks at the chemical structures of well-known medicines to find molecules with similar biological properties. Chemoinformatics software, chemistry databases like PubChem, ChEMBL, and DrugBank, and molecular fingerprints (Table 13.2) are all useful tools in finding chemical similarities [39–41]. Structural biology is the study of the three-dimensional shapes of biological macromolecules. Knowing the structures of drug targets, like proteins, helps

TABLE 13.2
Repositories for Drug Repurposing

Name of Data	Repositories Name	Name of Data	Repositories Name
Drug omics data	PubChem DrugBank FDA Label Drug Information Portal Drug Map Central Daily Med	Toxicity data	Tox 21 Tox Cast
Pathway omics data	MANTRA KEGG PID Pathway Commons Reactome	Genomic data	SRA GEO OMM dbSNP
Target omics data	PDB TTD UNIPROT BioGRID HPRD STRING Binding DB STITCH	Clinical trials and adverse effects data	Clinicaltrials.gov SIDER Adverse Drug Reaction Database FAERS

us understand how drugs work with these targets and also suggests new uses of drugs that are already on the market. A few examples are X-ray diffraction, cryo-electron microscopy, and molecular docking [40,42,43]. Combining structural biology and chemical resemblance is a worthy way to identify new uses for already approved drugs. Finding new drug–target interactions is possible when the information of biological targets and chemical structures is known, with the help of bioinformatics, ML, and HTS tools [44]. Some of the successfully repurposed drugs with the methodology used for repurposing are listed in Table 13.3.

ADHD, attention deficit hyperactivity disorder; EMA, European Medicines Agency; FDA, US Food and Drug Administration; MS, multiple sclerosis; SUI, stress urinary incontinence.

CHALLENGES IN DRUG REPURPOSING

Even though drug repurposing has great potential, there are many challenges and barriers in this process. These barriers lengthen the scientific, regulatory, and commercial process and affect the feasibility and timeline for bringing the repurposed drugs to the market.

Scientific Challenges

Action Mechanism Recognition: The process of repurposing drugs often involves the use of a treatment that is intended to address one condition but is actually intended to address another. Further research may be necessary to determine the precise mechanisms by which these drugs function in novel indications [8].

Lack of Predictive Models: Predictive models, such as computer algorithms or animal models, are frequently needed to determine the novel therapeutic applications of chemical compounds. Rigorous validation processes are mandatory; hence, these models might not always accurately predict the therapeutic outcomes or side effects of drugs in humans. The models are not accepted as predictive models if they are not properly validated [48].

Regulatory Challenges

Regulations for Off-Label Use: Reusing drugs for conditions other than those for which they were actually approved is called “off-label use”. Regulations are different in each country. Getting approval for new uses of already approved drugs might need more clinical trials and regulatory applications [49].

Intellectual Property Issues: When patents for the original drug are still valid, the repurposing of the drug is difficult due to intellectual property rights and patents. Finding a way for drug repurposing around patents and licencing deals can be difficult and time-consuming [50].

Financial Challenges

Investment and Funding: Due to budget constraints, pharmaceutical companies may find developing new drugs more economical than repurposing existing ones. The absence of market incentives and exclusivity assurances may also discourage investment in drug repurposing activities [8].

Market Competition: The profit margins of repurposed drugs may be reduced by increased competition from generic copies of the same drugs. This may reduce the inducements for industries to invest in drug repurposing and commercialization [51].

Limited Data Availability

It is not always easy to find detailed information about drugs on the market and their mechanisms of action. This can make it challenging to identify drug candidates that are suitable for repurposing.

Clinical Trial Design and Biomarker Identification

Clinical trial design for repurposed drugs could be tricky and needs careful attention in selecting appropriate patient demographics, dosage schedules, and objectives.

Identifying the relevant biomarkers to assess efficacy can be challenging, especially if the drug being repurposed has a different mode of action.

Patient Recruitment

It can be challenging to recruit patients for clinical studies of repurposed drugs, especially if the patient population for the new indication is small or geographically dispersed.

Resistance to Change

Despite the advantages of repurposed medications, clinicians and healthcare systems may hesitate to adopt them and prefer to adhere to traditional therapies. It is crucial to raise awareness among healthcare professionals about repurposed drugs, provide solid evidence of their efficacy, and encourage them to overcome their reluctance and adopt repurposed drugs.

Safety Concerns

Although repurposed drugs have well-established safety profiles, their use for new indications may result in safety concerns, especially when the patient population or dosage differs drastically from the original application.

Pharmaceutical businesses, university researchers, regulatory bodies, and healthcare practitioners should work together to address these challenges. Improved data-sharing

TABLE 13.3
Selected Successful Drug Repurposing Examples and the Repurposing Approach Employed [8,45–47]

Drug Name	Original Indication	New Indication	Date of Approval (FDA)	Repurposing Approach Used
Zidovudine	Cancer	HIV/AIDS	1987	*In vitro* screening of compound libraries
Minoxidil	Hypertension	Hair loss	1988	Retrospective clinical analysis (identification of hair growth as an adverse effect)
Sildenafil	Angina	Erectile dysfunction	1998	Retrospective clinical analysis
Thalidomide	Morning sickness	Erythema nodosum leprosum and multiple myeloma	1998 and 2006	Off-label usage and pharmacological analysis
Celecoxib	Pain and inflammation	Familial adenomatous polyps	2000	Pharmacological analysis
Atomoxetine	Parkinson's disease	ADHD	2002	Pharmacological analysis
Duloxetine	Depression	SUI	2004	Pharmacological analysis
Rituximab	Various cancers	Rheumatoid arthritis	2006	Retrospective clinical analysis (remission of coexisting rheumatoid arthritis in patients with non-Hodgkin lymphoma treated with rituximab)
Raloxifene	Osteoporosis	Breast cancer	2007	Retrospective clinical analysis
Fingolimod	Transplant rejection	MS	2010	Pharmacological and structural analysis
Dapoxetine	Analgesia and depression	Premature ejaculation	2012	Pharmacological analysis
Topiramate	Epilepsy	Obesity	2012	Pharmacological analysis
Ketoconazole	Fungal infections	Cushing's syndrome	2014	Pharmacological analysis
Aspirin	Analgesia	Colorectal cancer	2015	Retrospective clinical and pharmacological analysis
Tamoxifen	Breast cancer	Infertility treatment	1978	Mechanistic insight
Dexamethasone	Inflammation	COVID-19	2020	Clinical trial
Finasteride	Benign prostatic hyperplasia (BPH)	Male pattern baldness	1997	Mechanistic insight
Eflornithine	Cancer	African sleeping sickness	1990	Mechanistic insight
Propranolol	Hypertension	Anxiety and migraine	1984	Mechanistic insight
Methotrexate	Cancer	Rheumatoid arthritis	1988	Mechanistic insight
Colchicine	Gout	Familial Mediterranean fever	2009	Clinical observation
Bupropion	Depression	Smoking cessation	1997	Mechanistic insight
Clomiphene	Infertility	Male hypogonadism	1976	Mechanistic insight
Allopurinol	Gout	Cardiovascular disease prevention	2014	Clinical observation
Metformin	Type 2 diabetes	Polycystic ovary syndrome (PCOS)	2001	Clinical observation
Valproic acid	Epilepsy	Bipolar disorder	1995	Mechanistic insight
Ketamine	Anaesthesia	Treatment-resistant depression	2019	Clinical trial
Dimenhydrinate	Motion sickness	Hyperemesis gravidarum	2006	Clinical observation
Hydroxychloroquine	Malaria	Rheumatoid arthritis, lupus	1955	Mechanistic insight
Naltrexone	Opioid dependence	Alcohol dependence	1994	Mechanistic insight
Doxycycline	Bacterial infections	Rosacea	2006	Mechanistic insight
Spironolactone	Hypertension	Acne, hirsutism	1985	Clinical observation
Fluoxetine	Depression	Premenstrual dysphoric disorder	2000	Mechanistic insight
Topiramate	Epilepsy	Migraine prophylaxis	2004	Clinical trial
Prazosin	Hypertension	Post-traumatic stress disorder	1995 (off-label use)	Clinical observation
Aprepitant	Chemotherapy-induced nausea	Postoperative nausea and vomiting	2006	Mechanistic insight

(*Continued*)

TABLE 13.3 (*Continued*)

Selected Successful Drug Repurposing Examples and the Repurposing Approach Employed [8,45–47]

Drug Name	Original Indication	New Indication	Date of Approval (FDA)	Repurposing Approach Used
Dapoxetine	Antidepressant	Premature ejaculation	2009	Mechanistic insight
Levodopa	Parkinson's disease	Restless legs syndrome	2005	Clinical observation
Ritonavir	HIV/AIDS	COVID-19 (in combination therapy)	2020	Clinical trial
Thiazolidinediones	Type 2 diabetes	Nonalcoholic steatohepatitis (NASH)	2017 (investigational)	Mechanistic insight
Cyclosporine	Immunosuppressant	Dry eye disease	2003	Clinical trial
Dimethyl fumarate	Psoriasis	Multiple sclerosis	2013	Mechanistic insight
Everolimus	Organ transplant rejection	Various cancers	2009	Mechanistic insight
Miltefosine	Cancer	Leishmaniasis	2014	Mechanistic insight
Alendronate	Osteoporosis	Paget's disease	1995	Clinical observation
Liraglutide	Type 2 diabetes	Obesity	2014	Clinical trial
Azathioprine	Organ transplant rejection	Autoimmune diseases (e.g., Crohn's)	1982	Clinical observation
Methadone	Pain relief	Opioid dependence	1972	Mechanistic insight
Triamcinolone	Skin inflammation	Macular oedema	2014	Clinical trial
Cladribine	Hairy cell leukaemia	Multiple sclerosis	2019	Mechanistic insight
Tadalafil	Erectile dysfunction	Benign prostatic hyperplasia (BPH)	2011	Mechanistic insight
Carvedilol	Hypertension	Heart failure	1997	Clinical trial
Remdesivir	Ebola	COVID-19	2020	Clinical trial
Baricitinib	Rheumatoid arthritis	COVID-19	2020	Mechanistic insight
Tocilizumab	Rheumatoid arthritis	COVID-19	2021	Mechanistic insight
Fluvoxamine	Obsessive-compulsive disorder	COVID-19	2021 (emergency use)	Clinical observation
Famotidine	Acid reflux	COVID-19	2020 (under investigation)	Clinical observation
Colchicine	Gout	COVID-19	2021 (under investigation)	Clinical observation
Ivermectin	Parasitic infections	COVID-19	2020 (under investigation)	Empirical evidence
Dexamethasone	Inflammation	COVID-19	2020	Clinical trial
Anakinra	Rheumatoid arthritis	COVID-19	2021 (under investigation)	Mechanistic insight
Metformin	Type 2 diabetes	COVID-19 (prevention in high-risk cases)	2021 (under investigation)	Clinical observation
Ruxolitinib	Myelofibrosis	COVID-19	2021 (under investigation)	Mechanistic insight
Nintedanib	Idiopathic pulmonary fibrosis	Systemic sclerosis-associated interstitial lung disease	2019 (close to timeline)	Mechanistic insight
Apremilast	Psoriasis	Behçet's disease	2020	Mechanistic insight
Cannabidiol	Seizures in Dravet syndrome	Tuberous sclerosis complex	2020	Clinical trial
Ozanimod	Multiple sclerosis	Ulcerative colitis	2021	Mechanistic insight
Lopinavir/ritonavir	HIV/AIDS	COVID-19	2020 (under investigation)	Clinical trial

programmes, adaptive clinical trial designs, and innovative intellectual property management procedures can all contribute to the effective repurposing of drugs for novel therapeutic purposes, benefiting both patients and healthcare systems.

FUTURE TRENDS IN DRUG REPURPOSING

Concepts that make finding new medical uses for already approved drugs cheaper, more efficient, and faster will affect future trends in this field.

Advanced Computational Techniques

Sophisticated AI algorithms will be used in measuring drug–target interactions, discovering new indications, and optimizing pharmacological combinations for repurposing [52].

Network Pharmacology: Drug interactions and causes for illness can be explored using network pharmacology methods that leverage omics data and large-scale biological networks [53]. This leads to targeted drug repurposing techniques.

Integrating Big Data: Data from wearables, patient registries, and electronic health records can be used to assess drug reactions in various patient groups. This data is referred to as RWD. According to Sultana et al., this helps in finding new indications and ways to customize medicine [54].

Omics Technologies: Progress in genomics, proteomics, and metabolomics makes it easier to find drug candidates for reuse [55].

Collaborative Initiatives

Public–Private Partnerships: To speed up drug repurposing research, partnerships between universities, drug companies, and government agencies are essential, and promoting the sharing of data and resources and working together to raise funds are also crucial [56].

Open Innovation Platforms: It is faster to find and test possible drug repurposing candidates with the help of internet databases and platforms that let people from all over the world work together on repurposing ideas [57].

Regulatory Advancements

Adaptive Licencing and Drug Approvals: Regulators are investigating flexible approval pathways and adaptive licencing for repurposed pharmaceuticals to provide faster access to new indications based on RWD and surrogate endpoints [17].

Global Harmonization: As a result of attempts to standardize the regulatory norms and standards across nations, the repurposed drug clearance process is becoming simpler, and barriers in abroad marketing are being eliminated [58].

The future of drug repurposing is promising with advances in collaborative models, big data analytics, computational biology, and laws. These improvements are expected to change the drug discovery business, allowing for speedier and more cost-effective development of answers to unmet medical needs.

Recent technological developments, changes in research methods, and new regulatory frameworks are all contributing to the progress of drug repurposing. Innovations that increase productivity, decrease prices, and speed up the identification of new therapeutic uses for approved drugs will shape future trends in this area.

ETHICAL AND REGULATORY CONSIDERATIONS IN DRUG REPURPOSING

Although repurposing medications may speed up therapeutic results, there are significant moral and legal considerations. Some of these concerns include fair benefit sharing, patient safety, consent, data protection, and measures required to obtain government clearance.

Ethical Considerations

Patient Safety and Informed Consent: When people use repurposed drugs off-label, they may be exposed to unrecognized risks. To safeguard the well-being of patients, informed consent should be obtained, and any adverse effects should be closely monitored [59].

Benefit–Risk Assessment: It is crucial to be transparent with patients and other stakeholders when considering the potential benefits of repurposing drugs against the risks of their off-label use, particularly while their safety and effectiveness profiles are still being established [60].

Regulatory Considerations for Off-Label Use and Approval: Typically, repurposing involves using medications already approved for one purpose for another without undergoing standard clinical trials. However, the approval process may become more challenging if regulatory bodies require additional evidence of safety and efficacy [61].

Data Security and Privacy: Privacy and data security concerns about patients can arise when RWD and electronic health records are reused for research. To maintain patients' trust and compliance, strict adherence to data privacy regulations is essential [62].

Equity and Access

Ensuring equal access to repurposed drugs for diverse populations is essential for reducing healthcare inequities. Regulations should facilitate the swift approval of these medications and ensure they are available at an affordable cost [63].

Negotiations over market exclusivity and intellectual property rights can influence the availability and pricing of repurposed drugs. Implementing policies that ensure equitable pricing and availability is crucial for optimizing public health benefits [64]. The drug repurposing process raises ethical and legal issues that should be addressed to protect patient well-being, ensure fair access, and maintain public trust. Effective collaboration among academics, healthcare practitioners, regulators, and policymakers is essential to address these complex challenges and maximize the benefits of repurposed pharmaceuticals.

SUMMARY

Drug repurposing offers a viable means to expedite the drug discovery process by identifying new therapeutic uses for approved drugs. Utilizing various computational, experimental, and clinical approaches, this strategy successfully explores the potential of pharmaceuticals beyond their original indications. Two computational tools that aid in predicting drug–target interactions and disease processes are network pharmacology and molecular docking. Experimental approaches such as phenotypic screening and HTS are used to verify the effect of candidate drugs in biological systems. Clinical activities supporting new indications include retrospective studies and adaptive trials, which provide empirical data.

Despite its potential, drug repurposing faces several challenges. Scientifically, these include verifying hypotheses in clinical settings and understanding complex pharmacological mechanisms. Regulatory hurdles, such as addressing off-label usage regulations and obtaining new licenses, can complicate the progress of repurposed drugs to the market. Economic considerations, including market competitiveness and budgetary constraints, also impact drug repurposing.

Future advancements in drug repurposing will likely focus on sophisticated computational approaches, such as AI and extensive data integration. Regulatory reforms and collaborative initiatives will be essential to address ethical issues, ensure patient safety, and provide equitable access to repurposed therapies. By leveraging advanced technology to overcome these challenges, drug repurposing has the potential to revolutionize the pharmaceutical industry, offering innovative treatments in a more efficient and cost-effective manner.

REFERENCES

1. DiMasi, J. A.; Grabowski, H. G.; Hansen, R. W. Innovation in the Pharmaceutical Industry: New Estimates of R&D Costs. *J. Health Econ.* **2016**, *47*, 20–33. https://doi.org/10.1016/j.jhealeco.2016.01.012.
2. Ito, T.; Ando, H.; Suzuki, T.; Ogura, T.; Hotta, K.; Imamura, Y.; Yamaguchi, Y.; Handa, H. Identification of a Primary Target of Thalidomide Teratogenicity. *Science* **2010**, *327* (5971), 1345–1350. https://doi.org/10.1126/science.1177319.
3. Ekins, S.; Mestres, J.; Testa, B. In Silico Pharmacology for Drug Discovery: Methods for Virtual Ligand Screening and Profiling. *Br. J. Pharmacol.* **2007**, *152* (1), 9–20. https://doi.org/10.1038/sj.bjp.0707305.
4. Swinney, D. C.; Anthony, J. How Were New Medicines Discovered? *Nat. Rev. Drug Discov.* **2011**, *10* (7), 507–519. https://doi.org/10.1038/nrd3480.
5. Schneeweiss, S. Learning from Big Health Care Data. *N. Engl. J. Med.* **2014**, *370* (23), 2161–2163.
6. Beigel, J. H.; Tomashek, K. M.; Dodd, L. E.; Mehta, A. K.; Zingman, B. S.; Kalil, A. C.; Hohmann, E.; Chu, H. Y.; Luetkemeyer, A.; Kline, S.; Lopez De Castilla, D.; Finberg, R. W.; Dierberg, K.; Tapson, V.; Hsieh, L.; Patterson, T. F.; Paredes, R.; Sweeney, D. A.; Short, W. R.; Touloumi, G.; Lye, D. C.; Ohmagari, N.; Oh, M.; Ruiz-Palacios, G. M.; Benfield, T.; Fätkenheuer, G.; Kortepeter, M. G.; Atmar, R. L.; Creech, C. B.; Lundgren, J.; Babiker, A. G.; Pett, S.; Neaton, J. D.; Burgess, T. H.; Bonnett, T.; Green, M.; Makowski, M.; Osinusi, A.; Nayak, S.; Lane, H. C. Remdesivir for the Treatment of Covid-19 — Final Report. *N. Engl. J. Med.* **2020**, *383* (19), 1813–1826. https://doi.org/10.1056/NEJMoa2007764.
7. Group, R. C. Dexamethasone in Hospitalized Patients with Covid-19. *N. Engl. J. Med.* **2021**, *384* (8), 693–704.
8. Pushpakom, S.; Iorio, F.; Eyers, P. A.; Escott, K. J.; Hopper, S.; Wells, A.; Doig, A.; Guilliams, T.; Latimer, J.; McNamee, C. Drug Repurposing: Progress, Challenges and Recommendations. *Nat. Rev. Drug Discov.* **2019**, *18* (1), 41–58.
9. de Oliveira, E. A. M.; Lang, K. L. Drug Repositioning: Concept, Classification, Methodology, and Importance in Rare/Orphans and Neglected Diseases. *J. Appl. Pharm. Sci.* **2018**, *8* (8), 157–165.
10. Hay, M.; Thomas, D. W.; Craighead, J. L.; Economides, C.; Rosenthal, J. Clinical Development Success Rates for Investigational Drugs. *Nat. Biotechnol.* **2014**, *32* (1), 40–51.
11. Downing, N. S.; Aminawung, J. A.; Shah, N. D.; Braunstein, J. B.; Krumholz, H. M.; Ross, J. S. Regulatory Review of Novel Therapeutics—Comparison of Three Regulatory Agencies. *N. Engl. J. Med.* **2012**, *366* (24), 2284–2293.
12. Scannell, J. W.; Blanckley, A.; Boldon, H.; Warrington, B. Diagnosing the Decline in Pharmaceutical R&D Efficiency. *Nat. Rev. Drug Discov.* **2012**, *11* (3), 191–200.
13. Dresser, R. Ethical Issues in Embryonic Stem Cell Research. *JAMA* **2001**, *285* (11), 1439–1440.
14. Aronson, J. K. Medication Errors: Definitions and Classification. *Br. J. Clin. Pharmacol.* **2009**, *67* (6), 599–604.
15. Nosengo, N. Can You Teach Old Drugs New Tricks? *Nature* **2016**, *534* (7607), 314–316.
16. Pollak, M. Metformin and Other Biguanides in Oncology: Advancing the Research Agenda. *Cancer Prev. Res.* **2010**, *3* (9), 1060–1065.
17. Pantziarka, P.; Bouche, G.; Meheus, L.; Sukhatme, V.; Sukhatme, V. P. Repurposing Drugs in Oncology (ReDO)—Cimetidine as an Anti-Cancer Agent. *Ecancermedicalscience* **2014**, *8*, 485.
18. Dong, L.; Hu, S.; Gao, J. Discovering Drugs to Treat Coronavirus Disease 2019 (COVID-19). *Drug Discov. Ther.* **2020**, *14* (1), 58–60.
19. Richardson, P. G.; Schlossman, R. L.; Weller, E.; Hideshima, T.; Mitsiades, C.; Davies, F.; LeBlanc, R.; Catley, L. P.; Doss, D.; Kelly, K. Immunomodulatory Drug CC-5013 Overcomes Drug Resistance and Is Well Tolerated in Patients with Relapsed Multiple Myeloma. *Blood J. Am. Soc. Hematol.* **2002**, *100* (9), 3063–3067.
20. Chen, H.; Engkvist, O.; Wang, Y.; Olivecrona, M.; Blaschke, T. The Rise of Deep Learning in Drug Discovery. *Drug Discov. Today* **2018**, *23* (6), 1241–1250.
21. Zarate, C. A.; Singh, J. B.; Carlson, P. J.; Brutsche, N. E.; Ameli, R.; Luckenbaugh, D. A.; Charney, D. S.; Manji, H. K. A Randomized Trial of an N-Methyl-D-Aspartate Antagonist in Treatment-Resistant Major Depression. *Arch. Gen. Psychiatry* **2006**, *63* (8), 856–864.
22. Patten, S. A.; Aggad, D.; Martinez, J.; Tremblay, E.; Petrillo, J.; Armstrong, G. A. B.; La Fontaine, A.; Maios, C.; Liao, M.; Ciura, S.; Wen, X.-Y.; Rafuse, V.; Ichida, J.; Zinman, L.; Julien, J.-P.; Kabashi, E.; Robitaille, R.; Korngut, L.; Parker, J. A.; Drapeau, P. Neuroleptics as Therapeutic Compounds Stabilizing Neuromuscular Transmission in Amyotrophic Lateral Sclerosis. *JCI Insight* **2017**, *2* (22), e97152. https://doi.org/10.1172/jci.insight.97152.

23. Caly, L.; Druce, J. D.; Catton, M. G.; Jans, D. A.; Wagstaff, K. M. The FDA-Approved Drug Ivermectin Inhibits the Replication of SARS-CoV-2 in Vitro. *Antiviral Res.* **2020**, *178*, 104787.
24. Zappacosta, A. R. Reversal of Baldness in Patient Receiving Minoxidil for Hypertension. *N. Engl. J. Med.* **1980**, *303* (25), 1480–1481.
25. Inglese, J.; Johnson, R. L.; Simeonov, A.; Xia, M.; Zheng, W.; Austin, C. P.; Auld, D. S. High-Throughput Screening Assays for the Identification of Chemical Probes. *Nat. Chem. Biol.* **2007**, *3* (8), 466–479.
26. Lanz, J.; Biniaz-Harris, N.; Kuvaldina, M.; Jain, S.; Lewis, K.; Fallon, B. A. Disulfiram: Mechanisms, Applications, and Challenges. *Antibiotics* **2023**, *12* (3), 524. https://doi.org/10.3390/antibiotics12030524.
27. Amaravadi, R.; Kimmelman, A. C.; White, E. Recent Insights into the Function of Autophagy in Cancer. *Genes Dev.* **2016**, *30* (17), 1913–1930.
28. Quan, Y.; Wang, Z.-Y.; Xiong, M.; Xiao, Z.-T.; Zhang, H.-Y. Dissecting Traditional Chinese Medicines by Omics and Bioinformatics. *Nat. Prod. Commun.* **2014**, *9* (9), 1934578X1400900. https://doi.org/10.1177/1934578X1400900942.
29. Johnson, S. C.; Rabinovitch, P. S.; Kaeberlein, M. mTOR Is a Key Modulator of Ageing and Age-Related Disease. *Nature* **2013**, *493* (7432), 338–345.
30. Wang, P. S. HMG-CoA Reductase Inhibitors and the Risk of Hip Fractures in Elderly Patients. *JAMA* **2000**, *283* (24), 3211. https://doi.org/10.1001/jama.283.24.3211.
31. Powe, D. G.; Voss, M. J.; Zänker, K. S.; Habashy, H. O.; Green, A. R.; Ellis, I. O.; Entschladen, F. Beta-Blocker Drug Therapy Reduces Secondary Cancer Formation in Breast Cancer and Improves Cancer Specific Survival. *Oncotarget* **2010**, *1* (7), 628–638. https://doi.org/10.18632/oncotarget.197.
32. Chong, C. R.; Sullivan Jr, D. J. New Uses for Old Drugs. *Nature* **2007**, *448* (7154), 645–646.
33. Burn, J.; Gerdes, A.-M.; Macrae, F.; Mecklin, J.-P.; Moeslein, G.; Olschwang, S.; Eccles, D.; Evans, D. G.; Maher, E. R.; Bertario, L. Long-Term Effect of Aspirin on Cancer Risk in Carriers of Hereditary Colorectal Cancer: An Analysis from the CAPP2 Randomised Controlled Trial. *The Lancet* **2011**, *378* (9809), 2081–2087.
34. Bowden, C. L.; Brugger, A. M.; Swann, A. C.; Calabrese, J. R.; Janicak, P. G.; Petty, F.; Dilsaver, S. C.; Davis, J. M.; Rush, A. J.; Small, J. G. Efficacy of Divalproex vs Lithium and Placebo in the Treatment of Mania. *JAMA* **1994**, *271* (12), 918–924.
35. Bothwell, L. E.; Greene, J. A.; Podolsky, S. H.; Jones, D. S. Assessing the Gold Standard — Lessons from the History of RCTs. *N. Engl. J. Med.* **2016**, *374* (22), 2175–2181. https://doi.org/10.1056/NEJMms1604593.
36. Andronis, C.; Sharma, A.; Virvilis, V.; Deftereos, S.; Persidis, A. Literature Mining, Ontologies and Information Visualization for Drug Repurposing. *Brief. Bioinform.* **2011**, *12* (4), 357–368.
37. Xue, H.; Li, J.; Xie, H.; Wang, Y. Review of Drug Repositioning Approaches and Resources. *Int. J. Biol. Sci.* **2018**, *14* (10), 1232.
38. Brown, A. S.; Patel, C. J. A Standard Database for Drug Repositioning. *Sci. Data* **2017**, *4* (1), 1–7.
39. Keiser, M. J.; Roth, B. L.; Armbruster, B. N.; Ernsberger, P.; Irwin, J. J.; Shoichet, B. K. Relating Protein Pharmacology by Ligand Chemistry. *Nat. Biotechnol.* **2007**, *25* (2), 197–206. https://doi.org/10.1038/nbt1284.
40. Shoichet, B. K. Virtual Screening of Chemical Libraries. *Nature* **2004**, *432* (7019), 862–865.
41. Schneider, G.; Baringhaus, K.-H. *Molecular Design: Concepts and Applications*; John Wiley & Sons, New York, 2008.
42. Anderson, A. C. The Process of Structure-Based Drug Design. *Chem. Biol.* **2003**, *10* (9), 787–797.
43. Ferreira, L. G.; Dos Santos, R. N.; Oliva, G.; Andricopulo, A. D. Molecular Docking and Structure-Based Drug Design Strategies. *Molecules* **2015**, *20* (7), 13384–13421.
44. Moffat, J. G.; Vincent, F.; Lee, J. A.; Eder, J.; Prunotto, M. Opportunities and Challenges in Phenotypic Drug Discovery: An Industry Perspective. *Nat. Rev. Drug Discov.* **2017**, *16* (8), 531–543.
45. Yella, J. K.; Yaddanapudi, S.; Wang, Y.; Jegga, A. G. Changing Trends in Computational Drug Repositioning. *Pharmaceuticals* **2018**, *11* (2), 57.
46. *Drugs | FDA*. https://www.fda.gov/drugs (accessed 2024-06-26).
47. *Drug Repurposing Hub | Broad Institute*. https://www.broadinstitute.org/drug-repurposing-hub (accessed 2024-06-26).
48. Mestres, J.; Gregori-Puigjane, E.; Valverde, S.; Sole, R. V. Data Completeness—the Achilles Heel of Drug-Target Networks. *Nat. Biotechnol.* **2008**, *26* (9), 983–984.
49. Parvathaneni, V.; Kulkarni, N. S.; Muth, A.; Gupta, V. Drug Repurposing: A Promising Tool to Accelerate the Drug Discovery Process. *Drug Discov. Today* **2019**, *24* (10), 2076–2085.
50. Damschroder, L. J.; Aron, D. C.; Keith, R. E.; Kirsh, S. R.; Alexander, J. A.; Lowery, J. C. Fostering Implementation of Health Services Research Findings into Practice: A Consolidated Framework for Advancing Implementation Science. *Implement. Sci.* **2009**, *4* (1), 50. https://doi.org/10.1186/1748-5908-4-50.
51. Stoller, J. K. The Challenge of Rare Diseases. *Chest* **2018**, *153* (6), 1309–1314.
52. Ching, T.; Himmelstein, D. S.; Beaulieu-Jones, B. K.; Kalinin, A. A.; Do, B. T.; Way, G. P.; Ferrero, E.; Agapow, P.-M.; Zietz, M.; Hoffman, M. M.; Xie, W.; Rosen, G. L.; Lengerich, B. J.; Israeli, J.; Lanchantin, J.; Woloszynek, S.; Carpenter, A. E.; Shrikumar, A.; Xu, J.; Cofer, E. M.; Lavender, C. A.; Turaga, S. C.; Alexandari, A. M.; Lu, Z.; Harris, D. J.; DeCaprio, D.; Qi, Y.; Kundaje, A.; Peng, Y.; Wiley, L. K.; Segler, M. H. S.; Boca, S. M.; Swamidass, S. J.; Huang, A.; Gitter, A.; Greene, C. S. Opportunities and Obstacles for Deep Learning in Biology and Medicine. *J. R. Soc. Interface* **2018**, *15* (141), 20170387. https://doi.org/10.1098/rsif.2017.0387.
53. Yıldırım, M. A.; Goh, K.-I.; Cusick, M. E.; Barabási, A.-L.; Vidal, M. Drug—Target Network. *Nat. Biotechnol.* **2007**, *25* (10), 1119–1126.
54. Sultana, J.; Cutroneo, P. M.; Crisafulli, S.; Puglisi, G.; Caramori, G.; Trifirò, G. Azithromycin in COVID-19 Patients: Pharmacological Mechanism, Clinical Evidence and Prescribing Guidelines. *Drug Saf.* **2020**, *43* (8), 691–698. https://doi.org/10.1007/s40264-020-00976-7.
55. Dudley, J. T.; Deshpande, T.; Butte, A. J. Exploiting Drug–Disease Relationships for Computational Drug Repositioning. *Brief. Bioinform.* **2011**, *12* (4), 303–311.
56. Ashburn, T. T.; Thor, K. B. Drug Repositioning: Identifying and Developing New Uses for Existing Drugs. *Nat. Rev. Drug Discov.* **2004**, *3* (8), 673–683.

57. Papadatos, G.; Gaulton, A.; Hersey, A.; Overington, J. P. Activity, Assay and Target Data Curation and Quality in the ChEMBL Database. *J. Comput. Aided Mol. Des.* **2015**, *29* (9), 885–896. https://doi.org/10.1007/s10822-015-9860-5.
58. Mullard, A. 2017 FDA Drug Approvals. *Nat. Rev. Drug Discov.* **2018**, *17* (2), 81–86.
59. Rusz, C.M.; HOsz, B.E.; Jîtcă, G.; Miklos, A.; Bătrînu, M.G.; Imre, S. Off-Label Medication: From a Simple Concept to Complex Practical Aspects. *Int. J. Environ. Res. Public. Health* **2021**, *18* (19), 10447.
60. Chaloner, C. Research Ethics in the Real World: Issues and Solutions for Health and Social Care. *Nurse Res.* **2009**, *16* (3), 87–88.
61. Roessler, H. I.; Knoers, N. V. A. M.; Van Haelst, M. M.; Van Haaften, G. Drug Repurposing for Rare Diseases. *Trends Pharmacol. Sci.* **2021**, *42* (4), 255–267. https://doi.org/10.1016/j.tips.2021.01.003.
62. Zisis, K.; Pavi, E.; Geitona, M.; Athanasakis, K. Real-World Data: A Comprehensive Literature Review on the Barriers, Challenges, and Opportunities Associated with Their Inclusion in the Health Technology Assessment Process. *J. Pharm. Pharm. Sci.* **2024**, *27*, 12302. https://doi.org/10.3389/jpps.2024.12302.
63. Hwang, T. J.; Carpenter, D.; Lauffenburger, J. C.; Wang, B.; Franklin, J. M.; Kesselheim, A. S. Failure of Investigational Drugs in Late-Stage Clinical Development and Publication of Trial Results. *JAMA Intern. Med.* **2016**, *176* (12), 1826. https://doi.org/10.1001/jamainternmed.2016.6008.
64. Küry, P.; Kremer, D.; Göttle, P. Drug Repurposing for Neuroregeneration in Multiple Sclerosis. *Neural Regen. Res.* **2018**, *13* (8), 1366. https://doi.org/10.4103/1673-5374.235242.

14 Biodiversity Ecological and Immuno-Informatics

Shifa Khan[1], Akanksha gupta[2], Priya singh[3], and Varsha Kashaw[4]
[1]Department of Pharmaceutical Chemistry Division, Adina College of Pharmacy, Bhopal Road, Sagar, India
[2]Department of Pharmaceutical Chemistry Division, Sanskriti College of Higher Education & Studies, Bhognipur, Kanpur, India
[3]UPMC Hillman Cancer Center, Division of Malignant Hematology and Medical Oncology, Department of Medicine, University of Pittsburgh School of Medicine, Pittsburgh, Pennsylvania
[4]Department of Pharmaceutical Chemistry Division, Sagar institute of Pharmaceutical Sciences, Narsinghpur Road, Sagar M.P.

INTRODUCTION

The current face of drug discovery is being revolutionized by the coming together of artificial intelligence (AI), computational chemistry, and bioinformatics. In the midst of this revolution, the fusion of biodiversity informatics, ecological informatics, and immuno-informatics as a potential interdisciplinary frontier is conspicuous. These fields, previously dealt with in isolation, are increasingly being combined through technological developments that facilitate large-scale data integration and predictive modeling and hence facilitate drug discovery and therapeutic innovation breakthroughs.

Biodiversity, encompassing the vast variability of life forms on the Earth, holds immense potential for pharmaceutical research. Natural compounds derived from diverse organisms have historically been the cornerstone of drug discovery, with more than half of all approved drugs being either natural products or their derivatives. However, the full potential of global biodiversity has remained largely untapped due to challenges in cataloging, accessing, and analyzing complex biological datasets. Biodiversity informatics, the discipline that merges data science and computational methods with biological diversity, is key to overcoming these constraints. It methodically processes species occurrences, genetic properties, ecological associations, and chemical structures and, in doing so, provides a valuable resource for drug discovery and ecological modeling.

Ecological informatics layers on more complexity and applicability by situating species within their ecological niches. Ecosystem-level interactions, including symbiotic associations, competition, or predator–prey relationships, can regulate gene expression and metabolite formation, thereby determining the bioactivity profiles of organisms. Stress-induced secondary metabolites in plants or microbial responses in specialized ecological niches frequently result in novel chemical scaffolds with potential therapeutic applications. AI-based ecological modeling tools now have the capability to include these dynamics to predict potential "hotspots" for bioprospecting [1–10].

Parallelly, immuno-informatics makes use of computational approaches to analyze the immune system at a molecular level. This comprises modeling of antigens, epitope prediction, vaccine design, and host–pathogen interactions based on genomics, transcriptomics, and proteomics data. Since numerous natural products exert immunomodulatory effects, associating immuno-informatics with biodiversity data offers novel opportunities for the identification of natural immunotherapeutics. The immune system, modified by evolutionary pressures and microbial exposures, also offers a special interface where ecological and immunological data meet.

AI and machine learning (ML) are the overarching paradigms within these fields, which allow for automated curation of data, pattern detection, and hypothesis formation. Deep learning models forecast plant metabolites to natural language processing (NLP) technologies that scan biodiversity publications, and AI shortens cycles of discovery and connects disparate disciplines. Computational chemistry further supports the biome by modeling how bioactive compounds interact with immune receptors or metabolic enzymes to enable drug screening at an early stage.

The convergence of biodiversity, ecological, and immuno-informatics into the AI–bioinformatics–computational chemistry triad is therefore not just a theoretical breakthrough—it is an overdue one. It enables scientists to read nature's pharmacopoeia with unprecedented precision and inform the discovery of new therapeutics from ecological and immunological findings. This chapter explores the convergences between these fields and their sum in revolutionizing drug discovery, with digital and biological convergence [11–16].

DOI: 10.1201/9781003464679-14

THE USE OF AI AND BIOINFORMATICS IN BIODIVERSITY AND ECOSYSTEM DATA

The digitalization of ecological and biodiversity data has opened up a new frontier for environmental and pharmaceutical research. AI and bioinformatics are now indispensable tools for handling, processing, and interpreting the huge, complex datasets created from natural ecosystems. Their use in biodiversity informatics improves our ability to comprehend ecosystem functionality and detect bioactive organisms of pharmaceutical interest [17,18].

Integration of AI and Biodiversity Informatics

Biodiversity informatics offers the technical basis for the organization of information on species distribution, taxonomy, genetic characteristics, and ecological relationships. Starting from specimen holdings in herbaria and natural history museums, biodiversity informatics has developed with the establishment of the Global Biodiversity Information Facility (GBIF) and other online platforms for the digitization of millions of biodiversity records.

AI algorithms have significantly enhanced the curation and retrieval of biodiversity data. Image recognition tools using convolutional neural networks (CNNs) now automate species identification from photos and herbarium scans, reducing the burden on taxonomists and improving annotation accuracy. For example, citizen science platforms such as iNaturalist and eBird use AI-powered models to validate public submissions of species sightings, transforming lay observations into high-quality biodiversity data.

Aside from taxonomy, ML models have the potential to reveal underlying patterns in environmental and biological variables governing species distributions. Predictive models such as MaxEnt and GARP use occurrence data and environmental variables to predict potential ranges of species, such as invasive weeds and endangered species. These models can be combined with metabolomic databases to analyze environmental factors with regard to the production of bioactive compounds, informing bioprospecting in understudied areas [19].

Ecological Informatics and Data-Driven Modeling

Ecological informatics is the use of computational methods to analyze the structure, function, and dynamics of ecosystems. It combines abiotic and biotic data from long-term ecological research (LTER), citizen science, remote sensing, and experiments. AI makes this process more efficient by allowing large-scale modeling and simulation of ecological processes such as species migration, community turnover, and ecosystem responses to climate change.

Ecological forecasting models based on AI have been utilized to evaluate the effects of climate and land-use change on biodiversity, usually demonstrating that loss of diversity is not linear but threshold-dependent and, on occasion, irreversible. Such findings are crucial for conservation efforts and determining ecological stresses that promote the evolution of new chemical defenses in microbes and plants—traits that can be exploited for therapeutic development.

AI also aids in the mapping of ecosystem services such as pollination, water filtration, and soil health. Through correlating land-use data with species characteristics, AI enables the predictions of how alterations in biodiversity affect ecosystem functioning, which further influences natural sources of pharmacologically useful compounds [20–22].

Environmental and Metagenomic Data Bioinformatics Pipelines

Omics-led bioinformatics also deepens our knowledge of biodiversity at the molecular level. eDNA sequencing enables scientists to identify organisms in soil, water, and air samples without any observation. Metagenomic tools decipher this information to determine the taxonomic and functional diversity of microbial communities, most of which synthesize metabolites of drug interest.

AI algorithms maximize sequence assembly, taxonomic assignment, and function prediction of genes from environmental samples. Such steps are essential for the identification of microbial strains or gene clusters responsible for new antibiotics, anticancer compounds, or immunomodulators. As illustrated by the Barcode of Life initiative, coupling DNA barcoding with voucher specimen databases speeds up the identification of cryptic species and their corresponding metabolic characteristics.

In addition, bioinformatics pipelines currently accommodate the integration of data from remote sensing and ecological surveys and produce comprehensive biodiversity maps capable of guiding regional drug prospecting. For example, integration of species distribution models with satellite-derived chlorophyll or temperature data can predict marine biodiversity areas that are dense with bioactive algae or symbiotic bacteria [23].

Semantic Ontologies and Interoperability Standards

For facilitating data interoperability, biodiversity and ecological databases are increasingly becoming dependent on semantic ontologies and metadata standards. The Darwin Core standard (DwC) and Ecological Metadata Language (EML) enable uniform data presentation between repositories. AI tools assist in annotating and converting legacy datasets into these formats so that they become machine-readable and integrable with drug discovery platforms.

Ontologies such as the Biological Collections Ontology (BCO) and Environment Ontology (ENVO) enable standardized tagging of species attributes, habitats, and sample protocols to facilitate multidomain data integration necessary for computational drug discovery. These tools enable researchers to automatically link ecological properties to molecular profiles with accelerated hypothesis-driven compound screening.

FROM ECOLOGY TO IMMUNITY—INFORMATICS-DRIVEN MOLECULAR INSIGHTS

The overlap between ecology and immunology—formerly disparate scientific domains—is increasingly connected via sophisticated informatics. This convergence is transforming the face of immuno-ecological science and pharmacological discovery. Ecological environments determine the evolutionary histories of organisms and, significantly, the immune tactics they employ. These tactics, typically embedded in secondary metabolites or immune genes, are now available for discovery via computational tools that connect ecological information with immuno-informatics platforms [24].

Ecological Pressure as a Motor of Immunological Creativity

In nature, organisms are continuously exposed to multitudes of pathogens, competitors, predators, and environmental stresses. These elements exert selective pressures that result in the evolution of defense mechanisms—most of which are mediated by bioactive molecules that possess antimicrobial, antiviral, or immunomodulatory activity. For example, plants growing in pathogen-dense or predator-dense environments frequently produce effective secondary metabolites that repel invaders or inhibit microbial infections.

Insects, amphibians, marine sponges, and even microbes that live in soil have been found to harbor distinctive biochemical arsenals hammered under ecological stress. These compounds present a valuable reservoir for pharmacological discovery. Ecological informatics enabled by AI allows the identification of environmental parameters—such as soil type, elevation, and species interactions—that are associated with the synthesis of these immune-critical compounds.

Immuno-Informatics: Computational Insights into Host Defense

Immuno-informatics uses computational methods to decipher the immune system at levels ranging from gene expression to antigen–antibody interactions. It includes epitope prediction, structural modeling of immune proteins, and simulation of host–pathogen interactions. When combined with ecological information, immuno-informatics provides deeper insights into how environmental stimuli modulate immune evolution and diversity.

For instance, epitope mapping software like NetMHCpan, when applied with the aid of environmental metagenomic data, can make predictions of likely immunogenic peptides from novel organisms, leading to the development of peptide-based drugs and vaccines. Such predictions are particularly valuable for the identification of natural compounds that mimic or interfere with immune function, including Toll-like receptor (TLR) agonists from marine life or plant-derived immunostimulants.

The integration of ecological and immunological informatics enables a systems biology framework for immune surveillance. Microbial diversity of the human gut, conditioned by dietary and environmental exposure, has direct implications for immune homeostasis and disease susceptibility. AI-facilitated metagenomic analysis bridges these microbial patterns with immunological markers, providing new opportunities in immunonutrition and gut-derived therapeutic discovery [25].

Connecting Biodiversity to Immune-Relevant Bioactivity

Computational chemistry and AI are used to screen molecular scaffolds from biodiversity databases for immunological potential. QSAR models, molecular docking simulations, and pharmacophore mapping are being used more and more with natural compounds isolated from ecologically different sites. These methods enable the prediction of interaction with immune receptors, including cytokine receptors, HLA proteins, or viral antigens, providing a preclinical evaluation of bioactivity prior to expensive laboratory verification.

Significantly, the combination of DNA barcoding information with immuno-informatics databases has the potential to accelerate species-specific bioactive profile screening. For example, correlating barcode information of medicinal plants and transcriptomic profiles under stress conditions identifies compounds most likely to be involved in immune modulation. Such a rich data environment allows AI to reveal patterns that may not be apparent to conventional approaches [26].

Functional Ecology Confronts Immuno-Bioinformatics

Functional ecology addresses how traits affect ecosystem function. Such traits, usually encoded by genes, can now be investigated with high-resolution omics technologies. Genes that are responsible for the biosynthesis of immune-acting molecules such as alkaloids, terpenoids, and peptides can be purified from functional metagenomic screens and assessed through immuno-informatics pipelines.

AI tools such as deep neural networks can predict immunogenicity, allergenicity, and toxicity profiles of new compounds derived from ecological datasets. Such predictive functions reduce drug pipeline attrition rates and facilitate more specific *in vitro* testing. As more ecological samples are being sequenced and annotated, the potential for identifying immune-active leads from nature gets larger by the day.

AI FOR IMMUNO-ECOLOGICAL INTERACTIONS AND DRUG TARGET DISCOVERY

The intersection of AI, immuno-informatics, and ecological modeling is reshaping the landscape of drug discovery by elucidating complex immuno-ecological relationships that would otherwise go undetected. These relationships, frequently context-dependent and multifaceted, are crucial to how organisms develop defense mechanisms and generate bioactive compounds. AI supplies the computational power to look at and forecast these interactions at scale, so that researchers can scan for promising drug targets from nature [27].

Deciphering Host–Pathogen–Environment Triads with AI

Pathogens co-evolve with their hosts in particular environmental settings within ecological systems. The triadic interaction dictates not only disease epidemiology, but also immune evasion and resistance molecular mechanisms. AI models combining host immune genetics, microbial diversity, and environmental factors provide new avenues for understanding these interactions. Supervised learning algorithms have been used, for instance, to forecast zoonotic spillover risk by connecting viral mutation profiles with host ecology and immune system traits.

Such AI-based ecological immunology models are precious in drug discovery. They assist in the identification of conserved or hypervariable areas in microbial genomes that are predicted to interact with the host immune system, thus informing the rational design of immunotherapeutics or vaccines. In addition, they assist in predicting how environmental changes—such as climate change or land-use changes—might affect pathogen prevalence and virulence, which is important for anticipatory drug development [28].

AI-Accelerated Identification of Immune-Modulating Natural Compounds

One of the most exciting uses of AI within the immuno-ecological paradigm is the forecast and ranking of natural products with the ability to modulate immunity. Leveraging enormous datasets developed from biodiversity informatics and ecological databases, AI can identify candidate organisms such as endophytic fungi, marine invertebrates, and desert plants that are evolutionarily geared to yield immune-relevant compounds.

Unsupervised learning methods such as clustering and dimensionality reduction facilitate the clustering of organisms sharing comparable chemical defense mechanisms, whereas deep learning models that have been trained on known immunomodulators can be used for screening compound libraries for new leads. For example, AI models have been able to predict cytokine-inducing activities of fungal extract polysaccharides and alkaloids from rainforest plants.

These forecasts are then optimized with molecular docking and molecular dynamics simulations, which mimic the interaction of the predicted compound with immune receptors such as MHC complexes, TLRs, and cytokine-binding domains. In most instances, AI also assists in designing analogs of natural products with greater affinity and lower toxicity—a critical step in medicinal chemistry pipelines [29].

Integrative AI Platforms for Target Prediction and Screening

Increased access to integrative platforms makes possible cross-talk among biodiversity data, immuno-informatics resources, and AI-based screening pipelines. Such platforms as VoSeq and the Barcode of Life Data System (BOLD) enable taxonomically associated molecular information that can be integrated into AI pipelines for immunological screening. These platforms support:

- Target identification by relating taxa to established or predicted bioactivities.
- Compound screening by applying structure–activity relationship (SAR) models with training sets from ecological libraries of compounds.
- Pathway analysis by projecting compound interactions onto immune signaling pathways with systems biology models.

This systems-based strategy expedites drug discovery by detecting not only bioactive molecules but also biochemical targets and downstream pathways that they affect. AI algorithms learned on multi-omics data—such as metagenomics, transcriptomics, and metabolomics—can even forecast off-target actions or synergistic pairings, lowering downstream attrition levels [30,31].

Biodiversity-Informed AI in Emerging Therapeutic Areas

AI-based biodiversity prospecting is also particularly useful in new therapeutic categories such as cancer immunotherapy, antimicrobial resistance (AMR), and neuroimmunology. For instance:

- **Cancer Immunotherapy:** Natural checkpoint inhibitors or marine or microbial ecosystem-derived adjuvants are assayed by AI models trained against immune evasion signatures from tumor microenvironments.
- **Antimicrobial Resistance:** AI frameworks combine pathogen phylogenetics, resistance patterns, and ecological sources to predict new antimicrobial peptides (AMPs) from distant locales such as deep-sea hydrothermal vents or tropical rainforests.
- **Neuroimmunology:** Certain neuroactive agents, such as anti-inflammatory drugs against the blood–brain barrier, have been found via ecological screening of soil bacteria and fungi—organisms that thrive in competitive environments by modulating host immunity.

The combination of AI and ecological and immunological datasets is more than a tool for analysis—it is an enabling strategy for directed, ethical, and effective drug discovery from nature. It reduces ecological disruption while optimizing therapeutic innovation, satisfying both the objectives of precision medicine and sustainable bioprospecting [32].

OMICS DATA INTEGRATION—BRIDGING BIODIVERSITY WITH IMMUNO-INFORMATICS

The convergence of omics technologies with immuno-informatics and biodiversity has revolutionized the discovery platform for bioactives and immunomodulatory compounds. Omics—genomics, transcriptomics, proteomics, metabolomics, and metagenomics—enables scientists to decipher biological complexity from molecules to ecosystems. Combining these datasets in a single platform using AI and bioinformatics, scientists can relate ecological diversity with immune system modulation, thus providing a powerful new paradigm for drug discovery [33].

Genomics and Metagenomics in Natural Product Discovery

Genomic and metagenomic studies lie at the basis of learning about how biodiversity converts into biochemical potential. Plant, animal, and microbial genomes frequently harbor biosynthetic gene clusters (BGCs) that encode for the generation of secondary metabolites with pharmaceutical activities. Environmental metagenomics, especially of extreme or unusual environments, discovers gene signatures from uncultivated organisms that are likely to contain new immune-related compounds.

AI software supports mining large genomic datasets for structure predictions of encoded metabolites and functional annotations. Computer programs such as antiSMASH and PRISM predict BGCs, while ML classifiers evaluate their potential to yield compounds with immune significance—such as AMPs, immunosuppressants, and cytokine modulators.

Furthermore, ecological genomics—sequencing genomes in the context of environmental parameters—uncovers how selective pressures shape biosynthetic pathways. For instance, soil microbiomes from drought-prone areas tend to express stress-related metabolites, including molecules that modulate host immunity when applied in therapeutic contexts [34].

Transcriptomics and Epigenomics in Ecological Adaptation and Immune Regulation

Transcriptomics yields information on gene expression patterns under different ecological and immunological challenges. Plants and microbes overexpress certain gene sets upon exposure to pathogens or abiotic stress, which in turn is generally correlated with the production of bioactive compounds. Differential gene expression analysis using AI identifies candidate genes that potentially encode immune-modulating metabolites.

Epigenomic changes—e.g., DNA methylation and histone acetylation—contribute to environmental adaptation and affect secondary metabolite production. Through the integration of epigenomic and transcriptomic information, scientists can simulate regulatory circuits responsible for triggering bioactive molecule expression in ecological niches, providing a guide for lab-based compound induction or for synthetic biology applications.

These datasets are critical in immuno-informatics to comprehend how environmental exposure influences immune gene regulation in both host populations and microbial populations. For example, symbiotic microbes can epigenetically alter host immune responses, a process now trackable with combined omics platforms and predictive AI models [35].

Proteomics and Immunopeptidomics for Mechanism Elucidation

Proteomic profiling, particularly in combination with mass spectrometry, facilitates the description of bioactive proteins and peptides of natural origin. Immunopeptidomics, a branch of proteomics, aims to study peptides presented by MHC molecules or secreted in immune cascades. Such methods are being employed to an increasing extent in the discovery of therapeutic peptides from plants, fungi, marine sponges, and microbiota.

AI supports this procedure by detecting conserved protein domains, peptide–MHC binding affinity prediction, and antigen–antibody modeling. For instance, deep learning algorithms trained on immune epitope datasets can filter novel peptides from natural ecological sources as possible vaccines or adjuvants. These assist in speeding up early-phase immunotherapeutic research through structure–function relationships prior to clinical trials.

In conjunction with biodiversity information, proteomics assists in linking particular taxa or environmental conditions with the synthesis of immunologically active proteins. For instance, a particular marine snail species under predation stress will overexpress toxin proteins with anti-inflammatory functions that can be pharmaceutically exploited [36].

Metabolomics and Chemoinformatics in Compound Discovery

Metabolomics interprets the entire range of small molecules that organisms produce. These molecules tend to mediate immunity, stress responses, and ecological interactions. The combination of metabolomics, cheminformatics, and AI results in the rapid classification, annotation, and prediction of the activity of intricate chemical mixtures in ecological samples [37].

ML algorithms trained on metabolite–activity databases can predict immunomodulatory, anti-inflammatory, or anticancer activities of novel natural products. Principal component analysis and t-SNE visualization methods uncover chemical diversity patterns associated with ecological function and species-specific biosynthetic potential.

Furthermore, AI-aided dereplication methods prevent the rediscovery of established compounds, as research is directed toward truly novel entities. Coupling with biodiversity databases also ranks distinct species or habitats for focused bioprospecting, e.g., alpine plant endophytes and volcanic soil thermophilic bacteria [38].

Multi-Omics Platforms for Integrated Immuno-Ecological Insight

Multi-omics integration—integration of genomics, transcriptomics, proteomics, and metabolomics—gives a top-down vision of biological systems and their immune interfaces. AI-fueled systems biology makes the following possible:

- **Pathway Reconstruction:** Connecting genetic information with enzyme expression and metabolite production pathways.
- **Mechanism Elucidation:** Anticipating how environmental factors induce immune-active compound synthesis.
- **Drug Discovery Pipelines:** Linking ecological diversity to molecular targets and screening candidates via immuno-informatics.

Examples of such integration include AI models that combine marine metagenomics, proteomics of symbiotic bacteria, and host immunity profiles to discover new antibiotics and anti-inflammatory agents.

Omics integration also underpins emerging fields such as "eco-immunology," where the focus is not only on how organisms defend themselves but also how their immune systems evolve under ecological constraints—a fertile ground for therapeutic mining [38].

CASE STUDIES AND APPLICATIONS IN DRUG DISCOVERY

The intersection of immuno-informatics, ecological informatics, and biodiversity—facilitated by AI and omics technologies—has transitioned from theory to reality in drug discovery. Actual case studies now illustrate how unified data systems can detect new bioactive compounds, confirm drug targets, and accelerate the therapeutic pipeline. These examples highlight the applied potential of cross-disciplinary informatics in contemporary pharmaceutical development.

Identification of Immunomodulators from Marine Biodiversity

Marine ecosystems are one of the richest sources of pharmacologically active compounds. Secondary metabolites of varying bioactivities, such as immune modulation, are produced by sponges, corals, tunicates, and marine bacteria. One classical example is the discovery of the powerful analgesic ziconotide, isolated from the cone snail *Conus magus*. Its discovery was based on peptide sequencing, ecological behavior study, and neuroimmune profiling.

Researchers now use AI to screen metagenomic data from marine microbiomes to find BGCs linked to new non-ribosomal peptides and polyketides. Immuno-informatics pipelines then make predictions of MHC binding, cytokine modulation, or anti-inflammatory inhibitory potential.

In one such experiment, peptide toxins isolated from marine sponges were found to have high TLR-4 inhibition, a major pathway of innate immunity, which indicated their utility for autoimmune disease, and the findings were fueled by AI-directed docking and deep learning-based epitope prediction models [39].

Forest Ecosystem Metabolites and Cancer Immunotherapy

Biodiverse forest ecosystems, especially tropical rainforests and boreal biomes, contain a rich diversity of medicinal plants and fungi. Ranging from *Taxus brevifolia* (where paclitaxel is found) to *Camptotheca acuminata* (used to produce camptothecin), ecological species have provided life-saving anticancer drugs.

Based on this heritage, recent research utilized transcriptomics and AI modeling to search for plant species whose secondary metabolites complement immune checkpoint inhibition—a pillar of cancer immunotherapy. One AI platform employed molecular fingerprints and expression data to screen more than 3,000 plant extracts for compounds that are synergistic with PD-1/PD-L1 blockade.

In addition, researchers have used multi-omics profiling of forest fungi to identify new immunomodulatory β-glucans. AI has been used to correlate fungal phylogenetic characteristics with bioactivity, with a view to prioritizing strains for laboratory testing and identifying candidate molecules for adjunctive cancer immunotherapy [40,41].

Microbiome-Derived Metabolites and Anti-Inflammatory Drugs

Soil microbiomes and gut-associated bacteria are yet another frontier in the discovery of natural compounds. Advancements in AI and immuno-informatics have been instrumental in connecting microbial ecology to host immune modulation. For instance, *Faecalibacterium prausnitzii*, an important anti-inflammatory gut bacterium, was associated with the generation of metabolites that inhibit NF-κB signaling, an important inflammatory pathway.

Scientists employed metagenomic assembly and ML-based functional annotation to detect target gene clusters that account for such immune-suppressive activity. Predictive immuno-informatics tools validated the downregulation of IL-6 and TNF-α, supporting the anti-inflammatory activity of the compound. Such findings are of direct interest to diseases such as IBD and rheumatoid arthritis [42].

Citizen Science and AI in Drug Candidate Prioritization

Citizen science projects, when integrated with biodiversity informatics and AI, assist with new candidate identification. iNaturalist and eBird have become goldmines of geotagged biodiversity records. Researchers associated these records with biogeographic patterns of plant species with a history of traditional medicinal uses.

In a project, AI models learned from the metadata of citizen-provided images to prioritize understudied plants of South Asia for pharmacological screening. Its integration with immuno-informatics tools gave rise to the identification of flavonoid compounds of *Cajanus cajan* with the possibility of IL-17 inhibition—a target in autoimmune diseases like psoriasis.

This crowdsourced data, automated annotation, and AI-prioritized model offers a cost-effective framework for low-cost drug discovery, particularly in resource-poor biodiversity-dense areas [43].

Integrative Frameworks in Drug Target Discovery

There are now several integrated platforms that implement the synergy between immune science, ecology, and biodiversity:

- **The Global Natural Products Social Molecular Networking (GNPS):** It is an open-source knowledge base in which scientists upload and compare the spectra of natural compounds. AI supports the clustering and annotation of unidentified metabolites.
- **The Barcode of Life Data Systems (BOLD):** It relates taxonomic DNA barcodes to ecological and phenotypic information, allowing researchers to trace the bioactivity of individual species and their chemical products.
- **ImmunomeBrowser:** With the use of environmental genomics platforms, this software enables the superimposing of epitopes onto ecological peptide libraries to expedite immune-targeted discovery.

These platforms are the backbone of contemporary eco-immuno-informatics and are being increasingly utilized in pharmaceutical pipelines to discover and validate new compounds and targets.

The case studies above illustrate how ecological diversity, when run through AI, omics, and immuno-informatics pipelines, can lead to real-world therapeutic advances. These applications show the unexploited potential of global biodiversity as a strategic asset for combating emerging and resistant diseases [44].

DATA STANDARDS, REPOSITORIES, AND INTEROPERABILITY

To tap the full potential of biodiversity, ecological, and immuno-informatics in drug development, strong data standards and interoperating systems are essential. Merging omics data, ecological data, and immunological data demands effortless data exchange, standardized annotation, and machine-processability. This section describes the most important frameworks, databases, and guidelines that make this possible, comprising the digital infrastructure for eco-immuno-bioinformatics research.

The Role of Standards in Biodiversity and Ecological Data Sharing

The biodiversity research community has taken up a number of structured data formats for standardizing species information, taxonomic metadata, and ecological observations. Among the most notable is DwC, which establishes a consistent vocabulary for publishing species occurrence data, such as event date, location, scientific name, and catalog number. DwC is employed extensively by international biodiversity platforms such as GBIF and the Brazilian Marine Biodiversity Database.

Concurrently, EML supports the documentation of ecological datasets. It provides rich information describing data provenance, sampling, spatiotemporal location, and taxonomic coverage. Software such as Morpho and Metacat assist investigators in creating EML-conformant metadata, enabling reproducibility and interpretation of data.

These standards enable biodiversity and ecological data to be made interoperable with omics platforms and drug discovery tools. AI models that have been trained on standardized inputs can better evaluate species distribution patterns, forecast biosynthetic potentials, and direct bioprospecting campaigns [45].

Ontologies and Semantic Integration

Ontologies are important in the provision of a framework to ensure that data from heterogeneous origins is consistently understood by machines. BCO and ENVO offer a controlled vocabulary for the description of biological materials, sampling events, habitats, and ecological features.

Ontologies are utilized by AI-driven semantic integration software to connect biological samples with environmental context, phenotypic characteristics, and molecular profiles. For instance, a fungal sample represented with BCO terms can be computationally associated with proteomic data exhibiting immunomodulatory activities, facilitating systems-level discovery platforms.

By incorporating semantic annotations in actual data repositories, scientists can automate queries like: "Find all ocean-derived bacteria with predicted cytokine inhibitory peptides isolated from low-salinity environments." Intelligent search capability of this sort is feasible only using ontology-driven data structuring [46].

Cross-Domain Data Interoperability Supporting Repositories

Multiple open-access repositories allow the storage and integration of biodiversity, ecological, and immunology data:

- **GBIF:** It houses more than a billion species occurrence records, digitized specimens, and linked metadata, adheres to DwC standards, and has API access for programmatic integration.
- **BOLD:** It hosts DNA barcode sequence data and taxonomic, geographic, and ecological metadata to facilitate genotype linkage to ecological niches and traits.
- **EMBL-EBI and NCBI:** They host genomics, transcriptomics, and proteomics data with immune annotations, such as immune epitope databases like IEDB.
- **DataONE:** It is an open-access platform that aggregates ecological and environmental data with EML-compliant metadata, version control, and long-term data preservation.
- **GNPS:** It involves sharing and MS/MS spectra annotation from natural products, combined with ecological metadata for compound source tracking.

These repositories, once made accessible through semantic layers and AI-prepared APIs, allow multidimensional analysis cutting across ecology, immunity, and pharmacology.

AI-Powered Data Curation and Quality Control

Big data sharing becomes meaningful only if data quality is ensured. AI methods now participate centrally in the automated tasks of curation, including:

- **Species Identification:** Applying image recognition models to scans of herbarium sheets and citizen science photos.
- **Geo-Referencing Validation:** Verifying coordinate mismatches with tools such as BioGeomancer.
- **Reconciliation Naming:** Using fuzzy string matching (e.g., Levenshtein distance, TaxaMatch) to fix taxonomic misspelling.
- **Metadata Completeness Scoring:** Scoring EML or DwC datasets for missing essential attributes.

Platforms such as Kurator and the Data Quality Framework establish reproducible pipelines for data quality determination and reporting, embedded into biodiversity and ecological informatics platforms.

Ethical and Legal Considerations in Data Sharing

Cross-border bioprospecting and ecological exploration raise concerns regarding ownership, access to data, and benefit-sharing. The Nagoya Protocol and the Convention on Biological Diversity (CBD) offer legal frameworks for the fair access and utilization of biological resources. Data repositories now incorporate metadata fields for maintaining permits, provider nations, and terms of use, facilitating legal compliance and transparency.

Moreover, AI can assist in monitoring compliance by marking records with missing or incongruent ownership metadata. Data traceability systems based on blockchain technology are also being investigated to ensure integrity in the global utilization of bioresources.

The development and upkeep of open, interoperable, and high-quality databases that are supported by AI, ontologies, and global policies make up the digital backbone for the combination of biodiversity, ecology, and immunology in drug discovery. They ensure that new knowledge from nature is made accessible, verifiable, and ethically shared across borders and disciplines.

CHALLENGES, OPPORTUNITIES, AND THE ROAD AHEAD

The integrated use of biodiversity, ecological, and immuno-informatics driven by AI and omics technologies has extraordinary prospects in drug discovery. At the same time, the convergence also encounters enormous challenges in relation to data heterogeneity, limited accessibility, ethical issues, and the lack of standardized practices. Overcoming these challenges is of utmost importance for converting nature's molecular library full of therapeutic potential into a clinically effective arsenal of drugs.

Current Challenges in Eco-Immuno-Bioinformatics Integration

a. **Fragmentation and Inconsistency of Data:** Even with improvements in data digitization, biodiversity and ecological data continue to be fragmented over various institutions and formats. Separate repositories employ different taxonomic references, spatial resolutions, or data schemas, complicating integration without heavy cleaning and normalization.
b. **Incomplete or Biased Sampling:** Ecological and biological information tends to be biased toward well-surveyed areas or groups (e.g., mammals and flowering plants), with numerous ecosystems and microbial organisms understudied. This "Wallacean shortfall" hinders AI models from generalizing or identifying new patterns.
c. **Limited Availability of High-Quality Omics Data:** Omics technologies, even though powerful, are still costly and infrastructure-dense. Most biodiversity hotspots in the Global South cannot afford to carry out large-scale metagenomic, transcriptomic, or proteomic analyses, creating knowledge disparity.

d. **Ethical and Regulatory Challenges:** Global laws like the Nagoya Protocol, though important for fairness, build up intricate legal environments that hinder or slow down the exchange of biodiversity information and assets. Ethical issues also come into play in the utilization of indigenous knowledge, demanding open systems for benefit-sharing.
e. **Absence of Interdisciplinary Expertise:** Proper integration of ecology, biodiversity, and immunology needs to be done in collaboration with biologists, ecologists, chemists, immunologists, data scientists, and legal professionals. Interdisciplinary groups are uncommon and usually hampered by differences in language, methodology, and incentives.

Potential for Drug Innovation and Scientific Advances

a. **New Drug Leads from Unexplored Ecosystems:** Uncharacterized microbes, extremophiles, and orphan plants probably possess novel secondary metabolites with unexplored medicinal properties. AI can direct focused bioprospecting in undersampled environments such as deep seas, deserts, and high-altitude forests.
b. **Immunotherapy and Personalized Medicine:** Through the association of omics information with ecological metadata and patient-specific immune signatures, personalized immunotherapies can be constructed from natural products. This is particularly exciting in cancer, autoimmune disorders, and novel infections.
c. **Predictive Ecology and Climate Resilience:** AI algorithms integrating climate, ecological, and immunological information can forecast the impact of global changes on disease emergence, microbial resistance, and loss of biodiversity—offering pre-emptive drug targets and conservation strategies.
d. **Open Data Platforms and Citizen Science:** Validated crowdsourced biodiversity data, using AI, can underpin monitoring at the global scale of species, ecosystem shifts, and candidate therapeutic agents, particularly in remote or politically intractable areas.
e. **AI as a Scientific Equalizer:** AI can democratize drug discovery by cutting the demand for costly laboratory work during initial discovery. Through open-source tools and international datasets, researchers in low-resource environments can meaningfully contribute to natural product science.

Future Directions and Recommendations

1. **Establish Global Interoperability Protocols:** Efforts such as the Biodiversity Information Standards (TDWG) and Global Biodata Coalition need to be extended to cover immuno-informatics and omics metadata, providing genuine cross-domain integration.
2. **Invest in Global South Data Infrastructure:** Governments and funding agencies need to invest in the digitization of natural collections and omics capacity building in biodiversity hotspots to provide a level playing field for global discovery participation.
3. **Encourage Interdisciplinary Training and Collaboration:** Research institutions and universities should develop joint programs in ecology, bioinformatics, and immunology, giving birth to a new generation of "eco-pharma-informaticians."
4. **Harmonize Ethical and Legal Frameworks:** There is a need for clear global standards on data use, indigenous knowledge, benefit-sharing, and AI-driven bioprospecting in order to guarantee innovation without exploitation.
5. **Use AI to Monitor Real-Time Biodiversity and Health:** Edge-AI devices, drones, and sensor networks can be used to facilitate real-time ecological and disease surveillance to connect ecosystem health with pharmacological warning alerts directly.

CONCLUSION

The convergence of biodiversity, ecological, and immuno-informatics with AI, computational chemistry, and bioinformatics constitutes a revolutionary paradigm in drug discovery. Utilizing the enormous genetic and chemical diversity that exists in nature coupled with computational prowess, researchers can identify novel therapeutic candidates at an accelerated pace with higher precision. Ecological informatics improves our knowledge of environmental settings that influence biological functions, and immuno-informatics gives us essential insights into immune processes crucial for vaccine and immunotherapy development. The convergence of these two fields enables drug discovery pipelines by enhancing modeling, simulation, and prediction capacities. Looking ahead, adopting this multidisciplinary approach is necessary to combat the intricacies of human diseases and to speed up the progress toward creating safer, more effective medicines, ultimately enhancing health outcomes around the world.

REFERENCES

1. Osawa, T. (2019). Perspectives on biodiversity informatics for ecology. *Ecological Research*, 34(4), 446–456. https://doi.org/10.1111/1440-1703.12023.
2. Bisby, F. A. (2000). The quiet revolution: Biodiversity informatics and the internet. *Science*, 289, 2309–2312.
3. Edwards, J. L. (2000). Interoperability of biodiversity databases: Biodiversity information on every desktop. *Science*, 289(5488), 2312–2314. https://doi.org/10.1126/science.289.5488.2312.

4. Canhos, V., de Sausa, S., De Giovanni, R., & Canhos, D. (2004). Global biodiversity informatics: Setting the scene for a “new world” of ecological modeling. *Biodiversity Informatics*, 1, 1–13. https://doi.org/10.17161/bi.v1i0.3.
5. Soberón, J., & Peterson, A. T. (2004). Biodiversity informatics: Managing and applying primary biodiversity data. *Philosophical Transactions of the Royal Society B: Biological Sciences*, 359(1444), 689–698. https://doi.org/10.1098/rstb.2003.1439.
6. Berendsohn, W. G., Güntsch, A., Hoffmann, N., Kohlbecker, A., Luther, K., & Müller, A. (2011). Biodiversity information platforms: From standards to interoperability. *ZooKeys*, 150, 71–87. https://doi.org/10.3897/zookeys.150.2166.
7. Salski, A., & Recknagel, F. (2003). *Ecological Informatics: Understanding Ecology by Biologically-Inspired Computation*. Springer, Cham.
8. Osawa, T., & Totsu, K. (2017). Darwin core archive, the standard data format for biodiversity information and its extension for ecological data “sample-based data”. *Japanese Journal of Conservation Ecology*, 22, 371–381.
9. Troudet, J., Vignes-Lebbe, R., Grandcolas, P., & Legendre, F. (2018). The increasing disconnection of primary biodiversity data from specimens: How does it happen and how to handle it. *Systematic Biology*, 67(6), 1110–1119. https://doi.org/10.1093/sysbio/syy044.
10. Watanabe, K. (2016). Biodiversity information and natural history museum. *Japanese Journal of Ecology*, 66, 247–252.
11. Kaufmann, S. H. E. (2019). Immunology’s coming of age. *Frontiers in Immunology*, 10, 684. https://doi.org/10.3389/fimmu.2019.00684.
12. Kimbrell, D. A., & Beutler, B. (2001) The evolution and genetics of innate immunity. *Nature Reviews Genetics*, 2, 256–267.
13. Segel, L. A. (2001) Controlling the immune system: Diffuse feedback via a diffuse informational network. *Novartis Foundation Symposia*, 239, 31–40, discussion 40–51.
14. Gardy, J. L., Lynn, D. J., Brinkman, F. S. L., & Hancock, R. E. W. (2009) Enabling a systems biology approach to immunology: Focus on innate immunity. *Trends in Immunology*, 30, 249–262.
15. Davies, M. N., & Flower, D. R. (2007) Harnessing bioinformatics to discover new vaccine. *Drug Discovery Today*, 12, 389–395.
16. Lopez-Campos, G., Bermejo-Martin, J. F., Almansa, R., & Martin-Sanchez, F. (2014). Immunoinformatics and systems biology in personalized medicine. *Immunoinformatics*, 662, 457–475. https://doi.org/10.1007/978-1-4939-1115-8.
17. Gupta, R., Srivastava, D., Sahu, M., Tiwari, S., Ambasta, R. K., & Kumar, P. (2021) Artificial intelligence to deep learning: Machine intelligence approach for drug discovery. *Molecular Diversity*, 25, 1315–1360. https://doi.org/10.1007/s11030-021-10217-3.
18. Paul, D., Sanap, G., Shenoy, S., Kalyane, D., Kalia, K., & Tekade, R. K. (2021) Artificial intelligence in drug discovery and development. *Drug Discovery Today*, 26(1), 80–93. https://doi.org/10.1016/j.drudis.2020.10.010.
19. Balke, M., Schmidt, S., Hausmann, A., Toussaint, E., Bergsten, J., Buffington, M., & Hobern, D. (2013). Biodiversity into your hands – A call for a virtual global natural history “metacollection”. *Frontiers in Zoology*, 10(1), 55. https://doi.org/10.1186/1742-9994-10-55.
20. Chon, T.-S., & Park, Y.-S. (2006). Ecological informatics as an advanced interdisciplinary interpretation of ecosystems. *Ecological Informatics*, 1(3), 213–217. https://doi.org/10.1016/j.ecoinf.2006.02.007.
21. Recknagel, F. (2017) Ecological informatics. In: Gibson D (ed) *Oxford Bibliographies in Ecology*. Oxford University Press, New York. https://www.oxfordbibliographies.com/view/document/ obo-9780199830060/obo-9780199830060-0174.xm.
22. Porter, J., & Lin, C. C. (2017). Ecoinformatics: A new paradigm of ecology. In Chao JT, King HB (eds) *Ecology: Concept and Application Version 7 Chinese Translation*, pp. 555–568. McGraw Hill Edcation, New York.
23. Abraham, G., & Inouye, M. (2015) Genomic risk prediction of complex human disease and its clinical application. *Current Opinion in Genetics and Development*, 33, 10–16. https://doi.org/10.1016/j.gde.2015.06.005.
24. Nikolich-Žugich, J. (2018) The twilight of immunity: Emerging concepts in aging of the immune system. *Nature Immunology*, 19, 10–19. https://doi.org/10.1038/s41590-017-0006-x.
25. Vivona, S., Gardy, J. L., Ramachandran, S., Brinkman, F. S., Raghava, G. P., Flower, D. R., & Filippini, F. (2008) Computer-aided biotechnology: From immuno-informatics to reverse vaccinology. *Trends in Biotechnology*, 26(4), 190–200.
26. Ranjbar, M. M., Ebrahimi, M. M., Shahsavandi, S., Farhadi, T., Mirjalili, A., Tebianian, M., & Motedayen, M. H. (2019) Novel applications of immuno-bioinformatics in vaccine and bio-product developments at research institutes. *Archives of Razi Institute*, 74(3), 219–233.
27. Glenszczyk, M., Lis, A., Porc, W., Pacwa-Płociniczak, M., & Babczyńska A. (2005) The apple of discord: Can spider cocoons be equipped with antimicrobial factors?—a systematic review. *Frontiers in Zoology*, 22(1), 1–7.
28. Yakimovich, A. (2021) Machine learning and artificial intelligence for the prediction of host–pathogen interactions: A viral case. *Infection and Drug Resistance*, 14, 3319–3326.
29. Ogidi, O. I., Alfred-Ugbenbo, D., & Poripo, B. E. (2005) AI-driven drug discovery, design, and development in immunological disorders. In Tonika B, Chandra P (eds) *AI-Assisted Computational Approaches for Immunological Disorders*, pp. 143–180. IGI Global Scientific Publishing, New York.
30. Ali, H. (2023) Artificial intelligence in multi-omics data integration: Advancing precision medicine, biomarker discovery and genomic-driven disease interventions. *International Journal of Science and Research Archive*, 8(1), 1012–1030.
31. Ahmed, Z., Mohamed, K., Zeeshan, S., & Dong X. (2020) Artificial intelligence with multi-functional machine learning platform development for better healthcare and precision medicine. *Database*, 2020, baaa010.
32. Ullah, F., Saqib, S., & Xiong, Y. C. (2025) Integrating artificial intelligence in biodiversity conservation: Bridging classical and modern approaches. *Biodiversity and Conservation*, 34(1), 45–65.
33. Srivastava, P., & Garg, P. (eds.) (2024) *Omics Applications and Avenues*. Cambridge Scholars Publishing, Cambridge.
34. Medema, M. H., & Fischbach, M. A. (2015) Computational approaches to natural product discovery. *Nature Chemical Biology*, 11(9), 639–648.
35. Tuong, Z. K., Stewart, B. J., Guo, S. A., & Clatworthy, M. R. (2022) Epigenetics and tissue immunity—Translating environmental cues into functional adaptations. *Immunological Reviews*, 305(1), 111–136.
36. Olsson, N., Heberling, M. L., Zhang, L., Jhunjhunwala, S., Phung, Q. T., Lin, S., Anania, V. G., Lill, J. R., & Elias, J. E. (2021) An integrated genomic, proteomic, and immunopeptidomic approach to discover treatment-induced neoantigens. *Frontiers in Immunology*, 12, 662443.

37. Fernandes, M., Sanches, B., & Husi, H. (2019) Cheminformatics and computational approaches in metabolomics. *Computational Biology*, 9, 143–160.
38. Ndoja, S. (2022) The components of puzzle of existential fear: An integrated neuro-psycho-ecological model of understanding experiences during the covid-19 pandemics. *Acta Scientific Neurology*, 5(2), 11–22.
39. Cutolo, E. A., Campitiello, R., Caferri, R., Pagliuca, V. F., Li, J., Agathos, S. N., & Cutolo, M. (2024) Immunomodulatory compounds from the sea: From the origins to a modern marine pharmacopoeia. *Marine Drugs*, 22(7), 304.
40. Martinez-Outschoorn, U. E., Bartrons, M., & Bartrons, R. (2019) Cancer ecosystems. *Frontiers in Oncology*, 9, 718.
41. Li, Q., Morimoto, K., Kobayashi, M., Inagaki, H., Katsumata, M., Hirata, Y., Hirata, K., Suzuki, H., Li, Y. J., Wakayama, Y., & Kawada, T. (2008) Visiting a forest, but not a city, increases human natural killer activity and expression of anti-cancer proteins. *International Journal of Immunopathology and Pharmacology*, 21(1), 117–127.
42. Mafe, A. N., & Büsselberg, D. (2025) The effect of microbiome-derived metabolites in inflammation-related cancer prevention and treatment. *Biomolecules*, 15(5), 688.
43. Ponti, M., & Seredko, A. (2022) Human-machine-learning integration and task allocation in citizen science. *Humanities and Social Sciences Communications*, 9(1), 1–5.
44. Duarte, Y., Márquez-Miranda, V., Miossec, M. J., & González-Nilo, F. (2019) Integration of target discovery, drug discovery and drug delivery: A review on computational strategies. *Wiley Interdisciplinary Reviews: Nanomedicine and Nanobiotechnology*, 11(4), e1554.
45. Michener, W. K. (2015) Ecological data sharing. *Ecological Informatics*, 29, 33–44.
46. Walls, R. L., Deck, J., Guralnick, R., Baskauf, S., Beaman, R., Blum, S., Bowers, S., Buttigieg, P. L., Davies, N., Endresen, D., & Gandolfo, M. A. (2014) Semantics in support of biodiversity knowledge discovery: An introduction to the biological collections ontology and related ontologies. *PLoS One*, 9(3), e89606.

15 Quality by Design in Drug Formulation Design

Sakshi Soni, Arpana Purohit, and Vandana Soni
Department of Pharmaceutical Sciences, Dr. Harisingh Gour Vishwavidyalaya, Sagar, India

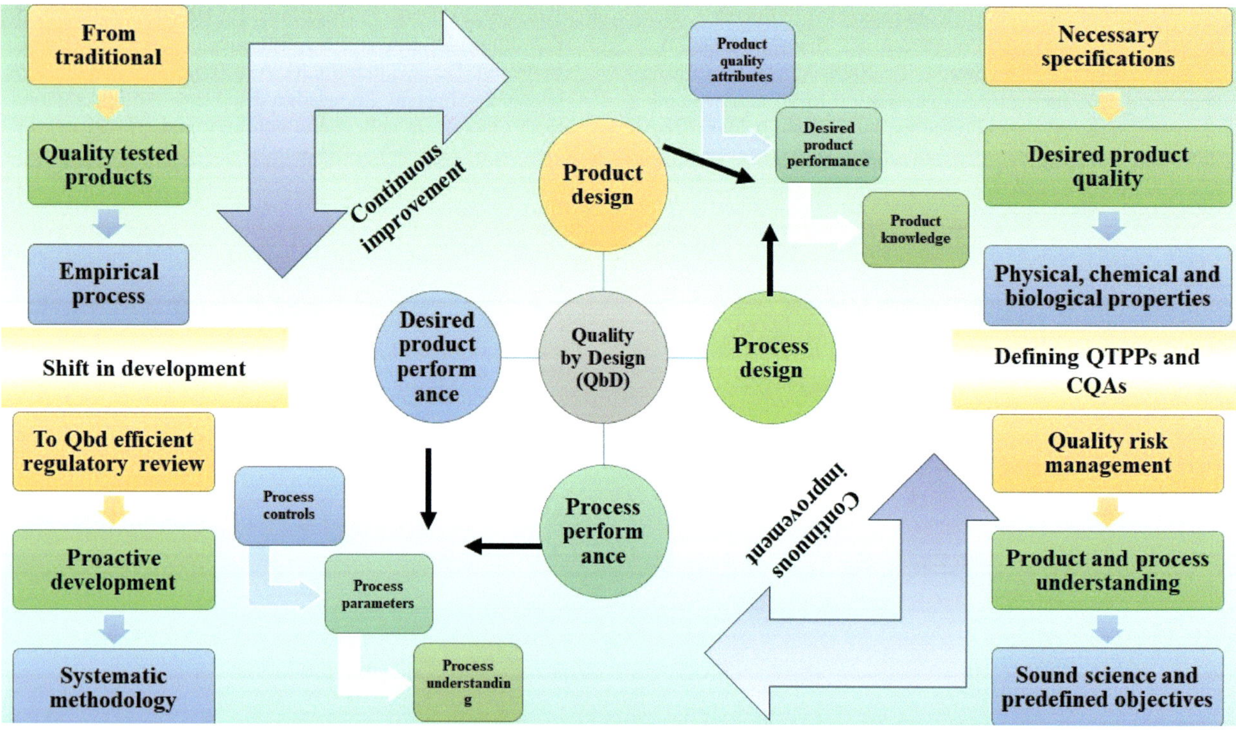

Graphical Abstract: Framework of qbd

INTRODUCTION

Quality by Design (QbD) is a systematic approach to drug formulation and development that emphasizes the design of products and processes to ensure consistent quality and performance. Unlike traditional methods, where quality is often tested in products at the end of the production process, QbD integrates quality considerations from the very beginning. This approach involves identifying critical quality attributes (CQAs) and critical process parameters (CPPs) that influence the final product's quality, safety, and efficacy. The importance of QbD in drug formulation lies in its ability to enhance the understanding and control of the manufacturing process, thereby minimizing variability, reducing costs, and ensuring regulatory compliance. In contrast, the traditional approach to drug formulation often relies on empirical methods, where changes in formulation and process are made in response to problems as they arise, rather than being proactively designed to meet predefined quality objectives. This often leads to a lack of understanding of how different factors impact the final product, resulting in potential risks to patient safety and increased regulatory scrutiny. The QbD approach, however, is data-driven and requires thorough scientific understanding, risk assessment, and a focus on quality from the outset. Regulatory bodies such as the US Food and Drug Administration (FDA) and the European Medicines Agency (EMA) have increasingly recognized the value of QbD and are encouraging its adoption by providing guidance and frameworks that help pharmaceutical companies align their processes with these principles. Adopting QbD in drug formulation design not only ensures compliance with regulatory requirements but also promotes innovation, accelerates development timelines, and ultimately leads to higher-quality pharmaceuticals that meet the needs of patients and healthcare providers [1–4].

DOI: 10.1201/9781003464679-15

KEY PRINCIPLES OF QbD IN DRUG FORMULATION

Defining Quality Target Product Profile

The quality target product profile (QTPP) is a foundational concept in QbD for drug formulation, serving as a strategic guide that outlines the desired quality characteristics of a final pharmaceutical product. It encompasses various attributes such as the intended use, dosage form, route of administration, strength, release profile, and stability criteria, which collectively define the product's quality standards. The QTPP acts as a blueprint for development, ensuring that the formulation and process design meet specific patient needs, therapeutic objectives, and regulatory requirements. Defining the QTPP is crucial because it establishes the starting point for identifying CQAs, which are the properties of the drug product that should be controlled within predefined limits to ensure the product's safety, efficacy, and quality. By clearly defining the QTPP at the early stages of development, manufacturers can systematically design and optimize their formulation and process parameters, reducing the risk of variability and ensuring consistent quality throughout the product lifecycle. This approach also facilitates more efficient regulatory submissions by demonstrating a thorough understanding of the product and process, aligning with the expectations of regulatory bodies such as the FDA and EMA. In essence, the QTPP serves as a critical reference point throughout the development process, guiding decision-making and ensuring that the final product meets its intended quality, safety, and efficacy standards [5].

Identifying Critical Quality Attributes of the Drug Product

Identifying CQAs is a key principle in the QbD framework for drug formulation, which is aimed at ensuring that the final product meets its predefined quality standards. CQAs are the physical, chemical, biological, or microbiological properties or characteristics of a drug product that should be maintained within specific limits to ensure its safety, efficacy, and quality. These attributes are directly linked to the QTPP and are crucial in determining the quality and performance of the final product. Identifying CQAs involves a thorough understanding of the drug substance, excipients, formulation, and manufacturing processes, as well as how each factor can impact the product's quality. For example, attributes such as particle size, dissolution rate, assay, impurities, pH, and moisture content can significantly influence a drug's bioavailability, stability, and therapeutic efficacy. Therefore, during the development process, a risk assessment is conducted to evaluate which attributes are critical and require stringent control.

The identification of CQAs enables manufacturers to focus on what is truly important for product quality, allowing them to design robust processes that consistently produce a product meeting these criteria. By understanding the relationship between CQAs and process parameters, manufacturers can develop strategies to control variability and maintain product quality within the desired range. This approach not only ensures compliance with regulatory standards but also reduces the risk of product recalls, enhances manufacturing efficiency, and promotes a deeper understanding of the product and process. Furthermore, documenting CQAs and their impact on the final product facilitates a more streamlined regulatory review process by demonstrating a well-defined, science-based approach to product development. Ultimately, the identification of CQAs is a critical step in the QbD framework that ensures a thorough understanding of the factors influencing product quality, leading to safer, more effective, and higher-quality pharmaceuticals [6–10].

Risk Assessment and Risk Management in Formulation Design

Quality risk management (QRM) is a systematic approach or technique for comprehending risks, their underlying causes, and how they affect quality. "Quality risk management is a systematic process for the identification, assessment, and control of risks to the quality of pharmaceutical products across the product lifecycle," according to the ICH Q9 guidelines. Risk assessment, mitigation, elimination, communication, and review are some of its components. The foremost step in creating a workable QRM system is risk assessment. The efficacy of the risk assessment process determines the measures for risk removal and mitigation that come after. As ICH puts it, "A systematic process of organizing information to support a risk decision to be made within a risk management process. It consists of the identification of hazards and the analysis and evaluation of risks associated with exposure to those hazards" [11–14].

Design Space: Establishing Acceptable Ranges for Process Variables

The multifaceted interaction and combination of process parameters and input factors ensure quality. According to ICH, "The relationship between the process inputs (material attributes and process parameters) and the critical quality attributes can be described in the design space." The risk assessment and process development experiments can help determine the linkage and effect of process parameters and material attributes on product CQAs and a variable, even helping in the identification of variables and their ranges within which consistent quality can be attained. Thus, these process parameters and material attributes may be chosen to be included in the design space. Operating inside the design area is not regarded as a modification, but movement outside of the design space could initiate a regulatory post-approval change procedure. Every proposed design space is subject to regulatory review and approval (ICH Q8). Design space may be thought of as a parameter adjustment or mathematical function. When a product operates inside its design space, it fulfills its quality standards [15–17] (Figure 15.1).

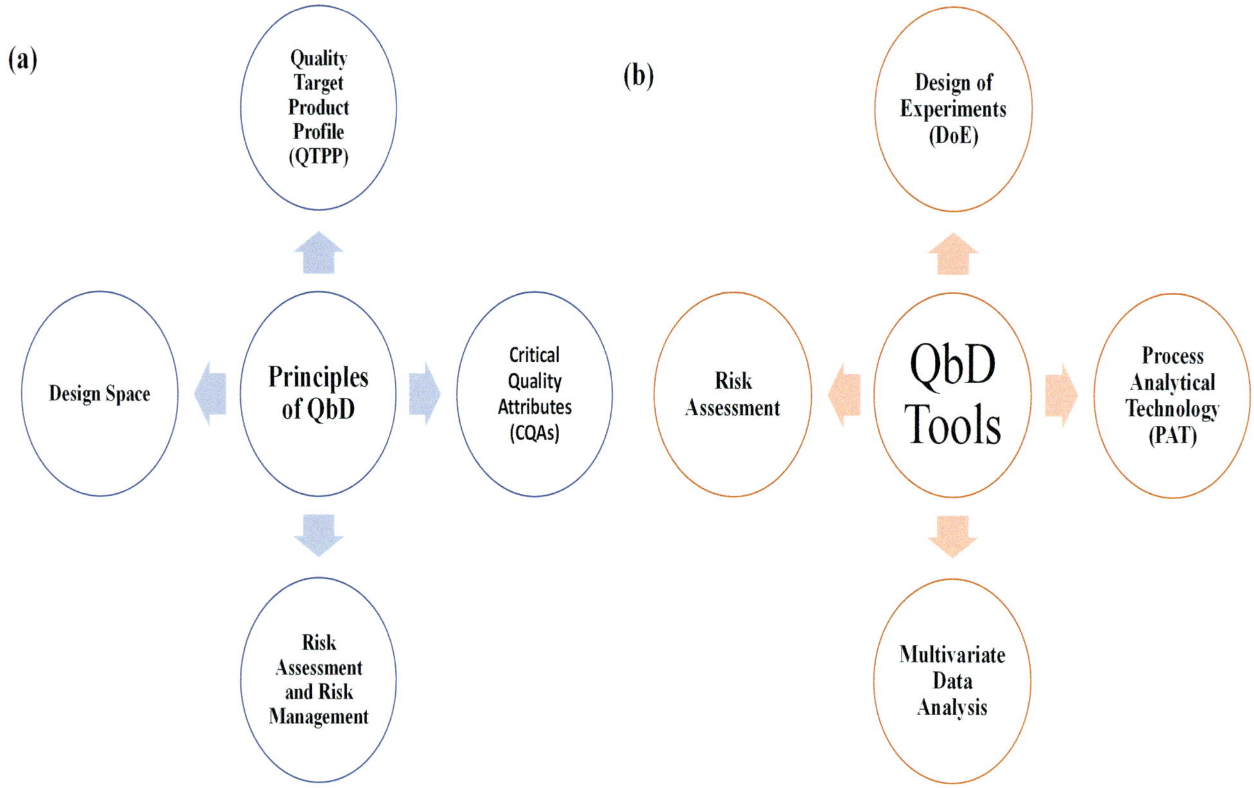

FIGURE 15.1 (a) Overview of the principles of quality by design (QbD) and (b) key tools utilized in QbD.

Control Strategy: Ensuring Consistent Quality Throughout the Process

A control strategy is required to ensure that a product with the desired quality is being generated consistently. Input material attributes (e.g., active pharmaceutical ingredients, excipients, primary packaging materials), unit operation, in-process or real-time release testing, and monitoring programs (e.g., full product testing at regular intervals) all are part of the control strategy. The components of the control strategy dossier should highlight how the quality of the finished product is affected by in-process controls as well as the controls of input materials, intermediates, container closing systems, and drug products. Control over important process parameters and material qualities should be based on an understanding of the product, formulation, and process [18].

TOOLS AND TECHNIQUES FOR QbD IMPLEMENTATION

Design of Experiments for Formulation Optimization

Experimental design is a statistical approach used to understand the relationship between product quality and factors such as CPPs. Design of experiments (DoE) employs various mathematical models, such as Box–Behnken, full and fractional factorial designs, Plackett–Burman, surface design, and Taguchi, to identify correlations between input factors and output responses. Pharmaceutical companies often use full and fractional factorial designs to optimize processes. The DoE process involves goal setting, factor selection, leveling, standard parameter establishment, design selection, experimentation, analysis, temporary application, statistical analysis, and final implementation of optimized parameters. Factorial and randomized designs are commonly used in DoE, with factorial design being particularly popular for conducting multiple runs of a single procedure [19,20].

Full Factorial Design: Full factorial design involves conducting experiments using a certain number of variables and levels. Equation L^F, where L is the level and F is the factor, forms the foundation of this design. Factors such as the names of the excipients (diluents, binder, disintegrants, lubricant, etc.) are often independent variables used to create dosage forms. The term "level" refers to the classification assigned to an independent variable, such as the amount of binder needed for the process, such as minimum, optimal, or maximum. The terms "minimum," "optimum," and "maximum" refer to levels. If the formulator has three levels and four components, according to the formula, total runs will be $L^F = 3^4 = 81$, and the process's best run is selected from these 81 runs [21].

Fractional Factorial Design: Equation L^{F-1} is the foundation of this design. According to the equation, if the formulator has three levels and four factors, then the total number of runs will be $L^{F-1}=3^{4-1}=33=27$. The process's best run is selected from these 27 runs [22].

Plackett–Burman Design: The Plackett–Burman design is utilized to examine studies closely. Because it incorporates the interactions of two elements, 11 factors can be operated simultaneously using the Plackett–Burman design [23].

Process Analytical Technology for Real-Time Process Monitoring

QbD acknowledges that while end-product testing may offer more assurance of product quality, a suitable combination of process controls, predefined material attributes, and intermediate quality attributes (IQAs) is required during processing. It is recommended to employ the effective analysis method to track the connection between process and quality. To analyze IQAs while creating regulatory specifications and supporting continuous manufacturing improvement, process analytical technology (PAT) was introduced.

The FDA published the PAT guidance for innovative pharmaceutical manufacturing and quality in September 2004 after the Center for Drug Evaluation and Research (CDER) discussed the need for FDA guidance to support PAT implementation. It is acknowledged as a significant paradigm shift in the processes of inspecting and approving the continuous process verification of pharmaceutical production processes.

PAT's enormous information collection capacity makes it possible to quickly identify defects, optimize processes, and solve problems. Furthermore, PAT can be used to find the underlying reasons for undesirable medicinal product quality problems in the case of unplanned process modifications. In light of this, proper PAT facilitates prompt process parameter change, guarantees high and consistent product quality, and reduces overall production time. These frameworks offer benefits that facilitate rapid and simple process control, and because they make a substantial contribution to the development of control technology, they have been steadily embraced and presented as a trend. Moreover, several studies have used PAT and QbD in pharmaceutical production processes [24–27].

Workflow of PAT Framework for the Pharmaceutical Manufacturing Process

Before implementing PAT, it is essential to comprehend the materials and methods as well as the features of the PAT tool. The PAT procedure begins with the selection and application of the most suitable PAT instrument. Considerations for this procedure include the measuring technique and the PAT tool's position. Three categories apply to the PAT measuring techniques used during the process: at-line, on-line, and in-line. An area extremely near to the process is sampled, separated, and analyzed using the at-line technique of measurement. Using an on-line measurement technique, a sample is measured during the production process, its appropriateness is assessed, and it is either rejected or reintroduced to the process. As an alternative to taking a sample from the process flow, in-line monitoring involves the use of software to monitor data in real time.

Following this, process monitoring is carried out, and statistical techniques are used to assess and analyze the data gathered. Three categories might be used to categorize statistical methods: data assessment, chemometric modeling, and preprocessing technologies. In the preprocessing stage, standard normal variate (SNV), multiplicative scatter correction (MSC), and derivatives are employed to minimize data interference and provide accurate data. To verify the relationship between CQAs, critical material attributes (CMAs), and CPPs, chemometric modeling techniques such as partial least squares (PLS), principal component analysis (PCA), multiple linear regression (MLR), etc., are used. Using metrics such as root mean square error of the prediction (RMSEP), root mean square error of the calibration (RMSEC), and so on, data assessment aims to quantify and improve data predictability.

PAT Tools for the Pharmaceutical Manufacturing Process

PAT tools for pharmaceutical manufacturing include near-infrared spectroscopy (NIRS), Raman spectroscopy, hyperspectral imaging (HSI), terahertz pulse imaging (TPI), mass spectrometry (MS), acoustic resonance spectrometry (ARS), spatial filter velocimetry (SFV), focused beam reflectance measurement (FBRM), X-ray fluorescence (XRF), and other advanced techniques.

Multivariate Data Analysis and Modeling

A crucial concern in all empirical fields, including pharmaceutical sciences, is identifying the pertinent information included within the data. The goals of DoE are to generate "good data," offer a limited number of experiments that are information-rich, and extract pertinent information from the measured data; hence, multivariate data analysis is employed. Complex pharmacological data may be made simpler and easier using multivariate projection methods. In addition, they enable processes such as sample categorization and result prediction.

When multivariate analysis and process analytical chemistry (PAC) instruments are combined, it becomes possible to identify multivariate correlations between many variables, including raw materials, process conditions, and final products. This allows for efficient process monitoring and management. Therefore, multivariate approaches can be extremely important for fault identification and diagnosis, process control, process scale-up, and multivariate statistical process control [28–31].

Quality by Design Software and Tools

Design-Expert Software for DoE

Stat-Ease Inc.'s statistical software is specifically developed to perform DoE. It is frequently employed in comparison tests, as well as in the examination, characterization, and optimization of various designs, including combination, mixed, and robust designs. About 50 factors may be tested and screened using Design-Expert. It also employs analysis of variance (ANOVA) for its statistical study. Graphical tools are utilized to determine the effect of every component on the final output. It also aids in determining the required number of test runs using a strong calculator. ANOVA can be used to assess statistical significance. In addition to this, a user can utilize a numerical optimizer to forecast the optimal values of each relevant element. By concurrently changing the values of the relevant elements, this program also makes it possible to determine the impacts of each component and the interactions between them. A design space may be mapped out with a response surface model and a significantly smaller number of experiments. It is common practice to utilize this optimization characteristic to determine the ideal operating parameters for a particular process [32,33].

PRACTICAL APPLICATIONS OF QbD IN DRUG FORMULATION

QbD is a methodical approach to developing pharmaceuticals that starts with predetermined quality targets and places a strong emphasis on process control, understanding of the product and process, and excellent science-based QRM. QbD principles are integrated across all stages of the formulation development process, starting from initial development phases to final commercial production as shown in Figure 15.2. This method focuses strongly on designing and developing pharmaceutical products systematically and proactively. Ensuring that quality is consistent throughout the product lifecycle requires having a thorough understanding of the product and process, identifying CQAs, determining CPPs, and establishing controls [5]. In other words, QbD ensures that quality is not tested or inspected into the product but is embedded in it from its inception. Product QbD has

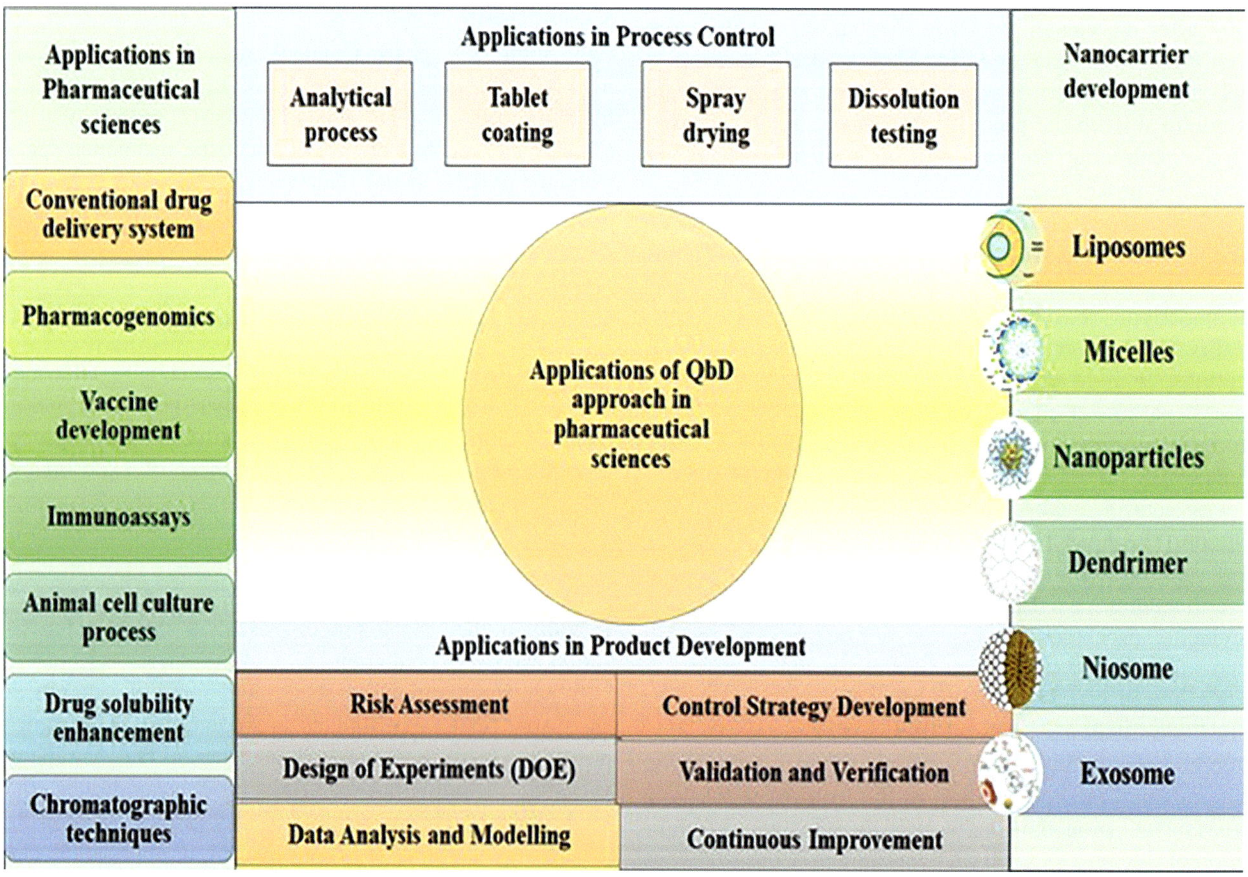

FIGURE 15.2 Applications of QbD in pharmaceutical sciences.

improved the efficiency of the product design process for regulatory compliance purposes. Its success can be seen in many areas as follows.

QbD for Oral Solid Dosage Forms (Tablets, Capsules)

The development and manufacture of oral solid dosage forms, e.g., tablets and capsules, are crucial in the pharmaceutical sector. Such dosage forms are found convenient by patients because they are easy to use and accurately dosed. However, personalizing these dosage forms has proven difficult as well as standardizing them, especially considering their complexity in manufacturing and tailor-made doses [34]. QbD is one approach that can be employed to solve this problem, which involves developing pharmaceutical products with pre-established quality attributes [35]. A bilayered tablet containing hydrochlorothiazide (HCTZ) was optimized using QbD methodology by Dholariya et al. The design included a loading dose of croscarmellose sodium and a maintenance dose of hydrophilic polymers. Risk assessment and factorial designs were used to optimize the disintegration time (DT) and medication release of the tablet. The final formulation using erosion as the major release mechanism had a DT of 70s and a drug release time (DRT95%) of 720 minutes. QbD was found effective for developing HCTZ bilayered tablets [36]. A gastroprotective bilayer tablet was developed by Singh et al. for lamivudine and zidovudine with controlled release and floating bioadhesion properties. Bioadhesive strength (BS), buoyancy time (Tb), and drug release characteristics were evaluated using a face-centered cubic design to optimize CMAs. A defined design space was achieved by balancing these quality attributes in selecting the optimal composition. Gamma scintigraphy studies in healthy volunteers showed that the optimized formulation maintained gastro-retentive behavior for up to 6 hours as compared to the marketed immediate-release formulation, which lasted for 1 hour only [37]. Critical quality parameters for different batches of empty hard gelatin capsules were studied by Stegemann et al. This study demonstrated that when considering DT measured with automatic endpoint detection, the variability of hard capsules remained highly consistent within the specified critical quality parameters. QbD principles were adhered to as this method offered consistent and reliable measurements; therefore, formulated hard capsules met the required quality standards [38]. Zhang et al. combined hot-melt extrusion (HME) with fused deposition modeling (FDM) to develop and evaluate 3D-printed tablets. They focused on the impact of design parameters on tablet weights, drug loadings, hardness at 0° and 45°, release at 180 minutes, and average drug release rates through the use of the Box–Behnken design for DoE. The mid-point dataset was reproduced three times to ensure reproducibility. Amorphous conversion of the crystalline drug was demonstrated by Differential Scanning Calorimetry (DSC), X-ray Diffraction (XRD), and Polarized Light Microscopy (PLM) characterization studies, while robust filament mechanical properties were confirmed by texture analysis. Drug release amounts and rates were found to be greatly influenced by the chosen variables during DoE [39]. Based on failure mode and effects analysis (FMEA), Bicouv et al. performed a quantitative evaluation of gência Nacional de Vigilância Sanitária (ANVISA) reported solid medicines' irregularities between 2017 to 2019 within the QbD framework. Out of the 421 indications about defective pharmaceutical items, 28.5% of the complaints are technical in nature, and 60.0% of the complaints refer to solid oral dosages. The evaluation showed that approximately 80% of nonconformities concern the aspect of appearance, dissolution, packaging, dosage, and/or purity. QbD was established as a suitable approach for identifying the risks of production and controlling the costs associated with the execution of measures designed to reduce the production of technical complaint cases of oral solid dosage form products [40]. Ansari et al. elucidated the application of the QbD approach in the formulation of oral dosage forms, which includes tablets and pellets concerning CQAs and CMAs. The evaluation of risks concerning CPPs in oral dosage forms and CMAs was carried out by applying the risk priority number (RPN) method. The RPN method focuses on three parameters, severity, occurrence, and detectability, at which the physical and chemical characteristics affect the CQAs, and assigns a numerical value to them. Data was analyzed using multivariate modeling techniques, including PLS and response surface methodology (RSM), to establish the association between input formulation parameters, process parameters, and attributes of tablets. These models assist in the comprehension of the effect of variables on the properties of tablets and assisted in their optimization [41].

QbD for Novel Drug Delivery Systems

The concept and technology of drug delivery systems (DDSs) have developed and evolved progressively in the past 60 years. Emerging between the 1960s and the 1980s, the field of controlled drug delivery featured the establishment of a large number of systems and devices for various administration routes and functions. These formulations employed natural and/or synthetic active agents, which sought to extend release, minimize side effects, increase safety, increase patient concordance, and increase efficacy. Numerous techniques were used in the preparation of DDSs to alter or provide a controlled release of the drug in solid, semisolid, or liquid state [42]. Due to today's development in science, the nanoscopic era of targeted nanocarriers has transformed the methodology of DDSs. Since the development of DDSs is related to pharmaceuticals, it is important to control formulation variables using QbD principles by providing a structural framework to ensure the adequate quality, efficacy, and safety of DDSs [43] [44]. Nanocarrier formulation involves multiple components to enhance the safety and efficiency of the formulation. These carriers are developed using bottom-up and top-down approaches, which involve a sequence of steps and are influenced by several factors such as surfactant concentration, time and speed of mixing, size reduction exposure, and temperature. Analyzing these interactions and responses to changes in the concentration of the constituents is critical for high-volume manufacturing with consistent quality [45].

ICH Q8 guideline (Q8r2 is the revised version) for pharmaceutical development describes QbD as a systematic method with set targets, where the main priorities and decision criteria are product and process understanding and control supported by scientific knowledge and QRM. This method is backed up by the FDA's description in 2004 of the risk-based approach to implementation of Current Good Manufacturing Practice (cGMP). Thus, for applying ICH Q8 to DDS development, consulting with relevant regulatory authorities is essential. This guideline entails the provision of regulatory and process flexibility meant for QbD submissions, patent protection, and quality of patents [46,47].

QbD involves the definition and management of formulation CPPs and manufacturing process variables, thus enhancing fast and efficient approval of nanocarrier-based formulations by the authorities. DoE in QbD investigates the relationships between material characteristics and process parameters, and it provides a range of knowledge about formulations and solves manufacturing and scale-up problems. Implementing consistency in processes is required since focusing on end products cannot guarantee quality. It is the method of risk assessment and control in products and processes that QbD supports and encourages. This scientific approach enables root cause analysis, product development, and manufacturing efficiency during minor modifications. It will be less time-consuming and cost-effective and also gives justification to the regulatory agencies concerning post-approval changes [48]. Some of the novel DDSs developed using QbD are listed in Table 15.1.

CASE STUDIES AND APPLICATIONS OF QUALITY BY DESIGN IN DRUG FORMULATION DESIGN

QbD is a rather systematic approach to the development of drugs that is oriented toward the process design and understanding to ensure the predetermined level of product quality. From practically implemented and tested case studies and applications of QbD in the formulation development of a drug, it has resulted in several benefits and enhanced the formulation process in numerous ways. Lee et al. used stepwise QbD to design a fixed-dose combination tablet of rabeprazole sodium and sodium bicarbonate for gastroesophageal reflux disease. Based on the evaluation and previous knowledge, they defined QTPP parameters and identified CQAs. By applying Preliminary Hazard Analysis (PHA), Failure Mode and Effects Analysis (FMEA), and Design of Experiments (DoE) methodologies, they managed to fine-tune binding and disintegration features of this formulation. Optimized formulation revealed more than 24 months of stability and a faster Tmax in beagle dogs than the reference drug, hence testifying QbD's ability to assure good quality and stability [70].

Rapalli et al. developed a topical hydrogel with cubosomes containing ketoconazole using the QbD approach. The research aimed at lower concentrations of a surfactant. The identified CQAs were the surfactant and stabilizer concentration impacting particle size, size distribution, encapsulation efficiency (EE), and DRT. Consequently, time and speed were established as the significant CPPs that affect the size and efficiency of entrapment of the pre-emulsion. The study employed a 2×3 factorial design with the help of Design-Expert® software; the optimized formulation disclosed the release profile commensurate with the Korsmeyer–Peppas model, releasing 67% ketoconazole within 24 hours. The *ex vivo* permeation of the gel through goat ear skin revealed a 92.73% release in 24 hours, which endorses the QbD strategy in the improvement of the solubility of the active component and the safety of the patient [71].

Patel et al. 2021 produced solid lipid nanoparticles (SLNs) using the QbD approach; QTPP was established, and CQAs were defined with the help of the Ishikawa diagram. Using FMEA, other high-risk factors were identified to be surfactant, oleic acid, and sonication amplitude. Optimization using the three-factor, two-level DoE arrived at the formulation of SLNs with 2.00 % poloxamer 188, 1.5% oleic acid, and 59.78% sonication amplitude, resulting in a particle size of 273.6 nm with a drug loading efficiency of 8.57%. This study illustrated how QbD enhances the design of DDSs with reduced iterations [72].

Pallagi et al. applied QbD methodology in the concept development stage of liposomes for nasal delivery for brain application. Quality by Design (QbD) and Risk Assessment (RA) approaches were applied concurrently in this study to define the QTPP characteristics and identification of CQAs. Some aspects influencing product quality were systematically evaluated using liposome preparation based on the factorial design and lipid film hydration technique. Some of the key factors such as concentration of phospholipid (Phospholipon 90G®) and ultrasonication temperature were adjusted to optimize the size of the vesicle, size distribution, and surface charge. This approach also helped improve the formulation process, especially the creation of liposomes for lipophilic drugs that have poor solubility [73].

Saha et al. implemented QbD for the creation of drug-containing nanoparticles, where patient-related QTPPs were defined and CQAs were determined by RA. CMAs and CPPs were discovered by the application of definitive screening design (DSD). The optimized formulation in terms of lecithin and drug concentration as per the central composite design (CCD) ensured that the developed nanoformulation is stable and has a very small particle size to provide a QTPP for ocular application. On a laboratory scale, the process was deemed to have worked well; however, issues

TABLE 15.1
QbD for Novel Drug Delivery Systems

Delivery System	Drug	Route of Administration	Disease	Formulation Design/Method and Parameters Used	Response Observed	References
Liposomes	Faeonol (PAE)	Transdermal	Atopic dermatitis	Box–Behnken design (3-factor 3-level) Independent variables studied were the DC-Chol concentration, the molar ratio of lipid/drug, and the polymer concentration, and the levels of each factor were low, medium, and high. The dependent variables studied were the encapsulation efficiency of PAE, flux of PAE, and viscosity of the gels.	Response surface plots were generated. The encapsulation efficiency (%EE) of PAE increased with a higher lipid/drug molar ratio but decreased with higher polymer concentration. The gel exhibited a non-Fickian diffusion release mechanism for PAE, fitting Higuchi's order model for *in vitro* release profiles.	[49]
Liposomes	Tenofovir	Parenteral	Antiviral	Placket–Burman and central composite design (CCD) Plackett–Burman method was used to screen eight high-risk variables: drug encapsulation efficiency, particle size, and physical stability. CCD was used to investigate the relationship between these variables and encapsulation efficiency.	The developed model established a design minimizing preparation variability and demonstrated stability for at least 2 years when stored at 4°C, offering time and cost savings and resulting in a more robust liposome preparation process.	[50]
Ligand-conjugated nanoparticles	Docetaxel trihydrate	Parenteral	Cancer	2-factor 3-level full factorial design The influence of the two independent variables was studied on the response size and percentage of encapsulated drugs.		[51]
Niosomes	Pioglitazone	Transdermal	Diabetes	Box–Behnken design (3-factor 3-level) Tween80, phospholipid, and cholesterol were selected as independent variables, whereas dependent variables were size, entrapment efficiency, and flux, with constraints applied in the formulation.	The quantitative impact of these factors at various levels on the responses could be predicted using polynomial equations, as there was a high degree of linearity observed between the predicted and actual response variable values.	[52]
Microemulsion	Itraconazole	Topical	Fungal infection	D-optimal mixture design A D-optimal mixture experimental design was conducted, incorporating three components: benzyl alcohol, Cremophor® EL, and Transcutol® P. The experiment revealed effects on various parameters, including globule size, pH, rheological properties, *in vitro* permeability, and drug release.	The optimized formulation exhibited the ideal globule size, pH, rheological properties, and *in vitro* permeation profile, maintaining stability for 6 months at 40°C and 75% relative humidity.	[53]

(Continued)

TABLE 15.1 (*Continued*)
QbD for Novel Drug Delivery Systems

Delivery System	Drug	Route of Administration	Disease	Formulation Design/Method and Parameters Used	Response Observed	References
Solid lipid nanoparticles	Curcumin	Parenteral	Cancer	Placket–Burman and Box–Behnken design The Plackett–Burman method identified three significant factors that affect encapsulation efficiency. These factors were then optimized using the Box–Behnken design to refine the formulation.	The optimized formulation exhibited a size of 135.3 ± 1.5 nm with an entrapment efficiency of 91.09% ± 1.23%, while also demonstrating a sustained release profile. The *in vivo* pharmacokinetic study showed a 12.27-fold increase in AUC (area under the curve) and a relative bioavailability of 942.53% compared to curcumin suspension.	[54]
Nanostructured lipid carriers	Iloperidone	Parenteral	Schizophrenia	Box–Behnken factorial design (3-factor, 3-level) The concentration of lipids, concentration of drug, and concentration of surfactant were selected as independent variables.	Particle size and entrapment efficiency were the selected responses, with particle size in the nanorange and entrapment efficiency ranging from 63% to 96%. The *in vivo* release study over 24 hours showed an 8.3-fold increase in bioavailability.	[55]
PLGA nanoparticles	Tacrolimus	Parenteral	Immunosuppressant	Placket–Burman design Independent variables included cyclosporine A, PLGA, emulsifier concentration, stirring rate, type of organic solvent, and organic-to-aqueous phase ratio.	The combination of factors resulted in an entrapment efficiency ranging from 10.17% to 93.01%, mean particle size from 41.60 to 372.80 nm, and zeta potential from 29.60 to 34.90 mV. Dissolution efficiency varied from 52.67% to 84.11%, with an initial burst release followed by sustained release.	[56]
Niosome	Lacidipine	Transdermal	Hypertension	Box–Behnken design (4-factor, 3-level) The selected independent variables for the experimental design were Span 60 concentration, cholesterol concentration, hydration time, and sonication time.	Particle size and percent entrapment efficiency were selected as responses. The particle size was within the nanorange, and entrapment efficiency ranged from 63% to 96%. An *in vivo* release study over 24 hours demonstrated an 8.30-fold increase in bioavailability.	[57]
Transferosomal gel	Miconazole nitrate	Transdermal	Fungal infection	2-factor 3-level full factorial design The independent variables selected were the type of surfactant, total lipid, and phospholipid-to-surfactant ratio. The response factors measured were entrapment efficiency, vesicle size, and flux.	The MIC transferosomes exhibited a high encapsulation efficiency (EE%) ranging from 67.98 ± 0.66% to 91.47 ± 1.85%, with small particle sizes between 63.5 ± 0.604 nm and 84.5 ± 0.684 nm. The *in vitro* release study indicated an inverse relationship between EE% and the release rate.	[58]

(*Continued*)

TABLE 15.1 (*Continued*)
QbD for Novel Drug Delivery Systems

Delivery System	Drug	Route of Administration	Disease	Formulation Design/Method and Parameters Used	Response Observed	References
Transferosomes	Raloxifene	Transdermal	Osteoporosis	Box–Behnken design (3-factor, 3-level) Phospholipon® 90G, sodium deoxycholate, and sonication time, each at three levels, were chosen as independent variables. The dependent variables were entrapment efficiency, vesicle size, and transdermal flux.	The optimized conditions had low polydispersity (0.08) with a particle size of 134±9 nm, an entrapment efficiency of 91.00%±4.90%, and a transdermal flux of 6.5±1.1 μg/cm²/h. Raloxifene hydrochloride-loaded transfersomes showed superior drug permeation and skin deposition.	[59]
Nanostructured lipid carrier	Silymarin	Transdermal	Skin cancer	2-factor 3-level full factorial design The independent variables were HHPH pressure, number of cycles, and stirring speed.	Photostability testing confirmed the formulation's stability in the presence of light. The NLC-based gel demonstrated significantly higher permeability, onset of action, prolonged activity up to 24 hours, and better efficacy.	[60]
Agarose-based liposomes	Sorafenib tosylate	Parenteral	Liver cancer	2-factor 3-level full factorial design	The formulations had a particle size of 150–180 nm and a zeta potential of 30–60 mV, depending on ligand and drug loading, with over 90% entrapment efficiency, and showed twice the cytotoxicity.	[61]
Liposomes	Citalopram hydrobromide	Parenteral	Brain targeting	Fractional factorial The formulations were assessed for drug encapsulation, vesicular size, size distribution, conductivity, and drug release characteristics.	Through a systematic screening of formulation and process variables in the optimization phase, improvements were made in drug loading, achieving fine vesicular size with a narrow size distribution, greater stability, and sustained release features.	[62]
Microemulsion-based hydrogel	Sertaconazole	Transdermal	Antifungal	Full factorial design The formulation was assessed by varying the oil type, oil concentration, surfactant type, and surfactant: cosurfactant ratio.	The optimized microemulsion formulations demonstrated higher skin retention (686.47 μg/cm²) in *ex vivo* studies. The antimycotic activity against *C. albicans* showed larger inhibition zones.	[63]
PEGylated lipidic nanocapsules	Paclitaxel	Parenteral	Cancer	Box–Behnken experimental design D-LNCs were characterized using dynamic light scattering, scanning and transmission electron microscopy, Fourier transform infrared spectroscopy, and differential scanning calorimetry.	*In vitro* release studies indicated a sustained drug release. *In vivo* pharmacokinetic studies showed prolonged circulation time and slower elimination of paclitaxel from D-LNCs compared to the marketed formulation.	[64]

(*Continued*)

TABLE 15.1 (*Continued*)
QbD for Novel Drug Delivery Systems

Delivery System	Drug	Route of Administration	Disease	Formulation Design/Method and Parameters Used	Response Observed	References
Nanoliposomes	α-galactosidase	Parenteral	Fabry disease	Full factorial design The influence of GLA concentration and lipid concentration was identified as the most important for the quality of nano-GLA obtained through compressed fluid processing.	The design space analysis predicted that to achieve a higher enzyme concentration, the lipid concentration in the DELOS-susp step could be reduced. This adjustment would result in a high-quality product in terms of particle size and uni-lamellarity while avoiding aggregation and sedimentation.	[65]
Ultrasmall nanostructured lipid carriers	Atorvastatin calcium and curcumin	Parenteral	Glioblastoma	Full factorial design The formulation composition and process parameters included the solid lipid ratio, type and concentration of liquid lipids and surfactants, and the type of production method.	Particles with an average diameter of approximately 50 nm and a polydispersity index lower than 0.3 were produced. These particles demonstrated high stability, scalability, drug protection, and sustained co-release properties.	[66]
Nanolipid vesicular system	N/A	Transdermal	Osteoarthritis	Full two-level factorial design The independent variables selected were the concentration of PC, the concentration of edge activator (EA), and the type of EA.	The factorial design revealed that the selected formulation variables significantly impacted the tested responses. The optimal permeation flux in the *ex vivo* study was 0.415 mg/cm²/h, following zero-order kinetics. Moreover, the *in vivo* study of the optimized formulations showed a significant reduction in inflammatory mediators associated with osteoarthritis (OA).	[67]
Microemulsion gel	Agomelatine	Transdermal	Depression	Design of experiment The mixture design's independent variables were the percentages of capriole 90, Cremophor RH40, and Transcutol HP. The dependent variables included globule size, optical clarity, cumulative amount permeated after 1 and 24 hours, and enhancement ratio (ER).	The gel significantly enhanced drug permeation through rat skin with an enhancement ratio of 37.30 compared to the drug hydrogel. The optimized microemulsion gel formula demonstrated significantly higher Cmax, AUC 0–24 hours, and AUC0–∞ than the reference agomelatine hydrogel and oral solution.	[68]
Lyotropic liquid crystals	Temozolomide	Parenteral	Cancer	Three-level Box–Behnken design The effect of lipid concentration, surfactant concentration, and co-surfactant concentration on response variables such as size, size distribution, zeta potential, entrapment efficiency, and drug loading was investigated using statistical data obtained.	The LLCs were obtained with sizes ranging from 53.15 to 186.50 nm and a PDI of less than 0.25. The optimized formulation showed a particle size of 97.70±0.481 nm and an entrapment efficiency of 36.46±1.48%. TMZ-loaded LLCs followed Korsmeyer's Peppas model and demonstrated sustained release for up to 72 hours.	[69]

of scalability were highlighted, which makes it pertinent to investigate its application in the industrial setup within the specified design space [74].

Several pharmaceutical organizations have realized that the application of QbD for developing products can be instrumental in the improvement of processes. AstraZeneca in 2005 undertook a significant implementation of the lean manufacturing concept as well as DMAIC (define, measure, analyze, improve, control). This initiative known as "pull manufacturing" stressed customer focus and customer satisfaction as opposed to just the product. Due to the visibility of all the gaps and removal of the unnecessary, it made a remarkable increase in AstraZeneca's revenues of \$60–\$70 million in the framework of the project, with a total investment of no more than \$0.1 million. This considerable return on investment proved the effectiveness of the application of lean methodologies in the firm in the elimination of waste and the improvement of profitability [75].

Similarly, a global six sigma program was launched in Capsugel, a division of Pfizer, in 2006, which had added to its continuous improvement process. This program was intended to take efficiency to higher levels and enhance the quality of products in the organization. Directors and managers were trained in six sigma with a more complex approach, including tools such as DoE, Voice of the Customer (VoC), and Pareto charts among others. These tools were useful in exposing gap-prone areas to be given top priority. It was also found that within a matter of one year, the company was able to cut down the cycle time of gelatin feed tanks by 50%, which in turn enhanced the company's production processes. Further, there was a 50% cut in waste, proving the efficiency of the program in waste management. A major learning was achieved within the context of the loading process, where there was a marked improvement and a nearly 88% decrease in cases of loss of capsules. These changes in turn coherently led to increased productivity and efficiency of operations [76].

To implement its integrated physical and logical structure, the PDO (Physically and Logically Integrated Product Development Organization), Dr. Reddy's Laboratories consolidated all aspects of R&D related to product development. They adopted the *Design for Six Sigma* principle through the DRDDI framework (Define, Research, Design, Develop, and Implement). This strategy improved due date performance, R&D efficiency, process stability, and the ability to achieve the desired final product price. In addition, Dr. Reddy's effectively applied methodologies such as CCPM and QbD, incorporating tools like FMEA, DoE, and control strategies. Together, these quality and process improvement approaches demonstrate how a comprehensive operating model and compliance standardization can be achieved across the pharmaceutical sector through systematic implementation [77].

Johnson & Johnson used lean manufacturing together with QbD principles at their facilities to improve their manufacturing processes, where they applied poka yoke, Value Stream Mapping (VSM) , DoE, statistics, and FMEA lean manufacturing tools to improve operator interface and get the best return on investments [78]. Similarly, in the early 1920s, Baxter Healthcare employed a novel strategy of applying QbD principles for organizational transformation, where every facility was managed as a separate entity of a small company. These focused on the use of DMAIC, lean methodologies, and SCOT analysis (strengths, challenges, opportunities, and threats) in the enhancement of different processes in Baxter. Applying all these ideas made a great difference in the organization's operational, quality, process, and compliance systems [79].

ADVANCEMENTS IN QbD TECHNOLOGIES: ARTIFICIAL INTELLIGENCE AND MACHINE LEARNING

Drug formulation and development have been greatly improved by advances in QbD technology, especially when it comes to the combination of *in silico* predictive modeling, molecular dynamics simulations, computational modeling and simulation, and the QbD approach in pharmacogenomics. Virtual screening platforms and quantitative structure–activity relationship (QSAR) models enable the rapid identification of lead compounds and estimations on drug interactions with the receptor, pharmacokinetics, and toxicity. Specifically, molecular dynamics simulation has also achieved significant advances, especially the usage of high-performance computing capability resources and improved sampling techniques to yield a detailed picture regarding molecular interactions, conformational dynamics, and drug binding routes. Physiologically based pharmacokinetic (PBPK) modeling and integration of pharmacogenomics provide a picture of how a drug or a biological compound will behave, the biological reactions it will cause, and more importantly, how it affects patient-specific dosing strategies [80].

Pharmacogenomics, in both its functional and structural aspects, is necessary in the process of drug discovery as it is based on the use of genomic information for the identification of potential drug targets and their further evaluation. This process includes lead optimization through high-throughput screening and the assessment of drug-metabolizing enzymes, drug transporters, and receptors using computer-aided techniques and bioinformatics [81].

Pharmacogenomics also offers a strong and stable background for assessing and enhancing dosage types, as well as offering chances to treat new diseases using repurposed drugs. This approach is particularly useful for different kinds of drugs, such as anticancer agents, vaccines, gene and DNA delivery systems, and immunological agents, as such drugs are evaluated according to the genetic indicators of a certain disease. Furthermore, various polymers used in formulations can evaluate the change in the pharmacokinetic and pharmacodynamic properties of drugs, which are essential in formulation development [82].

However, high cost and limited access to these sophisticated tools are still the major factors hindering the use of pharmacogenomics in drug discovery and formulation enhancement. Nonetheless, in future, developments in technologies are expected to reveal that the application of QbD in conjunction with pharmacogenomic applications can open a new avenue for the identification of better and safer formulations of drugs. This future outlook means that there are radical changes in the way treatment plans are set because they will depend on the genes of a given individual, and this will prove fruitful in terms of treatment returns with fewer side effects.

In recent years, the pharmaceutical industry has witnessed a transformative shift with the integration of artificial intelligence (AI) and machine learning (ML) into QbD approaches. It is significant to mention that, within the QbD approaches, AI and ML have appeared to be the enablers of the modern toolkit of pharmacological research and development, drug formulation, and manufacturing. In pharmaceutical production, these technologies enhance the quality of the product, enhance its production, and guarantee regulatory compliance by establishing complex algorithms and data insights. It has also provided new opportunities in terms of improving both the quality and the processes of manufacturing pharmaceutical products. Therefore, AI and ML have a significant and multifaceted influence on QbD in terms of risk assessment, modeling and simulation, PAT, drug discovery and development, submitting documents, and issues relevant to validation [83].

Predictive maintenance and quality control are two of the most critical elements that determine the efficiency of the operation and quality of manufactured pharmaceutical products. In this domain, supervised learning methods are used to address equipment failures, deviations in product quality, and other process abnormalities before they occur. Supervised learning models look into real-time data from manufactured part processes, equipment sensor data, or quality testing parameters to identify any signs of equipment failure or a shift from original manufacturability standards. Specifically, with this predictive capacity, pharmaceutical business organizations are in a position to have preventive plans of maintenance, solve quality problems early, and meet the high standards of quality assurance [84].

Neural networks, evolutionary algorithms, fuzzy logic, and other AI and ML approaches are essential for streamlining pharmaceutical production procedures. These sophisticated algorithms are capable of sorting through enormous amounts of data to find the best set of input variables that will enable the achievement of the required quality standards. This improvement results in cost savings and increased productivity by streamlining manufacturing operations and improving product quality [85].

Mazumder et al. worked on formulation and process variables for oral disintegrating films (ODFs) using the QbD approach, where lamotrigine (LMT) was used as the model drug. The factors that were categorized as formulation variations were the ratio of plasticizer to the film former, drying temperature, air flow rate in the chamber, drying time, and the wet coat thickness on the film. A definitive screening DoE was used to obtain information to help determine important factors that impacted CQAs. In addition, the creation of a predictive model of the process was also involved. The results highlighted the need to maintain a precise record of the plasticizer-to-film-former proportions and efficient control over the drying processes in the manufacture of LMT ODF that enhance CQAs [86].

Simoes et al. showcased the integration of QbD principles with artificial neural networks (ANNs) to develop a test drug product with defined material specifications correlated with its *in vivo* performance. Based on the CMAs identified, drug particle size distribution (PSD) was evaluated and managed through a three-tier specification strategy. The ANN model, employing a modest architecture with five hidden nodes in one hidden layer, demonstrated robust prediction capabilities validated with high R^2 values (>0.94) for training and validation datasets across three responses. The manufacturing process validation, which includes the production of exhibit batches coupled with the prediction of the *in vitro* dissolution with a minor error of about 5%, proved the efficiency of the ANN model. Further clinical research work vindicated bioequivalence to the reference listed drug, establishing the congruency of the formulation with *in vitro* performance and positive patient outcomes, which underscored the effectiveness of ANNs [87].

Kim et al. studied the effect of Microcrystalline Cellulose (MCC) variation on the quality of immediate-release tablets containing amlodipine besylate. The development of the formulation was carried out using a QbD concept, and the mixture was designed using a D-optimal mixture design. The design space was identified to change based on the type of MCC manufacturers and grades used in the research. This change of position in the design space provided a correlation between the physicochemical properties of MCC and the quality characteristics of tablets, namely the CQAs.

An ANN was then used to train a predictive model to establish the relationship between the physicochemical properties of MCCs and dissolution rates. High accuracy was recorded for the model by matching the actual profiles recorded for dissolution, and the overall absolute and relative errors were acceptably low [88]. Aksu et al. developed a correlation between formulation components and product characteristics. ANNs and genetic programming were used for the examination of ramipril tablets produced using the direct compression method. The data used in this study was systematically obtained using the principles of DoE. ANN and neuro-fuzzy logic programs were applied to analyze the optimization of the formulation process and product quality.

Using this integrated multivariate study design that was proposed by researchers for the first time, it was envisaged to obtain new knowledge about the interdependencies that are characteristic of the formulation and manufacturing of pharmaceuticals aimed at providing a good and improved pharmaceutical product [85].

Dawoud et al. established the applicability of AI alongside the QbD approach in formulating a compound that has poor solubility for potential use in carcinoma treatment. Silymarin, effective in liver cancer, was used in this study. Therefore, this study determined the influence of CPPs on particle size, size distribution, and lecithin/chitosan nanoparticle entrapment efficiency using an effective QbD approach. AI usage was strictly recognized for its high predictiveness of the drug release rate from the optimized formula in the investigated period using ANN models. In addition, the cytotoxicity of the optimized formula was determined, and the results depicted better activity than silymarin, as evident from the low IC50 value. The study provided the design space of the optimized formula, which pointed out the possibility of applying AI-based methodologies in terms of changing the nature of pharmaceutical formulations from the reliance on experiences to more evidence-based methodologies [89].

The progression of PAT systems in exercising pharmaceutical manufacturing through AI and ML has greatly advanced. These advanced systems can provide constant and continuous information on and control these valuable process parameters in real time, or as close to it as possible. Through providing opportunities for tweaking according to data generated by AI, PAT systems improve business knowledge and uniformity of processes, which in turn leads to the production of higher-quality products [90].

Kim et al. studied the PAT and the developments made in the past two decades in the pharmaceutical industry. PAT has an important application in the assessment of in-process quality attributes (IQAs) during manufacturing; in the development of actual benchmarking for the formulation of further adequate rules by the regulatory department; and in implementing constant improvement in the overall procedures in manufacturing.

These tools have played a major role in the continuous monitoring and control of parameters in all the steps of unit operations in the pharmaceutical industry. PAT should be implemented with QbD and Continued Process Verification (CPV), the basic approaches that would help in strengthening the quality of drug products and the improvement of the manufacturing process continuously [25,80].

Despite the myriad benefits offered by AI and ML in pharmaceutical manufacturing, the validation of AI-/ML-based systems remains a significant challenge. Ensuring the reliability, reproducibility, and consistency of AI-driven models and algorithms requires meticulous validation protocols. Moreover, addressing concerns related to data integrity, algorithm transparency, and cybersecurity is paramount to instilling confidence in AI-/ML-based systems within the regulatory framework. Table 15.2 provides a comprehensive overview of how AI and ML technologies are applied across various stages of QbD in drug formulation, ranging from predictive modeling to process optimization, real-time monitoring, regulatory compliance, and so on.

CONCLUSION

The QbD approach has revolutionized the drug formulation process, shifting the paradigm from reactive quality control to proactive quality assurance. By focusing on understanding the CQAs of a drug product and designing the formulation and manufacturing process to ensure consistent quality, QbD has significantly improved product safety, efficacy, and time-to-market. Key principles of QbD include defining the QTPP, identifying CQAs, conducting risk assessment and management, establishing design space, and implementing a control strategy. Advanced tools such as DoE, PAT, multivariate data analysis, and QbD software have facilitated the efficient application of QbD principles. Practical applications of QbD extend across various drug formulation types, including oral solid dosage forms, parenteral formulations, topical and transdermal formulations, and novel vesicular systems. Case studies demonstrate the successful implementation of QbD in drug development and manufacturing. Emerging technologies such as computational modeling and simulation, molecular dynamics simulations, *in silico* predictive modeling, pharmacogenomics-based QbD, AI, and ML are further enhancing the capabilities of QbD and driving innovation in drug formulation. QbD has become an indispensable approach for ensuring the quality and safety of drug products. By adopting QbD principles and leveraging advanced technologies, pharmaceutical companies can streamline their development processes, reduce risks, and deliver high-quality medications to patients.

TABLE 15.2
The Applications of Artificial Intelligence (AI) and Machine Learning (ML) in Quality by Design (QbD) for Drug Formulation

Application Area	Description	Outline	Inference	References
Predictive modeling	Predicts formulation and process parameters' impact on critical quality attributes (CQAs)	The significance of QbD, explains ICH guidelines, defines key QbD concepts, highlights ANNs' role, discusses experimental design, presents a ramipril tablet case study, and details methodology.	Integrating QbD with ANNs and DoE principles improves pharmaceutical formulations by understanding input interactions, optimizing processes, and ensuring high-quality product development and efficient manufacturing.	[85]
Process optimization	Using ML algorithms like neural networks, genetic algorithms, and fuzzy logic for optimizing manufacturing processes	Detailed QbD studies critical process parameters to optimize lecithin/chitosan nanoparticles. AI effectively predicts drug release rates, enhancing silymarin's cytotoxic effect with decreased IC50.	Integrating AI with QbD enhances the formulation process of poorly water-soluble drugs, demonstrating effective optimization, prediction, and cytotoxicity improvement, potentially revolutionizing pharmaceutical methodologies.	[91]
Real-time monitoring and control	Implementing AI-powered systems for real-time monitoring and control of critical process parameters (CPPs)	QbD in pharmaceuticals aims to meet patient needs with targeted quality. Process analytical technology (PAT) supports QbD by enabling real-time measurement and control of quality and performance criteria for raw and processed materials.	QbD and PAT together ensure consistent pharmaceutical quality, meeting patient needs efficiently and supporting the evolving industry with systematic, scientifically validated approaches.	[92]
Data analysis, pattern recognition, risk assessment, and decision support	Utilizing AI/ML for analyzing complex datasets, identifying patterns, and informing decision-making. Risk assessment, decision support, and optimizing QbD strategies for improved product quality	Developments in PAT, QbD, and material characterization in pharmaceuticals. The emerging trends in QbD can be useful in ensuring compliance with the regulatory requirement for the production of quality pharmaceutical products.	Advancements in the understanding and implementation of PAT real-time tools inclusive of Raman spectroscopy in tracking the manufacture of pharmacy products best illustrate the transition to enhanced production of pharmaceuticals with themes on efficiency and quality.	[93]
Virtual screening and drug design	AI-driven virtual screening of compounds, predicting drug–target interactions, and designing novel drug candidates	This study aims to develop a graph neural network model called PLANET to predict protein–ligand binding affinity accurately and efficiently. PLANET utilizes graph-represented 3D structures of protein binding pockets and 2D chemical structures of ligand molecules. The model was trained using multiobjective processes, including binding affinity, contact map, and ligand distance matrix.	PLANET's performance, tested on benchmarks like CASF-2016 and DUD-E, shows comparable accuracy to top deep learning models and conventional programs like Glide but with significantly reduced computation time. This highlights PLANET's potential as a valuable tool for large-scale virtual screening in drug design.	[94]
Regulatory compliance and quality assurance	Using AI/ML for automating regulatory compliance tasks, quality assurance, and ensuring adherence to standards	A variety of knowledge applications are employed, including heuristics, decision trees, correlation analysis, and first-principle models. These tools provide valuable insights that guide decision-making in production processes, aid in selecting excipients, and determine the appropriate size of equipment.	AI and its networks, technologies, and tools contribute to enhanced product quality, reduced waste, and higher profitability in manufacturing operations. The QbD approach guarantees enhanced quality standards in the produced goods.	[95]

REFERENCES

1. Nyarko S. A review on the application of quality by design methodology in the formulation, development, and evaluation of fixed-dose combination tablets. *London Journal of Medical and Health Research*. 2023 Dec 5;23(12):71–81.
2. Pramod K, Tahir MA, Charoo NA, Ansari SH, Ali J. Pharmaceutical product development: A quality by design approach. *International Journal of Pharmaceutical Investigation*. 2016 Jul;6(3):129.
3. Amrih P, Damayanti RW. Pharma 4.0 quality management challenge: A literature review. In *Proceedings of 12th Annual International Conference on Industrial Engineering and Operations Management (IEOM) Conference*; 2022. Istanbul, Turkey.
4. Jadhav JB, Girawale NN, Chaudhari RA. Quality by design (QBD) approach used in development of pharmaceuticals. *Indian Journal of Pure & Applied Biosciences*. 2014;2(5):214–23.
5. Yu LX, Amidon G, Khan MA, Hoag SW, Polli J, Raju GK, Woodcock J. Understanding pharmaceutical quality by design. *The AAPS journal*. 2014 Jul;16:771–83.
6. Aru PB, Gulhane MS, Katekar VA, Deshmukh SP. Quality by design (QbD) in pharmaceutical development: A comprehensive review. *GSC Biological and Pharmaceutical Sciences*. 2024;26(1):328–40.
7. Pawar SS, Mahale YS, Kalamkar PA, Satdive RA, Sonawane SK, Bhapkar SP. Quality by design (QbD): A comprehensive understanding and implementation in pharmaceuticals development. *Current Pharmaceutical Analysis*. 2023 Nov 1;19(9):677–86.
8. Beg S, Hasnain MS, Rahman M, Swain S. Introduction to quality by design (QbD): Fundamentals, principles, and applications. *Pharmaceutical Quality by Design*. 2019 Jan 1;2019:1–17.
9. Jupally P, Damagundam S, Domaraju P. Quality by design (QBD) tool for quality control in pharmaceutical industry. *International Journal of Pharmaceutical Sciences and Nanotechnology (IJPSN)*. 2023 May 1;16(2):6480–7.
10. Sreelekha S, Kumar KV, Mahammed N, Reshma T, Sree GU, Basha SS, Bhuvaneswari M. Quality by design approaches in pharmaceutical development and solid dosage forms production: A narrative overview. *Jordan Journal of Pharmaceutical Sciences*. 2023 Dec 25;16(4):770–84.
11. Rodríguez-Pérez J. *Quality Risk Management in the FDA-Regulated Industry*. Australia: Quality Press; 2017 Feb 21.
12. Tishko L, Dramaretska S, Kurbanov A, Yunger V. Risk management in medicine: The process of identification, assessment and control of risks in medical practice. *Revista de Gestão e Secretariado*. 2023;14(10):19211.
13. Wickramaratne KA. Risk based quality management: The way forward to improve quality in medical laboratories. *The Sri Lanka Journal of Haematology*. 2024 Apr 24;15(1):12–8.
14. Charoo NA, Ali AA. Quality risk management in pharmaceutical development. *Drug Development and Industrial Pharmacy*. 2013 Jul;39(7):947–60.
15. Kucherenko S, Sopittakamol N, Shah N. Design space identification of a coupled two-stage batch reactor system. *Computers & Chemical Engineering*. 2024 Aug 1;187:108727.
16. Oberleitner T, Zahel T, Herwig C. Identifying design spaces as linear combinations of parameter ranges for biopharmaceutical control strategies. *Computers & Chemical Engineering*. 2024 Apr 1;183:108555.
17. Geremia M, Bezzo F, Ierapetritou MG. Design space determination of pharmaceutical processes: Effects of control strategies and uncertainty. *European Journal of Pharmaceutics and Biopharmaceutics*. 2024 Jan 1;194:159–69.
18. van der Galien R, Langen AL, Jacobs LJ, Hagen B, Flahive K, Chatterjee SD, van Amsterdam MC. Setup of a contamination control strategy using the hazard analysis critical control point (HACCP) methodology. *PDA Journal of Pharmaceutical Science and Technology*. 2023 Jul 1;77(4):317–28.
19. Rodrigues MI, Iemma AF. *Experimental Design and Process Optimization*. Boca Raton, FL: CRC Press; 2014 Dec 11.
20. Li Z, Lu D, Gao X. Optimization of mixture proportions by statistical experimental design using response surface method-A review. *Journal of Building Engineering*. 2021 Apr 1;36:102101.
21. Gunst RF, Mason RL. Fractional factorial design. *Wiley Interdisciplinary Reviews: Computational Statistics*. 2009 Sep;1(2):234–44.
22. Collins LM, Dziak JJ, Li R. Design of experiments with multiple independent variables: A resource management perspective on complete and reduced factorial designs. *Psychological Methods*. 2009 Sep;14(3):202.
23. Vanaja K, Shobha Rani RH. Design of experiments: Concept and applications of Plackett Burman design. *Clinical Research and Regulatory Affairs*. 2007 Jan 1;24(1):1–23.
24. Reklaitis GV, Seymour C, García-Munoz S, editors. *Comprehensive Quality by Design for Pharmaceutical Product Development and Manufacture*. New York: John Wiley & Sons; 2017 Oct 9.
25. Kim EJ, Kim JH, Kim MS, Jeong SH, Choi DH. Process analytical technology tools for monitoring pharmaceutical unit operations: A control strategy for continuous process verification. Pharmaceutics. 2021 Jun 21;13(6):919.
26. Djuris J, Ibric S, Đurić Z. Quality by design in the pharmaceutical development. In: Djuris J. (eds) *Computer-Aided Applications in Pharmaceutical Technology*. Cambridge: Woodhead Publishing; 2024 Jan 1, pp. 1–21.
27. Gnoth S, Jenzsch M, Simutis R, Lübbert A. Process analytical technology (PAT): Batch-to-batch reproducibility of fermentation processes by robust process operational design and control. *Journal of Biotechnology*. 2007;132:180–6.
28. Chen Y. Multivariate data analysis for biotechnology & bio-processing. *Journal of Commercial Biotechnology*. 2022;27(4):197–204.
29. Dong Y, Yang T, Xing Y, Du J, Meng Q. Data-driven modeling methods and techniques for pharmaceutical processes. *Processes*. 2023 Jul 13;11(7):2096.
30. Esbensen KH, Guyot D, Westad F, Houmoller LP. Multivariate data analysis: In practice: an introduction to multivariate data analysis and experimental design. *Multivariate Data Analysis*. 2002;16(2):117–8.
31. Rajalahti T, Kvalheim OM. Multivariate data analysis in pharmaceutics: A tutorial review. *International Journal of Pharmaceutics*. 2011 Sep 30;417(1–2):280–90.
32. Kay P, Ingham B, Winterburn J. Statistical design of experiments for reaction modelling and optimisation. *Royal Society of Chemistry*. 2023;26:303–18.
33. Rhyne J, Bailey M, Morgan J, Lekivetz R. Assessing the usability of statistical software using designed experiments. In *International Conference on Human-Computer Interaction*; 2023 Jul 9 (pp. 681–688). Cham: Springer Nature Switzerland.

34. Wening K, Breitkreutz J. Oral drug delivery in personalized medicine: Unmet needs and novel approaches. *International Journal of Pharmaceutics*. 2011 Feb;404(1–2):1–9.
35. Zhang L, Mao S. Application of quality by design in the current drug development. *Asian Journal of Pharmaceutical Sciences*. 2017 Jan;12(1):1–8.
36. Ac-Di-Sol®. Pharma Solutions. [cited 2024 Sep 9]. Available from: https://pharma.iff.com/industry-segments/pharma-solutions/products/ac-di-sol
37. Karalia D, Siamidi A, Karalis V, Vlachou M. 3D-printed oral dosage forms: Mechanical properties, computational approaches and applications. *Pharmaceutics*. 2021 Sep 3;13(9):1401.
38. Stegemann S, Connolly P, Matthews W, Barnett R, Aylott M, Schrooten K, et al. Application of QbD principles for the evaluation of empty hard capsules as an input parameter in formulation development and manufacturing. *AAPS PharmSciTech*. 2014 Feb 14;15(3):542–9.
39. Zhang J, Thakkar R, Zhang Y, Maniruzzaman M. Structure-function correlation and personalized 3D printed tablets using a quality by design (QbD) approach. *International Journal of Pharmaceutics*. 2020 Nov; 590:119945.
40. Bicouv C, Pereira LNA, Giorgetti L. Application of FMEA tool to analyze notifications of technical complaints of solid dosage forms in the context of quality by design. *Brazilian Journal of Natural Sciences*. 2020 Dec 1;3(3):474.
41. Ansari MT, Alahmed TAA, Sami F. Quality by Design (QbD) concept for formulation of oral formulations for tablets. In: Jain NK, Bajwa N. (eds) *Introduction to Quality by Design (QbD)*. Singapore: Springer Nature Singapore; 2024 [cited 2024 Sep 9], pp. 161–84. Available from: https://doi.org/10.1007/978-981-99-8034-5_7
42. Park K. Controlled drug delivery systems: Past forward and future back. *Journal of Controlled Release*. 2014 Sep; 190:3–8.
43. Yu B, Tai HC, Xue W, Lee LJ, Lee RJ. Receptor-targeted nanocarriers for therapeutic delivery to cancer. *Molecular Membrane Biology*. 2010 Oct;27(7):286–98.
44. Din F, Aman W, Ullah I, Qureshi OS, Mustapha O, Shafique S, et al. Effective use of nanocarriers as drug delivery systems for the treatment of selected tumors. *International Journal of Nanomedicine*. 2017 Oct;12:7291–309.
45. Lan HR, Zhang YN, Han YJ, Yao SY, Yang MX, Xu XG, et al. Multifunctional nanocarriers for targeted drug delivery and diagnostic applications of lymph nodes metastasis: A review of recent trends and future perspectives. *Journal of Nanobiotechnology*. 2023 Aug 2;21(1):247.
46. Calvey M. *What You Need to Know about ICH Q8*. POS Nation. 2023 [cited 2024 Sep 10]. Available from: https://www.qualio.com/blog/ich-q8
47. *ICH Q8 (R2) Pharmaceutical Development - Scientific Guideline*. European Medicines Agency (EMA). Available from: https://www.ema.europa.eu/en/ich-q8-r2-pharmaceutical-development-scientific-guideline
48. Cunha S, Costa CP, Moreira JN, Sousa Lobo JM, Silva AC. Using the quality by design (QbD) approach to optimize formulations of lipid nanoparticles and nanoemulsions: A review. *Nanomedicine: Nanotechnology, Biology and Medicine*. 2020 Aug;28:102206.
49. Wang Y, Yue Y, Jia R, Liu X, Cheng Z, Cheng Y, et al. Design and evaluation of paeonol-loaded liposomes in thermoreversible gels for atopic dermatitis. *Gels*. 2023 Mar 5;9(3):198.
50. Spinks C, Zidan A, Khan M, Habib M, Faustino P. Pharmaceutical characterization of novel tenofovir liposomal formulations for enhanced oral drug delivery: In vitro pharmaceutics and Caco-2 permeability investigations. *Clinical Pharmacology: Advances and Applications*. 2017 Feb;9:29–38.
51. Jose S, Cinu TA, Sebastian R, Shoja MH, Aleykutty NA, Durazzo A, et al. Transferrin-conjugated docetaxel–PLGA nanoparticles for tumor targeting: Influence on MCF-7 cell cycle. *Polymers*. 2019 Nov 19;11(11):1905.
52. Haider Md F, Kanoujia J, Tripathi CB, Arya M, Kaithwas G, Saraf SA. Pioglitazone loaded vesicular carriers for anti-diabetic activity: Development and optimization as per central composite design. *Journal of Pharmaceutical Sciences and Pharmacology*. 2015 Mar 1;2(1):11–20.
53. Kumar NS. D-optimal experimental approach for designing topical microemulsion of itraconazole: Characterization and evaluation of antifungal efficacy against a standardized Tinea pedis infection model in Wistar rats. *European Journal of Pharmaceutical Sciences*. 2015 Jan;67:97–112.
54. Bhatt S, Sharma S, Sharma JB, Kumar M. Optimization and fabrication of curcumin loaded solid lipid nanoparticles using box-behnken design for nasal delivery. *Nanoscience & Nanotechnology-Asia*. 2022 Dec;12(6):7–18.
55. Mandpe L, Pokharkar V. Quality by design approach to understand the process of optimization of iloperidone nanostructured lipid carriers for oral bioavailability enhancement. *Pharmaceutical Development and Technology*. 2013 Dec 13;20(3):320–9.
56. *International Journal of Pharmaceutical Sciences and Research*. 2015 Oct 1;6(10).
57. Qumbar M, Ameeduzzafar A, Imam SS, Ali J, Ahmad J, Ali A. Formulation and optimization of lacidipine loaded niosomal gel for transdermal delivery: In-vitro characterization and in-vivo activity. *Biomedicine & Pharmacotherapy*. 2017 Sep;93:255–66.
58. Qushawy M, Nasr A, Abd-Alhaseeb M, Swidan S. Design, Optimization and characterization of a transfersomal gel using miconazole nitrate for the treatment of candida skin infections. *Pharmaceutics*. 2018 Feb 23;10(1):26.
59. Mandal U, Mahmood S, Taher M. Experimental design and optimization of raloxifene hydrochloride loaded nanotransfersomes for transdermal application. *International Journal of Nanomedicine*. 2014 Sep;9:4331–46.
60. Iqbal B, Ali J, Ganguli M, Mishra S, Baboota S. Silymarin-loaded nanostructured lipid carrier gel for the treatment of skin cancer. *Nanomedicine*. 2019 May;14(9):1077–93.
61. Dhawan V, Lokras A, Joshi G, Marwah M, Venkatraman M, Mohanty B, et al. Polysaccharide and monosaccharide guided liver delivery of Sorafenib Tosylate – A nanostrategic approach and comparative assessment of hepatospecificity. *International Journal of Pharmaceutics*. 2022 Sep;625:122039.
62. Kamal N, Cutie AJ, Habib MJ, Zidan AS. QbD approach to investigate product and process variabilities for brain targeting liposomes. *Journal of Liposome Research*. 2014 Oct 13;25(3):175–90.
63. Radwan SAA, ElMeshad AN, Shoukri RA. Microemulsion loaded hydrogel as a promising vehicle for dermal delivery of the antifungal sertaconazole: Design, optimization and ex vivo evaluation. *Drug Development and Industrial Pharmacy*. 2017 Apr 27;43(8):1351–65.

64. Anwar M, Akhter S, Mallick N, Mohapatra S, Zafar S, Rizvi MMA, et al. Enhanced anti-tumor efficacy of paclitaxel with PEGylated lipidic nanocapsules in presence of curcumin and poloxamer: In vitro and in vivo studies. *Pharmacological Research*. 2016 Nov;113:146–65.
65. Merlo-Mas J, Tomsen-Melero J, Corchero JL, González-Mira E, Font A, Pedersen JN, et al. Application of quality by design to the robust preparation of a liposomal GLA formulation by DELOS-susp method. *The Journal of Supercritical Fluids*. 2021 Jul;173:105204.
66. Mendes M, Basso J, Silva J, Cova T, Sousa J, Pais A, et al. Biomimeting ultra-small lipid nanoconstructs for glioblastoma treatment: A computationally guided experimental approach. *International Journal of Pharmaceutics*. 2020 Sep;587:119661.
67. El Afify MS, Zein El Dein EA, Elsadek BEM, Mohamed MA, El-Gizawy SA. Development and optimization of a novel drug free nanolipid vesicular system for treatment of osteoarthritis. *Drug Development and Industrial Pharmacy*. 2017 Dec 19;44(5):767–77.
68. Said M, Elsayed I, Aboelwafa AA, Elshafeey AH. Transdermal agomelatine microemulsion gel: Pyramidal screening, statistical optimization and in vivo bioavailability. *Drug Delivery*. 2017 Jan 1;24(1):1159–69.
69. Waghule T, Laxmi Swetha K, Roy A, Narayan Saha R, Singhvi G. Quality by design assisted optimization of temozolomide loaded PEGylated lyotropic liquid crystals: Investigating various formulation and process variables along with in-vitro characterization. *Journal of Molecular Liquids*. 2022 Apr;352:118724.
70. Lee SH, Kim JE. Quality by design applied development of immediate-release rabeprazole sodium dry-coated tablet. *Pharmaceutics*. 2021 Feb 14;13(2):259.
71. Rapalli VK, Banerjee S, Khan S, Jha PN, Gupta G, Dua K, et al. QbD-driven formulation development and evaluation of topical hydrogel containing ketoconazole loaded cubosomes. *Materials Science and Engineering: C*. 2021 Feb;119:111548.
72. Patel D, Patel M, Soni T, Suhagia B. Topical arginine solid lipid nanoparticles: Development and characterization by QbD approach. *Journal of Drug Delivery Science and Technology*. 2021 Feb;61:102329.
73. Pallagi E, Jójárt-Laczkovich O, Németh Z, Szabó-Révész P, Csóka I. Application of the QbD-based approach in the early development of liposomes for nasal administration. *International Journal of Pharmaceutics*. 2019 May;562:11–22.
74. Saha M, Saha DR, Ulhosna T, Sharker SM, Shohag MH, Islam MS, et al. QbD based development of resveratrol-loaded mucoadhesive lecithin/chitosan nanoparticles for prolonged ocular drug delivery. *Journal of Drug Delivery Science and Technology*. 2021 Jun;63:102480.
75. Thomas F. Operational excellence in pharmaceutical manufacturing. *Industry Immersion Learning*. 2009 Jan 28;2006:141–73.
76. O'Donnell ME. *Applying Six Sigma to Areas Outside of Manufacturing in the Space Industry. Lessons Learned from Applying Raytheon Six Sigma to PATRIOT Missile Software and Systems Engineering*. American Institute of Aeronautics and Astronautics, Reston, Virigina; 2005 [cited 2024 Sep 11]. Available from: https://doi.org/10.2514/6.2005-6764
77. Bloom D. *Achieving HR Excellence through Six Sigma*. Productivity Press, New York; 2017 [cited 2024 Sep 11]. Available from: https://doi.org/10.1201/b15357
78. Venables M. Striving for process excellence [process industry]. *Manufacturing Engineer*. 2004 Dec 1;83(6):40–3.
79. Tarantino A. Lean six sigma in the age of smart manufacturing. *Smart Manufacturing*. 2022 May 10;577:21–42.
80. *The Advent of Artificial Intelligence & Machine Learning in Medical Devices*. QbD Group. 2022 [cited 2024 Sep 11]. Available from: https://qbdgroup.com/en/blog/the-advent-of-artificial-intelligence-machine-learning-in-medical-devices/
81. Verch T. Application of quality by design and statistical quality control concepts in immunoassays. *Bioanalysis*. 2014 Dec;6(23):3251–60.
82. Gupta S, Jhawat V. Quality by design (QbD) approach of pharmacogenomics in drug designing and formulation development for optimization of drug delivery systems. *Journal of Controlled Release*. 2017 Jan;245:15–26.
83. Vora LK, Gholap AD, Jetha K, Thakur RRS, Solanki HK, Chavda VP. Artificial intelligence in pharmaceutical technology and drug delivery design. *Pharmaceutics*. 2023 Jul 10;15(7):1916.
84. Kavasidis I, Lallas E, Gerogiannis VC, Charitou T, Karageorgos A. Predictive maintenance in pharmaceutical manufacturing lines using deep transformers. *Procedia Computer Science*. 2023;220:576–83.
85. Aksu B, Paradkar A, de Matas M, Özer Ö, Güneri T, York P. Quality by design approach: Application of artificial intelligence techniques of tablets manufactured by direct compression. *AAPS PharmSciTech*. 2012 Sep 6;13(4):1138–46.
86. Mazumder S, Pavurala N, Manda P, Xu X, Cruz CN, Krishnaiah YSR. Quality by design approach for studying the impact of formulation and process variables on product quality of oral disintegrating films. *International Journal of Pharmaceutics*. 2017 Jul;527(1–2):151–60.
87. Simões MF, Silva G, Pinto AC, Fonseca M, Silva NE, Pinto RMA, et al. Artificial neural networks applied to quality-by-design: From formulation development to clinical outcome. *European Journal of Pharmaceutics and Biopharmaceutics*. 2020 Jul;152:282–95.
88. Kim JY, Choi DH. Control strategy for excipient variability in the quality by design approach using statistical analysis and predictive model: Effect of microcrystalline cellulose variability on design space. *Pharmaceutics*. 2022 Nov 8;14(11):2416.
89. Dawoud MHS, Mannaa IS, Abdel-Daim A, Sweed NM. Integrating artificial intelligence with quality by design in the formulation of lecithin/chitosan nanoparticles of a poorly water-soluble drug. *AAPS PharmSciTech*. 2023 Aug 8;24(6):169.
90. Dorota-Owczarek. *Process Analytical Technology (PAT) and Artificial Intelligence: The Expanding Impact of ML in Drug Manufacturing*. NEXOCODE. 2021. Available from: https://nexocode.com/blog/posts/process-analytical-technology-pat-with-ai-in-biopharmaceutical-manufacturing/
91. Sushma D. Machine learning applications in drug discovery and development. *Open Access Journal of Pharmaceutical Research*. 2023;7(4):000290.
92. Calhan SD, Eker ED, Sahin NO. Quality by design (QbD) and process analytical technology (PAT) applications in pharmaceutical industry. *European Journal of Chemistry*. 2017 Dec 31;8(4):430–3.

93. Stagner WC, Haware RV. QbD innovation through advances in PAT, data analysis methodologies, and material characterization. *AAPS PharmSciTech*. 2019 Aug 22;20(7):295.
94. Zhang X, Gao H, Wang H, Chen Z, Zhang Z, Chen X, et al. PLANET: A multi-objective graph neural network model for protein–ligand binding affinity prediction. *Journal of Chemical Information and Modeling*. 2023 Jun 15;64(7):2205–20.
95. Siepmann J, Siepmann F. Mathematical modeling of drug delivery. *International Journal of Pharmaceutics*. 2008 Dec;364(2):328–43.

16 In Silico Designing of Vaccines
Methods, Tools, and Their Limitations

Rakibul Islam[1], Abhishek Samanta[1], Sawan Kumar[1], Agniv Paul[1], Jasmeet Kaur[2], Utpal Mohan[2], and Devendra Kumar Dhaked[1]
[1]Department of Pharmacoinformatics, National Institute of Pharmaceutical Education and Research (NIPER), Kolkata, India
[2]Department of Medicinal Chemistry, National Institute of Pharmaceutical Education and Research (NIPER), Kolkata, India

INTRODUCTION

Vaccines are biological preparations intended to stimulate the body's immune system to identify and protect against disease-causing pathogens, such as viruses and bacteria, thereby preventing the disease. They typically contain weakened or inactivated forms of disease-causing agents or parts of them, such as proteins and sugars. Some vaccinations include well-defined antigens, such as the surface antigen of hepatitis B and the polysaccharide of *Haemophilus influenzae* type B. In contrast, other vaccines contain entire organisms, such as live attenuated viruses and dead *Bordetella pertussis*, that present several possible antigens on the cell surface [1]. Another category includes subunit-based vaccines, which contain fragments of a pathogen, not the entire organism. As they cannot infect, this qualifies them for use in populations including small children, the elderly, and immunocompromised individuals who should not receive vaccines based on live organisms. Existing vaccines are classified into multiple subtypes, as mentioned in Table 16.1. When vaccination is given, the immune system develops a defense against these invading foreign substances by producing antibodies and memory cells. The antibodies produced by the immune system are specific to the pathogen introduced by the vaccine. When the individual is exposed for the second time to the actual pathogen, the immune response enables the body to identify and combat the actual pathogen to avoid infection. Vaccines thus play a crucial role in preventing contagious diseases and significantly reducing morbidity and mortality globally [2].

The history of vaccination began in 1796 with the discovery of a vaccine against smallpox, a deadly sickness that had afflicted humans for generations. In the 18th century, a British physician named Edward Jenner observed

TABLE 16.1
Classification of Different Types of Vaccines with Description and Examples

Type of Vaccine	Description	Example
Live attenuated vaccine	Weakened versions of the live virus or bacterium	BCG vaccine
Inactivated vaccine	Destroyed versions of viruses or bacteria	Inactivated polio vaccine (IPV)
Recombinant vaccine	Viral or bacterial genome is inserted into plasmid to produce vaccine	Human papillomavirus (HPV) vaccine
mRNA vaccine	Newer forms of vaccines that function by directing cells in the body to make a protein that causes an immunological reaction	Pfizer-BioNTech and Moderna's COVID-19 vaccine
Toxoid vaccine	Contains inactivated toxins (toxoids) that stimulate the body's immune response to produce antibodies against the toxin	Tetanus vaccine
DNA vaccine	Unique approach to stimulate humoral and cellular immune responses by injecting genetically engineered DNA	Polyclonal and monoclonal antibodies
Multi-epitope-based vaccine	B-cell and T-cell epitopes are joined via suitable linkers	HIVACAT T-cell immunogen (HTI) vaccine [3]

DOI: 10.1201/9781003464679-16

that milkmaids with cowpox, a less severe disease, seemed immune to smallpox. This realization was the inspiration for the development of the first vaccination. Jenner proved that vaccinating a youngster against cowpox and then exposing him to smallpox would prevent the development of the more serious illness. Jenner's discoveries made possible immunizations against polio, measles, mumps, rubella, and other illnesses. Right now, vaccinations rank among the best methods for preventing infectious diseases [4]. The development of vaccines in human history with the timeline is shown in Figure 16.1a. When our body is invaded by pathogens such as viruses and bacteria, the T-cell receptor detects the pathogen's antigen by means of antigen-presenting cells and triggers a cell-based immune response. Now, the sensitized T-cells trigger sap-based immune response, and B-cells are activated, which read the epitopes on the presented antigen and start secreting immunoglobulin (an antibody) specific to the pathogen's antigen as an immediate response. T-cells have the ability to recognize antigens present on the surface of antigen-presenting cells through T-cell receptors complexed with the major histocompatibility complex (MHC) and the cluster of differentiation (CD) receptor. Figure 16.1b depicts how immune cells, e.g., B-cells and T-cells, recognize the entered antigen [5].

Antibodies are molecular players that are secreted during an immune response. Structurally, it has two fragments: Fab (stands for fragment along with the antigen-binding domain) and F_c (crystallizes readily but does not bind to the antigen). *Fc* plays an essential role in mediating responses, termed effector functions. Two polypeptide chains, a 50-kd heavy (H) chain and a 25-kd light (L) chain, make up immunoglobulin (IgG), which is the most abundant antibody in the serum. L2H2 can represent this antibody. The antibody structure is shown in Figure 16.1c. L chain and H chain are connected via a disulfide bridge [6].

Computational vaccine design has emerged as an interesting field that specifically focuses on the design of vaccine molecules. The success story of computational vaccine design has been observed in recent years; when the human population was suffering from the COVID-19 pandemic, the global research community came out with a vaccine that saved millions of lives with the help of computational tools [7].

BASIC STEPS OF VACCINE DESIGN

Selection of the Target for Epitope Identification

Epitope identification in the target protein involves a combination of experimental and computational methods. Experimental epitope identification is costly and challenging, requiring antibody production in animal models and epitope validation through crystallography. Identification of epitopes remains an interesting area for numerous reasons, such as identifying the causes of diseases, immune surveillance, diagnostic tests, and the development of vaccines based on epitopes [8]. The epitope identification process is time-consuming, and it requires experimental screening of a wide range of epitope candidates. However, there is a surge of a wide array of *in silico* techniques and tools designed to narrow down the number of possible epitope choices before testing them experimentally. Computational approaches guide experimental assays by selecting regions with a good probability of serving as useful epitopes. These approaches consider amino acid properties such as hydrophobicity and

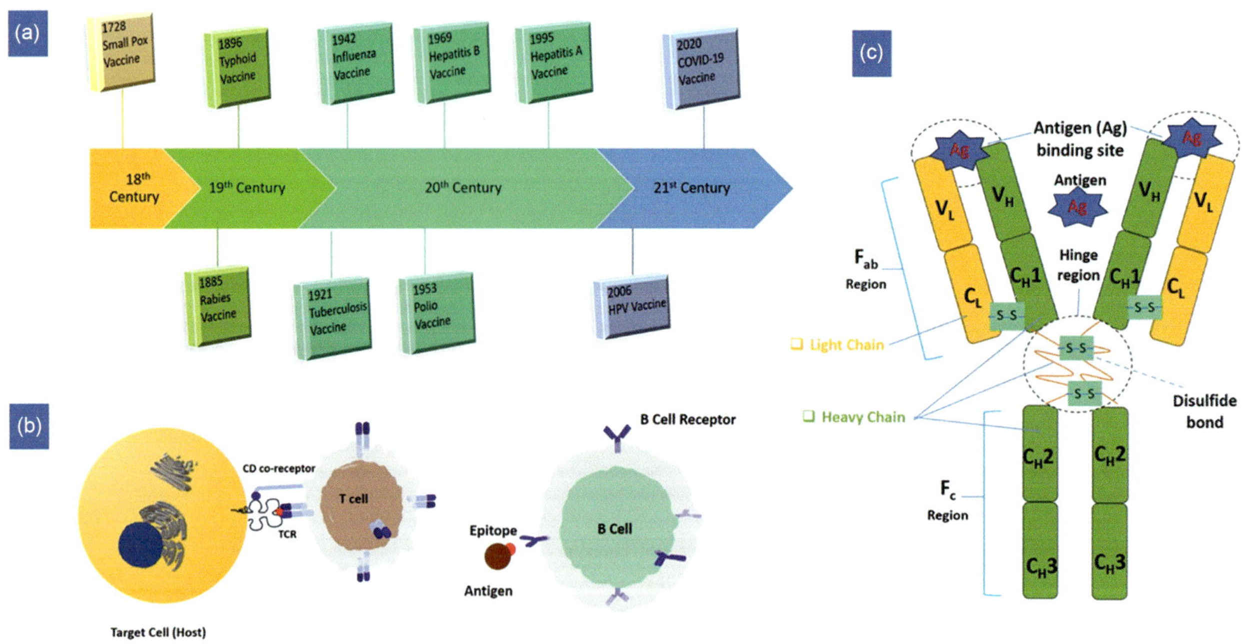

FIGURE 16.1 (a) Vaccine development timeline, (b) BCR and TCR complexes with the epitope of antigen, and (c) conventional representation of antibody structure.

hydrophilicity to predict epitope localization accurately. Computational techniques can accurately predict epitopes from the protein of interest, and they can foster the choice of valuable epitopes by evaluating some properties of epitopes, such as allergenicity, toxicity, and antigenicity [9]. Computational vaccine design requires extensive and in-depth knowledge of epitopes because epitopes are immediately recognized by immune cells, and then, they release antigen-specific antibodies to neutralize antigens [10].

Sequence Retrieval and Analysis

The National Center for Biotechnology Information [11] (NCBI) and the Universal Protein Resource [12] (UniProt) are well-known repositories for the sequences of millions of proteins. Protein Database [13] (PDB), NCBI, and UniProt are generally utilized to acquire amino acid sequences for a target protein. These platforms provide sequences of proteins in the fasta file format. Multiple fasta sequences can also be stored in a single file. Many web servers often use fasta sequences. Understanding how a computationally designed vaccine interacts with targets such as toll-like receptors (TLR3 and TLR4) in the body is necessary. It is also crucial to thoroughly study how the developed vaccine interacts with the MHC. The interactions between multi-epitope-based vaccines and targets at the molecular level can be estimated using molecular docking and molecular dynamics (MD) simulations. These techniques require three-dimensional (3D) structures of target proteins that are available in PDB [14].

Epitope Prediction

Immunological cells in the body, specifically the T- and B-cells, recognize different antigens. B-cells recognize epitopes that are highly accessible sites on the exposed surface of the antigen [15]. B-cell epitopes are found on soluble antigens and bind to the membrane-bound antibody on the B-cell membrane surface. On the other hand, T-cells only recognize peptides combined with MHC molecules. These MHC-bound peptides are found on antigen-presenting and altered-self cell surfaces. These T-cell epitopes cannot be considered apart from their associated MHC molecules [5]. Different methods and tools are used for epitope prediction: sequence-based, structure-based, and machine learning (ML)-based.

B-Cell Epitope

B-cell epitope prediction involves identifying specific protein regions that antibodies can recognize. B-cell epitopes are of two types: (i) linear epitope and (ii) conformational epitope. B-cell epitope prediction is crucial for various applications such as vaccine development and understanding immunological processes. Tools such as BepiPred and BCEPRED are commonly used to predict linear B-cell epitopes. Discotope and BEPro tools are used to predict conformational B-cell epitopes. Some B-cell epitope prediction tools are listed in Table 16.2.

BepiPred-3.0

This tool predicts B-cell epitopes based on sequence and uses protein language models in the background. It is the updated version of BepiPred-2.0 [17]. This tool is trained on a large dataset of protein sequences and structures. From amino acid sequences, it can reliably predict both local and global protein structural features. It can predict linear and conformational B-cell epitopes with improved accuracy and precision. This tool can be accessed at https://services.healthtech.dtu.dk/services/BepiPred-3.0/. It takes a fasta sequence of an antigen as the input. The fasta sequence for which epitopes need to be predicted should be pasted in the textbox under the submission tab, or the same can be uploaded by clicking the "choose file" button on the web server. It allows a maximum of 40 sequences of epitopes to be predicted simultaneously. A fasta sequence starts with a greater than (>) symbol followed by the names of the target protein and organism or species in the first line. The amino acid sequence with single-letter code is present in the next line. After submitting a fasta sequence to the BepiPred-3.0 web server, results are generated and can be found in a zip

TABLE 16.2
B-Cell Epitope Prediction Tools

Server	Function
BCEPRED [16]	Predicts continuous B-cell epitopes using a neural network algorithm
BepiPred [17]	Predicts sequence-based linear B-cell epitopes
Discotope [18]	Predicts B-cell epitopes from 3D structure
BEPro [19]	Predicts discontinuous B-cell epitopes
EpiSearch [20]	Automatically searches possible locations of conformational epitopes on the antigen surface
Pep-3D-Search [21]	Predicts conformational B-cell epitopes
Epitopia [22]	Predicts continuous and discontinuous B-cell epitopes
ElliPro [23]	Predicts continuous and discontinuous B-cell epitopes

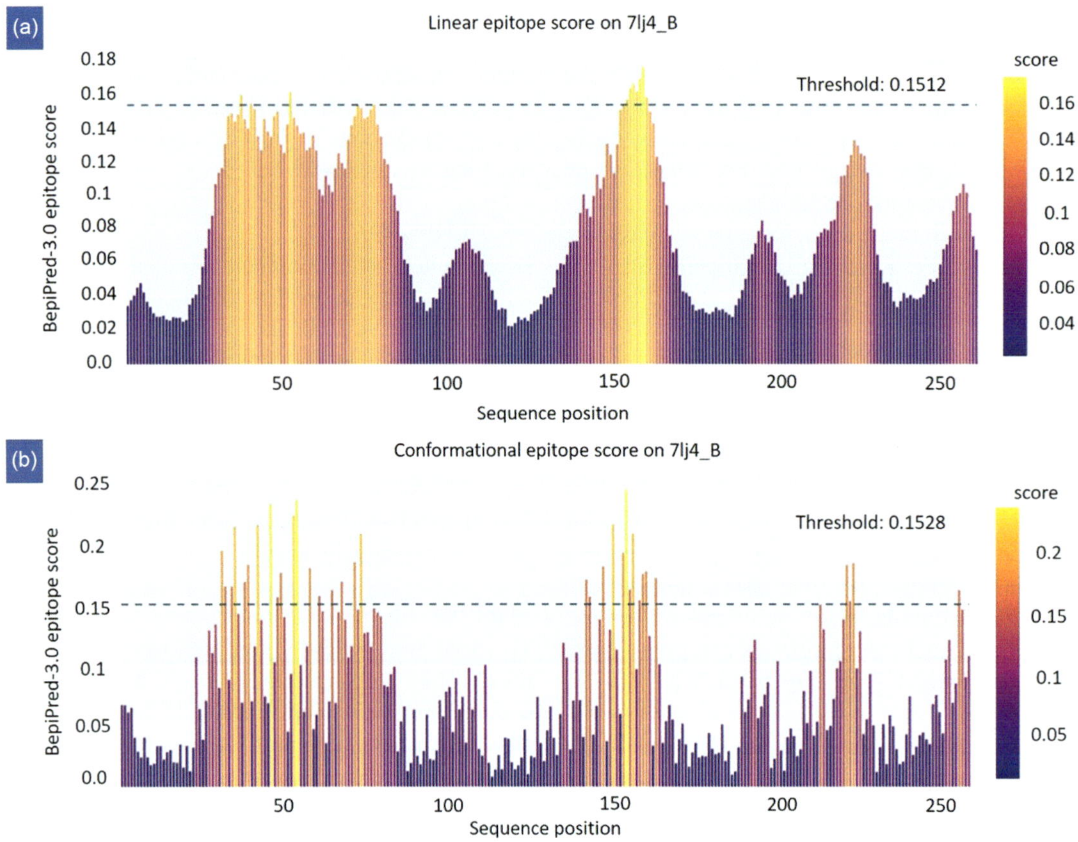

FIGURE 16.2 Graph of (a) B-cell linear epitope and (b) B-cell conformational epitope.

file consisting of four files, and a separate HTML file containing a graph (epitope score vs. sequence position) is displayed (Figure 16.2).

Limitation: It has memory limitations and can predict no more than 40 antigen sequences at a time. Overfitting of the model may lead to incorrect epitope prediction. Solving overfitting problems may enhance performance. High-quality training data is required to correctly predict the true positives of peptides. The independent benchmarking of three datasets shows the area under the curve (AUC) in the range of 0.66–0.77, indicating ample room for improvement.

Discotope-2.0

This tool predicts discontinuous B-cell epitopes from 3D structures of an antigenic protein. It takes the protein's PDB format as the input for which epitopes need to be predicted. It utilizes a unique method that calculates the epitope propensity score of amino acids as well as surface accessibility. The propensity scores of residues in close geographic proximity and contact numbers are combined to determine Discotope scores. This tool is super-user-friendly. The steps involved in Discotpe-2.0 are as follows: first, enter the PDB ID (four-letter code) of the respective protein or upload the protein structure in the PDB format; then, select a particular chain or multiple chains (for selecting multiple chains, the PDB file needs to be modified); and then, enter a threshold value for epitope prediction. Finally, results are displayed in the server in 3D view (epitope prediction) and Table view, which contains information about identifiers of chain and residue along with Discotope and propensity score. The contact number of a residue is defined as the number of alpha carbon atoms within a distance of 10 Å of the alpha carbon atom in that residue. A low contact number indicates that the residue is near the antigen's surface. The propensity score describes the probability of that residue being part of an epitope. In addition, it allows users to view 3D structures that are displayed using Jmol, with favorable predictions indicated in yellow. Discotope-2.0 server is accessible at https://services.healthtech.dtu.dk/services/DiscoTope-2.0/.

Limitation: The correctness of the prediction depends on the quality of the input structure (poor resolution may reduce the prediction quality). The error in the input structure may impact the prediction accuracy. The used training dataset does not cover all the antigen and antibody complexes. This server achieves an AUC performance of 0.737. The performance of prediction can be improved by considering the entire protein complex rather than a single unit and excluding the residues that do not interact with the antibody (like residues embedded in the transmembrane region).

ElliPro

ElliPro is a powerful tool that predicts epitopes of B-cells. It utilizes a residue clustering algorithm. It also has a molecule viewer option to display the predicted epitopes of the given antigen. ElliPro is trained on many antibody–protein complexes to predict epitopes accurately. One of the major advantages of using this tool is that it can predict both continuous and discontinuous epitopes of B-cells from the 3D structure of a protein. This tool can be accessed at http://tools.immuneepitope.org/tools/ElliPro. ElliPro accepts either a protein sequence or a 3D structure as the input. A fasta file or UniProt ID of the respective protein can be supplied. Then, this tool looks for a suitable template model for the submitted sequence by performing a BLAST search. The user can provide a PDB ID or even submit their protein in the PDB format. They will be asked to choose chain(s) if their protein has multiple chains. This tool has two parameters, protrusion index and distance, represented by S and R symbols. The minimum value of S and R can be specified. ElliPro uses the BLAST algorithm to search for homologous proteins in the database. One can develop homology models using the MODELLER software if BLAST cannot produce a satisfactory model.

ElliPro does three important tasks: first, it approximates the shape of the protein; second, it calculates the protrusion index of individual residues; and third, it performs cluster analysis of neighboring residues. ElliPro gives results of linear and conformational epitopes, epitope score, and sequence length.

Limitations: The quality of the prediction is solely dependent on the quality of the input. It considers protein as an ellipsoid shape, which may not be the appropriate representation for a multidomain protein. It only considers the whole geometrical characteristic of proteins, which may not be a sufficient feature to predict potential peptides. This is a structure-based method that depends on the 3D shape of a protein; hence, it cannot be used to predict epitopes from protein sequences alone. The performance is limited, with an AUC of 0.732. It lacks tools for antibody–antigen modeling or docking; users need to use other tools or servers.

Epitopia

Epitopia server is an online tool designed to predict regions in a protein's structure or sequence that are likely to trigger an immune response via B-cell activation. It utilizes the Naïve Bayes classifier algorithm, which is trained to identify antigenic properties within proteins, offering higher predictive accuracy than similar tools. Special attention has been given to creating a user-friendly interface for result visualization. Epitopia can predict epitopes from a protein 3D structure or a linear protein sequence input. Users can also specify relevant protein chains, choose all chains or specific ones, or mark checkboxes to include or exclude certain chains. All chains influence structural considerations for immunogenicity scores, but scores are computed only for the selected chains. Users can paste amino acid sequences or upload local sequence files in the fasta format for protein sequence inputs. Epitopia computes immunogenicity and probability scores for each input. Furthermore, immunogenicity scores are projected onto the protein and highlighted in color. Scores are also visually represented on the protein structure using FirstGlance in Jmol, offering various display options. In addition, a RasMol command script is available to view results.

Limitation: The AUC underestimates the true predictive potential of the prediction since the antigen is thought to harbor a far higher number of epitopes than are currently known. The performance of Epitopia is limited, with an AUC of 0.6, which is lower compared to other methods discussed. It cannot model antibody–antigen interactions.

AxIEM

It is an ML-based B-Cell epitope prediction tool. It is a revolutionary approach for epitope mapping. Traditional methods of epitope mapping are valuable but have limitations. Techniques such as X-ray crystallography and peptide binding assays often struggle with the dynamic nature of specific viral proteins, particularly fusion proteins. These proteins, crucial for viral entry into host cells, undergo significant structural changes during infection. These conformational shifts can render epitopes invisible to traditional methods, hindering our understanding of immune response and vaccine development.

Accelerated Class I Fusion Protein Epitope Mapping (AxIEM) is an ML-based approach for predicting discontinuous antibody epitopes while considering the flexibility of fusion proteins. AxIEM leverages a type of ML called low-complexity supervised learning methods to analyze amino acid sequences and protein structures.

The datasets, containing 46,710 residues, are built from seven class I fusion proteins from 34 PDB structures containing 82 unique epitopes. The prepared datasets are trained and tested on different supervised ML algorithms. Among these models, the linear regression model outperforms other epitope prediction tools such as ElliPro and Discotope in predicting tertiary and quaternary epitopes of class I viral fusion proteins. It requires less processing power and offers a valuable tool for vaccine development approaches, including finding new antigenic sites, contrasting homologs of fusion proteins, and using AxIEM in vaccine design platforms. The AxIEM repository is freely available at https://github.com/mfsfischer/AxIEM.

Limitation: Sequence identity may have an impact on the prediction. Other tools such as MD simulation may help study how protein flexibility influences antigenicity. This tool further needs to be improved in terms of accuracy and precision.

T-Cell Epitope

T-cell epitopes are short linear peptides separated from antigenic proteins and identified by T-cells. These are typically 8–11 and 13–17 residues long for MHC classes I and II, respectively. Predicting T-cell epitopes is crucial for understanding immune responses and designing vaccines or immunotherapies. T-cell epitope prediction tools are utilized to predict the ability of a peptide–MHC complex to elicit an immune response. Several tools and servers for predicting T-cell epitopes are listed in Table 16.3. The best strategy for preventing infectious diseases is vaccination, which has enormous advantages for human health. Accurately identifying T-cell epitopes and designing multi-epitope-based vaccines depend massively on the precise and reliable prediction of the interaction between the MHC protein and peptide. Up to now, the prediction of MHC class I or II peptide binding has been shown to be resistant and erratic. Here, we demonstrate the usefulness of currently available computational techniques for the prediction of peptide binding to MHC proteins [24].

IEDB

The Immune Epitope Database (IEDB) is a comprehensive compilation of tools designed to predict and analyze possible epitopes of T- and B-cells. Various tools are available in this database to analyze a given peptide sequence to determine possible T-cell and B-cell epitopes. The IEDB tool offers both linear and discontinuous epitope predictions. It provides MHC class I and II prediction using a sequence-based method. The prediction method can be chosen from a list of methods available on the web tool, which includes methods such as artificial neural network (ANN), consensus, and others. Most of these methods give predicted IC_{50} values as output. The consensus method, on the other hand, predicts the percentile rank. These values can be used to screen out the predicted epitopes. This website also contains a library of alleles from which we can choose the alleles of interest if the data on the allele is available. We can also work with the whole allele library if allele data is unavailable. Choosing an allele is very important for the prediction of possible T-cell epitopes. As previously mentioned, T-cell epitopes fall into two categories: MHC class I and II. Both can be predicted using the IEDB toolkit, and similar methods can be applied. The only difference is that the allele data in the library differs between these two classes.

EpiJen

EpiJen is an integrated multistep tool for T-cell epitope prediction. As we know, recognition by T-cells involves multiple steps. First, the target protein is degraded by the multimeric proteinase known as proteasome. The three active sites of the proteasome include one similar to trypsin, the second that functions similarly to chymotrypsin, and the third that functions similarly to peptidylglutamyl-peptide hydrolytic activity. Subsequently, the resultant peptides are transported via a transporter associated with antigen processing (TAP). TAP is a peptide transport protein that is ATP-dependent and a member of the ABC family of transporters. This family moves various molecules across membranes, including large polypeptides and small carbohydrates. According to experimental studies, TAP favors peptides with eight or more amino acids and basic or hydrophobic residues at the carboxy terminus. Most of the epitopes found in membrane or secretory proteins and cytosolic antigens are presented by TAP-mediated antigen presentation. Lastly, these peptides bind with MHC proteins and are presented for T-cell recognition. The EpiJen tool simulates all these steps in an integrated manner and

TABLE 16.3
T-Cell Epitope Prediction Tools or Servers

Server Name	Description
EpiJen [25]	It uses a multistep algorithm for the prediction of MHC-I molecules
SYFPEITHI [26]	Database that includes peptide motifs and MHC ligands for humans and other species, as well as for MHC classes I and II. T-cell epitopes, MHC ligands, and MHC peptide motifs are all included in the database
nHLAPred [27]	This tool is used for the prediction of MHC-I epitope; this server is divided into two parts, COMpred and ANNpred
ProPred1 [28]	This server is written in PERL programming and is used to predict T-cell epitopes and proteasomal cleavage sites
ProPred [29]	To predict MHC class II binding sites by utilizing quantitative matrices in an antigen sequence
MHCPred [30]	This server implements statistical techniques to estimate binding strength between peptide and MHC molecules
MHC2Pred [31]	This server predicts MHC class II binders by using matrix optimization and a supervised machine learning algorithm
NetMHC [32]	This server uses ANN-based method for the identification of peptide–MHC class I binding
RANKPEP [33]	This server is used to predict MHC I and II ligands using position-specific scoring matrices (PSSM)
SVMHC [34]	This tool can predict T-cell epitopes with high accuracy. This tool uses a model that was trained on SVM-based machine learning algorithm
IEDB [35]	IEDB is one of the most commonly used databases, and it has numerous tools to predict MHC binding epitopes
MMBPred [36]	This server applies a quantitative matrix algorithm for the prediction of MHC-I binders
SVRMHC [37]	This predicting tool uses a support vector-based regression model to calculate binding affinity between peptides and MHC molecules accurately

recognizes the top 5% of all peptide fragments of the given target as T-cell-recognizable antigens.

Limitation: This tool uses multiple filters and steps to predict top epitope candidates, making it difficult for non-experts to understand. One of the limitations is the filtering capability. Sometimes, it may not remove all the true negative peptides. There is still a chance to improve the optimization of the peptides. Fine-tuning the different filtering criteria may enhance its performance.

ProPred

The ProPred is a graphical server designed to predict the HLA-DR binding of an antigenic protein's primary structure. This server is based on another server, TEPITOPE, initially developed by Sturniolo and coworkers, based on a novel approach called Pocket Profiles for constructing matrices [38]. The ProPred server uses the ReadSeq program to read input sequences. So, fasta format files or plain text are accepted as input. This server generates graphics in the GIF format using GDPlot. The main advantage of ProPred is that, besides predicting the binding regions of a given primary structure, it can also predict the binding strength of those regions to the HLA-DR.

Limitation: Every matrix used in the server is taken from the literature and various servers. The selection of matrices is based not on performance but rather on availability from a single source. Thus, if more than one matrix is accessible in the literature, we do not necessarily need to use the optimal matrix for an allele. Only 9-mer peptide lengths are predicted in this service. Therefore, ProPred overlooks the probable 8-mer and 10-mer binders.

MHCPred

It is a server that predicts peptide–MHC binding quantitatively. MHCPred uses the partial least squares (PLS) method (a reliable multivariate statistical technique) to develop allele-specific QSAR models. Models of radioligand IC_{50} values are obtained from the literature by integrating the effects of individual amino acid side chains at each site, as well as the interactions between neighboring residues, specifically those that are 1–2 and 1–3 positions apart.

Limitation: It has computational constraints that do not use sequences with more than 1,000 residues. One potential limitation is that biased training data leads to less reliable predictions. Since it considers limited interactions between neighboring amino acids, specifically between 1–2 and 1–3 amino acids, it may not capture all possible interactions. It supports the prediction of only binders and no information on non-binders.

NetMHC-4.0

It is a sequence-based method for T-cell epitope prediction. The NetMHC version 4.0 server predicts the binding affinity between peptides and MHC class I molecules. This server was developed by the Department of Health Technology at DTU Health Tech. It employs an ANN architecture and a sequence alignment-based method that allows for insertions and deletions in the alignment. This innovative method can predict or identify potential epitopes, thereby reducing experimental efforts. The server is available at http://www.cbs.dtu.dk/services/NetMHC-4.0. The ANN model used in this tool is trained on 81 different human MHC alleles and 41 animal alleles. Users can type an amino acid sequence of a protein using a single-letter code or upload a fasta sequence or peptide file. They can customize several options such as specifying peptide length. By default, a 9-mer peptide is selected. Users can also choose species or loci and alleles (single allele or multiple alleles) and specify the threshold for strong and weak binders. Optionally, they can sort epitopes based on the score if they sort by affinity box. After job submission, they can monitor the status of the job, and the results of the submitted job are saved in the XLS file format.

Limitation: It is trained on hydrophobic peptides; it shows bias toward hydrophobic peptides and predicts them as a strong binder and hydrophilic peptides as a non-binder. NetMHC-4.0 is trained on limited MHC alleles, generating false positives and negatives in prediction for datasets different from the training set [39].

IEDB Tool for MHC Class II Binding Prediction

IEDB can predict and analyze epitopes for B- and T-cells. This database includes epitopes for infectious diseases, transplant epitopes, and autoimmune diseases. Sequences should be entered in the fasta format (upload files with a maximum size of 200 sequences or smaller than or equal to 10 MB). For MHC class II binding prediction, IEDB recommends different methods such as the consensus method, combinatorial library, neural network-based methods, scoring matrix method, and others. A list of available alleles will be generated by choosing a specific prediction method. Next, users can upload a file with a list of alleles or select a particular allele to use for prediction. In addition, users can choose one or more peptide sequences for prediction. Prediction results can be saved in a plain text format. Results are also shared with the users through registered email.

NetMHCII and NetMHCIIpan

These two tools are extensively and efficiently used to predict T-cell epitopes, specifically MHC class II binding peptides. The difference between these two methods is that NetMHCII is an allele-specific tool that restricts predictions to MHC molecules in its training data. Meanwhile, NetMHCIIpan is trained on an extensive experimental dataset, a pan-specific model. It can forecast binding affinities between any MHC molecule and antigen. The AUC–ROC curve is utilized to evaluate the prediction's performance. These tools are accessed separately at www.cbs.dtu.dk/services/NetMHCII-2.3 and www.cbs.dtu.dk/services/NetMHCIIpan-4.1. The standalone versions of the two tools can be downloaded and installed on Linux and Darwin platforms.

Limitation: These tools should be provided with an MHC amino acid sequence to predict binding affinity, limiting its applicability.

Vaccine Construction

The top candidates of cytotoxic T-lymphocyte (CTL), helper T-lymphocyte (HTL), and B-lymphocyte (BCL) epitopes identified in the previous steps are linked using variable-length linkers. Most of the time, KK, AAY, and GPGPG linkers join BCL, CTL, and HTL epitopes to enhance their immunogenicity and stability [40]. Besides AAY, GGGS is sometimes used to link CTL epitopes. To make sure that the designed vaccine does not cause autoimmunity, identified epitopes can be checked against a protein database using BLAST for identity. If epitopes are more than 35% similar to a human protein, it is considered too similar. So, the identified epitopes should have less similarity (degree of homology) to human proteins to be selected for final vaccine construction. The final vaccine includes an additional adjuvant component. Adjuvants with variable lengths for different biological purposes are commonly used in multi-epitope vaccine design. This is used to enhance the body's immune response or selectivity to the target. Using a linker like EAAAK, the adjuvant is joined to the multi-subunit sequence's N-terminal, which increases stability and provides an independent function for the construct. This linker is sometimes used to achieve sufficient distance between the domains to provide stability. Cholera toxin B subunit and porcine beta-defensin-2 are examples of adjuvants that enhance the immunogenicity of multi-epitope vaccines by acting as an agonist for TLR4. Together, CTL and HTL multi-epitope peptide regions make up the multi-subunit sequence [41]. The epitopes in the CTL region, which have shown superior results in CTL epitope prediction, are fused together using AAY linkers. On the other hand, the epitopes in the HTL region, which have demonstrated superior results in HTL epitope prediction, are fused together using GPGPG linkers. Vaccine design and evaluation require following a multistep procedure, and a complete *in silico* vaccine construction workflow is displayed in Figure 16.3. Here, we have discussed the vaccine construction workflow, including its evaluation in various steps.

Once the vaccine is designed, some essential properties, such as allergenicity, antigenicity, and toxicity, should be evaluated. In general, an ideal vaccine should not produce any detrimental effects on the host. The 3D structure of the vaccine can be determined using AlphaFold, homology modeling with MODELLER, etc. Some quality checkpoints, such as the Ramachandran plot and Z-score analysis, further evaluate the predicted structure of the vaccine. TLRs are pattern recognition receptors (PRRs) mediating various inflammatory pathways that are crucial to elicit the host's immune response. The compatibility of the vaccine–TLR complex is evaluated by employing computational techniques such as molecular docking, MD simulations, and MM-PBSA analysis.

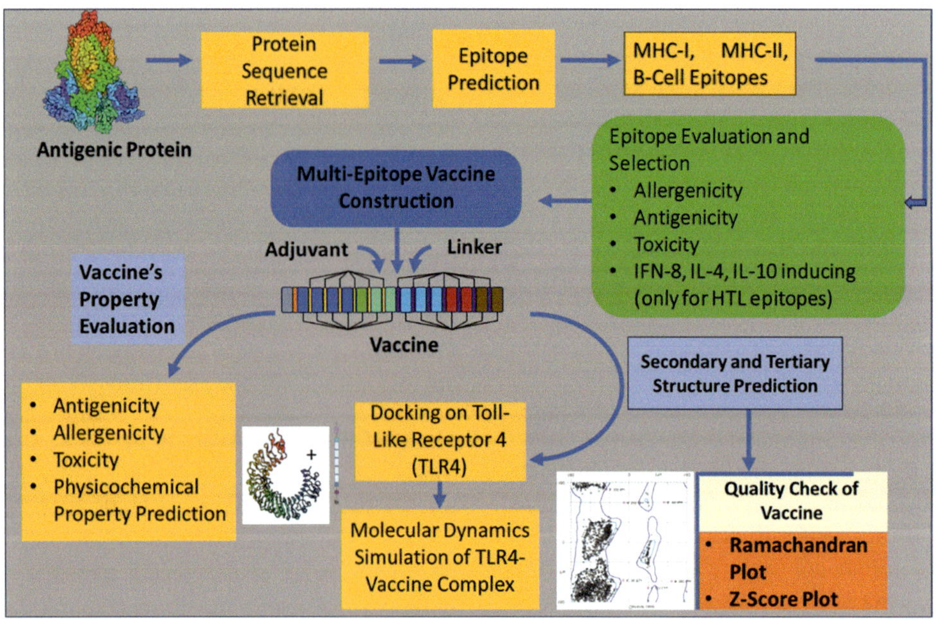

FIGURE 16.3 The steps of *in silico* vaccine construction workflow.

Tertiary Structure Prediction and Validation

The process of predicting 3D structures is usually done by homology modeling, which involves several steps, such as identifying a suitable template for protein structure modeling based on sequence homology. Several servers, such as SWISS-MODEL [42], I-TASSER-MTD [43], and RaptorX [44], can predict some of the best-ranking models. The quality of the models can be assessed using several metrics. For example, *P*-value can be calculated. *P*-value is the probability that the predicted model would perform worse than alternative random models. A low *P*-value indicates a high-quality model. An additional metric employed is the unnormalized global distance test (uGDT) score, which quantifies the absolute quality of the model. A good model for a protein with more than 100 amino acid residues is indicated by an uGDT score exceeding the threshold value. Once the model is of good quality, it can be used for further refinement and analysis.

Validation is an important step for evaluating the quality of crude-identified models, and it can be done using the ZLab server (https://zlab.umassmed.edu/bu/rama/), PROCHECK, MolProbity, and the ProSA-web server (https://prosa.services.came.sbg.ac.at/prosa.php). Ramachandran plot and ProSA-web server are essential tools in the analysis of protein structures. The two-dimensional distribution of the dihedral angles of amino acids (phi (φ) and psi (ψ)) on the protein backbone is given by the Ramachandran plot. This plot is crucial in understanding the spatial arrangement of the protein's structure. On the other hand, the ProSA-web server assesses the overall quality score of the protein structure. This score helps determine the accuracy of the protein structure, as a score outside the normal range indicates potential errors in the structure. Sometimes, the formation of disulfide bonds is essential to improve the stability of the vaccine construct by decreasing conformational flexibility. The Disulfide by Design 2.13 server (http://cptweb.cpt.wayne.edu/DbD2/) can be used to identify suitable locations to make a disulfide bond.

Epitope Evaluation for Physicochemical Properties

In the *in silico* vaccine design process, the number of proposed epitopes mainly depends on the length of the extracted sequence. These proposed epitopes with favorable properties should be thoroughly evaluated. This evaluation process can be significantly streamlined with online servers and software, which can screen for specific features. Some programs even automate this process, making it more efficient. The evaluation of epitopes considers several factors, such as safety, toxicity, antigenicity, solubility, immunogenicity, and immune response. Online tools such as AllerTOP v.2, AlgPred [45], and AllergenFP [46] are used to assess the feasibility of predicted epitopes to ensure the safety of the proposed vaccine. Among these tools, AllerTOP v.2 is known for its high accuracy in identifying allergic epitopes [47]. For optimal safety, these tools can be used in combination. Once the allergic portion is removed, the efficacy of the epitope can be determined using the VaxiJen 2.0 server by predicting the antigenicity of the selected sequence [48].

Population Coverage

T-cells recognize MHC and epitope complexes. MHC molecules are highly polymorphic. MHC alleles are highly variable in different populations. Population coverage analysis of the peptide-based vaccine is essential as MHC (HLA) allele types are expressed to varying frequencies in various ethnicities. The frequency of the HLA allele can be extracted from the allele frequency database, which can be accessed at www.allelefrequencies.net.

Conservancy Analysis

Conserved epitopes are critical in *in silico* epitope-based vaccine design because they provide more extensive protection across different species and strains than epitopes derived from highly variable regions. The measurement of epitope conservancy can ascertain the variability of epitopes within a particular collection of protein sequences. The conservancy of an epitope can be analyzed at http://tools.immuneepitope.org/tools/conservancy.

AntigenPro

AntigenPro is an alignment-independent and sequence-based server developed using a large antigenic and non-antigenic protein dataset. Antigenicity of the constructed vaccine can be measured by the tool AntigenPro server [49]. This tool is integrated with a scratch proteomics predictor (https://scratch.proteomics.ics.uci.edu/). This server will show the antigenicity of the vaccine with probability values. This tool is trained on a protein microarray dataset using several ML algorithms.

VaxiJen Server

This tool utilizes proteins' physicochemical properties and alignment-independent antigenicity prediction; hence, it can be used for proteins with no sequence similarity. It classifies a particular protein into antigenic or non-antigenic with high accuracy and precision. VaxiJen is a sequence-free antigenicity prediction tool [48]. This tool can be accessed at http://www.jenner.ac.uk/VaxiJen (Table 16.4).

The Protein-Sol server (https://protein-sol.manchester.ac.uk/) can be used to test the solubility of the multi-epitope vaccine, with a value higher than 0.45 indicating higher solubility than the average *E. coli* protein. The ProtParam tool on the Expasy server (https://web.expasy.org/protparam/) can be utilized to compute different physicochemical properties of the vaccine candidate. These properties include molecular weight, half-life, aliphatic index, instability index, hydropathicity, and amino acid count. Thus, the process of *in silico* vaccine design involves a careful and systematic evaluation of proposed epitopes to ensure their safety and efficacy.

TABLE 16.4
Epitope Evaluation Tools or Servers

Tools	Description
AllerTOP-v.2 [47]	Bioinformatics tool for antigenicity prediction
AlgPred [45]	A web server for predicting allergens and mapping of IgE epitopes
AllergenFP [46]	Alignment-free technique based on the primary characteristics of amino acids to predict allergenicity
VaxiJen [48]	This tool has been developed to predict antigenicity. It calculates the physicochemical properties of amino acids to classify antigens. This tool is free from alignment
Population coverage	Population coverage analysis of the peptide-based vaccine is very much essential as MHC (HLA) allele types are expressed at different frequencies in different ethnicities
Conservancy analysis	Epitope conservancy can be measured to find out how variable an epitope is within a particular set of protein sequences
AntigenPro [49]	Calculates antigenicity of vaccine with probability values

TABLE 16.5
Protein–Protein Docking and Molecular Dynamics Prediction Tools

Tools	Description
	Molecular Docking Tools
AutoDock CrankPep [51]	AutoDock CrankPep or ADCP is a tool for performing docking between a protein and peptide
HPEPDOCK [52]	Server for blind protein–peptide docking using a hierarchical approach
RosettaDock server [53]	Finds out the minimum energy state of protein–protein complexes by optimizing side chains
HDOCK [54]	Robust and fast protein–protein docking
ClusPro [50]	This server is free and used for the study of protein–protein docking
	Molecular Dynamics Tools
Gromacs [55]	Free and widely used MD engine with CPU and GPU supports
AMBER [32]	Biomolecular simulation programs for performing MD
Desmond	Commercial software for running MD, developed by Schrodinger
NAMD [56]	MD simulation program for a large biomolecular system with parallel computing facility
CHARMM [57]	CHARMM is a popular and very adaptable molecular modeling program

Molecular Docking and Molecular Dynamics Simulation

Molecular docking is one of the most popular computational techniques that predicts how a ligand interacts with its target protein. Nowadays, different commercial and free tools (software and servers) are available to predict the binding mode of drugs, peptides, proteins, nucleic acids, etc. After a vaccine is constructed, its stability and compatibility with the target immunoreceptor (TLR) need to be measured. ClusPro 2.0 is a protein–protein docking server that predicts the binding of vaccine constructs based on the generation of billions of conformations, followed by clustering the 1,000 lowest energy structures to get the best poses [50].

MD is a computational technique that estimates atomic movements with respect to time. This technique is widely used to determine the stability of protein–protein and protein–ligand complexes obtained from docking studies. The human TLR4 exists in monomeric and dimeric forms, and it is reported that the dimeric form is directly related to downstream signal transduction. This receptor also contains oligosaccharide moieties on the surface. The effects of these moieties should also be evaluated during MD simulation in monomeric and dimeric forms.

An MD study is often utilized to check the vaccine's stability, fluctuation, and binding affinity. The docking complex of the vaccine and TLR4 is submitted for MD simulations. It gives the interaction profile of the multi-epitope vaccine (bioligand) with the receptor (TLR4) at the molecular level. Many software are available for performing MD simulations, as mentioned in Table 16.5. MD simulation is generally carried out at 300K temperature and 1 bar pressure. The stability of the vaccine is measured by computing root mean square deviation (RMSD), root mean square fluctuation (RMSF), radius of gyration (RG), and solvent-accessible surface area (SASA) with reference to the starting structure of the vaccine or the average of all frames generated during simulation. The vaccine consists of multiple amino acids. RMSF is the fluctuation of individual residues present in the vaccine, which can be plotted against simulation time. Higher peaks of the RMSF plot represent higher flexibility of the regions of the vaccine. The RG and

SASA of the multi-epitope vaccine are computed to measure compactness and protein folding, respectively. MD generates different protein conformations with respect to time; these are called trajectory frames. The binding affinity of the vaccine and receptor is measured using molecular mechanics generalized Born surface area (MMGBSA) or molecular mechanics Poisson–Boltzmann surface area (MMPBSA) calculations, which are end-point-based binding free energy estimation methods. In addition, free energy perturbation (FEP) and thermodynamic integration (TI) are computationally more expensive methods that provide more accurate closer to experimental values of free energy binding between vaccine and receptor.

Codon Optimization and In Silico Cloning

Codon optimization is a technique that can improve the efficiency of foreign gene expression in organisms such as *E. Coli*. The Java Codon Adaptation Tool (JCat) can be used to reverse-translate and optimize codons and determine the codon adaptation index (CAI) value and GC content of the vaccination sequence. It can be accessed at http://www.jcat.de/. In practice, researchers commonly use vectors such as pET28a, pET30, and pET32a for *in silico* cloning due to their superior expression capabilities.

REFERENCES

1. Kroger, A. T.; Atkinson, W. L.; Marcuse, E. K.; Pickering, L. K.; Advisory Committee on Immunization Practices (ACIP) Centers for Disease Control and Prevention (CDC). General Recommendations on Immunization: Recommendations of the Advisory Committee on Immunization Practices (ACIP). *MMWR Recomm Rep* **1994**, *43* (2), 1–35.
2. Lahariya C. A Brief History of Vaccines & Vaccination in India. *Indian J Med Res* **2014**, *139* (4), 491–511.
3. Bailón, L.; Llano, A.; Cedeño, S.; Escribà, T.; Rosás-Umbert, M.; Parera, M.; Casadellà, M.; Lopez, M.; Pérez, F.; Oriol-Tordera, B.; Ruiz-Riol, M.; Coll, J.; Perez, F.; Rivero, À.; Leselbaum, A. R.; McGowan, I.; Sengupta, D.; Wee, E. G.; Hanke, T.; Paredes, R.; Alarcón-Soto, Y.; Clotet, B.; Noguera-Julian, M.; Brander, C.; Molto, J.; Mothe, B.; the AELIX002 Study Group; Barriocanal, A. M.; Benet, S.; Cobarsi, P.; Geleziunas, R.; Leselbaum, A. R.; Loste, C.; Meulbroek, M.; Miranda, C.; Muñoz, J.; Naval, J.; Nieto, A.; Pujol, F.; Puig, J. Safety, Immunogenicity and Effect on Viral Rebound of HTI Vaccines in Early Treated HIV-1 Infection: A Randomized, Placebo-Controlled Phase 1 Trial. *Nat Med* **2022**, *28* (12), 2611–2621. https://doi.org/10.1038/s41591-022-02060-2.
4. Stern, A. M.; Markel, H. The History of Vaccines and Immunization: Familiar Patterns, New Challenges. *Health Affairs* **2005**, *24* (3), 611–621. https://doi.org/10.1377/hlthaff.24.3.611.
5. Sanchez-Trincado, J. L.; Gomez-Perosanz, M.; Reche, P. A. Fundamentals and Methods for T- and B-Cell Epitope Prediction. *J Immunol Res* **2017**, *2017*, 1–14. https://doi.org/10.1155/2017/2680160.
6. Chiu, M. L.; Goulet, D. R.; Teplyakov, A.; Gilliland, G. L. Antibody Structure and Function: The Basis for Engineering Therapeutics. *Antibodies* **2019**, *8* (4), 55. https://doi.org/10.3390/antib8040055.
7. Parvizpour, S.; Pourseif, M. M.; Razmara, J.; Rafi, M. A.; Omidi, Y. Epitope-Based Vaccine Design: A Comprehensive Overview of Bioinformatics Approaches. *Drug Discov Today* **2020**, *25* (6), 1034–1042. https://doi.org/10.1016/j.drudis.2020.03.006.
8. Yurina, V.; Adianingsih, O. R. Predicting Epitopes for Vaccine Development Using Bioinformatics Tools. *Ther Adv Vaccines Immunother* **2022**, *10*, 251513552211002. https://doi.org/10.1177/25151355221100218.
9. Backert, L.; Kohlbacher, O. Immunoinformatics and Epitope Prediction in the Age of Genomic Medicine. *Genome Med* **2015**, *7* (1), 119.
10. Soria-Guerra, R. E.; Nieto-Gomez, R.; Govea-Alonso, D. O.; Rosales-Mendoza, S. An Overview of Bioinformatics Tools for Epitope Prediction: Implications on Vaccine Development. *J Biomed Inform* **2015**, *53*, 405–414. https://doi.org/10.1016/j.jbi.2014.11.003.
11. NCBI Resource Coordinators. Database Resources of the National Center for Biotechnology Information. *Nucleic Acids Res* **2016**, *44* (D1), D7–D19. https://doi.org/10.1093/nar/gkv1290.
12. The UniProt Consortium; Bateman, A.; Martin, M.-J.; Orchard, S.; Magrane, M.; Ahmad, S.; Alpi, E.; Bowler-Barnett, E. H.; Britto, R.; Bye-A-Jee, H.; Cukura, A.; Denny, P.; Dogan, T.; Ebenezer, T.; Fan, J.; Garmiri, P.; Da Costa Gonzales, L. J.; Hatton-Ellis, E.; Hussein, A.; Ignatchenko, A.; Insana, G.; Ishtiaq, R.; Joshi, V.; Jyothi, D.; Kandasaamy, S.; Lock, A.; Luciani, A.; Lugaric, M.; Luo, J.; Lussi, Y.; MacDougall, A.; Madeira, F.; Mahmoudy, M.; Mishra, A.; Moulang, K.; Nightingale, A.; Pundir, S.; Qi, G.; Raj, S.; Raposo, P.; Rice, D. L.; Saidi, R.; Santos, R.; Speretta, E.; Stephenson, J.; Totoo, P.; Turner, E.; Tyagi, N.; Vasudev, P.; Warner, K.; Watkins, X.; Zaru, R.; Zellner, H.; Bridge, A. J.; Aimo, L.; Argoud-Puy, G.; Auchincloss, A. H.; Axelsen, K. B.; Bansal, P.; Baratin, D.; Batista Neto, T. M.; Blatter, M.-C.; Bolleman, J. T.; Boutet, E.; Breuza, L.; Gil, B. C.; Casals-Casas, C.; Echioukh, K. C.; Coudert, E.; Cuche, B.; De Castro, E.; Estreicher, A.; Famiglietti, M. L.; Feuermann, M.; Gasteiger, E.; Gaudet, P.; Gehant, S.; Gerritsen, V.; Gos, A.; Gruaz, N.; Hulo, C.; Hyka-Nouspikel, N.; Jungo, F.; Kerhornou, A.; Le Mercier, P.; Lieberherr, D.; Masson, P.; Morgat, A.; Muthukrishnan, V.; Paesano, S.; Pedruzzi, I.; Pilbout, S.; Pourcel, L.; Poux, S.; Pozzato, M.; Pruess, M.; Redaschi, N.; Rivoire, C.; Sigrist, C. J. A.; Sonesson, K.; Sundaram, S.; Wu, C. H.; Arighi, C. N.; Arminski, L.; Chen, C.; Chen, Y.; Huang, H.; Laiho, K.; McGarvey, P.; Natale, D. A.; Ross, K.; Vinayaka, C. R.; Wang, Q.; Wang, Y.; Zhang, J. UniProt: The Universal Protein Knowledgebase in 2023. *Nucleic Acids Res* **2023**, *51* (D1), D523–D531. https://doi.org/10.1093/nar/gkac1052.
13. Zardecki, C.; Dutta, S.; Goodsell, D. S.; Voigt, M.; Burley, S. K. RCSB Protein Data Bank: A Resource for Chemical, Biochemical, and Structural Explorations of Large and Small Biomolecules. *J Chem Educ* **2016**, *93* (3), 569–575. https://doi.org/10.1021/acs.jchemed.5b00404.
14. Srinivasan, S.; Selvaraj, G.; Gopalan, V.; Padmanabhan, P.; Ramesh, K.; Govindan, K.; Chandran, A.; Dhandapani, P.; Krishnasamy, K.; Kitambi, S. Epitope Identification and Designing a Potent Multi-Epitope Vaccine Construct against SARS-CoV-2 Including the Emerging Variants. *J Global Infect Dis* **2022**, *14* (1), 24. https://doi.org/10.4103/jgid.jgid_96_21.
15. Lollier, V.; Denery-Papini, S.; Larré, C.; Tessier, D. A Generic Approach to Evaluate How B-Cell Epitopes Are Surface-Exposed on Protein Structures. *Mol Immunol* **2011**, *48* (4), 577–585. https://doi.org/10.1016/j.molimm.2010.10.011.

16. Saha, S.; Raghava, G. P. S. BcePred: Prediction of Continuous B-Cell Epitopes in Antigenic Sequences Using Physico-Chemical Properties. In *Artificial Immune Systems*; Nicosia, G.; Cutello, V.; Bentley, P. J.; Timmis, J.; Hutchison, D.; Kanade, T.; Kittler, J.; Kleinberg, J. M.; Mattern, F.; Mitchell, J. C.; Naor, M.; Nierstrasz, O.; Pandu Rangan, C.; Steffen, B.; Sudan, M.; Terzopoulos, D.; Tygar, D.; Vardi, M. Y.; Weikum, G.; (Eds.); Springer Berlin, Heidelberg, 2004; Vol. 3239, pp. 197–204. https://doi.org/10.1007/978-3-540-30220-9_16.
17. Jespersen, M. C.; Peters, B.; Nielsen, M.; Marcatili, P. BepiPred-2.0: Improving Sequence-Based B-Cell Epitope Prediction Using Conformational Epitopes. *Nucleic Acids Res* **2017**, *45* (W1), W24–W29. https://doi.org/10.1093/nar/gkx346.
18. Høie, M. H.; Gade, F. S.; Johansen, J. M.; Würtzen, C.; Winther, O.; Nielsen, M.; Marcatili, P. DiscoTope-3.0: Improved B-Cell Epitope Prediction Using Inverse Folding Latent Representations. *Front Immunol* **2024**, *15*, 1322712. https://doi.org/10.3389/fimmu.2024.1322712.
19. Sweredoski, M. J.; Baldi, P. PEPITO: Improved Discontinuous B-Cell Epitope Prediction Using Multiple Distance Thresholds and Half Sphere Exposure. *Bioinformatics* **2008**, *24* (12), 1459–1460.
20. Negi, S. S.; Braun, W. Automated Detection of Conformational Epitopes Using Phage Display Peptide Sequences. *Bioinform Biol Insights* **2009**, *3*, S2745. https://doi.org/10.4137/BBI.S2745.
21. Huang, Y. X.; Bao, Y. L.; Guo, S. Y.; Wang, Y.; Zhou, C. G.; Li, Y. X. Pep-3D-Search: A Method for B-Cell Epitope Prediction Based on Mimotope Analysis. *BMC Bioinform* **2008**, *9* (1), 538. https://doi.org/10.1186/1471-2105-9-538.
22. Rubinstein, N. D.; Mayrose, I.; Martz, E.; Pupko, T. Epitopia: A Web-Server for Predicting B-Cell Epitopes. *BMC Bioinform* **2009**, *10* (1), 287. https://doi.org/10.1186/1471-2105-10-287.
23. Ponomarenko, J.; Bui, H.-H.; Li, W.; Fusseder, N.; Bourne, P. E.; Sette, A.; Peters, B. ElliPro: A New Structure-Based Tool for the Prediction of Antibody Epitopes. *BMC Bioinform* **2008**, *9* (1), 514. https://doi.org/10.1186/1471-2105-9-514.
24. Dimitrov, I.; Garnev, P.; Flower, D. R.; Doytchinova, I. MHC Class II Binding Prediction—A Little Help from a Friend. *J Biomed Biotechnol* 2010, *2010*, 1–8. https://doi.org/10.1155/2010/705821.
25. Doytchinova, I. A.; Guan, P.; Flower, D. R. EpiJen: A Server for Multistep T Cell Epitope Prediction. *BMC Bioinform* **2006**, *7* (1), 131. https://doi.org/10.1186/1471-2105-7-131.
26. Rammensee, H.-G.; Bachmann, J.; Emmerich, N. P. N.; Bachor, O. A.; Stevanović, S. SYFPEITHI: Database for MHC Ligands and Peptide Motifs. *Immunogenetics* **1999**, *50* (3–4), 213–219. https://doi.org/10.1007/s002510050595.
27. Bhasin, M.; Raghava, G. P. S. A Hybrid Approach for Predicting Promiscuous MHC Class I Restricted T Cell Epitopes. *J Biosci* **2007**, *32* (1), 31–42. https://doi.org/10.1007/s12038-007-0004-5.
28. Singh, H.; Raghava, G. P. S. ProPred1: Prediction of Promiscuous MHC Class-I Bindingsites. *Bioinformatics* **2003**, *19* (8), 1009–1014. https://doi.org/10.1093/bioinformatics/btg108.
29. Singh, H.; Raghava, G. P. S. ProPred: Prediction of HLA-DR Binding Sites. *Bioinformatics* **2001**, *17* (12), 1236–1237. https://doi.org/10.1093/bioinformatics/17.12.1236.
30. Guan, P.; Doytchinova, I. A.; Zygouri, C.; Flower, D. R. MHCPred: Bringing a Quantitative Dimension to the Online Prediction of MHC Binding. *Appl Bioinform* **2003**, *2* (1), 63–66.
31. Lata, S.; Bhasin, M.; Raghava, G. P. S. Application of Machine Learning Techniques in Predicting MHC Binders. In *Immunoinformatics*; Flower, D. R.,; Walker, J. M; (Eds.); Humana Press: Totowa, NJ, 2007; Vol. 409, pp. 201–215. https://doi.org/10.1007/978-1-60327-118-9_14.
32. Lundegaard, C.; Lamberth, K.; Harndahl, M.; Buus, S.; Lund, O.; Nielsen, M. NetMHC-3.0: Accurate Web Accessible Predictions of Human, Mouse and Monkey MHC Class I Affinities for Peptides of Length 8–11. *Nucleic Acids Res* **2008**, *36* (suppl_2), W509–W512. https://doi.org/10.1093/nar/gkn202.
33. Reche, P. A.; Glutting, J.-P.; Zhang, H.; Reinherz, E. L. Enhancement to the RANKPEP Resource for the Prediction of Peptide Binding to MHC Molecules Using Profiles. *Immunogenetics* **2004**, *56* (6), 7. https://doi.org/10.1007/s00251-004-0709-7.
34. Donnes, P.; Kohlbacher, O. SVMHC: A Server for Prediction of MHC-Binding Peptides. *Nucleic Acids Res* **2006**, *34* (Web Server), W194–W197. https://doi.org/10.1093/nar/gkl284.
35. Dhanda, S. K.; Mahajan, S.; Paul, S.; Yan, Z.; Kim, H.; Jespersen, M. C.; Jurtz, V.; Andreatta, M.; Greenbaum, J. A.; Marcatili, P.; Sette, A.; Nielsen, M.; Peters, B. IEDB-AR: Immune Epitope Database—Analysis Resource in 2019. *Nucleic Acids Res* **2019**, *47* (W1), W502–W506. https://doi.org/10.1093/nar/gkz452.
36. Bhasin, M.; Raghava, G. P. S. Prediction of Promiscuous and High-Affinity Mutated MHC Binders. *Hybrid Hybridomics* **2003**, *22* (4), 229–234. https://doi.org/10.1089/153685903322328956.
37. Wan, J.; Liu, W.; Xu, Q.; Ren, Y.; Flower, D. R.; Li, T. SVRMHC Prediction Server for MHC-Binding Peptides. *BMC Bioinform* **2006**, *7* (1), 463. https://doi.org/10.1186/1471-2105-7-463.
38. Sturniolo, T.; Bono, E.; Ding, J.; Raddrizzani, L.; Tuereci, O.; Sahin, U.; Braxenthaler, M.; Gallazzi, F.; Protti, M. P.; Sinigaglia, F.; Hammer, J. Generation of Tissue-Specific and Promiscuous HLA Ligand Databases Using DNA Microarrays and Virtual HLA Class II Matrices. *Nat Biotechnol* **1999**, *17* (6), 555–561. https://doi.org/10.1038/9858.
39. Solanki, A.; Riedel, M.; Cornette, J.; Udell, J.; Vasmatzis, G. Hydrophobicity Identifies False Positives and False Negatives in Peptide-MHC Binding. *Front Oncol* **2022**, *12*, 1034810. https://doi.org/10.3389/fonc.2022.1034810.
40. Ayyagari, V. S.; Venkateswarulu, T. C.; Peele A. K.; Srirama, K. Design of a Multi-Epitope-Based Vaccine Targeting M-Protein of SARS-CoV2: An Immunoinformatics Approach. *J Biomol Struct Dyn* **2022**, *40* (7), 2963–2977. https://doi.org/10.1080/07391102.2020.1850357.
41. Hasan, M.; Mia, M. Exploratory Algorithm of a Multi-Epitope-Based Subunit Vaccine Candidate Against Cryptosporidium Hominis: Reverse Vaccinology-Based Immunoinformatic Approach. *Int J Pept Res Ther* **2022**, *28* (5), 134. https://doi.org/10.1007/s10989-022-10438-6.
42. Waterhouse, A.; Bertoni, M.; Bienert, S.; Studer, G.; Tauriello, G.; Gumienny, R.; Heer, F. T.; de Beer, T. A. P.; Rempfer, C.; Bordoli, L.; Lepore, R.; Schwede,

T. SWISS-MODEL: Homology Modelling of Protein Structures and Complexes. *Nucleic Acids Res* **2018**, *46* (W1), W296–W303. https://doi.org/10.1093/nar/gky427.
43. Zhou, X.; Zheng, W.; Li, Y.; Pearce, R.; Zhang, C.; Bell, E. W.; Zhang, G.; Zhang, Y. I-TASSER-MTD: A Deep-Learning-Based Platform for Multi-Domain Protein Structure and Function Prediction. *Nat Protoc* **2022**, *17* (10), 2326–2353. https://doi.org/10.1038/s41596-022-00728-0.
44. Källberg, M.; Wang, H.; Wang, S.; Peng, J.; Wang, Z.; Lu, H.; Xu, J. Template-Based Protein Structure Modeling Using the RaptorX Web Server. *Nat Protoc* **2012**, *7* (8), 1511–1522. https://doi.org/10.1038/nprot.2012.085.
45. Sharma, N.; Patiyal, S.; Dhall, A.; Pande, A.; Arora, C.; Raghava, G. P. S. AlgPred 2.0: An Improved Method for Predicting Allergenic Proteins and Mapping of IgE Epitopes. *Briefings Bioinform* **2021**, *22* (4), bbaa294. https://doi.org/10.1093/bib/bbaa294.
46. Dimitrov, I.; Naneva, L.; Doytchinova, I.; Bangov, I. Allergen FP: Allergenicity Prediction by Descriptor Fingerprints. *Bioinformatics* **2014**, *30* (6), 846–851. https://doi.org/10.1093/bioinformatics/btt619.
47. Dimitrov, I.; Bangov, I.; Flower, D. R.; Doytchinova, I. AllerTOP v.2—a Server for in Silico Prediction of Allergens. *J Mol Model* **2014**, *20* (6), 2278. https://doi.org/10.1007/s00894-014-2278-5.
48. Doytchinova, I. A.; Flower, D. R. VaxiJen: A Server for Prediction of Protective Antigens, Tumour Antigens and Subunit Vaccines. *BMC Bioinform* **2007**, *8* (1), 4. https://doi.org/10.1186/1471-2105-8-4.
49. Magnan, C. N.; Zeller, M.; Kayala, M. A.; Vigil, A.; Randall, A.; Felgner, P. L.; Baldi, P. High-Throughput Prediction of Protein Antigenicity Using Protein Microarray Data. *Bioinformatics* **2010**, *26* (23), 2936–2943. https://doi.org/10.1093/bioinformatics/btq551.
50. Kozakov, D.; Hall, D. R.; Xia, B.; Porter, K. A.; Padhorny, D.; Yueh, C.; Beglov, D.; Vajda, S. The ClusPro Web Server for Protein–Protein Docking. *Nat Protoc* **2017**, *12* (2), 255–278. https://doi.org/10.1038/nprot.2016.169.
51. Zhang, Y.; Sanner, M. F. AutoDock CrankPep : Combining Folding and Docking to Predict Protein–Peptide Complexes. *Bioinformatics* **2019**, *35* (24), 5121–5127. https://doi.org/10.1093/bioinformatics/btz459.
52. Zhou, P.; Jin, B.; Li, H.; Huang, S.-Y. HPEPDOCK: A Web Server for Blind Peptide–Protein Docking Based on a Hierarchical Algorithm. *Nucleic Acids Res* **2018**, *46* (W1), W443–W450. https://doi.org/10.1093/nar/gky357.
53. Lyskov, S.; Gray, J. J. The RosettaDock Server for Local Protein-Protein Docking. *Nucleic Acids Res* **2008**, *36* (Web Server), W233–W238. https://doi.org/10.1093/nar/gkn216.
54. Yan, Y.; Tao, H.; He, J.; Huang, S.-Y. The HDOCK Server for Integrated Protein–Protein Docking. *Nat Protoc* **2020**, *15* (5), 1829–1852. https://doi.org/10.1038/s41596-020-0312-x.
55. Abraham, M. J.; Murtola, T.; Schulz, R.; Páll, S.; Smith, J. C.; Hess, B.; Lindahl, E. GROMACS: High Performance Molecular Simulations through Multi-Level Parallelism from Laptops to Supercomputers. *SoftwareX* **2015**, *1–2*, 19–25. https://doi.org/10.1016/j.softx.2015.06.001.
56. Phillips, J. C.; Braun, R.; Wang, W.; Gumbart, J.; Tajkhorshid, E.; Villa, E.; Chipot, C.; Skeel, R. D.; Kalé, L.; Schulten, K. Scalable Molecular Dynamics with NAMD. *J Comput Chem* **2005**, *26* (16), 1781–1802. https://doi.org/10.1002/jcc.20289.
57. Brooks, B. R.; Brooks, C. L.; Mackerell, A. D.; Nilsson, L.; Petrella, R. J.; Roux, B.; Won, Y.; Archontis, G.; Bartels, C.; Boresch, S.; Caflisch, A.; Caves, L.; Cui, Q.; Dinner, A. R.; Feig, M.; Fischer, S.; Gao, J.; Hodoscek, M.; Im, W.; Kuczera, K.; Lazaridis, T.; Ma, J.; Ovchinnikov, V.; Paci, E.; Pastor, R. W.; Post, C. B.; Pu, J. Z.; Schaefer, M.; Tidor, B.; Venable, R. M.; Woodcock, H. L.; Wu, X.; Yang, W.; York, D. M.; Karplus, M. CHARMM: The Biomolecular Simulation Program. *J Comput Chem* **2009**, *30* (10), 1545–1614. https://doi.org/10.1002/jcc.21287.

17 Artificial Intelligence and Machine Learning in Drug Discovery

Jatin Malik, Sumit Kumar, and Gopal L. Khatik
Department of Medicinal Chemistry, National Institute of Pharmaceutical Education and Research-Raebareli, India

INTRODUCTION

A drug's development may cost up to 2.558 billion USD because it takes 10–15 years for the drug to become licensed for sale; however, the success rate is still nearly 13%. Clinicians have been using computer-assisted drug design (CADD) to increase the success rate. By creating lead compounds with ideal features and adjusting their molecular attributes (i.e. absorption, bioactivity, distribution, metabolism, selectivity, side effects, and excretion) at theoretical levels, the AI technique discovers novel therapeutic molecules. Preclinical attrition expenses can also be decreased by applying multiobjective optimization techniques [1].

Science and technology, high-speed computing, robotics, DNA sequencing, and chemical synthesis have allowed drug researchers to test thousands of compounds in a fraction of time and better understand the molecular mechanisms of diseases. The ability to search all published chemical libraries provides new possibilities to find drug-like candidates with desired biological effects and features. The plethora of data and tools may speed up drug discovery at a lower cost, but it is still laborious and expensive. Many drug developers use artificial intelligence (AI), particularly deep learning (DL), to understand diseases, identify and validate targets, discover predictive biomarkers, and study digital pathology. AI in drug discovery was demonstrated using unsupervised learning for endotypes in atopic dermatitis and the Bayesian additive regression test to create quantitative structure–activity relationship (QSAR) models to predict molecular activities [2].

AI and machine intelligence are terms used in computer science to describe the ability of machines, through either training or customization, to accomplish tasks like a human brain does. AI uses numerous algorithms for deciphering data and extracting insight from it. "Computational intelligence" refers to a broad range of disciplines, including probability theory, statistics, machine learning, fuzzy models, and neural networks. Several intricate applications involve AI strategies, such as classification, regression, predictions, and optimization techniques. Machine learning (ML) should be properly adjusted to extract and explore any information, such as a specific model's parameters, which must be defined. Thus, using trained data, machines can become proficient in the model with accessible parameters. In addition, the model can forecast future data and extract information from it [3]. Several factors have emerged recently due to increased interest in applying ML techniques in pharmaceutical sciences. These factors demonstrate the advances made in drug discovery disciplines using ML. Each step is carried out as a pipeline representing therapeutic potential. The phases mentioned above signify distinct iterations in terms of both time and cost. Using "omics" and "smart automation tools," medical data was mined and precisely assessed. The pharmaceutical sector will face new challenges and opportunities as these techniques expand into the biological field, given that many pharmaceutical companies seek to identify the most compelling clinical hypothesis. Practitioners or physicians can develop drugs based on the results received. The performance of any drug in the pharmaceutical industry can be evaluated using ML techniques. Now, improvements in datasets such as size and kind can serve as the basis for ML if they are integrated with infinite storage. It can acquire vast amounts of data from the pharmaceutical sector. Different data configurations are possible, including textual data, pictures, biometrics, assay information, and high-dimensional omics data [4]. As a result, the science of AI has progressed from theoretical understanding to facts. Much improved information is now used in personal computer (PC) hardware, such as graphical processing units (GPU), which speed up processing (i.e. computational procedures). One ML algorithm that has recently gained popularity is the DL model, which creates models to increase success in solving problems. Over the last two years, pharmaceutical companies have greatly expanded their use of ML algorithms.

In the clinical setting, natural drugs are the primary source of innovation for developing new drugs for chronic diseases. Recently, several medications have been developed to identify active ingredients in conventional therapies such as penicillin. Naturally occurring small molecules that help in treatments to identify substances such as cells and whole organisms are used in chemical laboratories, and this approach is known as "old-style pharmacology." The human genome has made cloning techniques and large-scale protein refinement possible, increasing high-throughput screening using various libraries. Reverse pharmacology is a condition that can be altered by biological targets being utilized as a screening tool for big molecules [5]. Several hits can be obtained from cell screening activities and further animal

DOI: 10.1201/9781003464679-17

studies to ensure the validity of results. Screening for possible hits and using optimization technologies to improve a drug's efficacy, affinity, and metabolic stability is necessary to discover novel medications. A particular medication will be created in clinical trials if the chemical meets all requirements and shows promise. In developing and discovering new drugs, lead optimization, target identification and validation, hit discovery, and clinical trials are all essential processes [6]. Further, we reviewed AI drug discovery approaches that employ DL techniques to tackle various drug development and discovery applications.

ARTIFICIAL INTELLIGENCE

Drug discovery entails a meticulous progression requiring a precise delineation of the problem space before employing a designated AI algorithm. The process of applying an AI algorithm in drug discovery involves defining the problem domain, designing an appropriate model architecture, training and evaluating the model, and understanding the results. Determining whether the problem belongs to discriminative or generative tasks is crucial. For discriminative tasks, random forests (RFs), support vector machines (SVMs), and artificial neural networks (ANNs) are frequently employed algorithms, while generative tasks often call upon models such as deep belief networks (DBNs), deep Boltzmann machines (DBMs), variational autoencoders (VAEs), generative adversarial networks (GANs), and adversarial autoencoders (AAEs) [1]. Hyperparameters, such as kernel type, kernel parameters, and error penalty, exhibit variability across different AI algorithms. After the architectural framework is decided upon, the dataset is carefully prepared, and then, the model is trained and evaluated to reduce prediction error. The final model needs to explain how molecular representations and the particular goals of practitioners might work together more effectively. Reinforcement learning, transfer learning, unsupervised learning, active learning, reinforcement learning (reiterated), semi-supervised learning, and multitask learning are the seven types of learning tasks and tactics used in drug discovery initiatives [7]. Each class harbours its unique set of advantages and drawbacks, necessitating a discerning selection of techniques commensurate with the exigencies of the task. The primary learning processes and principles are depicted schematically in Figure 17.1. Using the newest AI techniques, new AI developers in drug development projects frequently increase model performance. However, since training data provides the foundation for future development, focusing on it is more advantageous. Data preparation is difficult and labour-intensive; it necessitates a thorough comprehension of the entities, distribution, and origins of the data in chemical space. A preprocessing strategy and representation type should be considered if more data is needed. The initial step involves determining whether the task falls within the realm of generative or discriminative tasks. Subsequently, crafting an apt model architecture ensues, incorporating method selection and the establishment of pragmatic initial hyperparameter values. Each AI algorithm boasts unique hyperparameters, encompassing error penalties like kernel parameters such as γ for radial basis function (RBF) kernels in SVMs, and kernel selection. Architectural parameters for neural networks entail neuron and layer quantity and type selection, regularization parameters, learning rate and decay, and interlayer connectivity considerations. Upon finalizing a provisional architecture, meticulous data preparation ensues, where the quantity, quality, and representativeness of initial data profoundly influence model efficacy. Subsequent phases encompass model training and evaluation, entailing parameter optimization to minimize prediction errors. Failure to achieve this alignment may prompt practitioners to cultivate a tailored model "ecosystem" through individual case analyses to attain their objectives [1].

MACHINE LEARNING

Data mining epitomizes a multifaceted process aimed at distilling invaluable insights from huge databases or datasets through sophisticated search methodologies and algorithms to unravel intricate patterns and correlations latent within pre-existing databases. Central to this pursuit is developing models that delineate molecular descriptors regarding pivotal biological attributes, notably efficacy or absorption, distribution, metabolism, and excretion (ADME) properties [8]. These models provide insights into new compounds' structure–activity relationships (SARs) and function as prediction tools for identifying critical property values. This simplifies the prioritization of these compounds for further screening. Virtual screening (VS), a critical area in chemoinformatics, uses computational tools to thoroughly search large databases for new leads with higher-than-average chances of binding affinities to target proteins. Research endeavours in this field have yielded promising results, such as the identification of molecules that have a resemblance to natural ligands of certain targets or the discovery of completely new compounds. VS techniques are classified as structure-based (SBVS) and ligand-based (LBVS), depending on the availability of data on structure and bioactivity. The approaches used in LBVS are further divided into compound classification techniques and similarity search [9].

Using similarity metrics, the former pairs molecular fingerprints obtained from molecular graphs, 3D pharmacophore models, reduced molecular graph representations, or molecular shape queries with database substances. Contrarily, compound classification techniques are classified into ML and elementary classification techniques, including ANNs, k-nearest neighbours (k-NNs), decision trees (DTs), SVMs, and naïve Bayesian methods [10]. SVMs are the industry standard for ML techniques in CADD because of their broad range of applications, which include compound categorization, ranking, and regression-based property value prediction. SVMs are extremely good in predicting binary properties or activities, such as water

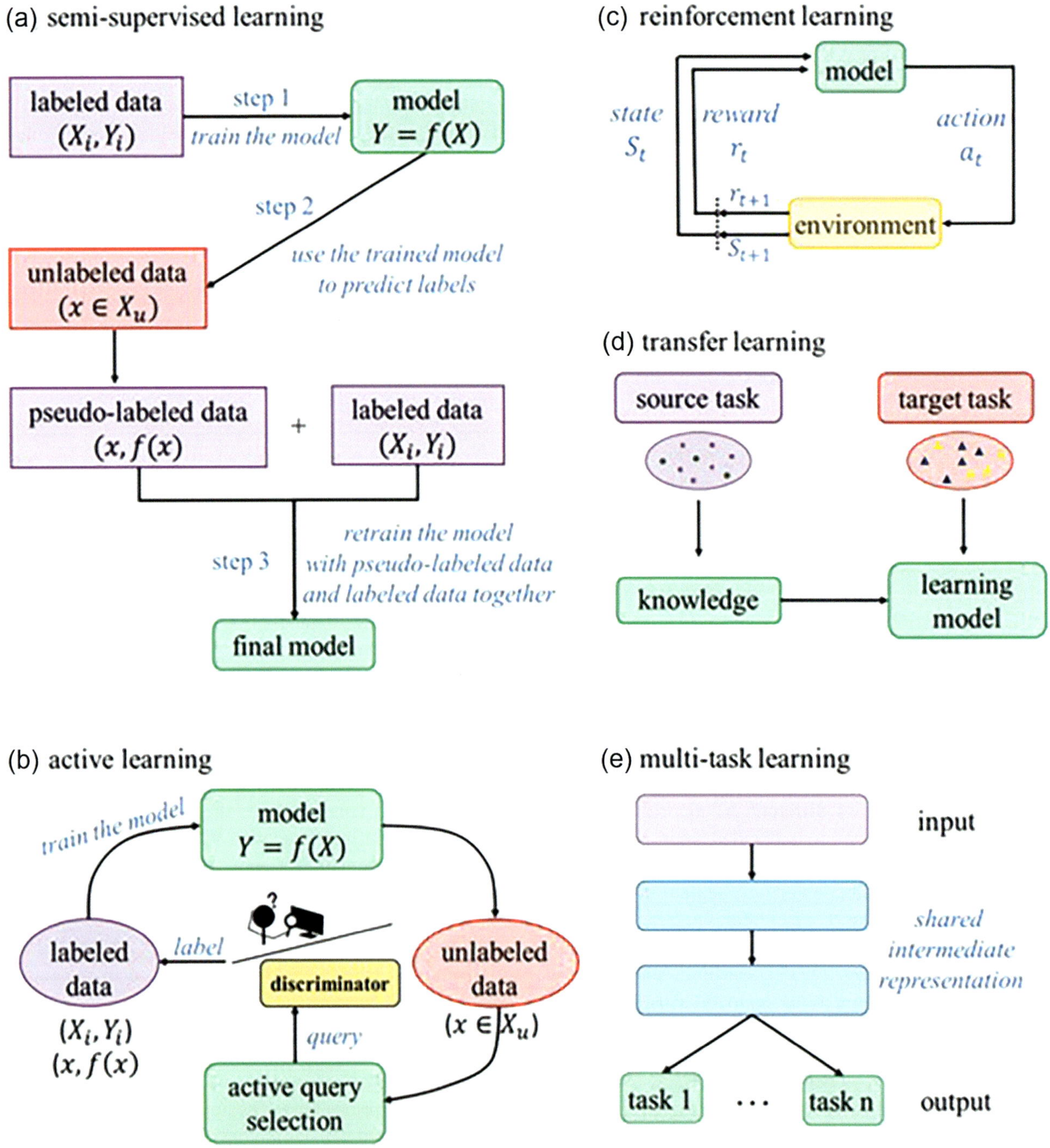

FIGURE 17.1 An illustration of different types of ML.

solubility, synthetic accessibility, and the ability to distinguish between drugs and nondrugs or molecules exhibiting particular activities. SVMs work in a high-dimensional feature space and are driven by kernel functions that include linear, polynomial, sigmoid, and RBF kernels. Of these, RBF-based SVMs perform better than the others. SVM kernel choice and setup are determined by empirical and experimental investigations, where support hyperplanes indicate the boundary between different compound classes in the feature space. Moreover, recent advancements in SVM research have ushered in novel kernel functions tailored to accommodate diverse applications, including graph and sequence kernels, which encapsulate disparate information for similarity assessment. Notably, the development of protein–ligand association prediction models employing SVMs underscores the nuanced interplay between ligand and target kernels, where refined strategies for target–ligand SVM kernel design yield increased discriminatory accuracy. Furthermore, the advent of hybrid techniques amalgamating multiple ML methodologies promises enhanced predictive fidelity, epitomizing a paradigm shift in predictive modelling endeavours. DTs occupy a central position within the domain of chemoinformatics, encapsulating a set of regulations that correlate specific molecular attributes or descriptor values with the desired activity or property. Renowned for their precision and ease of validation, DT models elucidate compound profiling data, prognosticate biological activities, and facilitate the design of combinatorial libraries. Nonetheless, their predictions are vulnerable to considerable variance owing to the process's hierarchical nature and the training dataset's scale, thus necessitating ensemble techniques such as RFs to ameliorate this constraint. In essence, the amalgamation of diverse ML methodologies, ranging from SVMs to DTs, represents a paradigm shift in predictive modelling endeavours within the domain of CADD, embodying the ceaseless quest for innovation and excellence in this multifaceted discipline [9].

APPLICATION OF AI/ML IN DRUG DISCOVERY

AI for Primary Drug Screening

The prowess of AI in discerning images featuring diverse objects or attributes surpasses human capabilities, particularly evident in the realm of eye examination, where the laborious and inefficient nature of manual analysis renders it unsuitable for extensive data scrutiny. Consequently, this domain presents an ideal arena for AI-driven computational methodologies [11]. The AI model necessitates rigorous training to swiftly and autonomously identify properties of cell types for classification or diagnosis. In the context of categorizing breast cancer cells, image contrast manipulation segregates them from the surrounding background. Subsequently, Tamura and wavelet-based texture features are extracted, followed by principal component analysis (PCA) to mitigate dimensionality. AI-based techniques undergo training to classify diverse cell types proficiently, leveraging statistical learning theory alongside regression; the least-square support vector machine (LS-SVM) approach emerges as the epitome of classification accuracy, achieving a remarkable accuracy of 95.34%. In order to achieve precise cell sorting, AI-enabled image analysis decision-making should exhibit rapidity commensurate with the requirements for the robot to discern various cell types within the sample (Figure 17.2) [12]. Modern advancements in image-activated cell sorting (IACS) technologies facilitate automated cell sorting by measuring optical, electrical, and mechanical parameters, thereby ensuring flexibility and scalability. These technologies use AI and deep neural network (DNN) algorithms to evaluate digital images quickly and make decisions in milliseconds. Analysis of high-content sorting of human platelets and *Chlamydomonas reinhardtii* highlights the remarkable sensitivity and specificity achieved with this method. In addition to cell identification and classification, AI is now being used to interpret computerized electrocardiography (ECG), a crucial step in clinical diagnostic and treatment procedures. For experienced practitioners, this has made the laborious manual checking procedure easier. The accuracy and scalability of automated ECG analysis are improved by DL algorithms and widely available digitized ECG data [13].

AI in Secondary Drug Screening

Forecasting Physical Attributes and Bioactivity

The meticulous selection of pharmaceutical candidates possessing the requisite attributes, encompassing bioavailability, bioactivity, and toxicity, is a pivotal endeavour in drug formulation. Pertinent physical parameters such as the melting point and partition coefficient (logP) profoundly influence drug bioavailability, warranting meticulous consideration in drug design. Melting point is a proxy for drug solubility in aqueous environments, whereas logP offers insights into water–oil solubility dynamics and aids in estimating cellular drug absorption rates. In AI-driven drug design, sophisticated algorithms leverage diverse molecular representations, including fingerprints, simplified molecular input line entry systems (SMILES) strings, potential energy measurements, graphs, Coulomb matrices, fragments, and electron density profiles. Various DNNs are adept at processing these inputs across generative and predictive stages during training, thereby facilitating reinforcement learning paradigms. In conventional approaches, the generative stage of a DNN operates on SMILES inputs to construct chemically viable molecular structures, while the predictive stage focuses on training molecular attribute classifiers. However, segregating these stages using supervised algorithms can inadvertently introduce bias through selective reward or penalty mechanisms during joint training. Matched molecular pair (MMP) analysis is a fundamental technique used in QSAR research to assess how localized changes to a drug candidate affect its molecular characteristics and bioactivity [14]. Retrosynthesis-based analytical techniques often produce MMPs for *de novo* drug discovery activities. A candidate molecule's chemical representation consists of two carefully encoded pieces that outline the transformation process and a static core. Using training data derived from IC_{50} values for five kinases and a protein containing a bromodomain, ML techniques such as RFs, gradient boosting machines (GBMs), and DNNs are used to predict novel transformations, fragments, and alterations based on the static core.

Interestingly, DNNs perform better in predicting chemical activity than RFs and GBMs. Since public databases such as ChEMBL and PubChem are becoming more and more prevalent, it is possible to predict bioactivity properties such as oral exposure, distribution coefficient (logD), intrinsic clearance, and ADME and clarify modes of action by combining MMP with ML techniques. Different new bioactivity prediction methods for drug candidates have emerged of late. For example, Tristan et al. embedded discrete chemicals into a continuous latent variable space (LVS) and used a graph convolutional network to create a pharmacological target site signature. This novel strategy makes it possible to optimize

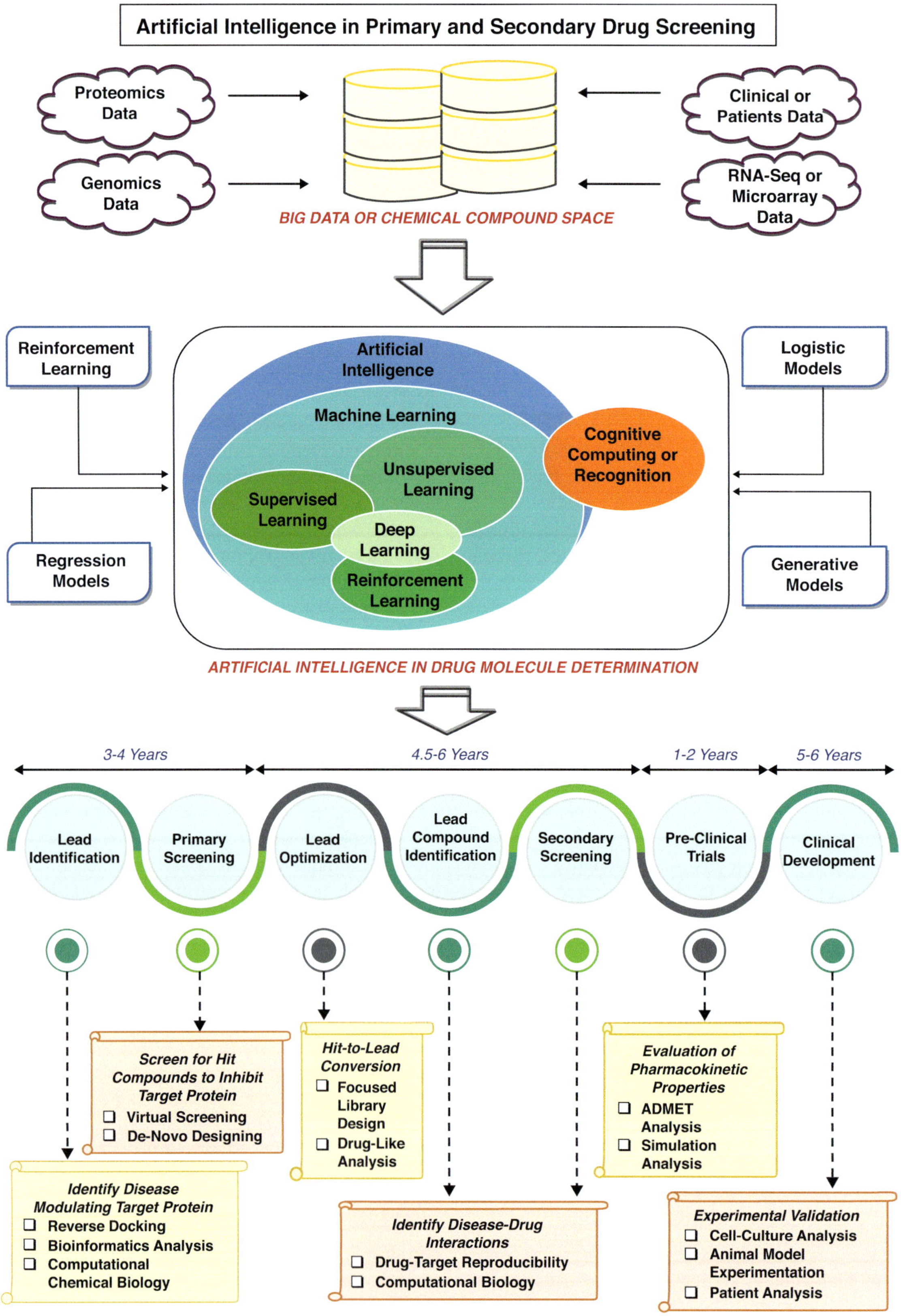

FIGURE 17.2 AI in primary and secondary drug screening.

gradient-wise in molecular space, making predictions based on prominent characteristics and differentiable binding affinities easier.

Toxicity Prediction

When developing new medications, a chemical's toxicity is a critical factor. In preclinical medication research, toxicity optimization is an expensive and time-consuming task. To develop a new drug, a precise chemical toxicity estimation is essential. In prepared experiments, the ML system DeepTox predicted 12 harmful effects for 12,000 pharmaceuticals and natural substances, demonstrating its superior performance in the Tox21 Data Challenge [15]. Chemical representations of substances are normalized using the DeepTox algorithm, which produces several chemical descriptors for ML algorithms. Descriptors can be either dynamic or static. Compounds' atom counts, surface areas, and predefined substructure presence are examples of static descriptors. In addition to other chemical parameters derived from conventional molecular fingerprint descriptors, the presence and absence of 2,500 predefined toxicophore characteristics are computed. Predefined computations are employed for dynamic descriptors. The approach restricts the dataset even if the number of dynamic features might be infinite. In common test scenarios, DeepTox predicts chemical toxicity with a high degree of accuracy [16].

Revolutionizing Drug Design through AI

Modelling 3D Protein Structure

Elucidating a target protein's three-dimensional (3D) configuration is pivotal in structure-based drug development, wherein novel drug compounds are engineered based on the intricate chemical environment surrounding the ligand-binding site. Homology modelling and *de novo* protein design represent established methodologies for this technique. The integration of AI-based technologies has markedly enhanced the precision and sophistication of predicting the 3D structure of target proteins. AlphaFold, an AI-powered tool, has demonstrated remarkable proficiency in forecasting the 3D structure of therapeutic target proteins, as evidenced by its performance in the recent Critical Assessment of Protein Structure Prediction competition [17,18]. Leveraging solely protein primary sequences, AlphaFold adeptly forecasted the structure of 25 out of 43 test sequences, outperforming its closest competitor by a considerable margin, which accurately predicted only three out of the 43 sequences. AlphaFold harnesses trained DNNs to extrapolate protein attributes from primary sequences, estimating the distances between amino acids and φ–ψ angles between adjacent peptide bonds. These probabilities are then utilized to generate a score, serving as a metric for evaluating the fidelity of a proposed 3D protein structure model. Employing scoring functions, AlphaFold meticulously assesses protein structures, discerning those that align with predictions [19].

Prediction of Drug–Protein Interactions

Quantum mechanics or molecular mechanics (QM/MM) methodologies play an essential role in drug development by elucidating protein–ligand interactions. These approaches account for quantum effects at the atomic scale, affording superior precision to conventional MM techniques, which rely on simplistic energy functions based on atomic coordinates [20]. QM-based methods offer increased accuracy but require higher computational time and cost. Integrating AI into QM calculations necessitates striking a delicate balance between accuracy and time-saving benefits inherent in MM models. AI models are adeptly trained to swiftly compute QM energies from atomic coordinates at a pace comparable to that of MM methods. Primarily employed for atomic simulations and electrical property predictions, AI, particularly DL, revolutionizes small molecule potential energy prediction by supplanting computationally intensive quantum chemistry calculations with rapid ML algorithms. Utilizing quantum chemistry-derived methods such as density functional theory (DFT), potential energies can be used for DNNs to be trained on extensive datasets, yielding remarkable accuracy. In a study encompassing 2 million elpasolite crystals, an ML model's accuracy demonstrated improvement with larger sample sizes, achieving a notable precision of 0.1 eV/atom for DFT formation energies trained on 10,000 structures; using this model, compositional options for diverse attributes can be efficiently screened. Predicting the intricate interplay between drugs and proteins represents a critical nexus in the realm of *in silico* drug development, where AI-driven methodologies hold immense promise. The vast predictions hinging upon myriad undisclosed interactions present formidable challenges in this domain.

Consequently, grappling with the complexities inherent in labelled and unlabelled data necessitates the implementation of semi-supervised training methodologies. Traditionally, it is recognized that superior outcomes are attainable solely through the utilization of labelled data. Furthermore, the semi-supervised paradigm integrates multifaceted data about chemical composition, genomic sequences, and intricate networks of drug–protein interactions. Ultimately, this can forecast well outcomes concerning the intricate interplay between drugs and proteins across diverse datasets encompassing ions, enzymes, and nuclear receptors.

The efficacy of pharmacological agents predominantly hinges upon their interactions with proteins, dictating therapeutic efficacy. However, our comprehension is primarily based on acknowledged therapeutic protein targets, leaving a chasm of understanding concerning non-target entities. Computational methodologies offer a potent avenue to bridge this information gap by identifying potential protein targets for various therapeutic compounds. Ain et al.'s investigation unveiled that an ensemble of 35 predictors markedly shapes the resemblance between protein and drug targets. These predictive variables were used to engender three distinct categories of similarity metrics grounded in drug structure, target sequence, and drug profiles. Appreciating the interrelations and repercussions among disparate database sources constitutes a key player in deciphering therapeutic mechanisms [21]. The serendipitous unearthing of drug–protein interactions assumes paramount importance in medication repurposing. Despite the inescapable incidence of pharmacological adversities, wherein many human proteins can instigate severe side effects, the repurposing of prevailing drugs retains substantial value. The entirety of an organism's proteome holds sway in this endeavour, wielding significant influence over therapeutic devices.

Planning Chemical Synthesis with AI

AI is a powerful tool that greatly increases efficiency in drug development. After a molecule has been virtually assessed for bioactivity and toxicity, the search for the best synthesis method for the therapeutic candidate begins. The synthesis of new drug candidates is frequently a challenging step. Inherent structural complexity or conflicting reactivities make it difficult to synthesize novel chemicals effectively, even with the large reservoir of known transformation processes [22]. Until more basic precursor molecules are identified, retrosynthesis analysis aims to pinpoint "backward" reaction pathways. The most popular technique for choosing branches in retrosynthesis route predictions is Monte Carlo tree search (MCTS), which requires repeatedly truncating the target molecule at different locations. Monte Carlo simulations search for optimum solutions without random step branching. Although computer-assisted synthesis planning (CASP) techniques have been created for retrosynthesis analysis, chemists have not widely adopted them. These methods need the incorporation of human knowledge into executable programs. Yet, chemical encoding by hand is impractical in the face of exponential information development, and reaction database outputs frequently lack chemical intelligence. ML techniques trained on empirical data present a potential way to guide the choice of random steps and predict the likelihood of a change at branching points for synthesis plans. Based on information from the literature on the costs and yields of transformation rules, AI systems can predict the most likely retrosynthesis pathway for a given molecule. A new technique called 3N-MCTS combines MCTS with three neural networks to provide a complete CASP procedure [23]. Every network carries out distinct duties, such as upgrading, rolling out, and expanding. Prospective retroactive transformations for molecules or intermediates are investigated in the expansion node, which incorporates an "in-scope" strategy to evaluate transformation feasibility based on 12.4 million criteria in the literature. Pathway selection is guided by neural networks that are taught to anticipate the best transformations for molecules and intermediates. The rollout node uses a similar "in-scope" policy, but it only applies transformation rules that are regularly reported. This allows for a comprehensive search for the best transformation possibilities during expansion and a quick review of position values during rollout. The update node incorporates route assessment into the search tree to facilitate iterative searches for high-scoring transformations and find probable antecedents for the full reaction pathway. Apart from pinpointing a reaction path, the duration required to arrive at a resolution is a crucial indicator of the algorithm's efficiency. The MCTS beats other algorithms on a test set of chemicals, solving 80% of retrosynthesis problems per molecule in less than 5 seconds and exceeding a 90% solution rate in less than 60 seconds. Remarkably, 3N-MCTS outperforms the traditional Monte Carlo method with a 20-fold quicker time per molecule. AI systems are quite good at forecasting yields and products of organic reactions based on the molecular characteristics of reactants. They also make the process of designing synthesis procedures easier. Many approaches in quantum chemistry, such as Hartree–Fock, semi-empirical techniques (AM1, PM3), and DFT, effectively depict experimental results *in silico*, thus overcoming the difficulty of anticipating the result of intricate chemical interactions [24]. AI techniques have been used recently to automate, improve, and generalize yield prediction; Doyle and Dreher's demonstration of ML's capacity to forecast Buchwald–Hartwig coupling reaction yields is an example. Using dipole moments and vibrational frequencies from quantum chemistry as descriptors in conjunction with high-throughput experimental syntheses, RF techniques can precisely predict the yields of additional expected products by examining the relationship between input descriptors and product yields [25].

Hit to Lead Identification

QSAR and Structure-Based Modelling with AI

Over more than half a century, Quantitative Structure-Activity Relationship/Quantitative Structure-Property Relationship (QSAR/QSPR) modelling has evolved substantially, marking profound strides in computational prowess. These models wield profound influence in drug development, showcasing remarkable ability in accurately prognosticating biological activity and essential pharmacokinetic attributes encompassing ADMET. Molecular descriptors, as they are commonly referred to, serve as conduits in this computational realm, translating intricate molecular structural nuances ranging from pharmacophore distribution to physicochemical attributes and functional moieties into machine-interpretable formats conducive to ligand-centric QSAR/QSPR modelling programs [26]. The lexicon of manually crafted molecular descriptors is vast and meticulously constructed, aimed at encapsulating manifold dimensions of underlying chemical architectures. The advent of more versatile ML methodologies, exemplified by SVM and GBM, has largely supplanted simpler models such as linear regression and k-NNs within QSAR/QSPR paradigms. This paradigmatic shift aims to grapple with intricate and often nonlinear interrelations between chemical constitution and its associated physicochemical or biological attributes, albeit potentially sacrificing interpretability. The first phase in the QSAR workflow is dataset compilation. It involves gathering and compiling data from public and literature databases and dividing it into various subsets for subsequent analysis. Then, data preprocessing is done as part of dataset processing. Curation is completed by computing molecular descriptors. Following the computation of the description, dataset processing includes data normalization and splitting into distinct sets. The third phase involves building the model, gathering datasets from internal and external sources, and using learning algorithms to model QSAR [27]. Furthermore, a statistical computation is performed to gauge the model's resilience. The last phase in the QSAR is model evaluation, which involves comparing the model to earlier benchmark models, identifying key features, assessing performance, and interpreting key aspects. Using the chosen modelling technique, a mathematical model predicting the biological activity or toxicity of novel drugs based on their molecular attributes is established. The curated dataset is used to train the model, while cross-validation techniques are used to assess its accuracy, reliability, and robustness. The validation process culminates with applying the QSAR model to an independent test set of chemicals absent from the training dataset. The comparison of anticipated and actual values facilitates the evaluation of the model's predictive prowess. Further refinement of the model's performance can be achieved through parameter adjustments, feature selection, and the incorporation of additional descriptors [28].

Machine Learning-Based QSAR Models

Using supervised learning approaches to uncover complex relationships between chemical structures and biological characteristics, the ML-based QSAR model is a crucial tool for drug design, screening, and property prediction. ML techniques such as RFs and SVMs have been crucial in the search for new drugs since the 1990s. A significant development has been incorporating ML into molecular docking-based scoring systems, enabling effective data mining using chemical graph analysis to extract two- or three-dimensional chemical descriptors. Prediction challenges are then increased to previously unheard-of levels when these descriptors are combined to create chemical fingerprints for different ML models. One of the most significant advances in this field is the combination of big data and ML to allow for the investigation of gene sequences, protein structures, interactions between multiple omics, and interactions between genes. This allows QSAR models to be applied to a wider range of biological phenomena. Conversely, generative models directly learn data distributions from samples and can generate new samples based on learned distributions, offering potential applications in generating candidate compounds. Notably, supervised learning encompasses various methods for discerning hidden patterns within data, including supervised, semi-supervised, unsupervised, and self-supervised learning. However, the QSAR model's focus on predictive ability necessitates a closer examination of discriminant models under supervised learning, emphasizing the pivotal role of feature selection, algorithm choice, and training set quality in model efficacy [29].

Conformation-related indicators have emerged as a crucial consideration in QSAR, with conformational calculations employing quantum chemistry, semi-empirical methods, and molecular dynamics simulations. These calculations, essential for assessing toxicity, enable predictions concerning conformational stability's impact on experimental outcomes, particularly under varying solvent environments. ML-based QSAR methods extend their purview to examining energy changes amidst conformational alterations, shedding light on solvent-induced conformational dynamics and their repercussions on molecular energy exemplified in imidazole scaffolds [30].

The significance of these features varies across different structure–activity prediction tasks, yet ML algorithms has the capacity to discern feature importance from data. Thus, the exhaustive inclusion of features pertinent to QSAR tasks and the application of traditional ML algorithms for training suffice to glean insights from the training set. ML-based QSAR methods excel in modelling systems that are too complex for physical modelling, efficiently computing elusive physical and chemical properties. While traditional QSAR models idealize compound characteristics in controlled environments, ML-based QSAR acknowledges the real-world complexity of solvent environments, necessitating consideration of hydrophilicity and lipophilicity for drug efficacy. ML-based QSAR model construction involves meticulous steps, including molecular coding, dataset selection, feature extraction, model training, and internal/external validation, culminating in the creation of robust predictive models (Figure 17.3). Continuous optimization, facilitated by advances and innovations in computing and algorithms, ensures the adaptability and versatility of ML-based QSAR models, thereby propelling drug discovery and development to unprecedented heights. Supervised learning models are employed in QSAR using labelled datasets, wherein chemical structures serve as input features, while related biological activities, toxicity, and other attributes are encapsulated in output labels. Furthermore, unsupervised learning approaches such as clustering can be used to uncover latent relationships or patterns within chemical datasets by comparing structural characteristics of comparable substances or by reducing dataset dimensionality. Traditional techniques, including RFs, multiple linear regression, Naïve Bayes, k-NNs, SVMs, and DNNs, are used in QSAR model construction. A comparative analysis revealed that the DNN method outperformed standard QSAR methodologies in identifying GPCR agonists and triple-negative breast cancer (TNBC) inhibitors [31].

Clarifying a mathematical framework that links molecular descriptors with quantitative activity measurements is the main goal of QSAR modelling in drug development. The AI algorithms included in QSAR models are structured. Amazingly, RF methods have been used in QSAR-based LBVS, exhibiting reliable prediction results and offering distinct insights into the contributions of features. Numerous

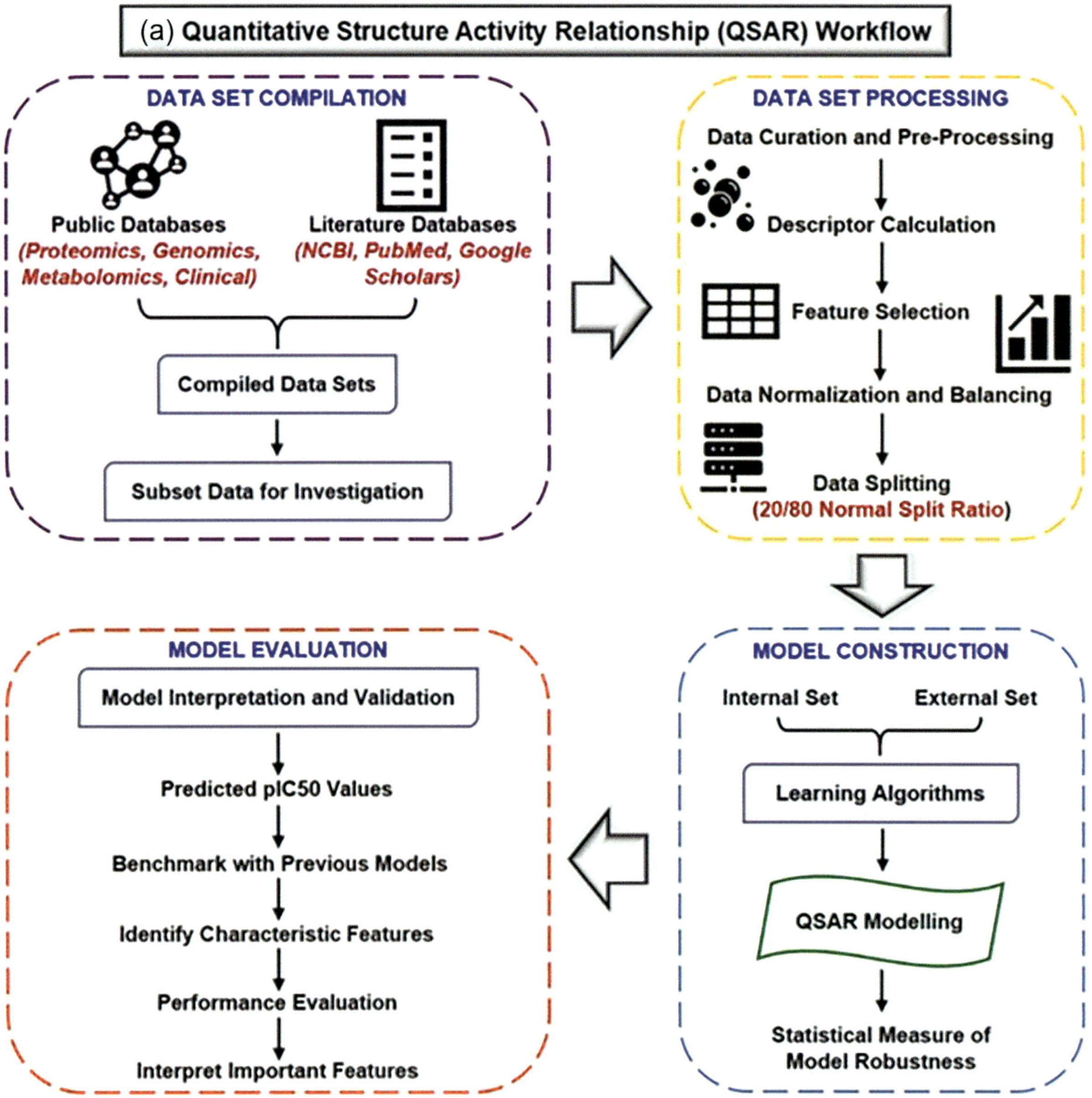

FIGURE 17.3 A QSAR workflow involving AI.

regression RF-based LBVS models have been created, such as Profile-QSAR (pQSAR), which facilitates a deeper understanding of the structural motifs and chemical descriptors that are crucial to the model. Potential medicinal pregnane X receptor activators, such as nimodipine, are identified using a Bayesian QSAR classification model trained on a dataset of 177 compounds. The identification of potent antibacterial 3-hydroxypyridine-4-one medications that target *Staphylococcus aureus* and the selection of novel antibacterial compounds are two significant applications of ANNs in QSAR research [32]. As demonstrated by the creation of an ANN-based multitask QSAR model with a 92% classification accuracy that can classify data from multiplexing experiments, multitask QSAR models have become effective tools for screening substances with various biological processes. More recently, efforts have been made to predict chemical immunotoxicity by building multifunctional QSAR-ANN models using several descriptors. Metrics such as sensitivity, specificity, recall, and accuracy are used to evaluate ML-QSAR models. External validation, conformal prediction, and VS are further methods that support these metrics and guarantee the dependability of the synthesized molecules. Due to computational constraints and intrinsic uncertainties, 2D QSAR models are a strong substitute for their 3D counterparts and frequently exceed them. The fundamental goal is still to create verified models that can precisely predict the biological activities of substances, with well-optimized measures such as R2 and QCV2 for evaluation [33].

De novo Evolution with AI

De novo design is a complex computational process that involves building unique molecules with desired pharmacological characteristics from scratch in a large chemical space containing drug-like compounds, which is thought to have between 10^{60} and 10^{100} different kinds of compounds. The complexity of creating new molecules stems from the enormous variety of atomic species and molecular structures that may be investigated. Depending on the data used, the *de novo* design method might be ligand-based, structured-based, or a combination of the two [34]. Rule-based and rule-free techniques are the two basic categories into which ligand-based strategies may be generally classified. Rule-based techniques, such as the Topliss strategy for sequential analogue production of active lead compounds to maximize potency, assemble molecules from a pool of "building blocks" (e.g. reagents or molecular fragments) according to predetermined construction rules. Modern techniques use predefined sets of chemical transformations to refine, such as matching molecular pairings or general rules for changing molecular structures and functional groups. Building block assembly and ligand manufacture synthesis principles are explicitly incorporated into synthesis-oriented methodologies, making it possible to create synthetically accessible libraries such as CHIPMUNK. Since the late 1990s, hybrid techniques have been developed to improve both chemical synthesizability and resemblance to known bioactive ligands while synthesizing new compounds. These tactics include TOPAS, DOGS, and DINGOS [35].

"Rule-free" approaches aim to produce desired molecules directly, eliminating the requirement for prescriptive standards controlling the production of molecules. Modern methods frequently depend on DL models, which extract fresh chemical samples from a learned latent molecular representation [36]. For *de novo* design, the idea of sampling from a numerical representation of molecules has been around for a while, having its roots in the "inverse QSAR" problem solved by Skvortsova and Zefirov in their seminal work back in the early 1990s. Using an existing QSAR model to determine the descriptor values corresponding to a desired property, inverse QSAR creates molecules. However, these approaches have several drawbacks, such as the possibility of several solutions for a certain feature and the difficulty of translating molecular descriptors into suitable structures. These problems are partially addressed by generative DL, which models the underlying distribution of a given collection of molecules and then uses sampling from the learned distribution to produce new compounds. Typically, SMILES from natural language processing are merged with the popular generative models [37]. Recurrent neural networks provide the basis for most of these systems, frequently in combination with transfer or reinforcement learning. Well-known DL-based generative models, such as VAEs, GANs, and variations based on graph convolutions, have been developed. Recently, conditional generative techniques have been presented with many design constraints, including drug-likeness, three-dimensional configuration, synthesizability, molecular descriptor values, and gene expression profiles. A major prospective problem in this regard will be creating balanced, objective functions that allow for intricate and limited multiparameter optimizations, such as those used in Pareto or desirability-based techniques, often required in drug discovery [38].

One reason for this exponential rise of ligand-based design strategies is the quick development of new generative neural network algorithms. More than forty new models have been developed in the past several years. Because there are so many possible drug design tools available, researchers are under increasing pressure to assess and benchmark generative methodologies in an unbiased and consistent manner. Recent attempts are best illustrated by initiatives such as the MOSES and GuacaMol platforms, which combine well-liked neural generative models with more conventional strategies such as genetic algorithms and provide a wide range of measures for comparison. Metrics such as scaffold and fragment variety, originality of matching molecules, resemblance to recognized compounds in terms of chemical and biological characteristics, and correctness of created molecular representations are frequently used. Reviewing *de novo* design tools after the fact is more difficult than reviewing predictive techniques [39].

Rule-based techniques rely on prior information, such as reaction laws and building blocks; they are good at synthesizing molecules with desired physicochemical properties and are often simple. Conversely, structure designs may be influenced by the choice of building block libraries and hardcoded criteria. Rule-free methods are unrestricted from hardcore structure and moreover allow exploring chemical space directly from data, but such molecules are difficult to synthesize, if not impossible, chemically. Combining both approaches might provide a viable medium for creating new bioactive and synthesizable molecules. Using a set of predefined virtual reactions, a new hybrid method shows promise for rule-free bioactive compound construction that ensures synthesizability in a microfluidic system. While ligand-based techniques have dominated much of the DL-based *de novo* design, investigating a structure-related generative strategy with an eye on orphan receptors and unidentified macromolecules offers a viable alternative. However, DL has not yet fully permeated these approaches, which usually use knowledge on ligand-binding site obtained via growing or fragment linking. However, some initial progress in ligand design has been made by considering the structure and characteristics of the binding pocket [40].

Other Applications of AI/ML

Prediction of Drug Efficacy

Assessing a drug's impact typically involves gauging its biological activity, yet effectively linking therapeutic efficacy with biological activity has proven challenging. To address this gap within the expansive realm of cellular assessments, a substantial corpus of data detailing the effects of cellular medications was amassed. Furthermore, despite being categorized as a psychoactive substance, leveraging random trees within a forest framework enables the derivation of a biomarker gene expression efficacy profile. This approach yielded a classification tree accuracy of 88.9% and an RF model accuracy of 83.3% in a medication therapy study. Consequently, generic genomic data proves instrumental in reconciling the effects of emerging physiological medications with clinical utility at the cellular level. In addition, *in vitro* gene expression data signatures serve as pivotal determinants of therapeutic efficacy, thereby facilitating medication development and target identification [41].

The ability to predict efficacy and pinpoint targets is imperative in drug development, enhancing profitability by validating novel medications. The juxtaposition of medical conditions can reduce treatment effectiveness, underscoring the importance of releasing medications endowed with beneficial therapeutic effects. To demonstrate the effectiveness of 238 drugs for 78 diseases, this study also scrutinized issues about gene efficacy and concomitant conditions. Furthermore, a network-based approach was used in crafting a drug–disease proximity metric, culminating in the development of a measure that evaluates the interplay between drug targets and diseases. Consequently, the proximity inherent within network-based systems facilitates the prediction of associations for novel drug disorders, thereby furnishing ample opportunities for conflict identification and drug repurposing [42].

Exploring the Safety of Biomarkers

In the realm of novel drug development, biomarkers serve as a bulwark against potential safety concerns, encompassing both biological nuances and analytical manifestations of the said biomarkers. A thorough safety evaluation process becomes imperative for biomarkers, facilitating the meticulous examination of experimental congruence. This process further enables the refinement of biomarkers' distinctive attributes and elucidates the methodologies underpinning their analysis, integration, and interpretation, alongside comparing improved biomarkers with traditional benchmarks [43]. Despite boasting extended clinical trial pedigrees, recent literature revealed a disquieting reality: contemporary medications offer no greater safety assurances than their predecessors. These medications swiftly transition into market circulation, often devoid of comprehensive inspections, thereby precluding the implementation of systematic safety verifications. Enhancing drug safety hinges on heeding prior signals associated with drug usage, which can unveil latent biomarkers predisposed to drug toxicity. However, safety indicators' efficacy may fluctuate depending on the targeted biological system. While no panacea exists to ensure absolute medication safety, cultivating a shared knowledge about benefit–risk evaluation emerges as a prudent strategy [44]. By using AI/ML technology, conventional and digital biomarkers may be used to enhance patient diagnosis and therapy. By removing the need for conventional sampling methods and in-person examinations by trained pathologists, real-time, passive/active, robust sensing of physiological and behavioural data from patients that can feed AI-based models can greatly improve diagnosis, treatment, and decision-making at the point of procedure [45]. AI-based therapies provide cutting-edge possibilities for individualized and superior treatment as shown by better care for Alzheimer's disease [46].

Drug Metabolism and Transportation

To mitigate the risks stemming from the metabolic characteristics of potential drugs, precise and reliable methodologies are imperative to forecast drug metabolism *in vitro*. Two primary research avenues in computer-aided metabolic prediction methods entail site of metabolism (SOM) prediction and metabolite structure analysis, both offering pivotal support and direction to experimental endeavours. SOM prediction techniques typically boast high predictive accuracy. The MetaSite software, for instance, employs protein structure data, GRID-derived molecular interaction fields (MIFs) of proteins and ligands, and molecular orbital calculations to assess the likelihood of metabolic reactions at specific atomic sites [47]. These methods use extensive databases housing metabolic data to deduce generalized rules for pinpointing regions within a molecule susceptible to metabolic modification. Furthermore, descriptor-based approaches are vital in elucidating a drug's metabolic trajectory, identifying pertinent enzymes, and delineating reaction pathways. For example, bioprint software includes a database of commonly used medications, reference compounds, and various biological and *in vitro* ADME assay data. This type of data is referred to as biological fingerprinting and allows for the projection of possible outcomes for novel compounds using proximity analysis and QSAR modelling. Comparably, huge data on substrate-related metabolic processes, including xenobiotic reactions, primary and secondary metabolites, and kinetic data pertinent to enzyme inhibition, may be found in the MetaDrug database. To create kernel partial least square models that could detect metabolic trends, a total of 317 compounds, parent drugs, and primary and secondary metabolites were randomly selected from this database [48].

CONCLUSION

The application of AI/ML technology spans diverse domains, including drug discovery, where its utilization is met with significant challenges. AI's potential remains promising despite the inherent complexity of the drug discovery process, compounded by the requisite interdisciplinary expertise spanning biology, chemistry, and medicine. Collaborative efforts between AI experts and domain specialists are imperative to overcome the significant disparity between the two realms. AI specialists should tailor algorithms specifically for drug discovery, prioritizing interpretability to elucidate mechanisms of action and facilitate informed decision-making. Experts in the field need to ensure that accurate biological and chemical data is generated simultaneously by reducing experimental mistakes and combining data into cohesive platforms for AI system improvement. Drug delivery methods are being revolutionized by AI, opening the door to personalized, adaptive, and targeted medicines. Pharmaceutical researchers and medical professionals may increase medication efficacy, reduce adverse effects, and improve patient outcomes using AI's data analysis, pattern identification, and optimization strengths. Numerous benefits come alongside using AI-driven computational approaches for pharmacokinetics and pharmacodynamics, such as anticipating drug distribution and clearance, optimizing dosing schedules, and reducing the need for human and animal experiments. Furthermore, big data and AI enable computational pharmaceutics, optimizing formulations, retrosynthesis, and drug administration procedures, eventually improving patient care. With ongoing developments in VS, predictive modelling, and drug design, the use of AI and ML in drug discovery is expected to grow, potentially leading to faster drug development and revolutionary changes in the pharmaceutical industry.

DECLARATIONS AND AUTHORS' CONTRIBUTIONS

G.L.K. and J.M. devised and designed the concept. J.M. and S.K. carried out the literature search and wrote the manuscript. All the authors have read and approved the final version of the manuscript.

CONFLICT OF INTEREST

The author(s) confirm that this article's content has no conflict of interest.

ACKNOWLEDGMENTS

The authors are highly grateful to the Department of Pharmaceuticals, Ministry of Chemical & Fertilizers, Government of India, for providing financial assistance.

REFERENCES

1. Yang, X.; Wang, Y.; Byrne, R.; Schneider, G.; Yang, S. Concepts of Artificial Intelligence for Computer-Assisted Drug Discovery. *Chem. Rev.* **2019**, *119* (18), 10520–10594. https://doi.org/10.1021/acs.chemrev.8b00728.
2. Dara, S.; Dhamercherla, S.; Jadav, S. S.; Babu, C. H. M.; Ahsan, M. J. Machine Learning in Drug Discovery: A Review. *Artif. Intell. Rev.* **2022**, *55* (3), 1947–1999. https://doi.org/10.1007/s10462-021-10058-4.
3. Levy, J.; Lu, Y.; Montivero, M.; Ramwala, O.; McFadden, J., Miles, C.; Diamond, A. G.; Reddy, R.; Reddy, R.; Hudson, T. Artificial Intelligence, Bioinformatics, and Pathology: Emerging Trends Part I—An Introduction to Machine Learning Technologies. *Adv. Mol. Pathol.* **2022**, *5* (1), e1–e24.
4. Reddy, B. M. Machine Learning for Drug Discovery and Manufacturing. In *AI and Blockchain in Healthcare*; Rai, B. K., Kumar, G., Balyan, V., Eds.; Springer: Singapore, 2023; pp. 3–30. https://doi.org/10.1007/978-981-99-0377-1_1.
5. Patwardhan, B.; Vaidya, A. D. B.; Chorghade, M.; Joshi, S. P. Reverse Pharmacology and Systems Approaches for Drug Discovery and Development. *Curr. Bioact. Compd.* **2008**, *4* (4), 201–212.
6. Gupta, R.; Srivastava, D.; Sahu, M.; Tiwari, S.; Ambasta, R. K.; Kumar, P. Artificial Intelligence to Deep Learning: Machine Intelligence Approach for Drug Discovery. *Mol. Divers.* **2021**, *25* (3), 1315–1360. https://doi.org/10.1007/s11030-021-10217-3.
7. Harnik, Y.; Milo, A. A Focus on Molecular Representation Learning for the Prediction of Chemical Properties. *Chem. Sci.* **2024**, *15* (14), 5052–5055.
8. Ferreira, L. L. G.; Andricopulo, A. D. ADMET Modeling Approaches in Drug Discovery. *Drug Discov. Today* **2019**, *24* (5), 1157–1165.
9. Lavecchia, A. Machine-Learning Approaches in Drug Discovery: Methods and Applications. *Drug Discov. Today* **2015**, *20* (3), 318–331. https://doi.org/10.1016/j.drudis.2014.10.012.
10. Pasrija, P.; Jha, P.; Upadhyaya, P.; Khan, M.; Chopra, M. Machine Learning and Artificial Intelligence: A Paradigm Shift in Big Data-Driven Drug Design and Discovery. *Curr. Top. Med. Chem.* **2022**, *22* (20), 1692–1727.
11. Singh, S.; Gupta, H.; Sharma, P.; Sahi, S. Advances in Artificial Intelligence (AI)-Assisted Approaches in Drug Screening. *Artif. Intell. Chem.* **2024**, *2* (1), 100039.
12. Vamathevan, J.; Clark, D.; Czodrowski, P.; Dunham, I.; Ferran, E.; Lee, G.; Li, B.; Madabhushi, A.; Shah, P.; Spitzer, M.; Zhao, S. Applications of Machine Learning in Drug Discovery and Development. *Nat. Rev. Drug Discov.* **2019**, *18* (6), 463–477. https://doi.org/10.1038/s41573-019-0024-5.
13. Siontis, K. C.; Noseworthy, P. A.; Attia, Z. I.; Friedman, P. A. Artificial Intelligence-Enhanced Electrocardiography in Cardiovascular Disease Management. *Nat. Rev. Cardiol.* **2021**, *18* (7), 465–478.
14. Yang, Z.; Shi, S.; Fu, L.; Lu, A.; Hou, T.; Cao, D. Matched Molecular Pair Analysis in Drug Discovery: Methods and Recent Applications. *J. Med. Chem.* **2023**, *66* (7), 4361–4377.
15. Cavasotto, C. N.; Scardino, V. Machine Learning Toxicity Prediction: Latest Advances by Toxicity End Point. *ACS Omega* **2022**, *7* (51), 47536–47546.
16. Bulusu, K.; Guha, R.; Mason, D.; Lewis, R.; Muratov, E.; Kalantarmotamedi, Y.; Cokol, M.; Bender, A. Modelling of Compound Combination Effects and Applications to Efficacy and Toxicity: State-of-the-Art, Challenges and Perspectives. *Drug Discov. Today* **2015**, *21* (2), 225–238. https://doi.org/10.1016/j.drudis.2015.09.003.
17. Jumper, J.; Evans, R.; Pritzel, A.; Green, T.; Figurnov, M.; Ronneberger, O.; Tunyasuvunakool, K.; Bates, R.; Žídek, A.; Potapenko, A. Highly Accurate Protein Structure Prediction with AlphaFold. *Nature* **2021**, *596* (7873), 583–589.
18. Kinch, L. N.; Pei, J.; Kryshtafovych, A.; Schaeffer, R. D.; Grishin, N. V. Topology Evaluation of Models for Difficult Targets in the 14th Round of the Critical Assessment of Protein Structure Prediction (CASP14). *Proteins Struct. Funct. Bioinforma.* **2021**, *89* (12), 1673–1686.
19. Zuluaga, F. H. G.; D'Arminio, N.; Bardozzo, F.; Tagliaferri, R.; Marabotti, A. An Automated Pipeline Integrating AlphaFold 2 and MODELLER for Protein Structure Prediction. *Comput. Struct. Biotechnol. J.* **2023**, *21*, 5620–5629.
20. Xia, S.; Chen, E.; Zhang, Y. Integrated Molecular Modeling and Machine Learning for Drug Design. *J. Chem. Theory Comput.* **2023**, *19* (21), 7478–7495. https://doi.org/10.1021/acs.jctc.3c00814.

21. Ain, Q. U.; Aleksandrova, A.; Roessler, F. D.; Ballester, P. J. Machine-Learning Scoring Functions to Improve Structure-Based Binding Affinity Prediction and Virtual Screening. *Wiley Interdiscip. Rev. Comput. Mol. Sci.* **2015**, *5* (6), 405–424. https://doi.org/10.1002/wcms.1225.
22. Coley, C. W.; Barzilay, R.; Jaakkola, T. S.; Green, W. H.; Jensen, K. F. Prediction of Organic Reaction Outcomes Using Machine Learning. *ACS Cent. Sci.* **2017**, *3* (5), 434–443. https://doi.org/10.1021/acscentsci.7b00064.
23. Segler, M. H. S.; Preuss, M.; Waller, M. P. Planning Chemical Syntheses with Deep Neural Networks and Symbolic AI. *Nature* **2018**, *555* (7698), 604–610.
24. Foster, M. E.; Sohlberg, K. Empirically Corrected DFT and Semi-Empirical Methods for Non-Bonding Interactions. *Phys. Chem. Chem. Phys.* **2010**, *12* (2), 307–322.
25. Chan, H. C. S.; Shan, H.; Dahoun, T.; Vogel, H.; Yuan, S. Advancing Drug Discovery via Artificial Intelligence. *Trends Pharmacol. Sci.* **2019**, *40* (8), 592–604. https://doi.org/10.1016/j.tips.2019.06.004.
26. Dahl, G.; Jaitly, N.; Salakhutdinov, R. Multi-Task Neural Networks for QSAR Predictions. *arXiv Prepr. arXiv1406.1231* **2014**.
27. Tsou, L. K.; Yeh, S.-H.; Ueng, S.-H.; Chang, C.-P.; Song, J.-S.; Wu, M.-H.; Chang, H.-F.; Chen, S.-R.; Shih, C.; Chen, C.-T.; Ke, Y.-Y. Comparative Study between Deep Learning and QSAR Classifications for TNBC Inhibitors and Novel GPCR Agonist Discovery. *Sci. Rep.* **2020**, *10* (1), 16771. https://doi.org/10.1038/s41598-020-73681-1.
28. Keyvanpour, M. R.; Shirzad, M. B. An Analysis of QSAR Research Based on Machine Learning Concepts. *Curr. Drug Discov. Technol.* **2021**, *18* (1), 17–30. https://doi.org/10.2174/1570163817666200316104404.
29. Mao, J.; Akhtar, J.; Zhang, X.; Sun, L.; Guan, S.; Li, X.; Chen, G.; Liu, J.; Jeon, H.-N.; Kim, M. S.; No, K. T.; Wang, G. Comprehensive Strategies of Machine-Learning-Based Quantitative Structure-Activity Relationship Models. *iScience* **2021**, *24* (9), 103052. https://doi.org/10.1016/j.isci.2021.103052.
30. Singh, K.; Ghosh, I.; Jayaprakash, V.; Jayapalan, S. Building a ML-Based QSAR Model for Predicting the Bioactivity of Therapeutically Active Drug Class with Imidazole Scaffold. *Eur. J. Med. Chem. Reports* **2024**, *11*, 100148.
31. Shi, L.; Yan, F.; Liu, H. Screening Model of Candidate Drugs for Breast Cancer Based on Ensemble Learning Algorithm and Molecular Descriptor. *Expert Syst. Appl.* **2023**, *213*, 119185.
32. Durrant, J. D.; Amaro, R. E. Machine-learning Techniques Applied to Antibacterial Drug Discovery. *Chem. Biol. Drug Des.* **2015**, *85* (1), 14–21.
33. Olier, I.; Sadawi, N.; Bickerton, G. R.; Vanschoren, J.; Grosan, C.; Soldatova, L.; King, R. D. Meta-QSAR: A Large-Scale Application of Meta-Learning to Drug Design and Discovery. *Mach. Learn.* **2018**, *107* (1), 285–311. https://doi.org/10.1007/s10994-017-5685-x.
34. Chávez-Hernández, A. L.; López-López, E.; Medina-Franco, J. L. Yin-Yang in Drug Discovery: Rethinking de Novo Design and Development of Predictive Models. *Front. Drug Discov.* **2023**, *3*, 1222655.
35. Blaschke, T.; Olivecrona, M.; Engkvist, O.; Bajorath, J.; Chen, H. Application of Generative Autoencoder in De Novo Molecular Design. *Mol. Inform.* **2018**, *37* (1–2), 123. https://doi.org/10.1002/minf.201700123.
36. Mouchlis, V. D.; Afantitis, A.; Serra, A.; Fratello, M.; Papadiamantis, A. G.; Aidinis, V.; Lynch, I.; Greco, D.; Melagraki, G. Advances in de Novo Drug Design: From Conventional to Machine Learning Methods. *Int. J. Mol. Sci.* **2021**, *22* (4), 676. https://doi.org/10.3390/ijms22041676.
37. Crucitti, D.; Pérez Míguez, C.; Díaz Arias, J. Á.; Fernandez Prada, D. B.; Mosquera Orgueira, A. De Novo Drug Design through Artificial Intelligence: An Introduction. *Front. Hematol.* **2024**, *3*, 5741. https://doi.org/10.3389/frhem.2024.1305741.
38. Nicolaou, C.; Kannas, C.; Loizidou, E. Multi-Objective Optimization Methods in de Novo Drug Design. *Mini Rev. Med. Chem.* **2012**, *12* (10), 979–987.
39. Bai, Q.; Liu, S.; Tian, Y.; Xu, T.; Banegas-Luna, A. J.; Pérez-Sánchez, H.; Huang, J.; Liu, H.; Yao, X. Application Advances of Deep Learning Methods for de Novo Drug Design and Molecular Dynamics Simulation. *WIREs Comput. Mol. Sci.* **2022**, *12* (3), e1581. https://doi.org/10.1002/wcms.1581.
40. Abe, K.; Ozako, M.; Inukai, M.; Matsuyuki, Y.; Kitayama, S.; Kanai, C.; Nagai, C.; Gopalasingam, C. C.; Gerle, C.; Shigematsu, H.; Umekubo, N.; Yokoshima, S.; Yoshimori, A. Deep Learning Driven de Novo Drug Design Based on Gastric Proton Pump Structures. *Commun. Biol.* **2023**, *6* (1), 956. https://doi.org/10.1038/s42003-023-05334-8.
41. Kasi, A.; Shetty, M.; Neil, R.; Murali, V.; Athri, P.; Srinivasa, G. Exploration and Comparison of Modern AI Algorithms to Predict Drug Efficacy. In *2020 IEEE International Conference on Electronics, Computing and Communication Technologies (CONECCT)*; IEEE, New York, 2020; pp. 1–5.
42. Badwan, B. A.; Liaropoulos, G.; Kyrodimos, E.; Skaltsas, D.; Tsirigos, A.; Gorgoulis, V. G. Machine Learning Approaches to Predict Drug Efficacy and Toxicity in Oncology. *Cell Rep. Methods* **2023**, *3* (2), 100413. https://doi.org/10.1016/j.crmeth.2023.100413.
43. Califf, R. M. Biomarker Definitions and Their Applications. *Exp. Biol. Med.* **2018**, *243* (3), 213–221.
44. Bates, D. W.; Levine, D.; Syrowatka, A.; Kuznetsova, M.; Craig, K. J. T.; Rui, A.; Jackson, G. P.; Rhee, K. The Potential of Artificial Intelligence to Improve Patient Safety: A Scoping Review. *NPJ Digit. Med.* **2021**, *4* (1), 54. https://doi.org/10.1038/s41746-021-00423-6.
45. Arya, S. S.; Dias, S. B.; Jelinek, H. F.; Hadjileontiadis, L. J.; Pappa, A.-M. The Convergence of Traditional and Digital Biomarkers through AI-Assisted Biosensing: A New Era in Translational Diagnostics? *Biosens. Bioelectron.* **2023**, *235*, 115387.
46. Subramanian, K.; Hajamohideen, F.; Viswan, V.; Shaffi, N.; Mahmud, M. Exploring Intervention Techniques for Alzheimer's Disease: Conventional Methods and the Role of AI in Advancing Care. *Artifi. Intell. Appl.* **2024**, *2*, 73–91.
47. Wang, D.; Liu, W.; Shen, Z.; Jiang, L.; Wang, J.; Li, S.; Li, H. Deep Learning Based Drug Metabolites Prediction. *Front. Pharmacol.* **2020**, *10*, 1586. https://doi.org/10.3389/fphar.2019.01586.
48. Testa, B.; Balmat, A.-L.; Long, A.; Judson, P. Predicting Drug Metabolism – An Evaluation of the Expert System METEOR. *Chem. Biodivers.* **2005**, *2* (7), 872–885. https://doi.org/10.1002/cbdv.200590064.

18 Estimation of Pharmacokinetic Parameters Using Computational Tools

Samyak Bajaj[1], *Shivam Kori*[2], *and Ratnesh Das*[3]
[1]Department of Pharmacy, Gyan Sagar College, Narsinghpur Road, Sagar, India
[2]Department of Pharmacy, Gyanveer University Rajghat Road Tili, Sagar, India
[3]Department of Chemistry, Dr. Harisingh Gour University, Sagar, India

INTRODUCTION

Historically, in the pharmaceutical industry, the development of new chemical entities (NCEs) was focussed on predicting the inherent biological activity of drug candidates, with optimization of potency being a prime concern. In this phase, attrition rates in late-stage development were high as candidate drugs were proven to lack *in vivo* efficacy despite exhibiting promising results *in vitro*. A seminal paper by [1] in 1988 demonstrated that unfavourable pharmacokinetics had resulted in attrition in 39% of late-stage failures. This moved the industry towards the concurrent optimization of both potency and pharmacokinetics, and from 1991 to 2000, attrition due to poor pharmacokinetics was reduced by 30% [2]. This is no cause for complacency as problems with safety and efficacy still result in high attrition rates, with a few NCEs being successfully introduced to the market every year. An investigation into reasons for attrition, across four major pharmaceutical companies between 2000 and 2010, demonstrated that failure due to poor pharmacokinetics remained a significant issue [3]. This is of prime concern to the sector because of the desire to save time, cost, and animal use in drug development, as well as the societal requirement to bring more effective medicines to the market more rapidly. Cooperative endeavours between pharmaceutical companies have attempted to address these issues by sharing data, facilitating the development of increasingly comprehensive databases and improved predictive models [4,5].

Absorption, Distribution, Metabolism, and Excretion (ADMET) data is regarded as a crucial component of drug discovery and development. Drug behaviour upon administration can be predicted using characteristics pertaining to ADMET qualities provided by both *in vitro* and *in vivo* models. Drug candidates are either advanced, retained, or terminated based on ADMET characteristics [6]. The evaluation of medication's post administration targeting is influenced by preclinical data on their ADMET capabilities, as this data can be used to determine pharmacokinetic profiles. Drug exposure in the targeted site of action is evaluated by taking into account factors such as the drug's metabolism, absorption rate, and deposition within the targeted organ [7]. A drug's pharmacokinetic characteristics, particularly its ADMET (Figure 18.1), should be ascertained in order to create medications with the necessary qualities and ideal dosage schedules. Owing to a number of risk factors related to drug development and discovery, as well as the lengthy procedures involved, *in vivo* models were used to reduce the anticipated undesirable characteristics of medications during the preclinical phase prior to their release to the market [8]. Drug developers face challenges in developing efficient and cost-effective drugs compared with existing therapies. To minimize adverse effects, they need to optimize physicochemical properties such as bioactivities and ADMET properties. Other factors to consider include dose size and frequency, drug bioavailability, oral absorption, clearance, volume of distribution (VD), and penetration through the blood–brain barrier (BBB). The relative solubility of drugs is crucial for toxicity and design, as poorly soluble drugs can cause issues during manufacturing and storage. The metabolic stability of drugs is also crucial, as the success of the development process relies heavily on their metabolism profile [9]. This chapter covers several computational tools to predict pharmacokinetics parameters.

OVERVIEW OF THE DEVELOPMENT OF ADMET PREDICTION MODELS

When developing an *in silico* ADMET prediction model, a number of elements need to be taken into account. The features of the model to be created are primarily determined by the model's intended use; for instance, the level of prediction accuracy changes according to the model's necessary prediction speed. Due to the high number of calculations needed for the prediction process, models with higher prediction accuracy tend to be more time-consuming [10]. When a large number of molecules need their ADMET properties (Figure 18.2) to be assessed quickly, simple models with quick computations are typically used; however, at more advanced stages of the drug development process, models with a high degree of prediction accuracy are needed for ADMET property assessment [11].

Choosing whether the model is global and used to forecast structurally diverse compounds or specific to a particular project where typically comparable molecules are present is another crucial consideration. The selection of an appropriate set of descriptors to be utilized in the model

DOI: 10.1201/9781003464679-18

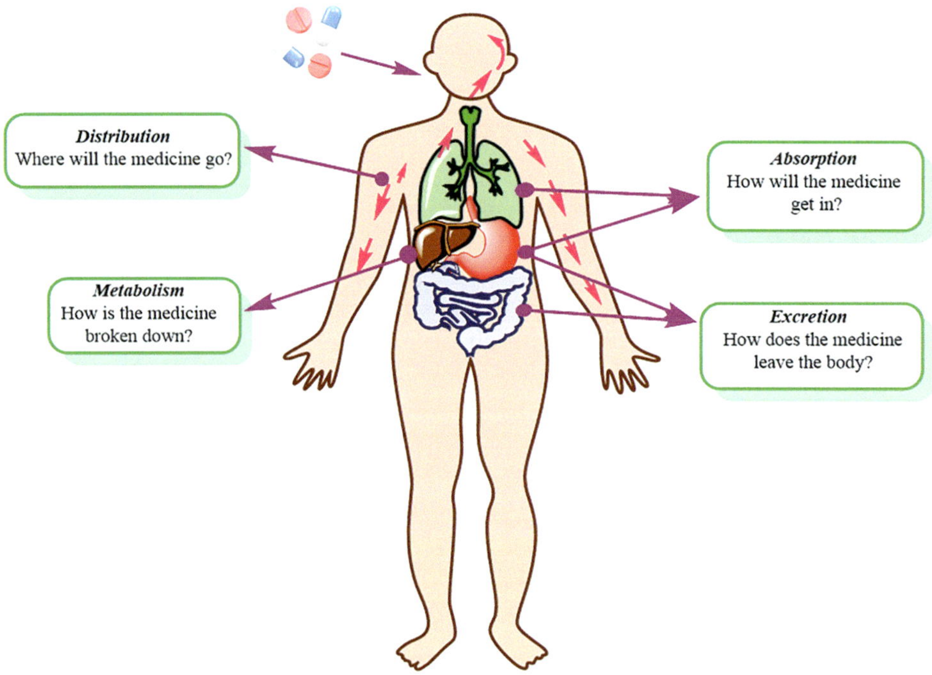

FIGURE 18.1 A scheme that represents ADME pharmacokinetics parameters of the drug.

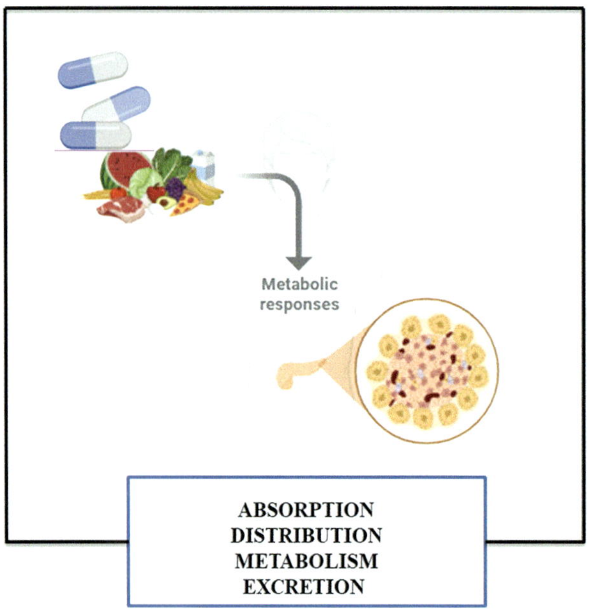

FIGURE 18.2 ADME pharmacokinetics parameters of the drug.

and the statistical techniques to employ can be decided to suit the goal of the model to be built based on the necessary features of the model [12].

A. **Descriptors:** Molecular descriptors are algorithm-generated mathematical representations of the characteristics of molecules. The physical and chemical properties of the molecules are quantitatively described by the numerical values of molecular descriptors. A quantitative indicator of a molecule's lipophilicity, LogP, is an example of a molecular descriptor. It is derived by measuring the molecule's partitioning between an aqueous phase and a lipophilic phase, which is typically composed of water and n-octanol. When conducting similarity searches in molecular libraries, molecular descriptors can be helpful since they can identify compounds with comparable chemical or physical characteristics depending on how similar their values are. In order to forecast the ADMET characteristics of molecules based on the values of their descriptors, ADMET prediction models employ molecular descriptors to correlate the structure–property association [13].

1D Descriptors	2D Descriptors	3D Descriptors
1D descriptors are the simplest type of molecular descriptors; they represent information that are calculated from the molecular formula of the molecule, which includes the number and type of atoms in the molecule and molecular weight.	2D descriptors are more complex than 1D descriptors; usually, they represent molecular information regarding the size, shape, and electronic distribution of the molecule.	3D descriptors mainly describe properties related to the 3D conformation of the molecule, such as intramolecular hydrogen bonding. Examples include polar and nonpolar surface area (PSA and NPSA, respectively), quantum mechanics calculations, and valence electron distribution.

B. **Datasets:** An accurate dataset that is appropriate for the model's application is necessary for the creation of a successful ADMET prediction model. Whether the model is constructed as a local model or a group model depends on the quantity of compounds and structural diversity in the datasets [14].

Local Model	Group Model
It is used for the prediction of a relatively similar class of compounds; they are generally used during advanced stages in the drug development process, because these models are specific to a class of similar molecules.	It can be used for the prediction of different classes of compounds; they are beneficial for use when a high number of compounds are present, such as in the initial stages of drug development.

C. **Statistical Methods:** Several statistical methods are usually used in the development of ADMET prediction models:
   - Multiple linear regression (MLR)
   - Multivariate data analysis (MVA)
   - Multiple nonlinear regression analysis (MNLR)
   - Partial least square (PLS)

D. **Model Validation:** It is crucial to confirm the ADMET prediction model's predictive power before implementing it. One technique that is frequently used in model validation is to purposefully leave out a subset of the dataset from the model and use it as a test to confirm the model's capacity for prediction after it has been developed. An alternative approach is to test the model's predictability using an external dataset, which will confirm that the model can accurately forecast external datasets that are not involved in the development process [12].

IN SILICO PREDICTION OF PHYSICOCHEMICAL AND PHARMACOKINETIC PROPERTIES

The association between the physicochemical characteristics and the biological behaviour of pharmacological therapies has been the subject of an increasing number of studies. Drugs' physicochemical characteristics can be altered by optimizing chemical structural modifications and their behavioural effects [15]. Water solubility, partition coefficient (LogP), melting point (MP), boiling point (BP), and bioconcentration factor (BCF) are some of the physicochemical characteristics that determine the drug's biological behaviour. Conversely, it has been demonstrated that these physicochemical characteristics influence a number of pharmacokinetic characteristics, including the drug's permeability, transfer, and bioavailability [16]. Indirectly, it is shown that the drug's physicochemical characteristics affect how its pharmacokinetic characteristics are interpreted in the early phases of development.

In Silico Prediction of Physicochemical Properties

A. **Lipophilicity:** The ratio of a compound's concentration between two phases, such as an oil phase and a liquid phase, at equilibrium is known as lipophilicity, or LogP [8]. It significantly affects a number of pharmacokinetic characteristics, including drug clearance pathways, permeability, distribution, and absorption [17]. Therefore, in the drug development process, automatic computational measures of water solubility, lipophilicity, and degree of ionization have been used and combined. Other prediction methods based on numerous fragments and atoms have been developed using models that predict LogP, which can facilitate the drug design process [18].

B. **Hydrogen Bonding:** The interaction between the H-bond donor chemical and the H-bond acceptor target, or vice versa, is reflected in hydrogen bonding [19]. It has been shown that the permeability of bioactive compounds across biological membranes is influenced by their hydrogen bonding capacity. In addition, it was demonstrated that a molecule should break hydrogen bonds with the drug's aqueous environment in order to pass a biological membrane. As a result, it is generally not a good idea for a molecule to form a lot of hydrogen bonds because this would negatively impact both its absorption and degree of permeability [20].

C. **Solubility:** Due to its involvement in determining drug uptake, transport, and removal from the body, aqueous solubility is a basic feature that is almost involved in every stage of drug development [21]. Since a drug's effectiveness is mostly determined by how soluble it is in water, medications with low solubility or low rates of dissolution will be removed from the bloodstream before they can provide the necessary pharmacological action [22,23].

MVA is typically used to create computational models for solubility prediction based on molecular structure. These models include artificial neural networks (ANNs), Support Vector Regression (SVR), Random Forest (RF), and Partial Least Squares (PLS). A recent *in silico* study has shown a practical computational algorithm that can forecast the impact of a material's solid state on its solubility. It is anticipated that this approach, along with others based on the same concepts of quantum and molecular mechanics, will have greater promise in the future and produce predictions that are more accurate. In addition, it is anticipated that this approach would precisely forecast crystal lattice characteristics and their effects on medication solubility and dissolution rate [18].

D. **Permeability:** The method of passive diffusion, in which compounds are transported by the action of a concentration gradient, is the primary way that permeable medications transit biological barriers, such as intestinal epithelium and BBB. To precisely forecast a drug's membrane permeability based on its lipophilicity profiles, molecular size, H-bonding capacity, and PSA, a number of *in silico* models were developed. It has been demonstrated that *in silico* prediction models reduce the amount of time required to generate novel medication candidates [24].

In Silico Prediction of Pharmacokinetic Properties

A. ***In Silico* Prediction of Drug Absorption**: The gastrointestinal absorption of pharmacological compounds is regarded as a complex mechanism because of a number of elements, which are primarily divided into formulation effects, physiochemical impacts, and physicochemical effects (Figure 18.3). Consequently, predicting the absorption of medications has historically been a significant challenge for academics and drug candidate developers [25].

In addition to models of *in vivo* and *in vitro* prediction technologies, a variety of *in silico* models have been employed to predict oral absorption. *In silico* models have been used to predict the bioavailability, stomach acid function, and associated hazards, as well as to assess the impact of gastrointestinal pH on the exposure of weak base medications [26].

B. ***In Silico* Prediction of Intestinal Permeation:** The intestinal permeability and lipophilicity of the drug substance have generally been found to have a sigmoidal relationship. This means that intestinal permeability, which is attained by the passive diffusion mechanism, typically determines the absorbability of a drug with a low molecular weight or no discernible metabolism [27].

The intestinal permeability of medications under development has been evaluated using computational modelling approaches prior to synthesis, a process that is thought to be quick and affordable. Intestinal permeability, a crucial component of drug discovery and development, is now predicted using a variety of models, including the basic models, the PSA, the fast PSA, and other sophisticated models [28].

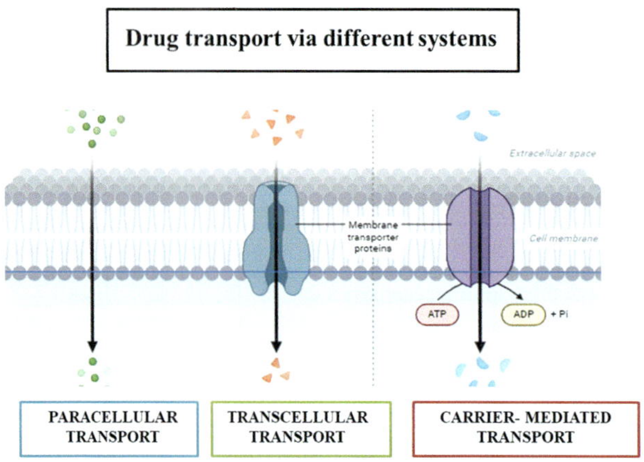

FIGURE 18.3 Drug transport across biological membranes including paracellular and transcellular passive diffusion in addition to carrier-mediated.

C. ***In Silico* Prediction of Drug Distribution:** BBB permeability, VD, and plasma protein binding (PPB) are the three primary areas of analysis that are used to predict the distribution of drugs throughout the body. Each of the three domains plays a discernible part in determining appropriate drug regimens, effective plasma concentrations, and permeability across the BBB, all of which aid in the prediction of CNS targets, adverse effects, and non-CNS treatments [29]. Drug tissue distribution is currently predicted using a variety of techniques, which include looking at tissue–plasma ratios or the volume of drug distribution at the steady state [25].

D. ***In Silico* Prediction of Drug Metabolism:** Among all pharmacokinetic characteristics, metabolism is the most unpredictable one since it is a very complex process involving a variety of enzyme activities that vary from one person to another due to genetic factors. Relative predictions about the metabolism of several medications have been estimated using a variety of computer (*in silico*) models.

    When evaluating a drug's metabolism profile in its early phases, several factors are optimized, including the kinetics of metabolizing enzymes as well as metabolic pathways, stability, and interactions [30]. Since cytochrome P450 (CYP) is thought to be the most important enzyme in drug metabolism, numerous models, including QSAR, have been developed to predict how the CYP enzyme will metabolize compounds.

E. ***In Silico* Prediction of Drug Excretion:** The process by which the body eliminates waste and harmful substances is referred to as excretion. Molecular weight is the most crucial determinant of the appropriate drug removal mechanism, with urine being the primary method of removal for drugs with relatively low molecular weights [31]. Several factors, including flow rate, lipophilicity, protein binding, and pKa value, can be used to predict passive excretion. A predictive model that gives a comprehensive model depicting the behaviour of the substance during the various stages of drug research and development should incorporate the information gathered after a drug's excretion profile has been predicted [25].

F. ***In Silico* Prediction of Toxicity Profile:** Traditionally, laboratory animals were used to test toxicity. New methods for toxicity optimization have been developed recently, and it has been stated that these methods reduce the dangers associated with animal toxicity testing by substituting safer alternatives [32]. Generally speaking, *in silico* toxicology is related to predictive science, and toxicology computational methods include toxicity databases that enable QSAR modelling. There are two types of *in silico* prediction approaches that are explicitly designed to forecast the toxicity of drugs: those that predict systemic toxicity and those that exclusively predict toxicity for a particular organ. Other *in silico* models, on the other hand, are thought to be more complex because they focus on both genotoxicity and carcinogenicity prediction [33].

TOOLS AVAILABLE FOR PREDICTION OF ADMET

As previously stated, the process of finding and creating new medications is thought to be expensive and time-consuming. In order to assess experimental results and offer appropriate recommendations for structures that should be synthesized, a number of computational models have been enhanced. Nowadays, *in silico* models are being used to quickly and affordably make appropriate cost-effective decisions on drug development prior to synthesis [34].

S. No.	Software	Developer	Applications
1.	ADMET Predictor	Simulation Plus, Inc.	ADMET prediction
2.	StarDrop	Optibrium, Ltd	ADMET prediction
3.	ADME Suite	Advanced Chemistry Development, Inc.	ADME Prediction
4.	Tox suite	Advanced Chemistry Development, Inc.	Toxicity prediction
5.	ADMEWORKS Predictor	Fujitsu FQS	ADMET prediction
6.	QikProp	Schrodinger, Inc.	ADMET prediction
7.	MetaDrug	GeneGo, Inc.	Metabolism and toxicity prediction
8.	TOPKAT	Accelrys, Inc.	Toxicity prediction
9.	PASS	Russian Academy of Medical Sciences	Toxicity prediction
10.	Bioclipse	Uppsala University, Sweden and European Bioinformatics Institute	Metabolism prediction
11	HazardExpert	CompuDrug, Ltd	Toxicity
12	SwissADME	Swiss Institute of Bioinformatics	ADMET prediction

Structure-based drug design (SBDD) and computer-aided drug design (CADD) are two popular computational technologies that offer a variety of computational tools for managing, analysing, modelling, and storing different substances. CADD offers a variety of computer programs that have made it easier to build compounds with desired physicochemical features and evaluate the potency and selectivity of potential candidates. One of the most widely used software examples is QikProp (Schrodinger, Inc.). To evaluate the drug-likeness of molecules, it can produce and forecast 50 molecular descriptors that specify the ADMET properties of molecules [35,36]. As discussed earlier, several tools are available for ADMET prediction.

CONCLUSIONS

One of the newest and fastest methods used in the drug discovery and development process is the use of *in silico* computational models. Because of the rapid advancements in the pharmaceutical industry, there is a growing need to develop more accurate methods for forecasting the pharmacokinetic characteristics of novel drug candidates in order to reduce the time and expense typically involved in developing new medications. Preclinical data is regarded as a crucial prerequisite when creating novel drug candidates, and the data needed varies depending on the stage of drug development. In addition to the drug's activity spectrum, transport, and toxicity, pharmacokinetic properties such as absorption, distribution, metabolism, and excretion are thought to be the most crucial aspects that should be anticipated early in the drug development process. Novel models of *in silico* techniques can predict a number of physicochemical properties, including permeability across biological membranes, hydrogen bonding, lipophilicity, and solubility. CADD, combinatorial chemistry, SBDD, virtual screening, and *in silico* ADMET screening are just a few of the many new commercial software tools that are currently available for prediction purposes. The lack of databases for recently created medications and the inability to use previously obtained data are still seen as obstacles for researchers and developers. It is anticipated that more precise and appropriate methods will be created in the future, offering faster and more accurate processes.

REFERENCES

1. Prentis RA, Lis Y, Walker SR. Pharmaceutical innovation by the seven UK-owned pharmaceutical companies (1964–1985). *British Journal of Clinical Pharmacology*. 1988;23:387–96.
2. Kerns EH, Di L. *Drug-Like Properties: Concepts, Structure Design and Methods: From ADME to Toxicity Optimization*. Elsevier, Burlington, NJ, 2008.
3. Waring MJ, Arrowsmith J, Leach AR, et al. An analysis of the attrition of drug candidates from four major pharmaceutical companies. *Nature Reviews Drug Discovery*. 2015;14:475–84.
4. Sanz F, Pognan F, Steger-Hartmann T, et al. Legacy data sharing to improve drug safety assessment: The eTOX project. *Nature Reviews Drug Discovery*. 2017;16:811–2
5. Pognan F, Steger-Hartmann T, Díaz C, Blomberg N, Bringezu F, Briggs K, Callegaro G, Capella-Gutierrez S, Centeno E, Corvi J, Drew P. The eTRANSAFE project on translational safety assessment through integrative knowledge management: Achievements and perspectives. *Pharmaceuticals*. 2021 Mar 8;14(3):237.
6. Zhang D, Luo G, Ding X, Lu C. Preclinical experimental models of drug metabolism and disposition in drug discovery and development. *Acta Pharmaceutica Sinica B*. 2012 Dec 1;2(6):549–61.
7. Zhuang X, Lu C. PBPK modeling and simulation in drug research and development. *Acta Pharmaceutica Sinica B*. 2016 Sep 1;6(5):430–40.
8. Bohnert T, Prakash C. ADME profiling in drug discovery and development: An overview. In Vishwanath G, Partha K, Ashit T (Eds.), *Encyclopedia of Drug Metabolism and Interactions*. IntechOpen, London, 2012, p. 135. https://doi.org/10.1002/9780470921920.edm021.
9. Peach M, Zakharov A, Liu R, Pugliese A, Tawa G, Wallqvist A. 2012. Computational tools and resources for metabolism-related property predictions. 1. Overview of publicly available (free and commercial) databases and software. *Future Medicinal Chemistry*. 2012;4(15):1907–32.
10. Bergström CA. Computational models to predict aqueous drug solubility, permeability and intestinal absorption. *Expert Opinion on Drug Metabolism & Toxicology*. 2005;1(4):613–27. https://doi.org/10.1517/17425255.1.4.613
11. Ekins S, Waller C, Swaan P, Cruciani G., Wrighton S, Wikel J. 2000. Progress in predicting human ADME parameters in silico. *Journal of Pharmacological and Toxicological Methods*. 2000;44(1):251272.
12. Hop C. Role of ADME studies in selecting drug candidates: Dependence of ADME parameters on physicochemical properties. In Vishwanath G, Partha K, Ashit T (Eds.), *Encyclopedia of Drug Metabolism and Interactions*. IntechOpen, London, 2012, p. 139. https://doi.org/10.1002/9780470921920.edm049.
13. Khan MZ, Jahan MR, Islam MN, Haque Z, Islam MR, Kon Y. Immunoglobulin (Ig)-containing plasma cells in the Harderian gland in broiler and native chickens of Bangladesh. *Tissue and Cell*. 2007 Jun 1;39(3):141–9.
14. Norinder U, Bergström CA. Prediction of ADMET properties. *ChemMedChem: Chemistry Enabling Drug Discovery*. 2006 Sep 11;1(9):920–37.
15. Wenlock MC, Barton P. In silico physicochemical parameter predictions. *Molecular Pharmaceutics*. 2013 Apr 1;10(4):1224–35.
16. Zang Q, Mansouri K, Williams AJ, Judson RS, Allen DG, Casey WM, Kleinstreuer NC. In silico prediction of physicochemical properties of environmental chemicals using molecular fingerprints and machine learning. *Journal of Chemical Information and Modeling*. 2017 Jan 23;57(1):36–49.
17. Tekade RK, Dutta T, Gajbhiye V, Jain NK. Exploring dendrimer towards dual drug delivery: pH responsive simultaneous drug-release kinetics. *Journal of Microencapsulation*. 2009 Jun 1;26(4):287–96.
18. Bergström C, Charman W, Porter C. Computational prediction of formulation strategies for beyondrule-of-5 compounds. *Advanced Drug Delivery Reviews*. 2016 Jun;101:6–21.
19. Schwöbel JA, Ebert RU, Kühne R, Schüürmann G. Prediction models for the Abraham hydrogen bond donor strength: Comparison of semi-empirical, ab initio, and DFT methods. *Journal of Physical Organic Chemistry*. 2011 Nov;24(11):1072–80.
20. Van De Waterbeemd H, Gifford E. ADMET in silico modelling: Towards prediction paradise?. *Nature Reviews Drug Discovery*. 2003 Mar;2(3):192–204.
21. Kurmi BD, Kayat J, Gajbhiye V, Tekade RK, Jain NK. Micro-and nanocarrier-mediated lung targeting. *Expert Opinion on Drug Delivery*. 2010 Jul 1;7(7):781–94.
22. Soni N, Soni N, Pandey H, Maheshwari R, Kesharwani P, Tekade RK. Augmented delivery of gemcitabine in lung cancer cells exploring mannose anchored solid lipid nanoparticles. *Journal of Colloid and Interface Science*. 2016 Nov 1;481:107–16.

23. Lüder K, Lindfors L, Westergren J, Nordholm S, Kjellander R. In silico prediction of drug solubility. 3. Free energy of solvation in pure amorphous matter. *The Journal of Physical Chemistry B*. 2007 Jun 28;111(25):7303–11.
24. Balimane PV, Chong S, Morrison RA. Current methodologies used for evaluation of intestinal permeability and absorption. *Journal of Pharmacological and Toxicological Methods*. 2000 Jul 1;44(1):301–12.
25. Boobis A, Gundert-Remy U, Kremers P, Macheras P, Pelkonen O. In silico prediction of ADME and pharmacokinetics: Report of an expert meeting organised by COST B15. *European Journal of Pharmaceutical Sciences*. 2002 Dec 1;17(4–5):183–93.
26. Saxena A, Shah D, Padmanabhan S, Gautam SS, Chowan GS, Mandlekar S, Desikan S. Prediction of pH dependent absorption using in vitro, in silico, and in vivo rat models: Early liability assessment during lead optimization. *European Journal of Pharmaceutical Sciences*. 2015 Aug 30;76:173–80.
27. Chaturvedi PR, Decker CJ, Odinecs A. Prediction of pharmacokinetic properties using experimental approaches during early drug discovery. *Current Opinion in Chemical Biology*. 2001 Aug 1;5(4):452–63.
28. Egan WJ, Lauri G. Prediction of intestinal permeability. *Advanced Drug Delivery Reviews*. 2002 Mar 31;54(3):273–89.
29. Lombardo F, Gifford E, Shalaeva MY. In silico ADME prediction: Data, models, facts and myths. *Mini Reviews in Medicinal Chemistry*. 2003 Dec 1;3(8):861–75.
30. Pelkonen O, Turpeinen M, Uusitalo J, Rautio A, Raunio H. Prediction of drug metabolism and interactions on the basis of in vitro investigations. *Basic & Clinical Pharmacology & Toxicology*. 2005 Mar;96(3):167–75.
31. Lamberti G, Cascone S, Marra F, Titomanlio G, d'Amore M, Barba AA. Gastrointestinal behavior and ADME phenomena: II. In silico simulation. *Journal of Drug Delivery Science and Technology*. 2016 Oct 1;35:165–71.
32. Alves VM, Muratov EN, Zakharov A, Muratov NN, Andrade CH, Tropsha A. Chemical toxicity prediction for major classes of industrial chemicals: Is it possible to develop universal models covering cosmetics, drugs, and pesticides?. *Food and Chemical Toxicology*. 2018 Feb 1;112:526–34.
33. Roncaglioni A, Toropov AA, Toropova AP, Benfenati E. In silico methods to predict drug toxicity. *Current Opinion in Pharmacology*. 2013 Oct 1;13(5):802–6.
34. Balakumar C, Lamba P, Kishore DP, Narayana BL, Rao KV, Rajwinder K, Rao AR, Shireesha B, Narsaiah B. Synthesis, anti-inflammatory evaluation and docking studies of some new fluorinated fused quinazolines. *European Journal of Medicinal Chemistry*. 2010 Nov 1;45(11):4904–13.
35. Deb PK, Sharma A, Piplani P, Akkinepally RR. Molecular docking and receptor-specific 3D-QSAR studies of acetylcholinesterase inhibitors. *Molecular Diversity*. 2012 Nov;16:803–23.
36. Dhingra MS, Deb PK, Chadha R, Singh T, Karan M. Synthesis, evaluation, and molecular docking studies of cycloalkyl/aryl-3, 4, 5-trimethylgallates as potent non-ulcerogenic and gastroprotective anti-inflammatory agents. *Medicinal Chemistry Research*. 2014 Jan;23:87–106.

19 Computational Genomics Concept and Its Utilization

Priyanshu Nema[1], *Bineet Kumar Mohanta*[1,2], *Sushil K. Kashaw*[1], and *Anshuman Dixit*[2]
[1]Integrated Drug Discovery Research Laboratory, Department of Pharmaceutical Sciences, Dr. Harisingh Gour University (A Central University), Sagar, India
[2]Computational Biology and Bioinformatics Laboratory, ILS, Bhubaneshwar, India

INTRODUCTION

The advent of computational genomics has revolutionized the biomedical landscape by enabling the interpretation of large-scale genomic data through algorithmic and statistical tools. This interdisciplinary field—at the intersection of molecular biology, computer science, and evolutionary theory—has become indispensable in modern drug discovery and development workflows [1–3]. With the exponential growth of sequencing technologies and the integration of high-throughput omics data, computational genomics has matured into a powerful framework for mechanism-based drug discovery, personalized medicine, and precision therapeutics [4,5].

Over the past two decades, genomics has transitioned from a passive cataloguing of genes to an active driver of target identification, prioritization, and validation. Resources such as the UK Biobank and GWAS Catalog have enabled population-scale genotype–phenotype correlation analyses, giving rise to the concept of "experiments of nature," where naturally occurring genetic variants can model drug responses or resistance [1,2]. These datasets, enriched by transcriptomic technologies such as bulk and single-cell RNA sequencing, have enhanced the ability to dissect disease-specific gene expression across tissues using databases such as GTEx and Human Cell Atlas [1,2].

CRISPR-based functional genomics has further strengthened target discovery by enabling gene-level screening using CRISPR knockout (CRISPRko), interference (CRISPRi), and activation (CRISPRa) approaches. These technologies allow the systematic perturbation of gene function, revealing essential and druggable genes implicated in diverse diseases, including cancer, diabetes, and infectious diseases [1,2]. Tools such as the Open Targets Platform integrate data from genetics, transcriptomics, and small molecules to enhance target validation and predict off-target effects [1].

Additional resources such as mutation constraint metrics (e.g., pLI and observed/expected variant ratios) from gnomAD help assess gene intolerance to loss-of-function mutations, informing therapeutic risk–benefit analysis [1]. While these developments are powerful, interpreting non-coding variants and integrating multi-omics data remain ongoing challenges [1,2].

The reduction in the cost of genome sequencing, now below $1,000 per genome, has expanded access to genomic data, enhancing our ability to conduct large-scale, ethnically inclusive studies. Tools powered by next-generation sequencing (NGS) platforms such as Illumina NovaSeq, which can generate trillions of base pairs in a single run, support deep genomic profiling across populations and timeframes, from ancient to modern DNA [3,4].

Through the application of advanced computational methods—such as Bayesian inference, likelihood models, and machine learning (ML)—scientists can explore genetic variation, selection pressures, recombination, and disease evolution with unprecedented depth [5–7]. This allows for better elucidation of complex traits and polygenic diseases, revealing pathways that may be missed by traditional methods; evolutionary modeling using tools such as coalescent theory, hidden Markov models, and ancestral recombination graphs offers insights into population-specific disease risks. This supports the development of therapeutics tailored to genetically distinct populations, advancing the goals of precision medicine [7].

The high dimensionality of genomic data, model assumptions, and the steep learning curve of computational tools still pose barriers to non-specialists. However, increasingly user-friendly platforms and integrated pipelines are bridging this gap [8].

Parallel to these developments, the integration of artificial intelligence (AI) with genomic sciences has further accelerated discovery. AI enables high-precision interpretation of complex datasets, reducing sequencing errors and unveiling cryptic disease-associated variants. The application of deep learning (DL) in genome annotation, variant prediction, and drug–gene interaction analysis is redefining genomic research and drug innovation [9–14].

Platforms integrating ML and clinical data enable AI to predict drug responses and disease outcomes at the individual level, forming the backbone of precision therapy [15–19]. AI has proven useful in mapping regulatory networks, identifying essential genes, and even enabling the development of synthetic gene circuits and DNA-based computing systems, which open new avenues in next-generation therapeutics [17–22].

DOI: 10.1201/9781003464679-19

Historical milestones in genomics—from the Watson–Crick DNA model to modern NGS technologies—have paved the way for AI-powered analysis, enabling tools that interpret omics layers across genomics, transcriptomics, and proteomics [12,18]. Initiatives such as the Human Functional Genomics Project and AI models developed by Coudray et al. demonstrate practical applications of neural networks in predicting disease-linked mutations and stratifying patients for treatment [21–31].

Furthermore, the fusion of genomics with computer-aided drug design (CADD) is shifting the landscape from trial-and-error discovery to rational, structure-guided drug development. This transformation is supported by a growing repository of structural biology data, AI-based target prediction, and the availability of virtual chemical libraries comprising billions of compounds [32]. These innovations enable giga-scale virtual screening, ligand optimization, and Absorption, Distribution, Metabolism, Excretion and Elimination modeling, drastically reducing the cost and time of drug development.

Platforms such as GEO, CCLE, GDSC, and CMap offer researchers access to curated genomic and pharmacogenomic data, supporting novel drug identification, repurposing, and therapeutic response modeling [33]. With the advent of AI-powered platforms such as Deep Learning-based Efficacy Prediction System (DLEPS) and generalist medical AI (GMAI), the integration of multi-omics data is becoming more efficient and predictive [33].

Finally, the conceptual roots of computational genomics—laid by early pioneers such as Alfred Sturtevant, R.A. Fisher, Pauling, and Zuckerkandl—have evolved into today's comparative genomics and systems biology approaches. These fields interpret biomolecular sequences as evolutionary documents, offering clues not just about ancestry but also about biological function, structural roles, and regulatory networks [34,35].

In essence, computational genomics represents a transformative convergence of biological insight and computational power. By uniting multi-omics data, evolutionary models, and AI-based analytics, it offers unprecedented opportunities for target discovery, therapeutic innovation, and individualized treatment. As we continue to advance into this data-rich era, computational genomics will remain a cornerstone of the biomedical sciences, reshaping the future of drug discovery and human health (Tables 19.1–19.3).

TABLE 19.1
Genomic Data Impacting Target Identification and Drug Development

Query	Representative Sources	Expected Output	Implication for Drug Development
Relevant population data for a given target	UK Biobank, GWAS Catalog	Genetic evidence of association between gene and target (similarity between clinical trait and drug indication)	Target identification, druggability
Genetic diseases	OMIM	Evidence for severe consequences of genetic variants	Druggability, consequences on long-term drug action and safety
Null individuals	gnomAD	Identification of individuals in the general population that tolerate heterozygous or homozygous loss of function	Druggability, consequences of long-term drug action and safety
Relevant tissue expression	GTEx	Target is pertinent to the disease tissue	Target identification, validation
Relevant cell expression	Human Cell Atlas	Target is pertinent to the cell implicated in pathogenesis	Target identification, validation
Expression perturbation	LINCS	The target responds to relevant perturbation(s)	Target identification, validation, mechanism of action
Target relevance and triage	CRISPR KO via DepMap	The target is relevant to *in vitro* or *in vivo* experimental endpoints	Target identification, validation
Gene-to-drug matching and precedent	Open Targets Platform	The target genetic perturbation matches the putative drug perturbation endpoints	Druggability, repurposing, chemical matter

TABLE 19.2
Selected Examples of Genetic Conditions Supporting the Indication of Approved Drugs

Gene (Protein)	Genetic Defect/ Variant	Human Phenotype	Drug: Indication	Mechanism of Action
PCSK9	GoF (deleterious), LoF (protective)	Familial hypercholesterolemia and CHD	Evolocumab, Alirocumab: Hypercholesterolemia	Inhibits LDL receptor degradation, enhancing LDL clearance
NPC1L1	GoF (deleterious), LoF (protective)	LDL reduction, reduced CV risk	Ezetimibe: Hypercholesterolemia	Inhibits intestinal cholesterol absorption
ANGPTL3	LoF (protective)	Familial hypolipidemia	Evinacumab: Hypercholesterolemia	Inhibits ANGPTL3, reducing lipids
LPA	GoF/LoF	High Lp(a) → CV disease	AKCEA-APO(a)-LRx	Antisense inhibition of Lp(a) production
LEPR	LoF	Severe early-onset obesity	Metreleptin, REGN4461	Leptin receptor agonist
MC4R	LoF	Early-onset obesity	Setmelanotide	MC4R agonist enhancing satiety
PPARG	LoF	Partial lipodystrophy, insulin resistance	Thiazolidinediones (e.g., pioglitazone)	Increases insulin sensitivity
SOST	LoF	Sclerosteosis, high bone density	Romosozumab	Neutralizes sclerostin to promote bone formation
SLC22A12	LoF	Uric acid imbalance	Lesinurad	Inhibits uric acid reabsorption
XDH	LoF	Xanthinuria	Allopurinol	Inhibits xanthine oxidase
IL4, IL13, IL4Ra	eQTL/GoF	Asthma, elevated IgE	Dupilumab	Blocks IL-4/IL-13 binding to IL-4Ra
NLRP3	GoF	CAPS syndrome	Canakinumab, Anakinra, Rilonacept	Inhibits IL-1β signaling
F10	LoF	Hemophilia	Rivaroxaban, Apixaban	FXa inhibitors; Andexanet Alfa: antidote
CFTR	Missense/LoF	Cystic fibrosis	Ivacaftor, Tezacaftor, etc.	Potentiators and correctors of CFTR
HCRTR2	LoF	Narcolepsy	Lemborexant, Suvorexant	Dual orexin receptor antagonists
SGLT2	LoF	Renal glucosuria	Dapagliflozin, etc.: Diabetes	Inhibits glucose reabsorption
JAK1	LoF	Immune deficiency	Tofacitinib, Baricitinib, Upadacitinib	JAK pathway inhibition
HCN4	GoF/LoF	Cardiac arrhythmias	Ivabradine	Selective HCN channel blocker

TABLE 19.3
Key Genetic and Genomic Data Sources

Database	Relevant Publication	Data Type	Web Page
GenBank (NIH genetic sequence database)	Benson et al. [36]	Various sequences including most single loci and mtDNA	ncbi.nlm.nih.gov/genbank
Thousand Genomes Project	1,000 Genomes Project Consortium 2015	Whole genomes aligned to hg19 (GRCh37)	ftp.1000genomes.ebi.ac.uk
Simons Genome Diversity Panel	Mallick et al. [37]; Watkins et al. [38]	Whole genomes aligned to hg19 (GRCh37)	reich_lab/sgdp/ phased_data2021
Human Genome Diversity Project	Bergström et al. [39]	Whole genomes aligned to hg38 (GRCh38)	internationalgenome.org
Human Pangenome Reference Consortium	Liao et al. [40]	Whole genomes including structural variants	ensembl.org/hprc

(Continued)

TABLE 19.3 (*Continued*)
Key Genetic and Genomic Data Sources

Database	Relevant Publication	Data Type	Web Page
Max Planck Institute Neanderthal and Denisovan Genomes	Various	Whole genomes aligned to hg19 (GRCh37)	eva.mpg.de/genetics
Estonia Biocenter Genome Diversity Panel	Pagani et al. [41]	Whole genomes aligned to hg19 (GRCh37)	evolbio.ut.ee
Allen Ancient DNA Resource	Mallick et al. [42]	SNP array data aligned to hg19 (GRCh37)	reich.hms.harvard.edu
GenomeAsia	Genome Asia Consortium [7]	Whole genomes aligned to hg19 (GRCh37)	genomeasia100k.org
LASI-DAD (Diagnostic Assessment of Dementia)	Lee et al. [43]; Kerdoncuff et al. [44]	SNP array data and whole-genome imputation	dss.niagads.org

THEORY OF GENOME SEQUENCING IN DRUG DEVELOPMENT

Genome sequencing technologies have revolutionized the landscape of drug discovery by providing unprecedented insights into the genetic underpinnings of diseases [45]. By analyzing the human genome, researchers can uncover novel drug targets, identify genetic variants associated with diseases, and optimize therapeutic strategies for precision medicine [46]. The three primary genomic approaches used in drug development—genome-wide association studies (GWAS), exome sequencing, and whole-genome sequencing (WGS)—each contribute unique value to the discovery and validation of therapeutic targets [47].

1. **GWAS**: GWAS have been instrumental in linking common genetic variants to complex diseases such as cardiovascular diseases, diabetes, and autoimmune disorders [48]. These studies scan the genome for single-nucleotide polymorphisms (SNPs) that are associated with specific phenotypes [49]. While the majority of GWAS findings are located in non-coding regions, which may not directly alter protein function, they provide valuable clues about regulatory elements that influence gene expression [50]. By analyzing large-scale populations, GWAS can pinpoint genetic variations that alter gene activity, thus revealing potential drug targets and mechanisms for disease treatment [51].
2. **Exome Sequencing and Gene Essentiality**: Exome sequencing focuses on the protein-coding regions of the genome, capturing rare variants that could have significant functional impacts [52]. This approach is especially valuable in identifying disease-associated mutations that are less frequent in the population but may play a pivotal role in disease development [53]. Exome sequencing allows the identification of essential genes—genes whose loss leads to severe consequences such as embryonic lethality or disease [54]. Such genes present promising drug targets for inhibition or activation, depending on whether their loss results in a protective effect (e.g., *PCSK9* for cholesterol regulation) or a deleterious outcome (e.g., *SCN9A* for pain perception) [55].
3. **WGS**: WGS offers comprehensive insights by analyzing both coding and non-coding regions of the genome [56]. It captures genetic variants across the entire genome, providing a deeper understanding of how genetic alterations in non-coding regions—such as promoters, enhancers, and other regulatory elements—can influence disease [57]. Although the druggability of non-coding regions is still a developing area, the identification of mutations in transcriptional regulators, chromatin remodeling pathways, and non-coding RNAs (e.g., miRNAs, lncRNAs) offers new opportunities for drug development, particularly in complex diseases such as cancer and neurodevelopmental disorders [58].
4. **The Concept of "Experiments of Nature"**: One of the key principles driving genomic drug discovery is the identification of "experiments of nature"—naturally occurring mutations that affect the function of a protein target [59]. These mutations can provide insights into the therapeutic potential of targeting a specific gene [60]. For example, individuals with loss-of-function mutations in genes like *PCSK9* exhibit lower cholesterol levels and a reduced risk of cardiovascular diseases, making them a prime target for drug development [61]. Similarly, the study of gain-of-function mutations in genes like *LPA* or *NPC1L1* can help identify potential inhibitors to reduce cardiovascular risk [62].
5. **Druggability of Non-Coding Regions**: The majority of the genome does not code for proteins but contains regulatory regions that control gene expression and protein function [63,64]. While traditionally considered "undruggable," recent advancements

in technology have made it possible to study the non-coding genome and identify potential therapeutic targets [65]. Non-coding RNAs, such as miRNAs and lncRNAs, play critical roles in gene regulation and disease development [66]. By targeting these molecules, it is now possible to modulate gene expression and potentially treat disorders that were once thought to be untargetable through traditional small-molecule drugs [67].

TRANSCRIPTOMICS—BULK AND SINGLE-CELL SEQUENCING

Transcriptomics plays a fundamental role in understanding gene expression changes associated with disease and drug response, making it a powerful tool in drug discovery [68]. It helps in elucidating the mechanism of action of compounds, identifying therapeutic targets, and discovering biomarkers for disease diagnosis and treatment stratification [69]. One of the key applications of transcriptomics is the analysis of gene expression after drug perturbation, also known as connectivity mapping [70]. This approach is used to identify drugs that induce or reverse disease-associated transcriptional signatures, enabling both mechanistic insights and opportunities for drug repurposing [71]. High-throughput transcriptomic profiling methods such as the L1000 assay have been developed to measure approximately 1,000 landmark genes while computationally inferring the rest, significantly reducing costs [72]. Other cost-effective transcriptome-wide approaches such as PLATE-seq, DRUG-seq, and BRB-seq have emerged to support large-scale screening efforts [73]. However, one of the persistent challenges in transcriptomic studies is the presence of batch effects, which can obscure true biological signals and complicate comparisons across datasets [74]. Resources such as the LINCS Connectivity Map 2.0 (https://clue.io/) provide extensive datasets and tools to explore drug-induced transcriptional signatures and to compare them with disease signatures, facilitating drug repositioning [75]. For example, the compound celastrol was identified through such approaches as a potential treatment for obesity [76].

Traditional transcriptomic studies have relied on bulk RNA sequencing, which measures the average gene expression of all cells in a sample [77]. Though informative, this approach cannot resolve the heterogeneity of cellular responses, which is especially critical in complex tissues or disease contexts such as cancer and inflammation. Single-cell RNA sequencing (scRNA-seq) has revolutionized transcriptomics by enabling gene expression profiling at the level of individual cells [78]. This allows researchers to identify rare cell types, track lineage trajectories, and study cell-type-specific drug responses. Recent innovations such as microwell-based approaches and split-pool barcoding have made it possible to scale up scRNA-seq to thousands or even millions of cells. Techniques such as sci-Plex, which combine nuclear hashing and chemical indexing, now allow for single-cell drug screening across multiple conditions in parallel [79]. In addition, the Human Cell Atlas (https://www.humancellatlas.org/) provides a comprehensive reference for understanding gene expression across different human tissues and cell types. However, even with scRNA-seq, interpreting transcriptional states can be challenging due to the convergence of different pathways on similar gene expression profiles, emphasizing the importance of combining transcriptomics with perturbation-based designs for more causal insights [80].

Transcriptome data can also be harnessed for biomarker discovery. By profiling patient-derived samples such as peripheral blood mononuclear cells (PBMCs) or tissue biopsies, transcriptomics can help identify gene expression signatures predictive of drug response or disease progression [68]. ML algorithms are often employed to classify patients or predict treatment outcomes based on these signatures [69]. However, many transcriptomic studies in the biomarker space are underpowered due to small sample sizes, limiting their clinical applicability [70]. To address this, approaches such as weighted gene co-expression network analysis (WGCNA) are used to reduce data dimensionality and group genes into biologically meaningful modules, enhancing interpretability [71]. Meta-analyses that combine data from multiple small-scale studies can also improve statistical power and robustness of biomarker findings [72].

Finally, integrating transcriptomic data with genomic information further enhances our ability to identify causal genes and understand disease mechanisms [73]. Projects such as the Genotype-Tissue Expression (GTEx) initiative enable the mapping of expression quantitative trait loci (eQTL), linking genetic variation to tissue-specific gene expression [74]. This integrative strategy, initially developed for rare disease diagnostics, is increasingly being used to identify novel drug targets by highlighting genes whose dysregulation is genetically associated with disease phenotypes [75]. Overall, transcriptomics—especially when combined with emerging single-cell and spatial technologies, advanced computational methods, and large-scale data integration—has become indispensable in modern drug discovery and precision medicine [76].

Genomic Data Analysis as a High-Dimensional Computational Challenge

Population genetics stands out due to its strong foundation in mathematical modeling, which allows researchers to explain and predict patterns of genome variation shaped by evolutionary forces such as genetic drift, mutation, migration, and natural selection. Central to this is the Wright–Fisher model, which provides a robust framework for understanding genetic drift in an idealized population. However, real-world populations deviate from these ideals due to ongoing influences from other evolutionary processes. Despite the complexities introduced by these factors, the Wright–Fisher model and its extensions form the basis for highly efficient simulation tools such as *msprime* and *SLiM*. The interplay of multiple forces in real-world populations creates intricate and sometimes chaotic patterns of genetic variation. Recombination adds an additional layer of complexity, as it affects the independence of genetic markers depending on their location in the genome. Although genome alignments may seem like simple two-dimensional data structures, they contain layered information derived from numerous evolutionary processes. This makes genomic data inherently high-dimensional, requiring sophisticated computational methods for meaningful interpretation and analysis [81].

Varieties of Genetic Data and Their Applications

Genetic and genomic data come in a variety of forms, and the nature of the data strongly influences which analytical tools and interpretations are appropriate. Traditionally, much of the available data has been generated through targeted, single-locus sequencing using methods like Sanger sequencing. This includes mitochondrial DNA (mtDNA), which is favored for its simplicity due to maternal inheritance, lack of recombination, and rapid mutation in hypervariable regions. Although useful for short-term evolutionary studies and species barcoding, mtDNA provides limited insights into the broader genomic landscape due to its small size and linkage structure. In contrast, autosomal data offers a richer source of genetic variation, especially when analyzed at the level of the whole genome. However, WGS remains computationally intensive and costly, particularly for large-scale studies. To manage this, genotyping arrays that target known SNPs are often used for human populations, providing cost-effective but somewhat limited datasets. For non-model organisms or populations lacking reference genomes, methods such as RADSeq provide an alternative, capturing random genomic fragments to detect variation. However, these approaches may offer uneven genome coverage and are often limited to within-species or population-specific comparisons. Overall, the selection of data type plays a crucial role in the feasibility, scope, and depth of population genomic research [81] (Table 19.4).

Artificial Intelligence-Driven Genome Sequencing

The integration of AI with genome sequencing represents a major advancement in biomedical science and healthcare, revolutionizing our ability to analyze complex genetic information with speed, accuracy, and scale. Traditionally, genome sequencing has been a time-consuming and expensive endeavor, relying on labor-intensive methods and vast computational resources. However, AI technologies—particularly ML and DL models—have significantly accelerated sequencing workflows, enhanced variant detection, and improved the interpretation of multidimensional biological data.

AI contribution spans multiple facets of the genome sequencing pipeline. For example, it enhances the precision of variant calling, reduces sequencing errors, identifies structural variations, and allows seamless integration of genomic data with electronic health records (EHRs) for personalized medicine applications. In particular, AI enables predictive modeling based on patient genetic profiles, offering clinicians a powerful tool to anticipate disease progression, treatment response, and even patient transfer to intensive care units. This capability goes beyond conventional clinical intuition, instead relying on learned patterns from vast patient datasets.

Furthermore, AI has enabled the real-time analysis of transcriptomes through RNA-seq and has contributed to our understanding of gene expression, alternative splicing, and transcriptional complexity in eukaryotic systems. In conjunction with proteomics, AI models are uncovering intricate protein networks and interactions, offering new pathways for biomarker discovery and therapeutic targeting. The progress in protein sequencing, peptide quantification, and structural prediction has broadened the clinical applications of AI across neuroscience, oncology, cardiology, and immunology.

High-throughput technologies such as NGS and RNA-seq, when paired with AI, have shifted the research paradigm toward scalable, high-resolution, and high-accuracy data generation. Acquisition strategies such as strategic alliances with sequencing platforms, in-house data generation, open-source collaborations, and citizen science initiatives have further expanded the scope of available genomic datasets for AI training and validation.

Ethical considerations and regulatory compliance remain important topics in the AI–genomics interface. As AI systems increasingly handle sensitive patient genomic data, questions related to data privacy, algorithmic bias, and equitable access should be addressed. Moreover, regulatory frameworks need to evolve to support AI-driven genomic applications without compromising safety, efficacy, and transparency.

In the context of drug discovery, AI is rapidly becoming indispensable. From identifying essential genes and therapeutic targets to accelerating drug repurposing and lead optimization, AI has outpaced many traditional computational tools in both performance and interpretability. For example, ML algorithms such as random forests and artificial neural networks (ANNs) are being used to predict patient outcomes, classify disease states, and simulate gene expression circuits. Advanced techniques such as DNA strand displacement (DSD) and synthetic gene circuits are laying the groundwork for programmable, intelligent therapeutics that can adapt to environmental conditions.

The growing body of evidence supporting the clinical utility of AI in genomics is exemplified by projects such as the Human Functional Genomics Project (HFGP) and applications in cancer genomics, such as predicting mutations in non-small-cell lung cancer. These examples demonstrate how the synergy between DL algorithms and genomic data can lead to robust diagnostic and prognostic models [81].

In conclusion, the fusion of AI with genomics has ushered in a new era of biological discovery and personalized healthcare. Although challenges remain—particularly in standardization, interpretability, and bioethics—the progress made thus far underscores the immense potential of AI to decode the complexity of the genome, optimize therapeutic strategies, and ultimately transform patient care on a global scale (Tables 19.5 and 19.6).

TABLE 19.4
Popular Computational Tools for Anthropological Genomics

Application Area	Tool Name	References	GitHub/Resource Page
Demographic reconstruction/ simulation	BEAST2	Bouckaert et al. [82]	beast2.org
	fastsimcoal2	Excoffier et al. [83]	cmpg.unibe.ch/software/fastsimcoal2
	SMC++	Terhorst et al. [84]	github.com/popgenmethods/smcpp
	CHIMP	Upadhya and Steinrücken [85]	github.com/steinrue/chimp
	diCal2	Steinrücken et al. [86]	github.com/popgenmethods/diCal2
	ARGweaver-D	Hubisz et al. [87]	compgen.cshl.edu/ARGweaver
	TSinfer	Kelleher et al. [88]	github.com/tskit-dev/tsinfer
	Relate	Speidel et al. [89]	myersgroup.github.io/relate
	Msprime	Baumdicker et al. [90]	tskit.dev/msprime
	SLiM	Haller and Messer [91]	messerlab.org/slim
Testing for selection in phylogenies	RELAX	Wertheim et al. [92]	datamonkey.org/relax
	PAML	Yang [93]	abacus.gene.ucl.ac.uk/software/paml.html
	HyPhy	Pond et al. [94]	github.com/veg/hyphy
Genome-wide selection scans	Selscan	Szpiech and Hernandez [95]	github.com/szpiech/selscan
	hapbin	Maclean et al. [96]	github.com/smilefreak/hapbin
	iSAFE	Akbari et al. [97]	github.com/alek0991/iSAFE
	CLUES	Stern et al. [98]	github.com/35ajstern/clues
Population admixture	ADMIXTURE	Alexander et al. [99]	dalexander.github.io/admixture
	NGSadmix	Meisner and Albrechtsen [100]	github.com/ANGSD/NgsAdmix
	qpAdm / qpGraph	Patterson et al. [101]	reich.hms.harvard.edu/software
	RFMix	Maples et al. [102]	github.com/slowkoni/rfmix
	ELAI	Guan [103]	github.com/GuangchuangYu/ELAI
Admixture visualization	pong	Behr et al. [104]	github.com/ramachandran-lab/pong
	CLUMPAK	Kopelman et al. [105]	clumpak.tau.ac.il
Kinship and relatedness estimation	KING	Manichaikul et al. [106]	people.virginia.edu/~wc9c/KING
	READ	Monroy Kuhn et al. [107]	bitbucket.org/egeland/read
	lcMLkin	Lipatov et al. [108]	github.com/molpopgen/lcMLkin
	Ngskit	Meisner and Albrechtsen [109]	github.com/ANGSD/NgsRelate

TABLE 19.5
AI-Powered Genome Sequencing Aspects

Aspect	Description
1. Accelerated sequencing	AI reduces the time and expense associated with genome sequencing.
2. Error reduction	AI decreases errors, increasing the accuracy of genome sequencing.
3. Variant identification	AI swiftly and correctly pinpoints genetic variations related to diseases or traits.
4. Personalized medicine	AI uses genomic data analysis to personalize medicines based on each patient's genetics.
5. Population studies	AI analyzes large-scale datasets to provide insights about population-level variants.
6. Structural variation analysis	AI detects large-scale genomic rearrangements and structural changes.
7. Data integration	AI integrates clinical, environmental, lifestyle, and genomic data for comprehensive insights.
8. Scalability	AI enables genome sequencing to scale and handle massive datasets efficiently.
9. Ethical considerations	AI raises concerns regarding sensitive genomic data storage and sharing.
10. Regulatory compliance	AI ensures sequencing processes meet legal and regulatory standards while safeguarding data.

TABLE 19.6
Acquisition Strategies in AI-Powered Genome Sequencing

Acquisition Strategy	Description
Partnerships and collaborations	Collaboration with academia, biotech, and healthcare sectors enables access to large, diverse genomic datasets for AI model development and validation.
Mergers and acquisitions	Companies merge or acquire to integrate AI with genome sequencing tools and foster innovation.
Data licensing and sharing	Licensing large curated datasets to AI companies strengthens model training and genetic insight accuracy.
In-house data generation	Companies create and control their own genomic datasets, ensuring high-quality input for AI algorithms.
Strategic alliances with sequencing platforms	AI algorithms are embedded into sequencing hardware for real-time data analysis and efficiency.
Crowdsourcing and citizen science	Public participation enhances data diversity and accessibility, expanding genetic databases.
Investment in R&D	Funding internal innovation promotes proprietary AI solutions in genomics.
Open-source collaboration	Engaging in open-source communities accelerates AI development for genome sequencing through shared resources.
Clinical trial collaborations	Partnering with pharma for clinical trial data aids in testing AI tools on real-world datasets.
Global expansion and market access	Expanding globally improves AI model generalizability by incorporating diverse genetic backgrounds.

The Role of Structural Genomics in Advancing Computational Drug Innovation

The advent of structural genomics has significantly advanced the field of computational drug discovery by providing detailed three-dimensional structures of proteins at an unprecedented scale. This wealth of structural data has catalyzed innovation in drug discovery, particularly through the integration of genomic information with *in silico* methodologies. By linking gene sequences to protein structures and biological functions, structural genomics facilitates the rational design of novel therapeutics with improved specificity, efficacy, and safety [110].

Genomics enables the identification of novel drug targets, including those involved in disease-specific mutations, genetic polymorphisms, and protein isoforms. Structural insights into these genomically defined targets allow researchers to explore binding sites, allosteric pockets, and protein–protein interaction (PPI) interfaces that were previously inaccessible. For example, hot-spot analyses guided by sequence data and structural modeling now support the development of small-molecule inhibitors for challenging targets such as PPIs—traditionally considered undruggable [111].

Furthermore, genomics-driven drug discovery leverages computational tools to conduct virtual screening against structurally characterized targets. Structure-based virtual screening benefits from high-resolution macromolecular models derived from genomic data, enabling more accurate docking and scoring of potential ligands [112,113]. Combined approaches, such as integrating ligand based methods with receptor-derived pharmacophores, allow for scaffold hopping and optimization of lead compounds tailored to disease-specific genetic profiles [114].

Beyond individual targets, systems pharmacology—built upon genomic and proteomic networks—provides a holistic view of drug action. The integration of genomic data helps map the interaction landscape of drugs across multiple proteins and pathways, offering insights into polypharmacology and off-target effects [115,116]. This network-based understanding is essential for addressing complex, multifactorial diseases such as cancer, diabetes, and neurodegenerative disorders.

Pharmacogenomics, a direct outcome of genomic research, further strengthens the drug development process by predicting patient-specific responses to drugs. SNPs and gene expression profiles can be computationally modeled to forecast efficacy and toxicity, paving the way for personalized medicine [117,118]. Tools such as ChemProt, PharmMapper, and STITCH utilize these datasets to associate chemical structures with target ensembles, facilitating drug repurposing and safety evaluation [119,120].

Despite these advancements, the application of structural genomics in drug discovery is not without challenges. Limitations include the difficulty in modeling receptor flexibility, refining docking scores, and capturing dynamic biological processes *in silico* [121,122]. Moreover, the effective use of genomic data requires not only computational power but also standardized, interoperable databases and collaborative efforts across disciplines.

In summary, structural genomics serves as a bridge between molecular biology and computational drug discovery, accelerating the identification and development of innovative therapeutics. By enabling a deeper understanding of the structural and functional consequences of genetic variation, it empowers researchers to design drugs that are more targeted, efficient, and safe—marking a significant leap forward in the era of precision medicine.

Transformative Impact of Genomics-Based Tools in Drug Discovery and Development

The evolution of genomics technologies, from microarrays to NGS, has revolutionized the field of drug discovery and development [123]. Early tools such as GEO, CCLE,

GDSC, and the original CMap marked a pivotal shift in understanding the relationships between gene expression, disease pathology, and drug mechanisms [71,124–126]. These first-generation resources laid the groundwork for identifying potential drug targets, repurposing existing drugs, and predicting therapeutic efficacy by providing gene expression signatures across various experimental conditions [127].

The transition to second-generation platforms introduced high-resolution multi-omics datasets, deeper biological annotations, and scalable analytic capabilities [128]. Platforms such as the next-generation Cancer Cell Line Encyclopedia (CCLE) and the L1000-based Connectivity Map (CMap) significantly enhanced the detection of molecular patterns associated with drug responses [129,130]. These tools enabled researchers to decipher phenotypic effects and mechanism-of-action insights at an unprecedented scale, strengthening the foundation for precision medicine [131].

Moreover, newly developed platforms such as the Drug Repurposing Hub, NCI Transcriptional Pharmacodynamics Workbench (NCI TPW), CancerRxTissue, and CancerTracer serve as critical resources for mapping drug–target interactions, tumor heterogeneity, and treatment sensitivity across patient cohorts [132–135]. These databases not only facilitate hypothesis generation but also support clinical decision-making and early-phase drug development [136].

DL-based tools, such as DLEPS, represent a paradigm shift in pharmacogenomics, offering novel approaches to predict drug efficacy from transcriptomic data [137]. By analyzing gene expression perturbations from drug-treated cell lines, DL models can infer complex biological relationships, identify synergistic effects, and forecast therapeutic responses *in silico* [138]. Despite limitations in data quality and interpretability, these methods are unlocking the potential of integrating AI into all stages of the drug development pipeline [43,139,140].

The discussion extends to the application of AI more broadly. GMAI models, spatial omics atlases such as MOSAIC, and AI-driven prediction platforms offer unprecedented capabilities for handling multimodal datasets—ranging from genomics and histopathology to clinical metadata [141,142]. These models can support tasks such as drug toxicity prediction, protein–ligand interaction mapping, and multitarget drug discovery.

Despite these advances, significant challenges persist. Data fragmentation, inconsistency in standards, limited sharing due to intellectual property concerns, and the "black-box" nature of AI models hinder their translation into clinical practice [136]. However, ongoing efforts in open-source tools such as PharmacoGx and community-shared databases continue to address these gaps, promoting interoperability and reproducibility [137,138].

In conclusion, genomics-based tools—amplified by DL and AI—have profoundly reshaped drug discovery, enabling data-driven strategies for identifying new drugs, repurposing existing ones, and personalizing treatments. The convergence of genomics, computational power, and AI promises to drive future innovations in both drug development and clinical pharmacology [139] (Tables 19.7 and 19.8).

TABLE 19.7
Examples of Drugs Obtained Using Genomics Databases

Database	Compound(s)	Disease/Indication	Theoretical Basis from Manuscript
CMap	Tomatidine	Skeletal muscle atrophy	CMap1 uses microarray-derived gene expression profiles and GSEA to connect small molecules to phenotypic states.
CMap	Parbendazole	Osteoporosis	Leveraging CMap's expression-based pattern recognition reveals osteogenic potential of existing small molecules.
GEO and CMap	Trichostatin A, Vorinostat, HC toxin, sodium phenylbutyrate, etc.	Colorectal cancer	GEO provides comprehensive cancer gene expression data; CMap links it with drug-induced transcriptional responses.
Human Metabolome+CMap	Diflunisal, Nabumetone, niflumic acid, Valdecoxib, phenoxybenzamine, Idazoxan	Type 2 diabetes	Multi-database integration of expression changes in disease states and compound bioactivity from CMap reveals potential therapies.
GEO and CMap	KM-00927, BRD-K75081836	Cancer	Gene expression datasets from GEO and perturbation signatures from CMap were used to correlate drug efficacy in tumor profiles.
GEO and CMap	CGP-60474	Endotoxemia	LINCS-derived expression signatures in CMap help identify CGP-60474 as an anti-endotoxemic using a similarity matching framework.
GEO and CMap	PP-2 (Src-kinase inhibitor)	Cystic fibrosis	Differential gene signatures from cystic fibrosis models matched PP-2 expression patterns, suggesting functional correction.

(Continued)

TABLE 19.7 (*Continued*)
Examples of Drugs Obtained Using Genomics Databases

Database	Compound(s)	Disease/Indication	Theoretical Basis from Manuscript
GEO and CMap	Cinnarizine, Digitoxigenin, Clofazimine	Metastatic uveal melanoma	Using CMap's GSEA with GEO cancer data revealed transcriptional antagonism relevant for targeting melanoma proliferation.
Drug Repurposing Hub+DL	Halicin	Broad-spectrum antibiotic use	Deep learning applied to transcriptional data in the Drug Repurposing Hub identified halicin's antimicrobial profile.
Drug Repurposing Hub	BRD4780	Mucin-1 kidney disease	Compound–protein mapping with clinical annotations helps identify BRD4780's target for lysosomal degradation.
Drug Repurposing Hub+CCLE	Disulfiram, Tepoxalin	Cancer	Gene dependency and mutation data from CCLE support the repurposing of these compounds via transcriptomic correlation.
GEO and CMap	Oxytocin, Fluoxetine, Saquinavir, Ribavirin	SARS-CoV-2	Drug–gene expression overlap is calculated using CMap and GEO COVID-19-related profiles.
TCGA+PharmacoGx	Guanidine hydrochloride	Triple-negative breast cancer	PharmacoGx enables integrative cross-dataset analysis to correlate gene sensitivity markers with guanidine hydrochloride.
Drug Repurposing Hub	Obatoclax	SARS-CoV-2	Drug perturbation data and known mechanisms within the Drug Repurposing Hub aid in identifying antiviral potential.
Drug Repurposing Hub+CCLE	CR8 (CDK inhibitor)	Cancer	CCLE dependency mapping links CR8's effects to cyclin K degradation via transcriptomic profiling.
DLEPS	Ataluren	Osteoporosis	DLEPS trained on L1000 identifies Ataluren through feature similarity scoring.

TABLE 19.8
Overview of Genomics-Based Databases and Tools

Database/Tool	Description	Datasets/Key Metrics	URL Link
Gene Expression Omnibus (GEO)	Public repository for microarray, NGS, and high-throughput genomics data. Includes GEO2R tool for comparative analysis.	Datasets: 4,348 Series: 211,302 Platforms: 25,476 Samples: 6,762,989	GEO
Genomics of Drug Sensitivity in Cancer (GDSC)	Integrates cancer genomic datasets with anticancer drug responses. GDSCTools identifies genomic markers of drug response.	Compounds: 621 Dose–response curves: 576,758 Genomic associations: 722,057	GDSC
Gene Expression Profile Compendium	Tool for studying drug effects via gene expression changes; builds drug feature libraries using expression profiles.	Not listed	–
Next-generation Cancer Cell Line Encyclopedia (CCLE)	Provides detailed multi-omics and genetic profiling of 1,100+ cancer cell lines.	329 Genetic datasets 1,019 RNA expression lines Fusion calls: 1,019 Epigenetics, proteomics, metabolomics, DNA methylation, chromatin profiling	CCLE
Drug Repurposing Hub	Repository of FDA-approved, clinical, and tool compounds with annotations on structure, targets, and indications.	Samples: 16,826 Protein targets: 2,183 Unique compounds: 7,934 Drug indications: 670	Repo Hub
NCI Transcriptional Pharmacodynamics Workbench (NCI TPW)	Visualization and analysis of NCI-60 cell lines' transcriptional response to 15 anticancer drugs.	NCI-60 cell lines 15 anticancer drugs	NCI TPW
CancerRxTissue	Predicts drug sensitivity in normal and tumor tissues based on genomic data.	Drug sensitivity predictions: 272	CancerRxTissue

(*Continued*)

TABLE 19.8 (*Continued*)

Overview of Genomics-Based Databases and Tools

Database/Tool	Description	Datasets/Key Metrics	URL Link
CancerTracer	Tracks tumor heterogeneity and evolution using genomic data across 6,000+ tumor samples.	Tumor samples: 6,000+	CancerTracer
Catalogue of Somatic Mutations in Cancer (COSMIC)	Catalog of coding/non-coding mutations, gene fusions, CNVs, and resistance mutations in cancer.	Variants: 2.38M+ Samples: 1.52M Exonic mutations: 5.07M CNVs: 1.2M Fusions: 19,428 Gene expression: 9.2M Methylated CpGs: 7.9M	COSMIC
Carcinogenic Potency Database (CPDB)	Database of 6,540 long-term animal tests for carcinogenicity across 1,547 chemicals.	Animal cancer tests: 6,540	CPDB
PharmacoGx	Open-source R package for pharmacogenomic dataset analysis and comparison across studies.	Not listed	PharmacoGx
CMap: L1000 Platform	L1000-based perturbation-driven gene expression profiles for drug mechanism discovery and repurposing.	Profiles: 3.02M Compounds: 33,609 Perturbagens: 81,979 Signatures: 1.16M 240 cell contexts (12 primary)	CMap
Deep Learning-based Efficacy Prediction System (DLEPS)	Deep learning tool using chemical structure, gene expression, and targets to predict drug efficacy.	Not listed	DLEPS
Generalist Medical AI (GMAI)	AI model capable of multimodal medical data interpretation with minimal task-specific training.	Not listed	–
Multi Omic Spatial Atlas in Cancer (Mosaic)	Integrates spatial and single-cell omics with pathology and clinical data for 7,000 patients across seven cancer types.	Indications: 7 Patient samples: 7,000 Data modalities: 6	Mosaic

REFERENCES

1. Holbein ME. Understanding FDA regulatory requirements for investigational new drug applications for sponsor-investigators. *J Investig Med*. 2009;57:688–694.
2. Spreafico R, Soriaga LB, Grosse J, Virgin HW, Telenti A. Advances in genomics for drug development. *Genes*. 2020;11(8):942. https://doi.org/10.3390/genes11080942
3. Witt KE, Villanea FA. Computational genomics and its applications to anthropological questions. *Am J Biol Anthropol*. 2025;186(Suppl. 78):e70010. https://doi.org/10.1002/ajpa.70010
4. Mullin E. 2022. "The Era of Fast, Cheap Genome Sequencing Is Here." WIRED. https://www.wired.com/story/the-era-of-fast-cheap-genome sequencing-is-here/.
5. Illumina. 2024. "Novaseq X and Novaseq X Plus Sequencing Systems Specification Sheet." Accessed October 14 2024. https://www.illum ina.com/content/dam/illumina/gcs/assembled-assets/marketing-literature/novaseq-x-series-spec-sheet-m-us-00197/novaseq-x-series-specification-sheet-m-us-00197.pdf.
6. Sohail M, Palma-Martinez MJ, Chong AY, et al. Mexican biobank advances population and medical genomics of diverse ancestries. *Nature*. 2023;622:775–783.
7. GenomeAsia 100K Consortium. The GenomeAsia 100K project enables genetic discoveries across Asia. *Nature*. 2019;576:106–111.
8. Jain A, Bhoyar RC, Pandhare K, et al. IndiGenomes: A comprehensive resource of genetic variants from over 1000 Indian genomes. *Nucleic Acids Res*. 2021;49:D1225–D1232.
9. Heather JM, Chain B. The sequence of sequencers: The history of sequencing DNA. *Genomics*. 2016;107:1–8. https://doi.org/10.1016/j.ygeno.2015.11.003
10. Garcia-Vidal C, Sanjuan G, Puerta-Alcalde P, Moreno-García E, Soriano A. Artificial intelligence to support clinical decision-making processes. *eBioMedicine*. 2019;46:27–29. https://doi.org/10.1016/j.ebiom.2019.07.019
11. Rajkomar A, Dean J, Kohane I. Machine learning in medicine. *N Engl J Med*. 2019;380:1347–1358. https://doi.org/10.1056/NEJMra1814259
12. Wirth T, Parker N, Ylä-Herttuala S. History of gene therapy. *Gene*. 2013;525:162–169. https://doi.org/10.1016/j.gene.2013.03.137
13. Wang Z, Gerstein M, Snyder M. RNA-Seq: A revolutionary tool for transcriptomics. *Nat Rev Genet*. 2009;10:57–63. https://doi.org/10.1038/nrg2484
14. Gupta R, Srivastava D, Sahu M, Tiwari S, Ambasta RK, Kumar P. Artificial intelligence to deep learning: Machine intelligence approach for drug discovery. *Mol Divers*. 2021;25:1315–1360.
15. Narula S, Shameer K, Salem Omar AM, Dudley JT, Sengupta PP. Machine-learning algorithms to automate morphological and functional assessments in 2D echocardiography. *J Am Coll Cardiol*. 2016;68:2287–2295. https://doi.org/10.1016/j.jacc.2016.08.062
16. Al-Amrani S, Al-Jabri Z, Al-Zaabi A, Alshekaili J, Al-Khabori M. Proteomics: Concepts and applications in human medicine. *World J Biol Chem*. 2021;12:57–69. https://doi.org/10.4331/wjbc.v12.i5.57

17. Vatansever S, Schlessinger A, Wacker D, Kaniskan HÜ, Jin J, Zhou MM, Zhang B. Artificial intelligence and machine learning-aided drug discovery in central nervous system diseases: State-of-the-arts and future directions. *Med Res Rev*. 2021;41:1427–1473. https://doi.org/10.1002/med.21764
18. Aromolaran O, Aromolaran D, Isewon I, Oyelade J. Machine learning approach to gene essentiality prediction: A review. *Brief Bioinform*. 2021;22:1–19. https://doi.org/10.1093/bib/bbab128
19. Mann M, Kumar C, Zeng WF, Strauss MT. Artificial intelligence for proteomics and biomarker discovery. *Cell Syst*. 2021;12:759–770. https://doi.org/10.1016/j.cels.2021.06.006
20. Hokken TW, Ribeiro JM, De Jaegere PP, Van Mieghem NM. Precision medicine in interventional cardiology. *Interv Cardiol*. 2020;15:e03. https://doi.org/10.15420/icr.2019.23
21. Xu J, Yang P, Xue S, et al. Translating cancer genomics into precision medicine with artificial intelligence: Applications, challenges and future perspectives. *Hum Genet*. 2019;138:109–124.
22. Heo J, Yoon JG, Park H, Kim YD, Nam HS, Heo JH. Machine learning-based model for prediction of outcomes in acute stroke. *Stroke*. 2019;50:1263–1265. https://doi.org/10.1161/STROKEAHA.118.024293
23. Hebbring S. Genomic and phenomic research in the 21st century. *Trends Genet*. 2019;35:29–41. https://doi.org/10.1016/j.tig.2018.09.007
24. Subramanian M, Wojtusciszyn A, Favre L, et al. Precision medicine in the era of artificial intelligence: Implications in chronic disease management. *J Transl Med*. 2020;18:472. https://doi.org/10.1186/s12967-020-02658-5
25. Racovita A, Jaramillo A. Reinforcement learning in synthetic gene circuits. *Biochem Soc Trans*. 2020;48:1637–1643. https://doi.org/10.1042/BST20200008
26. Nesbeth DN, Zaikin A, Saka Y, Romano MC, Giuraniuc CV, Kanakov O, Laptyeva T. Synthetic biology routes to bio-artificial intelligence. *Essays Biochem*. 2016;60:381–391. https://doi.org/10.1042/EBC20160014
27. Fil J, Dalchau N, Chu D. Programming molecular systems to emulate a learning spiking neuron. *ACS Synth Biol*. 2022;11:2055–2069. https://doi.org/10.1021/acssynbio.1c00625
28. Rauschert S, Raubenheimer K, Melton PE, Huang RC. Machine learning and clinical epigenetics: A review of challenges for diagnosis and classification. *Clin Epigenetics*. 2020;12:51. https://doi.org/10.1186/s13148-020-00842-4
29. Sohail A. Genetic algorithms in the fields of artificial intelligence and data sciences. *Ann Data Sci*. 2023;10:1007–1018. https://doi.org/10.1007/s40745-021-00354-9
30. Martin AR, Daly MJ, Robinson EB, Hyman SE, Neale BM. Predicting polygenic risk of psychiatric disorders. *Biol Psychiatry*. 2019;86:97–109. https://doi.org/10.1016/j.biopsych.2018.12.015
31. Uddin M, Wang Y, Woodbury-Smith M. Artificial intelligence for precision medicine in neurodevelopmental disorders. *NPJ Digit Med*. 2019;2:112. https://doi.org/10.1038/s41746-019-0191-0
32. Telenti A, Pierce LC, Biggs WH, et al. Deep sequencing of 10,000 human genomes. *Proc Natl Acad Sci USA*. 2016;113:11901–11906.
33. Fang J, Chen Q, Wu G. Genomics-based tools for drug discovery and development: From network maps to efficacy prediction. *J Holistic Integr Pharm*. 2023;4(3):199–209. https://doi.org/10.1016/j.jhip.2023.11.001
34. Mushegian AR. (2007). *Foundations of Comparative Genomics*. Academic Press, Boston.
35. Sadybekov AV, Katritch V. Computational approaches streamlining drug discovery. *Nature*. 2023;616(7957):673–683. https://doi.org/10.1038/s41586-023-05905-z
36. Benson DA, Cavanaugh M, Clark K, et al. GenBank. *Nucleic Acids Res*. 2012;41(D1):D36–D42.
37. Mallick S, Li H, Lipson M, et al. The simons genome diversity project: 300 genomes from 142 diverse populations. *Nature*. 2016;538(7624):201–206. https://doi.org/10.1038/nature18964.
38. Watkins WS, Feusier JE, Thomas J, Goubert C, Mallick S, Jorde LB. The Simons genome diversity project: A global analysis of mobile element diversity. *Genome Biol Evol*. 2020;12(6):779–794. https://doi.org/10.1093/gbe/evaa086.
39. Bergström A, McCarthy SA, Hui R, et al. Insights into human genetic variation and population history from 929 diverse genomes. *Science*. 2020;367(6484):eaay5012.
40. Liao WW, Asri M, Ebler J, et al. A draft human pangenome reference. *Nature*. 2023;617(7960):312–324.
41. Pagani L, Lawson DJ, Jagoda E, et al. Genomic analyses inform on migration events during the peopling of Eurasia. *Nature* 2016;538(7624):238–242.
42. Mallick S, Micco A, Mah M, et al. The Allen Ancient DNA Resource (AADR) a curated compendium of ancient human genomes. *Sci Data*. 2024;11(1):182.
43. Tsubaki M, Tomii K, Sese J. Compound-protein interaction prediction with end-to-end learning of neural networks for graphs and sequences. *Bioinformatics*. 2019;35(2):309–318.
44. Kerdoncuff E, Skov L, Patterson N, et al. 50,000 Years of evolutionary history of India: Insights from ~2,700 whole genome sequences. 2024. https://doi.org/10.1101/2024.02.15.580575.
45. Plenge RM, Scolnick EM, Altshuler D. Validating therapeutic targets through human genetics. *Nat Rev Drug Discov*. 2013;12:581–594.
46. Giral H, Landmesser U, Kratzer A. Into the wild: GWAS exploration of non-coding RNAs. *Front Cardiovasc Med*. 2018;5:181.
47. King EA, Davis JW, Degner JF. Are drug targets with genetic support twice as likely to be approved? Revised estimates of the impact of genetic support for drug mechanisms on the probability of drug approval. *PLoS Genet*. 2019;15:e1008489.
48. Rentzsch P, Witten D, Cooper GM, Shendure J, Kircher M. CADD: Predicting the deleteriousness of variants throughout the human genome. *Nucleic Acids Res*. 2019;47:D886–D894.
49. Bartha I, di Iulio J, Venter JC, Telenti A. Human gene essentiality. *Nat Rev Genet*. 2018;19:51–62.
50. Lek M, Karczewski KJ, Minikel EV, et al. Analysis of protein-coding genetic variation in 60,706 humans. *Nature*. 2016;536:285–291.
51. Rausell A, Luo Y, Lopez M, et al. Common homozygosity for predicted loss-of-function variants reveals both redundant and advantageous e ects of dispensable human genes. *Proc Natl Acad Sci USA*. 2020;117:13626 13636.
52. Karczewski KJ, Francioli LC, Tiao G, et al. The mutational constraint spectrum quantified from variation in 141,456 humans. *Nature*. 2020;581:434–443.
53. Cohen JC, Boerwinkle E, Mosley TH, Jr., Hobbs HH. Sequence variations in PCSK9, low LDL, and protection against coronary heart disease. *N Engl J Med*. 2006;354:1264–1272.

54. Clarke R, Peden JF, Hopewell JC, et al. Genetic variants associated with Lp(a) lipoprotein level and coronary disease. *N Engl J Med*. 2009;361:2518–2528.
55. Crosby J, Peloso GM, Auer PL, et al. Loss-of-function mutations in APOC3, triglycerides, and coronary disease. *N Engl J Med*. 2014;371:22–31.
56. Jorgensen AB, Frikke-Schmidt R, Nordestgaard BG, Tybjaerg-Hansen A. Loss-of-function mutations in APOC3andrisk of ischemic vascular disease. *N Engl J Med*. 2014;371:32–41.
57. Altmann SW, Davis HR, Jr., Zhu LJ, et al. Niemann-Pick C1 like 1 protein is critical for intestinal cholesterol absorption. *Science*. 2004;303:1201–1204.
58. Musunuru K, Pirruccello JP, Do R, et al. Exome sequencing, ANGPTL3 mutations, and familial combined hypolipidemia. *N Engl J Med*. 2010;363:2220–2227.
59. Dewey FE, Gusarova V, Dunbar RL, et al. Genetic and pharmacologic inactivation of ANGPTL3 and cardiovascular disease. *N Engl J Med*. 2017;377:211–221.
60. Robinson MK, Caminis J, Brunkow ME. Sclerostin: How human mutations have helped reveal a new target for the treatment of osteoporosis. *Drug Discov Today*. 2013;18:637–643.
61. Cirulli ET, Guo L, Leon Swisher C, et al. Profound perturbation of the metabolome in obesity is associated with health risk. *Cell Metab*. 2019;29:488–500.e2.
62. Long T, Hicks M, Yu HC, et al. Whole-genome sequencing identifies common-to-rare variants associated with human blood metabolites. *Nat Genet*. 2017;49:568–578.
63. Berger MF, Mardis ER. The emerging clinical relevance of genomics in cancer medicine. *Nat Rev Clin Oncol*. 2018;15:353–365.
64. Nogrady B. How cancer genomics is transforming diagnosis and treatment. *Nature*. 2020;579:S10–S11.
65. Fresard L, Montgomery SB. Diagnosing rare diseases after the exome. *Mol Case Stud*. 2018;4:a003392.
66. di Iulio J, Bartha I, Wong EHM, et al. The human noncoding genome defined by genetic diversity. *Nat Genet*. 2018;50:333–337.
67. Wells A, Heckerman D, Torkamani A, Yin L, Sebat J, Ren B, Telenti A, di Iulio J. Ranking of non-coding pathogenic variants and putative essential regions of the human genome. *Nat Commun*. 2019;10:5241.
68. Perenthaler E, Yousefi S, Niggl E, Barakat TS. Beyond the exome: The non-coding genome and enhancers in neurodevelopmental disorders and malformations of cortical development. *Front Cell Neurosci*. 2019;13:352.
69. Ning B, Yu D, Yu AM. Advances and challenges in studying noncoding RNA regulation of drug metabolism and development of RNA therapeutics. *Biochem Pharm*. 2019;169:113638.
70. Keenan AB, Wojciechowicz ML, Wang Z, Jagodnik KM, Jenkins SL, Lachmann A, Ma'ayan A. Connectivity mapping: Methods and applications. *Annu Rev Biomed Data Sci*. 2019;2:69–92.
71. Subramanian A, Narayan R, Corsello SM, et al. A next generation connectivity map: L1000 platform and the first 1,000,000 profiles. *Cell*. 2017;171:1437–1452.e17.
72. Bush EC, Ray F, Alvarez MJ, Realubit R, Li H, Karan C, Califano A, Sims PA. PLATE-Seq for genome-wideregulatorynetworkanalysisofhigh-throughputscreens. *Nat Commun*. 2017;8:105.
73. Ye C, Ho DJ, Neri M, et al. DRUG-seq for miniaturized high-throughput transcriptome profiling in drug discovery. *Nat Commun*. 2018;9:4307.
74. Alpern D, Gardeux V, Russeil J, Mangeat B, Meireles-Filho ACA, Breysse R, Hacker D, Deplancke B. BRB-seq: Ultra-a ordable high-throughput transcriptomics enabled by bulk RNA barcoding and sequencing. *Genome Biol*. 2019;20:71.
75. Leek JT, Scharpf RB, Bravo HC, Simcha D, Langmead B, Johnson WE, Geman D, Baggerly K, Irizarry RA. Tackling the widespread and critical impact of batch e ects in high-throughput data. *Nat Rev Genet*. 2010;11:733–739.
76. Csermely P, Korcsmaros T, Kiss HJ, London G, Nussinov, R. Structure and dynamics of molecular networks: A novel paradigm of drug discovery: A comprehensive review. *Pharmacol Ther*. 2013;138:333–408.
77. Vamathevan J, Clark D, Czodrowski P, et al. Applications of machine learning in drug discovery and development. *Nat Rev Drug Discov*. 2019;18:463–477.
78. Zou J, Huss M, Abid A, Mohammadi P, Torkamani A, Telenti A. A primer on deep learning in genomics. *Nat Genet*. 2019;51:12–18.
79. Vilhekar RS, Rawekar A. Artificial intelligence in genetics. *Cureus*. 2024 Jan 10;16(1):e51599. doi:10.7759/cureus.51599
80. Taboureau O, Baell JB, Fernández-Recio J, Villoutreix BO. Established and emerging trends in computational drug discovery in the structural genomics era. *Chem Biol*. 2012;19(1):29–41.
81. Fernández-Recio J. Prediction of protein binding sites and hot spots. *Comput Mol Sci*. 2011;1:680–698.
82. Bouckaert R, Heled J, Kühnert D, et al. BEAST 2: A software platform for bayesian evolutionary analysis. *PLoS Computat. Biol*. 2014;10(4):e1003537.
83. Excoffier L, Marchi N, Marques DA, Matthey-Doret R, Gouy A, Sousa VC. Fastsimcoal2: Demographic inference under complex evolutionary scenarios. *Bioinformatics*. 2021;37(24):4882–4885.
84. Terhorst J, Kamm JA, Song YS. Robust and scalable inference of population history from hundreds of unphased whole genomes. *Nat Genet*. 2017;49(2):303–309.
85. Upadhya G, Steinrücken M. Robust inference of population size histories from genomic sequencing data. *PLoS Comput Biol*. 2022;18(9):e1010419.
86. Steinrücken M, Kamm J, Spence JP, Song YS. Inference of complex population histories using whole-genome sequences from multiple populations. *Proc Natl Acad Sci*. 2019;116(34):17115–17120.
87. Hubisz MJ, Williams AL, Siepel A. Mapping gene flow between ancient hominins through demography-aware inference of the ancestral recombination graph. *PLoS Genet*. 2020;16(8):e1008895.
88. Kelleher J, Wong Y, Wohns AW, Fadil C, Albers PK, McVean G. Inferring whole-genome histories in large population datasets. *Nat Genet*. 2019;51(9):1330–1338.
89. Speidel L, Forest M, Shi S, Myers SR. A method for genome - wide genealogy estimation for thousands of samples. *Nat Genet*. 2019;51(9):1321–1329.
90. Baumdicker F, Bisschop G, Goldstein D, et al. Efficient ancestry and mutation simulation with Msprime 1.0. *Genetics*. 2022;220(3):iyab229.
91. Haller BC, Messer PW. SLiM 4: Multispecies eco-evolutionary modeling. *Am Nat* 2023;201(5):E127–E139.
92. Wertheim JO, et al. RELAX: Detecting relaxed selection in a phylogenetic framework. *Mol Biol Evol*. 2015;32:820–832.
93. Yang Z. PAML 4: Phylogenetic analysis by maximum likelihood. *Mol Biol Evol*. 2007;24(8):1586–1591.
94. Pond SLK, Frost SDW, Muse SV. HyPhy: Hypothesis testing using phylogenies. *Bioinformatics*. 2005;21(5):676–679. https://dx. doi.org/10.1093/bioinformatics/bti079.

95. Szpiech ZA. Selscan 2.0: Scanning for sweeps in unphased data. 2021. https://doi.org/10.1101/2021.10.22.465497.
96. Maclean CA, Chue Hong NP, Prendergast JGD. Hapbin: An efficient program for performing haplotype-based scans for positive selection in large genomic datasets. *Mol Biol Evol.* 2015;32(11):3027–3029. https://dx. doi.org/10.1093/molbev/msv172.
97. Akbari A, Barton AR, Gazal S, et al. Pervasive findings of directional selection realize the promise of ancient DNA to elucidate human adaptation, 2024.
98. Stern AJ, Wilton PR, Nielsen R. CLUES: A method to infer selection and allele frequency trajectories using genealogies, 2021.
99. Alexander DH, Novembre J, Lange K. Fast model-based estimation of ancestry in unrelated individuals. *Genome Res.* 2009;19(9):1655–1664.
100. Meisner J, Albrechtsen A. Inferring population structure and admixture proportions in low-depth NGS data. *Genetics.* 2018;210(2):719–731. https://dx. doi.org/10.1534/genetics.118.301336.
101. Patterson N, Moorjani P, Luo Y, et al. Ancient admixture in human history. *Genetics.* 2012;192(3):1065–1093.
102. Maples BK, Gravel S, Kenny EE, Bustamante CD. RFMix: A discriminative modeling approach for rapid and robust local-ancestry inference. *Am J Hum Genet.* 2013;93(2):278–288.
103. Guan Y. Detecting structure of haplotypes and local ancestry inference, 2014. GitHub repository: https://github.com/GuangchuangYu/ELAI.
104. Behr AA, Liu KZ, Liu-Fang G, Nakka P, Ramachandran S. Pong: Fast analysis and visualization of latent clusters in population genetic data. *Bioinformatics.* 2016;32(18):2817–23. https://dx. doi.org/10.1093/bioinformatics/btw327.
105. Kopelman NM, Mayzel J, Jakobsson M, Rosenberg NA, Mayrose I. Clumpak: A program for identifying clustering modes and packaging population structure inferences across K. *Mol Ecol Resour* 2015;15(5):1179–1191. https://dx. doi.org/10.1111/1755-0998.12387.
106. Manichaikul A, Mychaleckyj JC, Rich SS, Daly K, Sale M, Chen WM. Robust relationship inference in genome-wide association studies. *Bioinformatics.* 2010;26(22):2867–2873. https://dx. doi.org/10.1093/bioinformatics/btq559.
107. Monroy Kuhn JM, Jakobsson M, Günther T. Estimating genetic kin relationships in prehistoric populations. *PLoS One.* 2018;13(4):e0195491. https://dx. doi.org/10.1371/journal.pone.0195491.
108. Lipatov M, Sanjeev K, Patro R, et al. Maximum likelihood estimation of biological relatedness from low coverage sequencing data. bioRxiv 023374, 2015.
109. Meisner J, Albrechtsen A. Inferring population structure and admixture proportions in low-depth NGS data. *Genetics.* 2018;210(2):719–731.
110. Cavasotto CN, Phatak SS. Homology modeling in drug discovery: Current trends and applications. *Drug Discov Today.* 2009;14(13–14):676–683.
111. Tanrikulu Y, Schneider G. Pseudoreceptor models in drug design: Bridging ligand- and receptor-based virtual screening. *Nat Rev Drug Discov.* 2008;7(8):667–677.
112. Wu C, Peng YH, Wu JM, et al. Discovery of SGLT2 inhibitors using ligand-based pharmacophore and shape-based virtual screening. *J Med Chem.* 2010;53(2):272–281.
113. Berger SI, Iyengar R. Network analyses in systems pharmacology. *Bioinformatics.* 2009;25(19):2466–2472.
114. Taboureau O, Nielsen SK, Audouze K, et al. ChemProt: A disease chemical biology database. *Nucleic Acids Res.* 2011;39:D367–D372.
115. Mah LJ, Capriotti E, Daneshjou R, Karczewski KJ, Altman RB. Bioinformatics challenges for personalized medicine. *Bioinformatics.* 2011;27(13):1741–1748.
116. Thusberg J, Vihinen M. Bioinformatics methods for predicting the function of amino acid substitutions. *Hum Mutat.* 2009;30(5):703–714.
117. Liu X, Ouyang S, Yu B, et al. PharmMapper server: A web server for potential drug target identification using pharmacophore mapping. *Nucleic Acids Res.* 2010;38(Web Server issue):W609–W614.
118. Kuhn M, Szklarczyk D, Franceschini A, Campillos M, von Mering C, Jensen LJ, Beyer A, Bork P. STITCH 2: An interaction network database for small molecules and proteins. *Nucleic Acids Res.* 2010;38:D552–D556.
119. Bottegoni G, Rocchia W, Rueda M, Abagyan R, Cavalli A. Systematic exploitation of multiple receptor conformations for virtual ligand screening. *PLoS One.* 2011;6(5):e18845.
120. Sperandio O, et al. Rationalizing the chemical space and designing multitarget ligands in the context of polypharmacology. *Curr Pharm Des.* 2010;16(26):3185–3197.
121. Barrett T, Troup DB, Wilhite SE, et al. NCBI GEO: Mining tens of millions of expression profiles–database and tools update. *Nucleic Acids Res.* 2007;35:D760–D765.
122. Barretina J, Caponigro G, Stransky N, et al. The cancer cell line encyclopedia enables predictive modelling of anticancer drug sensitivity. *Nature.* 2012;483(7391):603–607.
123. Yang W, Soares J, Greninger P, et al. Genomics of drug sensitivity in cancer (GDSC): A resource for therapeutic biomarker discovery in cancer cells. *Nucleic Acids Res.* 2013;41:D955–D961.
124. Lamb J, Crawford ED, Peck D, et al. The connectivity map: Using gene-expression signatures to connect small molecules, genes, and disease. *Science.* 2006;313(5795):1929–1935.
125. Ghandi M, Huang FW, Jane-Valbuena J, et al. Next-generation characterization of the cancer cell line encyclopedia. *Nature.* 2019;569(7757):503–508.
126. Corsello SM, Bittker JA, Liu Z, et al. The drug repurposing hub: A next-generation drug library and information resource. *Nat Med.* 2017;23(4):405–408.
127. Monks A, Zhao Y, Hose C, et al. The NCI transcriptional pharmacodynamics workbench: A tool to examine dynamic expression profiling of therapeutic response in the NCI-60 cell line panel. *Cancer Res.* 2018;78(24):6807–6817.
128. Li Y, Umbach DM, Krahn JM, et al. Predicting tumor response to drugs based on gene-expression biomarkers of sensitivity learned from cancer cell lines. *BMC Genomics.* 2021;22(1):272.
129. Wang C, Yang J, Luo H, et al. CancerTracer: A curated database for intrapatient tumor heterogeneity. *Nucleic Acids Res.* 2020;48(D1):D797–D806.
130. Zhu J, Wang J, Wang X, et al. Prediction of drug efficacy from transcriptional profiles with deep learning. *Nat Biotechnol.* 2021;39(11):1444–1452.
131. LeCun Y, Bengio Y, Hinton G. Deep learning. *Nature.* 2015;521(7553):436–444.
132. Xu Y, Dai Z, Chen F, Gao S, Pei J, Lai L. Deep learning for drug-induced liver injury. *J Chem Inf Model.* 2015;55(10):2085–2093.
133. Kuenzi BM, Park J, Fong SH, et al. Predicting drug response and synergy using a deep learning model of human cancer cells. *Cancer Cell.* 2020;38(5):672–684.e6.

134. Moor M, Banerjee O, Abad ZSH, et al. Foundation models for generalist medical artificial intelligence. *Nature*. 2023;616(7956):259–265.
135. Lehar J, Madissoon E, Chevallier J, et al. MOSAIC: Multi-omic spatial atlas in cancer, effect on precision oncology. *J Clin Oncol*. 2023;41(16_suppl):e15076.
136. Lewis SM, Asselin-Labat ML, Nguyen Q, et al. Spatial omics and multiplexed imaging to explore cancer biology. *Nat Methods*. 2021;18(9):997–1012.
137. Chen W, Chen G, Zhao L, Chen CY. Predicting drug-target interactions with deep-embedding learning of graphs and sequences. *J Phys Chem A*. 2021;125(25):5633–5642.
138. Mayr A, Klambauer G, Unterthiner T, Hochreiter S. DeepTox: Toxicity prediction using deep learning. *Front Environ Sci*. 2016;3:80.
139. Lee I, Keum J, Nam H. DeepConv-DTI: Prediction of drug-target interactions via deep learning with convolution on protein sequences. *PLoS Comput Biol*. 2019;15(6):e1007129.
140. Li S, Wan F, Shu H, Jiang T, Zhao D, Zeng J. MONN: A multi-objective neural network for predicting compound-protein interactions and affinities. *Cell Syst*. 2020;10(4):308–322.e11.
141. Smirnov P, Safikhani Z, El-Hachem N, et al. PharmacoGx: An R package for analysis of large pharmacogenomic datasets. *Bioinformatics*. 2016;32(8):1244–1246.
142. Smirnov P, Kofia V, Maru A, et al. PharmacoDB: An integrative database for mining in vitro anticancer drug screening studies. *Nucleic Acids Res*. 2018;46(D1):D994–D1002.

Index

3D arrangement, 16, 89
3D molecular visualization, 6, 89, 99
3D protein structure, 43, 60, 67, 70, 71, 217
3D representation, 104
3D structure, 14, 16, 19, 20, 25, 41, 44, 57, 67–69, 71, 72, 79, 84, 89, 93, 94, 108, 140, 194, 201–203, 207, 217
3D-QSAR, 79, 96, 104, 107, 112, 113
3D-QSAR method, 107

A

Academia, 92, 239
ACEMD software, 6, 142
Achilles, 167
Active pharmaceutical ingredients, 182
Active sites, 42, 67, 71, 82, 109, 110, 204
Adenine, 88
Adjuvant selection, 199–210
ADMET parameter, 56
Adverse effects, 7, 78, 126, 155, 158, 159, 161, 165, 223, 225, 229
AI analysis, 238
AI/ML applications, 212–223
Algorithm, 2, 5, 7, 13, 22, 26, 28, 29, 31, 32, 44–52, 54, 56, 59–61, 68, 70, 71, 78–80, 85, 90–93, 95, 106, 109, 114, 120, 122, 125, 127–130, 137–141, 148, 152, 158–162, 165, 170, 173, 177, 192–194, 201, 203, 204, 207, 212, 213, 215, 217–219, 221, 228, 232, 236, 237, 239
Alignment, 2, 3, 5, 15, 21–23, 29, 42, 43, 67–70, 72, 82, 92, 94, 106, 107, 109, 110, 112, 114, 115, 128, 129, 205, 207, 208, 213, 237
Alpha helices (α), 14, 25
Amino acid residues, 42, 85, 87, 114, 137, 207
Amino acid sequences, 2, 5, 6, 17, 19, 72, 129, 201, 203
Amino acid sequencing, 1–6
Amino acids, 2, 5, 6, 13–17, 19–26, 32, 34, 39, 42, 44, 68–72, 79, 82, 85, 87, 114, 116, 129, 137, 143, 200–208, 217
AMMP VE software, 6
Analysis of mutation, 6
Ancient DNA sequencing, 1, 3, 5, 13, 170, 212
Anomaly detection, 7, 48
Antigen presentation, 204
Antigen-specific response, 169–178
Antiparallel, 16
Aqueous solubility, 84, 228
Aromatic chemistry, 148–160
Artificial neural networks, 192, 213, 237
Assembly, 5, 14, 16, 25, 68, 70, 82, 170, 174, 221
Atomic interactions, 84
AutoDock software, 6, 80, 81, 85, 138, 140, 208, 211
AutoDock Vina, 6, 85, 138
Automated protein sequencing, 2
Automatic genome annotation (KAAS), 3, 9

B

Backbone chain, 14–18, 24–26, 56, 57, 68–70, 86, 93, 143, 175, 176, 207, 232
Bacteriophage, 1, 2, 8
Balance in structure, vii, 28, 33, 81, 83, 144, 150, 221, 234
Base pair complementarity, 78
Base pairing, 152, 232
Bases, 21, 43, 44
Basic local alignment search tool (BLAST), 3, 22, 128
Basis sets, 152
Benzene rings, 116
Beta (β) domains, 16
Beta sheets (β), 14, 25
Big data challenges, 3, 39, 61, 131, 158, 159, 165, 176, 216, 219, 223
BiGG database, 10
Binding affinity, 45, 57, 71, 80, 81, 85, 86, 95, 112, 137, 140, 141, 159, 173, 194, 204–206, 208
Binding modes, 71, 93, 107
Binding pockets, 78, 194
Bioavailability, 88, 181, 188, 215, 225, 227
BioCarta database, 4
Biodiversity, vi, 169–177
BioGRID database, 4, 121, 161
Bioinformatics, v, vii, vii, viii, 1–36, 39–42, 44, 46, 48, 50, 56, 58, 60, 61, 64, 67, 68, 72, 82, 83, 104, 120, 124, 126–128, 130, 149, 158, 159, 161, 162, 169–171, 173, 175–177, 191, 208, 216, 229, 232
Bioinformatics analysis, 5, 216
Bioinformatics integration, 176
Bioinformatics resources, 39–55
Bioinformatics software, 60
Bioinformatics tools, 2, 3, 61, 68, 149, 158, 159, 161, 209
Bioinformatics, Computational Chemistry and AI in Drug Innovation, vii, viii, 1, 2, 8, 10, 12, 14, 16, 18, 20, 22, 24, 26, 30, 32, 34, 36, 38, 40, 42, 44, 46, 48, 50, 52, 54, 56, 58, 60, 62, 64, 66, 68, 70, 72, 74, 76, 78, 80, 82, 84, 86
Biological activity prediction, 4, 5, 46, 71, 89–92, 94, 103, 105–107, 109, 111, 112, 114, 116, 150, 159, 218, 219, 221, 225
Biological complexity, 70, 173
Biological data, v, vii, 1, 3, 7, 20, 39–53, 57, 60, 61, 92, 107, 108, 119, 120, 128, 129, 131, 161, 169, 237
Biological data management in, vii, 1
Biological databases, 39–55, 161
Biological molecules, 3, 41
Biological networks, 30, 127, 165
Biological pathways, 30, 44, 119
Biological processes, 2, 5, 47, 59, 63, 120, 123–126, 128, 130, 141, 161, 220
Biological standards, 56–66
Biological systems, 4, 5, 35, 46, 51, 57, 106, 124, 127–130, 137, 148, 160, 166, 174
Biological target discovery, 13, 47, 57, 58, 67, 79, 89–91, 93, 95, 158, 161, 162, 212
Biomarker discovery, 237
Biomarkers, 27–40, 48, 52, 119, 120, 126–128, 162, 212, 222, 236
  predictive, 225–231
Birmingham Metabolite Library, 5, 10
BLAST applications, 3, 5, 22, 60, 68, 120–122, 128–130, 190, 203, 206
BLAST database search, 3–6, 63–65
Blood-brain barrier, 225–231
Body systems computational, 46
Bonding interactions, 110, 152
Bootstrap analysis, 111
Both strands, 13, 15–17
Bottleneck solutions, 119, 130
Brain diseases, 84, 98, 172, 186, 189, 212, 225

C

CADD (computer-aided drug design), 91, 92, 148, 151, 212, 213, 215, 229, 230, 233
Calcium channels, 190
Cancer, viii, 4, 7, 30–33, 42, 71, 91, 94, 123, 125, 126, 130, 153, 156, 159, 160, 163, 164, 169, 170, 172, 174, 177, 188, 191, 193, 215, 219, 232, 235–237, 239–242
Cancer biology, 1, 125, 130
Cancer drug, 71, 153, 174, 241
Cancer treatment, 38, 123
Canonical structures, 75
Capacity analysis, 45, 48, 51, 52, 130, 150, 159, 177, 183, 192, 218, 219, 227, 228
Carbon chemistry, 14, 15, 18, 85, 110, 114, 186, 202
Carcinogenic compounds, 33, 85, 229, 242
Cardiac proteins, 134, 234
Case studies, vii, 174, 175, 186, 193
Categories, 39, 40, 47, 62, 70, 81, 83, 105, 139, 140, 172, 183, 204, 218, 221
Cell membrane transport, 151, 201
Central Drug Research Institute (CDRI), viii
Channels, 125
Characterization techniques, 59, 112, 119, 184, 185, 194
CHARMM force field, 6, 84, 114, 138, 142, 208
Chemical databases, 45, 47, 81, 103
Chemical descriptors, 84, 106, 217, 219, 220
Chemical features, 89, 90, 229
Chemical interactions, 149, 218
Chemical structure, 32, 46, 48, 57, 59–61, 82, 83, 85, 94–98
Cheminformatics, vii, 56, 61, 120–122, 159, 173
Chemometrics, 28
Chiral molecules, 14
Clearance kinetics, 165, 215, 223, 225, 227, 234

Clinical applications, 35, 37, 237
Cloud computing, 5, 62
ClusPro, 208
CLUSTAL algorithm, 2, 3, 5, 23, 68
CLUSTALW software, 3
Clustering algorithms, 203
CNS (central nervous system) drugs, 229
Coding regions, 1, 5, 41, 42, 63, 90, 125, 170, 171, 218, 219, 232, 235–237, 242
Coherent analysis, 56, 191
Columnar databases, 44
CoMFA (comparative molecular field analysis), 79
Commercial software platforms, 29, 208
Comparative genomics, 64, 233, 243
Comparative Molecular Field Analysis. *See* CoMFA
Competency assessment, 79, 107, 110–113, 115
Complementary DNA (cDNA), 5
Complex diseases, 91, 119, 120, 124, 126, 130, 132, 135, 235
Complexity of, 20, 32, 53, 61, 219, 221, 222
Composite, 19, 44, 69, 114, 187
Composite databases, 44
Composition of, 15, 28
COMPROTEIN software, 2
Computational analysis, 29, 40, 57
Computational analysis, 1–12, 119–135
Computational approaches, vii, 1, 45, 52, 108, 119, 121, 123, 125, 127, 129, 131, 133, 135, 155–168, 200, 212–223
Computational approaches for drug target identification, 119, 121, 123, 125, 127, 129, 131
Computational biology, viii, 3, 41, 78, 127, 165, 216
Computational chemistry, vii, viii, 2, 4, 6, 12, 14, 16, 18, 20, 22, 24, 26, 28, 30, 32, 34, 36, 38, 40, 42, 44–46, 48, 50, 52, 54, 56, 58, 60, 62, 64, 67, 70, 74, 76, 78, 80, 82, 84, 86, 88–90, 92, 94, 98, 100—246
Computational drug discovery, viii, 48, 85, 119, 170, 239
Computational efficiency, 81, 151
Computational facilities, 6
Computational genomics, 232, 233, 235, 237, 239, 241, 243, 245
Computational modeling, 120, 128, 191, 193
Computational models, 67, 120, 128, 191, 228, 229
Computational platforms, 129
Computational Protein Modelling, Refinement, Validation, and Its Application, 67, 69, 71, 73, 75, 77
Computational resources, 51, 61
Computational tools, vii, 3, 6, 27, 28, 45, 92, 95, 99, 166, 171, 200, 213, 225, 227, 229–232, 238, 239
Computed tomography (CT) scans, 7
Computer-aided (CADD), 91, 92, 100, 148, 212, 229, 231
CoMSIA (comparative molecular similarity indices analysis), 107, 112
Concentration gradient, 228
Concurrent analysis, 32, 112, 175, 184, 186, 225
Conformational changes, 6, 81, 111
Conformational ensembles, 139
Conformational flexibility, 207
Conformational sampling, 24, 116
Consensus sequences, 69, 70, 81, 140, 204, 205
Conservation patterns, 19
Controlled vocabulary, 175
Core-ligand complexes, 14
Cost-effectiveness, 3, 6, 237
Covalent, 14, 15, 17
Crick, Francis, 1
CRISPR arrays, 2
CRISPR-Cas systems, 1–12, 39–41, 232–245
Cross-disciplinary, 1, 39
Cross-examination, 15
Cross-linking experiments, 137–152
CRX format files, 67–76
Crystal structures, 57, 59, 71
Cufflinks software, 1–5
CYP450 metabolism, 84–93
CYP-mediated, 84–93
Cysteine residues, 14, 15

D

Data access, 45, 61, 62
Data alignment, 2, 3, 5, 15
Data analysis, 1, 3, 28
Data architecture, 15, 16
Data availability, 162
Data collection, 13, 28, 29
Data complexity, 15, 20, 32, 35
Data compression, 193
Data consistency, 62, 176
Data curation, 1, 127, 167
Data duplication, 44, 62
Data integration, 1–12, 26, 39–66
Data interpretation, 1, 2, 7, 28–30
Data management, 1, 56, 57, 61
Data mining, 1, 39
Data modeling, 37
Data organization, 56–66
Data preprocessing, 50, 219
Data quality, 29, 44, 51, 61, 62
Data repositories, 64, 175, 176
Data retrieval, 57, 59, 60
Data sharing, 53, 175, 176
Data storage, 56, 57, 61
Data transformation, 117, 218
Data type diversity, 66, 170
Data visualization, 62, 125
Database accessibility, 45, 48, 50, 62
Database analysis, 1, 2
Database annotation, 62, 63
Database architecture, 15–17
Database design, 56–66
Database integration, 30, 35, 39–55
Database maintenance, 55
Database queries, 62, 125
Database redundancy, 44, 62, 111
Database resources, 7–9
Database scalability, 51, 56, 62
Database search, 9, 22, 80, 82
Database structure, 1, 44, 56
Database updating, 40, 54
Databases bioinformatics, 1, 2
Dayhoff contribution, 2
Dayhoff, Margaret Oakley, 2
DDBJ (DNA Data Bank of Japan), 1, 3, 40, 44
De novo sequencing, 20, 23, 28, 82
Decision trees, 49, 213
Density functional theory (DFT), 46, 148
Descriptive, 42, 64
Descriptive annotations, 64
Descriptor generation, 150
Design of experiments (DoE), 182, 186
Design space, 181, 184, 192
DFT geometry optimization, 153
DFT models, 149
Diagnostic markers, 30, 31, 34
Diagram interpretation, 1, 2, 7, 28
Diamond lattice, 107, 110, 112
Diastereoisomers, 103–118
Differential gene expression, 232–245
Differential scanning calorimetry, 173
Digital tools, 131, 169, 175
Dimensionality reduction, 47, 49, 172
DINIES (drug-target interaction network inference), 3
Discovery by Ishino, 2
Discovery process, 48, 57, 59, 71, 83, 92
Disease diagnosis, 1, 3, 34, 236
Disease drug targets, 30–32, 37, 47
Disease mechanisms, 3, 30, 35, 127
Disease pathways, 64, 128, 129, 161
Disease prevention, 163
Disease progression models, 34, 48, 80, 164
Disease treatment, 148, 235
Disordered regions, 15
Disulfide bonding, 15, 200, 207
Divergent sequences, 44
Diversity monitoring, 27, 32, 34
DNA analysis, 2
DNA cloning, 1
DNA Data Bank of Japan. *See* DDBJ
DNA double-helix, 17
DNA mapping, 5
DNA methylation, 173, 241
DNA repair, 125
DNA sequences, 1, 22, 60
DNA synthesis, 232–238
Docetaxel, 187–196
Docking, 6, 13, 45
Docking accuracy, 145
Docking algorithms, 137–139
Docking predictions, 145
Docking results, 137–152
Docking software. *See* AutoDock Vina; GOLD; SwissDock; ClusPro; HDOCK; ZDOCK; HAD-DOCK
Docking tools, 80, 138, 208
Docking validation, 78–88
Document stores, 44
Domain architecture, 13–26, 67–76
Domain classification, 70, 72
Domain interactions, 70
Double-helix, 17
Double-helix model, 1–6
Draft genome, 232–245
Drug absorption, 215
Drug candidate prioritization, 175
Drug clearance, 165, 227
Drug combinations, 50
Drug combinatorial strategies, viii
Drug delivery, 212
Drug design, 45, 78, 140
Drug development, 1, 13, 29, 30, 34
Drug discovery, 1–12, 78–200
Drug distribution, 223, 239
Drug efficacy, 32, 221, 240, 242
Drug elimination, 225–231
Drug excretion, 229
Drug formulation, 1, 119, 180
Drug formulation design, 180, 181, 183
Drug interaction, 57, 59, 93, 165
Drug lead, 71, 117, 177

Drug lead optimization, 61, 78, 86, 91
Drug lipophilicity, 57, 219, 227
Drug metabolism, 120, 222
Drug permeability, 227, 228
Drug properties, 150, 161
Drug repositioning, 6, 155
Drug repurposing, 155–168
Drug resistance, 144
Drug safety, 222
Drug safety prediction, 103–118
Drug solubility, 215, 230, 231
Drug structures, 127
Drug target interaction, 120
Drug targets, 1–12, 27, 30, 31, 78–102, 119–135
Drug toxicity, 31
Drug transporters, 131
Drugs administration, 41, 54, 78
Drugs availability, 155–168
Drug-target interaction network interference. *See* DINIES
Drug-target interactions, 120
Drug-target recognition, 78–102
Dynamic aspects, 1, 6, 13
Dynamic loading, 78—102
Dynamics simulations. *See* molecular dynamics simulation
DynaMut software, 6, 11
Dysregulation gene, 236

E

Edman, Pehr, 2
Electronic structure, 46, 148–150, 160
Elimination, 83, 180, 189, 233
EMBL, 1, 3, 5, 15
Enzymes, 1, 25, 42, 44, 45
Essential, 1, 2, 7, 19, 35
Evolutionary analysis, 244
Excipient selection, 180–198

F

FASTA comparison, 3, 5, 10, 22, 201
Flexible docking, 81, 138, 139
Force fields, 6, 11, 46, 69, 81, 140
Formulation optimization, 182
Foundations, 149, 213
Functional, 148
Functional annotation, 64, 66, 173, 174
Functional databases, 1

G

GAATTC recognition, 1
GenBank, 1, 3, 23, 40, 44
Gene networks, 3, 66
Gene silencing, 232–245
Genetic, 1–3, 33
Genome analysis, 3
Genome editing, 2, 125, 134
Genomic, 1–4, 7, 27, 30, 33, 47, 48, 232–238
Genomic data, 7, 9, 41, 43, 47, 48, 51, 52, 64, 120, 161, 222, 232–239, 241, 242
Glucuronidation, 225–231
Goals, 1, 3, 148
GOLD, 6, 26, 80, 81, 84, 138, 140
GPU, 208, 212
GPU acceleration, 208, 212
Graph databases, 44
Greek key motif, 16
GROMACS, 6, 142, 208
GWAS studies, 33, 232, 235

H

Hardware, 151, 212, 239
HDOCK, 6, 139, 208
Hepatotoxicity, 84
Hierarchical, 49, 123, 208, 215
High-throughput, 2, 3, 27, 30–32, 35, 40, 47, 57, 61, 71, 72, 90, 92, 95, 124, 125, 128, 130, 131, 138, 218, 237
High-throughput screening, 57, 61, 71, 72, 104, 125, 159, 191, 212
Historical, 1, 2, 27, 67, 68, 155, 233
History, 1, 2, 7, 23, 41, 99, 170, 175, 199, 200
Hit-to-lead, 216
Homologous, 21, 22, 42–44, 67, 70, 71, 109, 129
Homology, 5, 13, 17, 21, 23, 25, 43, 67–72, 93, 109, 114, 140, 203, 206, 207, 217
Human genome, 2, 3, 212, 235, 237
Hydrogen, 14–17, 24, 25, 57, 80–85, 92, 95, 109, 110, 112
Hydrogen bonding, 15, 24, 25, 57, 81, 85, 95, 110–113, 137, 150, 152, 227, 230
Hydrophilic, 16, 185, 205
Hydrophobic, 16

I

IC50 values, 204, 205, 215
Identification, 4–6, 13, 29–33, 35, 42, 47–50, 71, 90–92, 94, 99, 109, 117, 119, 123, 125, 128–131, 158, 165, 169
Immunology, 42, 171, 176, 177, 237
Induced fit model, 81
Information management, 64
Interdisciplinary approach, 127
Intermolecular interactions, 15
Ionic, 13, 31, 32, 81, 109
Isoform, 5, 239

J

Jelly roll barrels, 16

K

KEGG, 1, 4, 5, 7, 43, 121, 126, 161
Kinases, 215

L

Large-scale, 124
Lead compound, 49, 71, 83, 91, 119, 150, 191
Lead molecules, 6
Lead optimization, 61, 78, 81, 91, 141, 216
Library design, 216
Ligand, 6, 13, 25, 44, 45
Ligand-receptor, 85, 110
M
Machine learning, 7, 13, 23, 31, 45, 46
Machine learning algorithm, 50, 61, 68, 125, 204
Macromolecular 3D structures, 44
Macromolecular structures, 13, 41, 59
Management, 2, 3, 7, 34, 56
MD simulations, 45, 46, 80, 92, 141, 144
MD trajectory data, 143
Medical imaging AI, 7
Membrane proteins, 16
Metabolic pathways, 5, 27, 28, 34, 35, 44, 61, 229
Metabolism, 4, 6, 31, 33, 44, 229, 230
Methods, 51, 56, 64, 67–71
MINT, 4
Missing values, 61
ML, 7, 39, 45
Modeling, 45
Models, 45
Molecular descriptors, 49, 90, 106, 122, 150
Molecular dynamics, 6, 70, 172, 191, 208
Molecular geometry, 113
Molecular graphics, 70
Molecular interactions, 5, 6, 45, 110, 144
Molecular modeling, 41, 46, 78, 81
Molecular weight, 32, 79, 84, 207, 228
Motifs, 3, 14–17, 20
Multiple sequence alignment, 42, 68
Mutation analysis, 6

N

NAMD, 142, 208
NCBI, 1–3, 5, 40, 58, 201
Network analysis, 47, 52, 120, 127, 161
Network pharmacology, 120, 127, 128
Network-based, 30, 31, 69
New indications, 126, 160, 165, 166
NGS data, 245
Noncoding, 10
Normalization, 61, 62, 176
NoSQL databases, 61
Novel therapeutics, 6, 72, 92, 126
Nucleotide sequences, 3, 5, 21, 40

O

Omics data, 30, 47
Optimization, 45, 59, 69, 78, 93
Oral absorption, 228
Oral bioavailability, 84, 88
Oral dosage form, 185

P

PAML, 238
Parallel, 15–17, 93, 110, 232
Passive diffusion, 228
Patents, 54, 57, 58, 61, 126
Pathway analysis, 34
Pattern detection, 169
PCR, 105, 111
PDB database, 6
pepATTRACT, 6
Personalized medicine, 7, 31, 48, 177, 238
Pharmaceutical research, 150
Pharmacogenomics, 35, 48, 191, 239, 240
Pharmacokinetic modeling, 191, 225
Pharmacophore modelling, 71, 91, 93, 96
Phase I reactions, 158
Phase II, 119, 158
Phylogenetic analysis, 5
Polypeptide, 14–19, 152
Positioning, 18, 19, 78, 112
Potency, 71, 85, 225, 229, 242
Predicting, 103, 107, 114, 120, 129, 137, 150
Prediction, 150, 169–174, 183, 192
Preprocessing, 30, 50, 129
Primary databases, 1, 40, 41, 44
Primary structure determination, 2, 14, 205

Principal component analysis, 49, 79, 105, 143, 174
Process development, 181
Prognosis, 72
Promising candidates, 96, 129
Properties, 129
Protein architecture, 15, 16
Protein backbone, 18, 25
Protein complex, 81, 203
Protein expression in, 126, 130
Protein flexibility, 203
Protein folds, 6, 14, 16, 19, 25, 209
Protein interactions, 26, 47, 71
Protein involvement, 200
Protein networks, 4, 237
Protein sequences, 15, 22
Protein similarity, 26
Protein stability, 144
Protein structure, 6, 8, 202, 207, 217
Protein synthesis, 8
Protein-ligand interactions, 137, 144
Protein-protein, 71

Q

QbD implementation, 182
QbD in design, 180, 184–186, 189
QSAR assessment, 79, 84
QSAR datasets, 106
QSAR descriptors, 79, 104
QSAR model development, 116
QSAR MODELS, 205, 215
QSAR research, 215, 220
Quality assurance, 11, 192, 193
Quality by design, 180, 183, 194
Quality control, 51, 64, 192, 196
Quality metrics, 62
Quantum mechanics, 84, 108, 146
Quaternary structure, 17, 70
Query, 90, 233
Query language, 60
Query optimization, 62
Query sequence, 5, 21, 129

R

Rational design, 79, 92, 172
Reactive, 193
Reactome, 1, 4, 5, 126, 161
Receptors, 151, 159, 169, 172, 191
Recognition sites, 1
Redundancy, 44, 62
Regularized, 106, 185
Regulatory approval, 7, 126, 156, 158
Relational, 61
Relational databases, 40, 44, 61, 63
Renal glucosuria, 234
Replication, 3, 17, 54
Repository systems, 39, 43, 59, 203
Repurposing opportunities, 159
Restriction enzymes, 1, 42
Reverse transcriptase, 5
Rigid docking, 138, 139
Risk assessment, 165, 180, 181, 192
RMSD, 69, 85, 143, 208
RNA sequencing, 126, 232, 236
Routes, 141, 185, 191

S

Sampling, 20, 24, 78, 120, 191, 221
Sanger sequencing, 3, 237
Scalability, 51, 56, 62, 191, 215
Schrodinger, 6, 46, 95, 138, 148
SCOP database, 1
Scoring functions, 69, 78, 80, 81, 140, 141
Screening, 155, 158, 166, 172
Search algorithms, 71, 139
Secondary databases, 1, 41
Secondary structure, 14–17, 25, 43
Selectivity, 45, 65, 83, 91–92, 115, 119, 128, 137, 206, 212, 229
Semantic, 56, 64, 170, 175–176, 179
Semantic annotation, 64, 175
Sequence alignment, 2, 5, 8–10, 21–22, 26, 42, 67–69, 72–73, 205
Sequence analysis, 2, 3, 5, 21
Sequence patterns, 42, 44
Sequence similarity, 21–22, 25, 109, 129, 207
Signal transduction, 4, 9, 208
Signaling, 4, 17, 47, 126, 172, 174, 234
Similarity indices, 112, 117
Small molecules, 4, 13, 25, 27, 34–35, 37, 58–59, 65, 71, 84, 86, 93, 116, 123, 132, 137, 144, 148, 212, 232, 240, 245
Software platforms, 89, 90, 95
Specificity, 13, 22, 30–34, 42, 48, 57, 73, 89, 110, 115, 117, 159, 196, 215, 220
Spectroscopic methods, 28
Stability, 6, 11, 14, 16–17, 23, 25, 29, 57, 69, 109–112, 141, 144, 152–153, 181, 186, 187, 189–191, 206–208, 213, 219, 225, 229
Statistical analysis, 4, 30, 39, 62, 78, 84, 134, 182, 197
Statistical approaches, 79
Statistical methods, 128, 183, 227
Stereochemistry, 26
Storage and retrieval, 5, 56–57, 59, 61, 63–65
Structure analysis, 69, 70, 222
Structure representation, 57
Substitution matrices, 21
Success stories, 101
Swiss Dock, 6, 11
Synergistic effects, 50
Systems biology, 9–10, 26–27, 36, 65, 120, 127–128, 131, 133–135, 171, 174, 178, 233

T

Target prediction, 91, 120–133, 172, 233
Target selectivity, 92
Taxonomy, 170
Text annotation, 64
Therapeutic applications, 155, 159–169
Therapeutic effect, 89, 128, 222
Therapeutics, 2, 6, 7, 13, 58, 59, 70–72, 77, 92, 126–134, 136, 144, 166, 169, 172, 209–210, 232, 239–244
Timelines, 156–157
Tools and methods, 145
Tools and resources, 230
Toxicity assessment, 33
Toxicity prediction, 50, 84, 217, 223, 229, 231, 240, 246
Training sets, 92, 115, 172
Trajectory analysis, 143
Transcript assembly, 5
Transcription, 3, 10, 34, 235–238, 240–246
Transcriptomics, 3–5, 10, 30, 33, 35–36, 47, 51, 83, 124, 127–128, 130, 132, 169, 172–173, 176, 232–233, 236, 242, 244
Transdermal delivery, 196
Translation, 3, 5, 41–42, 44, 79–81, 114–115, 126, 129, 131, 134, 138, 178, 224, 230
Transporters, 144, 191, 204

V

Validation, 4–6, 11, 31–32, 47, 51, 54, 57, 61, 64, 67–69, 75–77, 84, 88, 95, 97, 106–107, 111, 116, 119–120, 125–131, 135, 150, 158–160, 162, 176, 192–193, 200, 207, 213, 217, 219–220, 227, 232
Variant calling, 237
Variants, 19, 26, 35, 70, 104, 116, 118, 209, 232–235, 238, 242–244
Version control, 176
Virtual screening, 6, 11, 45, 47, 50, 55, 59, 71–72, 76, 81, 89–92, 94–102, 129, 132–133, 137–138, 141, 145–146, 158–159, 166, 167, 191, 194, 213, 216, 224, 230, 233, 239, 245–246
Volume of distribution, 225

W

Water solubility, 213, 227, 228
Web servers, 201
Workflow, 7, 67, 71, 90, 135, 150, 232, 237

Z

ZDOCK, 6, 11, 84

For Product Safety Concerns and Information please contact our
EU representative GPSR@taylorandfrancis.com Taylor & Francis
Verlag GmbH, Kaufingerstraße 24, 80331 München, Germany